Aquaculture

Aquaculture

The Farming and Husbandry
of Freshwater
and Marine Organisms

John E. Bardach

Hawaii Institute of Marine Biology

John H. Ryther

Woods Hole Oceanographic Institution

and

William O. McLarney

Woods Hole Oceanographic Institution

WILEY-INTERSCIENCE, A DIVISION OF JOHN WILEY & SONS, INC.

NEW YORK · LONDON · SYDNEY · TORONTO

Library of Congress Cataloging in Publication Data:

Bardach, John E
Aquaculture; the farming and husbandry of freshwater and marine organisms.

Includes bibliographies.
1. Aquaculture. I. Ryther, John H., joint author.
II. McLarney, William O., joint author. III. Title.

SH135.B37   639'.3   72-2516
ISBN 0-471-04825-9

Printed in the United States of America

10 9 8 7 6 5 4 3 2

Preface

Much of the success of the green revolution lies in the planned systems approach that was followed in developing certain key crops. Yet husbandry of land animals in the developed world had already taken a similar approach, including the development of automation. Chicken, for example, are a highly efficient source of animal protein production. On a modern chicken farm, perhaps best called an industry, adequate fixed and variable inputs permit one man to produce 500 tons or more of meat per year. He obtains his chicks from a hatchery and only tends to feeding, watering, and cleaning devices. It is relevant here that part of the variable cost inputs, namely feed, is spent for fish meal, that is, cheap aquatic animal protein.

The question has been asked if one could rear aquatic animals in a manner similar to these chicken production-line assemblies. This question is pertinent, not only to obtain, for example, the fish meal protein base, but also because there are large areas available in the developing world where water is or can be supplied and where intensive *aquaculture**
might well be practiced. Also, it has been said that it may be cheaper, pound per pound, to grow aquatic rather than land animals for food because they are of the same density as the medium in which they live and require only containers for their water rather than supporting and sheltering structures, such as stables and huts. Since man is a land dweller, it is logical that he applied scientific and engineering skills to land animal production first and foremost, especially since he could, and still can, gather, rather than grow and tend, fish from the sea and inland waters. That process—traditional fisheries—rather than aquaculture has

---

* Even though often called aquiculture—in consonance with agriculture—aquaculture is the etymologically correct term.

v

received significant technological inputs so that the weight of sea fish harvested per man per year in the well-mechanized fisheries for cheap sources of fish meal, the previously mentioned chicken feed, is easily twice or more that attained on a mechanized chicken farm.

Aside from the lack of attention to aquaculture so far, there exist in it biotechnical and economic obstacles of peculiar dimensions to efficiency upgrading: water rather than air is the environment of the animals we wish to rear; in part, this is the reason why animal husbandry in water is far less developed than the corresponding branch of agriculture on land. For instance, unused chicken feed and excreta—even at high densities—are easily disposed of, at least in the space where the animals grow. That poultry farms present off-site pollution problems is another matter, but at least the air in the chicken house is not full of feed or droppings. In contrast, in the water unused feed as well as metabolic products surround the animals, befoul the medium, and impose the need for continuous flushing, which requires either an overabundance of water or costly recycling schemes. Furthermore, young chickens—and calves and piglets after weaning—eat the same food as their parents; not so most cold-blooded aquatic animals. Their larvae undergo several transformations from hatching to their adult stages (e.g., six in the case of the shrimp) during which they neither resemble the parents nor consume the same food. In the life cycle, fish and invertebrates alike recapitulate their evolutionary history in the water, instead of warm-blooded animals *in ovo* or *in utero*. Growing fish and invertebrates requires, therefore, attention to larval survival and larval feeding and imposes constraints on successfully rearing them from the egg. At the same time, the large number of eggs they produce affords certain advantages, even for the selective rearing of desirable strains.

Man has tried for millennia to overcome these obstacles, and with some animals he has been very successful indeed; others hold considerable promise. In a few cases, helped by specially favorable conditions, man has been able to surpass the efficiency attained with chicken, producing more than 100,000 lb of flesh per man per year (e.g., mussels in the Bays of Galicia, Spain, and sewage-stream-fattened carp in Indonesia). In other cases, though not growing as much bulk, aquacultural entrepreneurs do a lucrative business (e.g., shrimp in Japan under rather unique socioeconomic conditions, trout in Denmark and the Snake River Valley of the United States, and milkfish and Chinese carp in Southeast Asia). Even for them, but especially considering the world's need for nutrition, it is clearly worthwhile to apply more technological and engineering principles to aquaculture than has been done in the past. The result would then be substantial improvements in production efficiency at various levels of intensity. One aim of this book is to provide a useful baseline for such endeavors by describing what is being done in aquaculture now in many parts of the world.

The division of the world into nutritionally and otherwise rich countries and poor and needy ones is a reality, much as one may deplore it, and aquaculture develops under somewhat different premises in each region. In the technically advanced countries it is strictly a question of producing mostly luxury food commodities at competitive costs for a diversified, often protein-glutted food market (or products for high monetary return or recreation). In the developing world the predominant problem is one of producing additional animal proteins, which may be so scarce there that any meat, unless excessively cheap, is a luxury commodity available only to the relatively wealthy few. The corollary here is that especially in developing nations herbivores or plankton filter feeders are most suitable for aquaculture, producing the most per surface or volume of water from the more-or-less natural amenities, such as solar energy, existing standing or flowing waters, and natural or man-enhanced fertility.

The rearing of herbivores as well as carnivores modifies the natural ecology of land or water by passing through the system energy and materials at a faster rate than that set there by natural evolution—faster than nature intended. This practice requires labor and installations and is likely to accelerate natural dysfunctions—waste accumulations and the like—which must be coped with through technical, hence economic, inputs. Thus an aquacultural product is likely to cost more per pound than many if not most that can be gathered through fisheries from natural aquatic populations. Aquaculture will not replace fisheries. However, it may, and in our opinion is likely to, supplement them increasingly as the natural production limit is reached in one after another of the large fishing areas of the world. True, there are presently unexploited fishing grounds in the Indian Ocean, the Indonesian Archipelago, Australian waters, and elsewhere that eventually could permit an expansion of traditional fishery products to perhaps double the present level. In addition there are such unused resources as the Antarctic krill where exploitation may become technically and economically feasible. But these areas have natural limits beyond which their yields cannot be sustained.

Seafaring, including the quest for fish, and the domestication of land plants and animals have enabled man to spread over the globe. We now wish to explore what species can be domesticated in the sea or in lakes and rivers, especially with the development of some technical mastery over the once alien, liquid portion of our biosphere. It is not surprising that aquaculture has advanced farther in fresh than in salt water and that mariculture is still in its infancy. Thus the only truly domesticated aquatic animals are carp and trout rather than saltwater creatures.

Whether we wish to grow crab or mullet, lobster or shrimp and whether the first consideration is to make money or to supply additional animal proteins in a country's diet, we should have complete manipulative mastery over the entire life cycle of the animal. Provided that ecological considerations and economic reasons prove (e.g., trout) or suggest (e.g., shrimp) that an animal can and should be reared the following should be considered. (1) *Control over reproductive biology* of the species in question is necessary. One should be able to produce offspring at will and at predetermined times. Lacking this level of control one should at least be able to gather the young in sufficient abundance (e.g., oysters) more than once a year, especially in the tropics. (2) Another set of problems common to all aquaculture concerns *nutrition* and *diseases*. As mentioned, different larval stages require different foods, and the exact nutritional requirements of but few species are known. The rearing of animals in close proximity in the water leads easily to their infection with viruses, bacteria, fungi, and multicellular parasites, many of which are only cursorily known. As in human populations, animals under crowding and stress, even if they are fish, are more likely to succumb to diseases than are calm, well-nourished ones, be it because of poor nutrition or for psychological reasons. Thus success in aquaculture requires an understanding of nutrition and diseases of a species and of their interplay with aspects of the animals' behavior. (3) Economically sound technology—*aquacultural engineering*—must be utilized. Technology here ranges from the extremely simple, such as the proper construction of ponds or of hand-operated sluice gates that mix fresh with salt water, to the more complex, such as sophisticated larval rearing schemes that employ pumps, filters, and ultraviolet sterilization of the water, all more or less automated. The book treats these three general areas of concern and involvement in sequence. Rearing practices of animals are arranged by species according to their occurrence in fresh, brackish, and salt water.

The book is an outgrowth of our longstanding interest in aquaculture, with individual practical experiences in cold, temperate, and tropical regions. A report prepared by two of us for the American Institute of Biological Sciences, under contract with the President's Council on Marine Resources and Engineering Development sparked the idea in 1968. Although that report may be considered as the nucleus of the present effort, a very considerable amount of new information has been added. Most of the new material and, in fact, virtually the entire book in its present guise is eminently the product of the writing talents and organizational abilities of the third author.

Our purpose in the book is to give an overview of present, and to some extent past, practices of food aquaculture the world over, within our language and subject to the following constraints. We do not wish to speculate excessively about what might be done if certain incipient technologies and scientific hunches were advanced and perfected. We can, of course, only report what is available in the literature and what workers in the field are willing to tell us. It must be understood that some aquaculture operations are highly competitive, purely profit-oriented ventures and that even scientists in them consider some facets of what they do their secrets, which are not to be published or otherwise divulged. Their practices are often experimental, however, and as likely to be scrapped as they are to be operationally perfected; these then would not represent "the state of the art" and would hardly contribute to the conservative report that it is our intent to give. In addition one must note that many practices, especially in the developing countries, are not reported in print. Thus we may well have missed some important items here and there. We can only add to our apology for this shortcoming that we were unable to prevent it. We have restricted ourselves to food species and to some mention of their foods, respectively. For reasons of space and unity we have omitted in our treatment ornamental or industrial living aquatic resources, such as pearls or aquarium fishes or marine colloids from red and brown algae.

News stories of the last decade show substantial preoccupation with the oceans and their use and abuse; the growing of crops in the sea is prominent in the former category. Much has been said in this context, often in glowing terms, that on closer examination has made little ecological or economic sense. We hope this book provides a basis on which such speculations can be measured. We have compiled a welter of information from literatures of various countries that should be useful throughout the world, especially in regions where it is difficult to come by much comparative material. But we realize that in a field as fast growing as this one we cannot be as up to date as we would like to be. We offer this caution to our prospective readers who, we hope, can acquire through our book a "wet green thumb."

The information given in this book comes from many sources, the most valuable and timely of which are interviews and correspondence with individuals working in the various areas of specialization. We acknowledge these contributions at the end of each chapter, but we also wish to express here our grateful appreciation for their generous assistance, as well as that of any others whose names we may have inadvertently omitted. The drawings were prepared by Ilyse Rosenthal and Ann Hinds, to whose skill and patience we are indebted, and we wish par-

ticularly to acknowledge the invaluable assistance of Anita Gunning
with many of the tasks involved in the preparation of the manuscript.

JOHN E. BARDACH
JOHN H. RYTHER
WILLIAM O. McLARNEY

Kanehoe, Hawaii
Woods Hole, Massachusetts

Contents

Aquaculture

1

General Principles and Economics

Husbandry of aquatic organisms, though a novelty to much of the world, has been practiced through the ages. Oyster culture, for instance, thrived in ancient Rome and Gaul. There are earlier, less certain reports of artificial propagation of fish; the legendary Chinese Croesus, Fan-Li, of the fifth century B.C., is said to have reared carps in ponds. Although the description of their arrangement is reminiscent of the legendary well-field system of ancient Chinese social organization, its authenticity is not established. It has been speculated that aquaculture may have even more remote roots in the highly organized ancient water-oriented civilizations of the Near East, in which fish were an important dietary component.

No matter the antiquity of aquaculture, the contribution of the world's waters to man's diet still stems largely from the hunting and gathering of fish and shellfish from untended stocks. There has been a spectacular increase in the production of world fisheries, but wild stocks of aquatic organisms are limited, and ecological reasoning suggests that we must eventually reach a ceiling on the harvest of wild aquatic organisms. Rec-ognition of these facts, coupled with the increasing efficiency of communi-cations and the establishment of international technical agencies, such

1

as the Food and Agricultural Organization of the United Nations, has led to a nearly worldwide interest in the last two decades in the potential of aquatic husbandry.

Neither aquaculture nor any other method of food production will be a panacea for human nutritional problems, but all can and must contribute if the specter of hunger is to be banned. Aquaculture,* that is, the growing of aquatic organisms under controlled conditions, can make a unique contribution to nutrition in many parts of the world by virtue both of its extremely high productivity in many situations and the fact that aquatic crops are primarily protein crops rather than sources of starchy staple foods. In this regard, it should be noted that certain aquatic organisms may be better converters of primary foods than ruminants, fowl, or even pigs. Some, such as filter feeding fishes and mollusks, feed on microscopic plankton, which cannot be used directly by man.

Whereas many existing aquaculture enterprises, including some of the most successful, rely on supplying high-protein feed to produce a luxury product for human consumption, aquaculture has the potential of producing large quantities of lower-cost protein-rich food. This has been done in parts of the Orient, but elsewhere applied scientific, technological, and managerial skills must be improved and substantial seed funds must be provided if aquaculture is to assume comparable importance.

Aquaculture is by no means restricted to food production. Sport fishermen have for centuries relied on hatcheries to supplement wild stocks and will increasingly do so in the future as recreational needs among developed nations grow in the face of much environmental degradation. Bait organisms are cultured for both sport and commercial purposes. Propagation of ornamental fish and plants constitutes an important industry in some areas. Pearls are cultured in appropriate molluscan species, and goldfish (*Carassius auratus*) and other species are commercially reared for use as laboratory animals. We have, however, limited ourselves to those organisms which are raised for use as human food,† which must be judged as the most important function of aquaculture at this point in history.

Aquaculture also contributes to human nutrition through the production of unicellular algae for use in animal feeds. Human and other animal

---

* Even though the term aquiculture is often used—in consonance with agriculture—it must be stressed that aquaculture is etymologically correct.

† As the methods for growing certain fish (e.g., salmon, pay-as-you-go pond trout, and catfish) are the same whether they be used for market or for sport, the use of portions of the fish crop, the rearing of which we treat, may not be primarily but only secondarily for food.

waste is sometimes used in this process in a manner that is ecologically sound and, under certain conditions, perhaps less costly than traditional sewage disposal methods. Attempts have been and are being made to apply similar methods to the production of human food items. Obstacles to large-scale adoption of such methods are sociocultural as well as technical, in that people would have to learn to accept new and unusual foods. We believe that practical aquaculture for such purposes is relatively far in the future, whereas production of species presently used for food, in some cases involving the use of organic wastes, is imminent and in some cases has been accomplished. For this reason, for the sake of unity, and because the culture of fish, shellfish, and multicellular plants comprises a vast subject in itself, we shall treat only organisms raised to be consumed more or less directly by man.

We have attempted to bring together information on cultured organisms from all over the world. The necessity for such a treatment is pointed up, for example, by the present simultaneous experimentation with mullet culture in Israel, Taiwan, Hawaii, and Great Britain, the widespread adoption of hybrid tilapia, and the worldwide practice of oyster culture. It is hoped that this book will make some contribution to the coordination of such research and management practices and also help researchers to avoid costly and time-consuming duplication. In this regard, however, planned international and regional exchanges of information are at least equally important.

Following a discussion, in this chapter, of general principles of aquaculture, including biological and some economic considerations, the book is organized by groups of organisms: true freshwater fish, fish which can adapt to varying salinities, true marine fish, invertebrates (mostly marine), and plants. An appendix treats some principles of construction and management of ponds.

RANGES OF AQUACULTURAL PRACTICES

Aquaculture is akin to terrestrial agriculture in that it cannot economically be carried out just anywhere. A site for aquaculture must present certain natural amenities, particularly an ample supply of water of suitable temperature, salinity, and fertility. It is also necessary that the culturist exercise control through ownership, lease, or other means of secure holding; this consideration is problematical for marine and brackish water aquaculture in many parts of the world, including much of the United States, where the traditional view is that the sea, its shores, and its resources are common property, available to all. Where this attitude prevails, aquaculture is effectively thwarted. Elsewhere, aquaculture in coastal waters is fostered by protective grants or leases. In Japan, where some of the most advanced and the greatest variety of aquacultural enterprises are carried out and are strongly encouraged by all levels of government, the prefectural governments (comparable to states or provinces elsewhere) designate the areas to be used for aquaculture, and the local fishermen's cooperative associations, unique and highly effective organizations, allocate subareas to individual aquaculturists at no charge.

However the site is held, it must usually be modified to greater or lesser extent; the amount of time and effort expended on management varies considerably. In a general way, aquaculture practices may be characterized by the relative intensity of human effort applied to them. Such a treatment permits the following arrangement according to increasing inputs of capital and labor, often with corresponding increases in yields.

1. Transplantation of organisms from poor to better growing grounds, not considered aquaculture in this book, is sometimes classified as the least intensive form of aquaculture. It is most prevalent in the Soviet Union, where by 1965 more than 50 species of fish had been acclimatized in 1225 lakes, 80 rivers, and nearly 100 reservoirs.

2. Transplantation often involves hatchery-reared fish, and transplanted stocks may be partially or totally dependent on hatcheries for their maintenance. A well-known recent example is the successful introduction of the coho salmon (*Oncorhynchus kisutch*) to Lake Michigan. Hatchery propagation has also often been employed to augment stocks of naturally occurring species. This practice was particularly common in the United States and Canada during the late nineteenth and early twentieth centuries. At that time, large numbers of fish fry and very early larvae of invertebrates were released, almost invariably with no discernible benefit to the fishery in question. More recently, it has been shown in a number of instances that if young animals can be reared to a later stage before release they may ultimately contribute significantly to fisheries.

3. Fish and invertebrates may be induced to enter special enclosures where they are trapped and held until ready for harvest. This technique, with virtually no further labor input, is used successfully to grow shrimp and various euryhaline fishes in the Malay Peninsula and the Mediterranean area, respectively.

4. The foregoing technique may be intensified by fertilization and/or the installation of devices to control the rate of exchange of water. Such schemes are widely applied in the culture of milkfish (*Chanos chanos*) and shrimp in southeast Asia.

5. More complete control of stocking may be achieved through the use of artificial enclosures so constructed as not to permit entry of wild fish. Earthen ponds are the oldest and most common of such enclosures. In classical freshwater fish culture, as best exemplified by the polyculture of cyprinids in China, food for the fish is produced "naturally" by fertilizing the pond. It is to be noted that this level of aquaculture represents the dividing line between what might still be called subsistence aquaculture, albeit with a high application of labor, and endeavors demanding higher capital as well as intensive labor inputs. The maximum yields which can be achieved with labor-intensive subsistence aquaculture are around 5000 to 8000 kg/ha and occur in tropical and subtropical regions.

6. Pond culture may be further intensified by feeding the stock directly, a practice which is usually essential in enclosures constructed of cement or wood, in the increasingly popular floating net cages, or wherever carnivorous animals are raised. Examples of such high-intensity methods in practice include catfish farming in the southern United States, trout farming in the United States and Europe, and the culture of common carp (*Cyprinus carpio*), eels (*Anguilla japonica*), yellowtail (*Seriola quinqueradiata*), kuruma shrimp (*Penaeus japonicus*), and other animals in Japan.

7. Another culture method that merits special treatment is raft culture of sessile invertebrates and macroscopic algae. Although the source of stock may be natural production (hatcheries also play a role, notably in oyster culture in the United States) and food is not artificially provided, the construction and management of the growing facilities usually involve considerable labor and expense, so that raft culture must be considered an intensive form of aquaculture. Probably the highest yields ever attained through aquaculture have involved the use of rafts; for example, 300,000 kg/ha of mussels (*Mytilus edulis*) have been raised in the Galician bays of Spain.

Yields obtained by this method, or through culture in floating cages, depend, however, on a much larger area or volume of water than the one in which the installations are found; they depend also on adequate tidal or current exchanges whereby food is carried to and/or wastes are removed from the sessile (oysters or mussels) or enclosed (e.g., yellowtail) animals.

With the preceding categories in mind, the reader may judge the approximate intensity, in terms of labor and other costs, of any aquaculture enterprise. Unfortunately, however, there is much more information available on yields per unit of water surface or volume than on the

TABLE 1. SELECTED EXAMPLES OF AQUACULTURAL YIELDS ARRANGED BY ASCENDING INTENSITY OF CULTURE METHODS

CULTURE METHOD	SPECIES	YIELD [KG/(HA)(YEAR)] OR ECONOMIC GAIN
Transplantation	Plaice (Denmark, 1919–1957)	Cost: benefit of transplantation, 1:1.1–1.3 in best years (other social benefits)
	Pacific salmon (U.S.)	Cost: benefit, based on return of hatchery fish in commercial catch, 1:2.3–5.1
Release of reared young into natural environment	Pacific salmon (Japan)	Cost: benefit 1:14–20, on above basis
	Shrimp, abalone, puffer fish (Japan)	Not assessed; reputed to increase income of fishermen
	Brown trout (Denmark, 1961–1963)	Maximum net profit/100 planted fish: 163%
Retention in enclosures of young or juveniles from wild populations, no fertilization, no feeding	Mullet	150–300
	Eel, miscellaneous fish (Italy)	1,250
	Shrimp (Singapore)	
Stocking and rearing in fertilized enclosures, no feeding	Milkfish (Taiwan)	1,000
	Carp and related spp. (Israel, S.E. Asia)	125–700
	Tilapia (Africa)	400–1,200
	Carp (Java, sewage streams) (1/4–1/2 of water area used)	62,500–125,000
Stocking and rearing with fertilization and feeding	Channel catfish (U.S.)	3,000
	Carp, mullet (Israel)	2,100
	Tilapia (Cambodia)	8,000–12,000
	Carp and related spp. (in polyculture) (China, Hong Kong, Malaysia)	3,000–5,000
	Clarias (Thailand)	97,000

Intensive cultivation in running water; feeding	Rainbow trout (U.S.)	2,000,000 [170 kg/(liter)(sec)][b]
	Carp (Japan)	1,000,000–4,000,000 [about 100 kg/(liter) (sec)]]
	Shrimp (Japan)	6,000
Intensive cultivation of sessile organisms, mollusks, and algae	Oysters (Japan, Inland Sea)[a]	20,000
	Oysters (U.S.)	5,000 (best yields)
	Mussels (Spain)[a]	300,000
	Porphyra (Japan)[a]	7,500
	Undaria (Japan)[a]	47,500

[a] Raft-culture calculations based on an area 25% covered by rafts.
[b] See text for volume of flow versus surface as basis of yield.

fixed and variable costs of producing these yields. The selective tabulation of Table 1 should be perused with this caution in mind.

BIOLOGICAL PRINCIPLES UNDERLYING THE PRACTICE OF AQUACULTURE

Aquatic animals possess a number of advantages for use in husbandry. Since the body density of fish and swimming crustaceans is nearly the same as that of the water they inhabit, they are spared the chore of supporting their weight, and thus may devote more food energy to growth than terrestrial animals. In addition, fish and invertebrates, being cold-blooded animals, expend no energy on thermoregulation. (Tuna and other fast swimmers are an exception here.) This property would further enhance their potential growth rate, which is far more plastic than that of higher vertebrates. Russian sources aver that accumulation of flesh in the body of carp, per unit of assimilated food, is one and one-half times as rapid as in swine or chickens and twice as rapid as in cattle or sheep. Repetition of these measurements would be of practical as well as theoretical interest.

Sessile shellfish achieve economies of energy by replacing the active search for food with a highly efficient method of filter feeding. The rate of filtration varies, but it may amount to 450 liters/day in large, healthy, favorably situated oysters.

Nor should it be overlooked that a body of water is a three-dimensional growing space. Many of the highest aquaculture yields have been achieved through polyculture of fish inhabiting different strata of the water column or by hanging strings of mollusks from floating structures, both methods which make use of the entire water column. Strictly speaking, such methods are not unique to aquaculture. Three-dimensional or multistory gardening incorporating trees, bushes, and low-growing plants has been developed, but to date aquaculture is alone in widespread commercial application of three-dimensional growing systems.

The principal disadvantages of the aquatic medium for production of human food have to do with the general properties of liquids and the specific property of water as the universal solvent. These two characteristics render physical and chemical contamination of bodies of water much more difficult to prevent or control than is the case for expanses of land. We need not cite specific instances of water pollution which have been detrimental or lethal to aquatic organisms or their consumers.

Of course some "pollutants," most notably human and animal metabolites, may be turned to the culturist's advantage; a number of instances of use of sewage or animal wastes as fertilizers are described in the text.

However, the application of organic wastes should be under the strict control of the culturist if consistent results are to be achieved. The same applies to heat; most of the large energy-generating stations, which are becoming increasingly prevalent, use water as a coolant. The result is often accurately termed "thermal pollution" but may, under carefully controlled conditions, be a boon to the aquaculturist.

Other pollutants, such as pesticides, heavy metal compounds, and polychlorinated biphenyls, are in no way beneficial to aquatic organisms. Yet they are present in ever-increasing concentrations in many waters. They present a particularly severe problem in heavily populated or industrialized areas or where large amounts of chemicals are employed in terrestrial agriculture. From such areas, they may be widely disseminated by water currents. Thus the aquaculturist or fisherman who relies on the sea, a large lake, or a river as a source of water or fish is often confronted with dangerous pollutants which are neither of his making nor subject to his control. While the majority of current research and development efforts in fisheries and aquaculture are directed at marine and brackish water resources, we feel that expansion of freshwater aquaculture should be at least equally encouraged. With proper management and intelligent site selection, freshwater culturists may largely avoid the pollution problem.

DESIRABLE CHARACTERISTICS IN A CULTURED ORGANISM

Since aquaculture developed during times when water pollution was not a serious problem, resistance to pollutants was not a consideration in species selection. Today it might reasonably be added to the following list of desirable attributes. Even with pollution resistance off the list, these properties limit choice greatly and explain why, among the 25,000 or so species of fish and the many thousand invertebrates, only a very few have thus far been successfully employed in intensive and commercially feasible aquaculture. In addition to such obvious factors as size, availability, and nutritive or gustatory value, the following biological attributes should be taken into account in considering any aquatic organism for culture:

1. Reproductive habits. Although it is highly desirable that man be capable of breeding the species in question in captivity, it is not strictly necessary. For example, the important milkfish and mullet industries of southeast Asia are solely dependent on natural reproduction to maintain their stocks, as are most shellfish culture enterprises. In such cases, the

reproductive habits of the culture species lead to the availability of adequate numbers of young where they may be captured for stocking.

Culture systems based on capture of wild stock are, of course, ultimately limited by the success of the natural population in reproducing itself. In some cases, for instance mullet, milkfish, and shrimp culture in the Philippines and Taiwan, culturists now experience limitations in the harvest of fry. Further expansion of the industry would be made possible by the development of means of breeding the animals in captivity. Success in such an endeavor would also do much to stabilize the fry market and would permit the genetic selection of stocks, as has been done for virtually every important terrestrial food organism.

Many ingenious methods have been devised toward the end of controlled reproduction of aquatic animals, but the most far-reaching advance was the development in Brazil in 1934 of the process of hypophysation of fish, which appears to have near-universal applicability. In hypophysation, female, and occasionally male breeders are injected with suspensions or extractions of pituitary gland material. The treatment raises the concentration of the sex hormones in the bloodstream of the recipient and facilitates maturation and shedding of the sex products. In addition to permitting the controlled breeding of hitherto unspawned species, hypophysation permits the culturist to exercise some control over the time of spawning.

2. Requirements of the eggs and larvae. The hardier the eggs and larvae, the easier the culturist's task. In general, it may be stated that animals which produce fewer and larger eggs have larger, and therefore less delicate larvae. Such animals, which include most of the intensively cultured species, usually make some provision to protect their eggs (e.g., trout bury their eggs, tilapia hatch them in their mouths, shrimp carry them on the body), but the culturist may often successfully apply artificial means of preventing predation or damage.

In contrast, many marine fish and most aquatic invertebrates produce small eggs which hatch into tiny, delicate larvae. Such species depend for their survival not on hardiness but on sheer reproductive capacity. With many of these animals, particularly the invertebrates, the situation is further complicated by the fact that the larvae may go through many developmental stages, each with its own distinct environmental and nutritional requirements, before assuming the adult form. Thus it is not surprising that culture of most of this second group of animals has succeeded only on a laboratory scale, if at all.

3. Feeding habits. There are two general approaches to feeding cultured aquatic animals. One is to raise animals which are low on the food chain, supply them with a low-cost feed, if any, and aim to produce a

protein product that can be sold in quantity at a low price or consumed at the subsistence level. The second is to select a species high on the food chain, which itself requires a high-protein diet. Food for such an animal will ordinarily be relatively expensive, thus the culturist's product will be a high-priced "luxury" food. Both approaches have often been success-ful; the point is that in order to determine whether he can economically grow a particular species, the culturist must know its feeding habits and nutritional requirements and the cost of satisfying them.

Usually fertilization of the water, rather than direct feeding, is em-ployed in rearing the first group of animals. Though fertilization of fish ponds is an ancient practice in the Orient, knowledge of the effects of fertilizers in aquatic systems lags decades behind what is known for ter-restrial communities. As is the case on land, fertilization of waters is as much a matter of art as science, and in the Orient, where pond fertiliza-tion is most extensively and successfully employed, local variation of dos-age and application is great.

Linked with the process of fertilization is the concept of polyculture —the growing together of different species or age groups. Almost all fertile bodies of water produce a variety of food organisms; for the most efficient utilization of these organisms it is essential that a variety of species be present to crop them. The more completely the culturist can fill the available feeding niches, the greater the total weight of flesh he can produce.

Since the culturist who chooses the second approach—direct feeding of an animal high on the food chain—usually supplies only one or two feeds, he is more likely to restrict himself to monoculture. However, any body of water produces a certain amount of natural food, and where carnivorous fish are cultured, this supply may be enhanced by the fer-tilizing effect of excess food and the metabolites of the stock. Thus the culturist again has the opportunity to increase total fish production through intelligent polyculture.

The grower of carnivorous species is sometimes able to reduce costs by taking advantage of inexpensive sources of food. For example, one of the reasons for Denmark's preeminence in commercial trout culture is the ready availability of trash fish at Danish fishing ports. Another sort of economy may be achieved by the use of pelleted feeds, which are much more easily handled than many unprocessed feeds. An increasing number of livestock feed concerns, particularly in the United States, are adding such fish feeds to their catalogs.

A word of caution is in order regarding feeding: The fish culture lit-erature contains more than a few reports of conversion efficiencies of 1:1 or only slightly more. Such reports, which appear to defy the second law of thermodynamics, are based on the dry weight of food and wet weight of fish. The best verified conversion efficiencies are on the order of 3 to 4:1, which is comparable to or slightly better than the best results obtained on land, where intensive feeding of animals is a much older process.

Caution is also in order in interpreting the reported yields of filter-feeding mollusks. Although it is true that higher yields may be obtained by culture of herbivores than carnivores and that the highest yields ever recorded for aquaculture have been achieved by growers of essentially herbivorous shellfish, it should be borne in mind that these yields depend not on local production of food organisms but on the action of tides and currents in transporting food organisms grown in a far wider area than that inhabited by the shellfish. Production figures from a few strategically located rafts cannot be extrapolated, on an areal basis, to an entire coastal or estuarine region.

4. Adaptability to crowding. It is obvious that the more individuals of a given size that can be confined in a given space, the greater the potential production of that space. Crowding, however, creates a host of problems, many of them unique to the aquatic environment. Growth of some species of fish has been shown to depend on population density, yet this most certainly is not true of the common carp. Far too little is known of the behavioral adaptations of fish to crowding. For example, when channel catfish (*Ictalurus punctatus*), which are normally territorial in the wild, are kept in ponds at very high population densities, their territorial behavior breaks down but their appetites do not suffer. Other territorial species, for example, some of the centrarchids, do not appear capable of this adaptation.

Concentration of waste products in standing water increases directly with population density, but this problem and the associated problem of oxygen depletion may be largely compensated for by increasing the exchange rate of water; raceways and net cages may be stocked at densities unthinkable in pond culture. Even such species as the rainbow trout (*Salmo gairdneri*), which has quite a high oxygen requirement, have thus been grown at extremely high densities.

Another concomitant of high population density is the facilitation of transmission of disease. This problem is not usually alleviated by rapid circulation of the water but calls for specific preventive or curative measures. The oldest disease control method is the fallowing of ponds, with or without lime treatment. More recently, chemical treatments have been developed for a number of diseases, and antibiotics are often incorporated in fish diets as a prophylactic measure, though this practice is of doubtful advisability. Present emphasis with many species is on the selective breeding of disease-resistant stock.

Yet another crowding effect sometimes observed is cannibalism, particularly in the early life stages. This problem may be alleviated to some extent by provision of ample amounts of food and shelter, but with some animals, notably pikes, pike-perches, and lobsters, it is impossible to stock them above a certain density.

THE GENERAL ECONOMICS OF AQUACULTURE

The success of aquatic farming ventures depends largely on the marketability of the product and the wise and efficient use of certain natural factors which render a particular site suitable for a particular form of aquaculture. For instance, the success of trout farming in southern Idaho hinges on the presence of an abundant supply of water at an optimal temperature for growth of salmonid fishes, and the farming of sessile mollusks originated, and remains concentrated, in areas where the seas are fertile and tidal exchange considerable. It is, of course, often possible to modify extensively natural environments or provide artificial habitats, but it is obviously most conducive to profitable operation to make maximum use of naturally favorable conditions.

It is often debated whether precedence should be given in aquaculture to development and improvement of programs designed to improve subsistence diets, or whether a more conventional, profit-oriented course should be pursued. In the latter case, it is often hoped that persons other than entrepreneurs and consumers will benefit initially through increased employment opportunities, and that eventually mass production will permit the price of the crop to drop until it is within the reach of all. It seems obvious, at least from a global vantage point, that high priority should be accorded to programs which show some promise of contributing to the alleviation of protein deficiencies. To discuss the relative efficiency in achieving this end of the two approaches just described would lead us into a discussion of economic development and far away from the subject of aquaculture. Certainly both approaches have been and will be explored, regardless of our opinions, so we will confine ourselves to analyzing the economics of a few selected aquacultural ventures.

It must be remembered that the term "luxury food" is relative and is defined within the context of a particular national or regional economy. Further, reliable information on the economics of aquaculture is scarce and difficult to obtain, both because of the understandable reluctance of many successful operators to reveal their economic secrets and because the situation, particularly in areas like the United States, where aquaculture is relatively new, is in a state of flux.

If our examples are drawn more from commercial enterprises than from subsistence aquaculture, it is only because the latter is even harder to analyze than the former. Not only are subsistence operations more diffuse and accurate data more scarce, but no one has yet ventured to quantify in dollars and cents the difference between a healthy, well-nourished and a sick, malnourished child.

Probably no form of aquaculture has been subjected to more intensive economic analysis than the young catfish culture industry in the United States (Table 2). It seems solidly entrenched as part of the "agribusiness" complex, but it is obvious that many of the early expectations of large and easy profits were extremely unrealistic. It is said that, at present, less than 10% of the American catfish farmers are making a profit, yet individual operators have realized returns on investment as high as 55%.

The crucial factor, as in most agricultural enterprises, is the degree of technical and managerial skill possessed by the operator. In fact, according to a publication of the United States Bureau of Commercial Fisheries, "Fish farming generally requires a higher level of management than conventional agriculture, in the sense that the technology as yet lies mainly in the realm of art rather than science." In catfish farming the required skills find expression in the ability of the best culturists to grow large fingerlings, to produce larger fish in a growing season than their competitors, to buy or produce cheap feed, and to harvest the fish more cheaply and efficiently.

It is also interesting to look at the price sensitivity of the catfish industry. In the last few years, the price of catfish has dropped considerably and, as might be predicted from Table 3 and Fig. 1, the percentage of operators realizing a profit has declined. Interestingly, most of the successful growers have been those who have emphasized a single product, usually live or fresh, dressed fish and have restricted themselves to local or regional markets. Attempts to diversify with packaged or prepared products and/or to seek a national market often have preceded economic failure. This information may not have wide applicability since catfish has historically been only regionally popular in the United States.

Looking into the future, it appears likely that the remaining catfish growers will continue to improve their yields, for a few years at least. The net result may be a further thinning of the ranks until the field is restricted to a few large producers, or emphasis may revert to small, low-profit operations serving local markets. It is conceivable that, with further application of cage and raceway culture, improvement of feeding methods, and so on, the evolution of American catfish culture could parallel that of chicken farming—from a luxury-food industry to a staple food

TABLE 2. POTENTIAL ANNUAL PROFIT (PER HA) OF CHANNEL CATFISH CULTURE

	UNDER AVERAGE MANAGEMENT		UNDER SUPERIOR MANAGEMENT	
Growing expense			Growing expense	
Fingerlings (1,200 @ 4¢)	$ 48		Fingerlings (1,200 @ 2¢)	$ 24
Chemicals	25		Chemicals	30
Feed (180 days, 22.5 kg/ha, $95/ton)	214		Feed (150 days, 33 kg/ha, $85/ton)	191
Labor	40		Labor (self-feeders)	50
Water pumping	8		Water pumping	8
Fuel and miscellaneous supplies	4		Fuel and miscellaneous supplies	4
Harvesting (4.6¢/kg)	30		Harvesting (2.2¢/kg)	14
Maintenance and taxes	25		Maintenance and taxes	25
Depreciation	13		Depreciation (self-feeders)	15
Interest on working capital	14		Interest on working capital	12
Total	$421		Total	$373
Income (93% survival, 0.56 kg ave. wt., 84¢/kg)	530		Income (93% survival, 0.68 kg ave. wt., 84¢/kg)	636
Profit (before tax)	109		Profit (before tax)	263
Return on investment (before tax)	23%		Return on investment (before tax)	55%
Cost of production			Cost of production	
Not including interest on investment	66¢/kg		Not including interest on investment	48¢/kg
Including interest on investment	73		Including interest on investment	53
Including interest on investment and 20% tax on profit	75		Including interest on investment and 20% tax on profit	73

SOURCE: U.S. Department of the Interior, Bureau of Sport Fisheries and Wildlife.

TABLE 3.   PRICE SENSITIVITY OF THE CATFISH INDUSTRY

PRICE ($/KG)	PROFIT ($/HA)	RETURN ON FIXED INVESTMENT (%)
1.09	324	28
1.04	272	23
0.88[a]	219	19
0.94[b]	170	14
0.89	119	10
0.84	67	6
0.79	15	1
0.74	−37	−3
0.69	−89	−8
0.64	−141	−12
0.59	−183	−16

[a] 1969 average price to producers.
[b] 1968 average price to producers.
SOURCE: U.S. Department of the Interior, Bureau of Sport Fisheries and Wildlife.

source. For this change to take place, however, would most certainly require a massive promotion campaign to increase the acceptance of catfish on American tables.

As if the situation were not confusing enough, the entire industry in the United States is threatened by foreign competition. Since its inception in the early 1960s, the center of the industry has shifted from Arkansas to southern Mississippi and Louisiana as a consequence of the longer

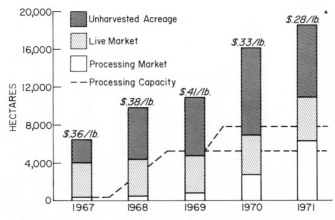

* To transform lb ⟶ kg multiply by 2.2

FIG. 1.   An estimate of the status of the catfish industry in the U.S.

growing season enjoyed by culturists in the latter locations. Further southward shifts would take the industry out of the United States. Already a farm in Honduras is capable of shipping frozen, dressed channel catfish to New Orleans at a price lower than that asked by Louisiana growers.

We have earlier asserted that animals which are low on the food chain may be produced more cheaply than carnivores. In making this claim, we assumed that other factors were equal. This, unfortunately, is seldom the case and we cannot offer comparative analyses of the culture of a carnivore and an herbivore within the same economy. A comparison of the American catfish industry with another fairly well analyzed aquaculture industry—milkfish culture in the Philippines—will, however, serve to illustrate the relative importance of biological, technical, and local economic factors.

Annual yields of good catfish and milkfish farms are roughly the same, as are annual profits to capital (Table 4). But the reasons for this similarity are quite different in the two cases.

Milkfish do not require extraneous feeding but subsist on a natural community of algae and associated microinvertebrates known locally as "lab-lab." However, to obtain maximum production of this food and hence the best yield of milkfish, the ponds must be fertilized at a cost which, relative to gross income, closely approaches that of feed in the best-managed catfish farms.

A major cost in Philippine milkfish farming, as in other forms of coastal aquaculture in southeast Asia, is that involved in the construction of ponds, often from virgin mangrove swamps. The large capital investment required for pond construction and related costs relative to the general level of economy of the Philippines and the high interest rate on loans for such purposes prevents the industry from enjoying much higher profits that would otherwise be indicated by yields and operational costs. Government-sponsored long-term, low-interest loans for pond construction could conceivably lower the price of milkfish and other species grown in the same manner to the point where they could represent a significant source of low-cost, high-quality protein for the Philippines and other developing countries. As matters now stand, milkfish is, in fact, more truly a luxury product than channel catfish; it fetches about one-quarter the price of catfish in a country where the average income is less than 5% of that in the United States.

Another group of cultivated organisms still closer to the base of the food chain are the bivalve mollusks. These animals feed directly on the unicellular algae or phytoplankton suspended in the water. No established commercial shellfish practice to date has involved the use of arti-

TABLE 4.   ECONOMICS OF A TEN-HECTARE MILKFISH PROJECT ($U.S.)[a]

Gross annual income	
15,000 kg of milkfish at $.22/kg	$3,300
Miscellaneous fish, shrimp, crabs, etc.	150
	3,450
Operating expenses	
Annual rental @ $1.50/ha	15
Purchase of 120,000 milkfish fry at $1.50/1,000	180
Salary of one caretaker, $18/month	216
Emergency labor	15
Supplies and material	22
30 tons of agricultural lime	9
2,500 kg organic fertilizers and 10 bags chemical fertilizers	787
Miscellaneous	15
Depreciation on equipment	
2% on concrete gate	8
20% on wooden gates	20
50% on nets	28
15% on flatboats	27
Interest (10% on capital)	726
Sales charges (brokers commission and other expenses)	158
Fish containers	6
	2,232
Annual net income for 1 year	1,240
Annual profit to capital	17%

[a] Expenses are based on the assumption that (1) the area is virgin mangrove swamp with second growth forest, (2) there are at least two creeks to close, (3) the desired elevation is one foot lower than the tide for the area, (4) labor is imported and workers get at least $0.75/day, and (5) area must be leveled and 50% excavated. Conversion from pesos at PS 1.00 = $0.15.

SOURCE: Esso Agroservice Bull. No. 8, Jan.–Feb., 1967

ficial food, artificially grown algae, or even the fertilization of natural waters to enhance the growth of the phytoplankton. Thus the yields of cultivated shellfish depend, among other things, upon the concentration of food organisms naturally present in the water and rate of movement of water carrying the suspended food to the animals. Greatly increased yields per unit of area under cultivation over those obtained by the traditional bottom culture of shellfish have been achieved by using the Japanese technique of three-dimensional culture, in which mollusks are suspended from rafts on ropes or wires 10 m or more in length.

Grown in this way, the bivalves have access to all of the food organisms suspended in the entire column of water from surface to bottom. A single raft 16 × 25 m in size and suspending 600 10-m strings of oysters is capable of producing over 4 tons of oyster meat per year, though average yields are probably about half that figure.

An economic breakdown of the Japanese oyster industry as practiced in the Inland Sea (Table 5) shows that profits and return on investment are high, as would be expected in a form of aquaculture dependent entirely on natural food. The capital investment is also modest, consisting only of the rather simple rafts (about $400 each), one or more small workboats, and a shed for shucking the oysters. An attempt was made to estimate capital outlay, but the figures may be very generous judging from the small annual cost of interest on capital, and the annual return on investment may therefore also be underestimated.

A distinct economic advantage to the industry is that the area under cultivation is designated by the prefectural government and assigned by the local fishermen's cooperative association to the individual grower at no cost. Raft culture is, however, a labor-intensive form of aquaculture. Labor costs in the Japanese oyster industry may account for as much as

TABLE 5.   ECONOMICS (IN $U.S.)[a] OF THE JAPANESE OYSTER INDUSTRY[b] (FOUR EXAMPLES RANKED BY SIZE OF THE OPERATION.)

Rank of grower	1	2	3	4
Area of growing field (ha)	3	6	12	24
Number of rafts	5	9	18	33
Capital investment (est.)[c]	10,000	14,000	19,000	27,000
Labor (hours/year)	5,121	10,963	27,106	28,343
Annual production, shucked meat (kg)	12,019	21,632	52,550	63,511
Annual gross income	6,000	10,000	17,400	32,600
Annual expenditures	2,200	6,500	13,700	23,700
Labor	540	1,800	5,700	9,300
Maintenance	265	930	2,950	3,450
Depreciation	615	1,120	3,100	4,200
All other[d]	780	2,650	1,950	16,750
Interest on capital investment	70	520	440	1,080
Annual profit[d]	3,730	2,980	4,260	7,820
Annual return on investment (%)[c]	38	25	25	33

[a] Conversion at 360 yen = $1.00 U.S.
[b] Data provided by Nansei Regional Fisheries Research Laboratory, Hiroshima, Japan, except where noted.
[c] Estimate of J. H. Ryther, based on personal visit and interviews with growers.
[d] Includes cost of seed oysters, fuel, rent, and charges.

50% of gross income, in contrast to about 10% for both U.S. catfish and Philippine milkfish farming. In countries where the cost of labor is appreciably higher than in Japan, raft culture is not economical for that reason alone.

It is clear from these few examples that the economic viability of a given aquacultural practice is dependent upon a large number of complex and interacting factors, often more sociological than technical, that are peculiar to the region if not unique to the individual enterprise. Broad generalizations and principles concerning the economics of aquaculture thus have little meaning. The blueprint of a profitable operation in one location is no guarantee of its successful application in another physical environment or in a different cultural or political setting.

Commercial aquacultural enterprises can also be evaluated and compared in terms of the tonnage produced per unit of human labor input. On the basis of the data presented in the preceding tables it can be calculated that approximately 4 tons of Japanese oysters (excluding shell weight), 15 tons of Philippine milkfish, and 30 tons of U.S. catfish may be produced per man-year of labor. Carp culture in the sewage ponds of the Bavaria Power Company near Munich, West Germany, yields over 30 tons per man-year (100 tons of carp from 200 ha of ponds tended by 3 men). Trout culture in Denmark yields about 16 tons of fresh fish per man-year, whereas on the more efficient trout farms in Idaho, the labor of one man may annually produce over 100 tons of fish. The latter figure is also closely approached if not exceeded in the high labor and moderately capital intensive mussel industry of northern Spain. It is to be noted that the Spanish mussel industry makes full use of extremely favorable natural amenities (high fertility of the water, protected natural sites, three-dimensional use of the water and its contained food organisms) and that the product therefore becomes relatively cheap.

Yields per unit of labor input in aquaculture compare favorably with medium-intensity pig and chicken farming, but the most advanced and mechanized high-capital input methods of rearing hogs and poultry produce considerably more animal meat per man-year of effort than commercial fish farming enterprises. Lack of adequate cost accounting and the difficulty in obtaining reliable information make such comparisons tenuous, to say the least. But it is not unreasonable to assume that the far greater application of research and development to land-animal husbandry, as compared to the rearing of aquatic animals, is largely responsible for the relative disadvantage in weight produced per man-year of aquacultural effort. There is no doubt that commercial application of newly developed, highly automated methods of feeding and harvesting can greatly improve aquacultural production efficiencies.

The foregoing information casts some doubt on the ability of intensive commercial aquaculture to contribute significantly either to the improvement of nutrition or the expansion of the economy in low-income countries. These two potential aims ought to be considered separately, though. Hundreds of hectares of ponds are now being brought under eel culture in Taiwan; the products so raised, small eels for further culturing, and others reared to minimal consumption size, are exported to Japan. Mullet culture is being perfected in that same country not so much because the mullet is a herbivore with a good growth rate and palatable flesh but because its sun-dried roe is a highly prized delicacy on the Japanese market. These enterprises provide employment and foreign exchange, both assets of great importance in the economy of an agricultural country of high population growth and low per capita income.

These assets of export aquaculture make it easy to find the capital necessary to achieve even moderately high production. Capital is less readily available for the production of low-priced fish for home consumption. Its supply must rely on the availability of government loans, which should result from a clearly stated policy of diversification of food production and the upgrading of nutritional levels. All too often governments lack the necessary experience to assess the security offered by proposed aquaculture enterprises. Thus aquaculture in low-income countries tends to remain a part of mixed subsistence or near-subsistence farming which defies conventional economic analysis.

Even where governments have correctly assessed the potential contributions of aquaculture, the incentives necessary to implement a successful program may be lacking, as illustrated by an example from Cambodia, a country in most respects naturally well-suited to aquaculture.

Cambodia's large fish consumption [about 20 kg/(person)(year)] has traditionally been supplied by largely unmanaged freshwater fisheries. By the mid-1950s, however, changes in land use and overfishing had resulted in a reduction in fishery yields. This, coupled with an annual net population increase of over 2%, made it apparent to Western technical advisers that more intensive fishery management, augmented by fish culture, would be necessary if the consumption of fish was to be maintained at its high level. Ginnelly (1962), who documented the successes and failures of fish culture in Cambodia, maintains that the failure to establish tilapia culture, attempted in 1955, resulted because: (1) conditions were not right for the acceptance of raising fish in this manner, that is, the government officials themselves were probably not convinced that it was a good thing, (2) the forestry division chiefs who were responsible for the ponds were neither interested nor trained in the management of fish ponds, and (3) an unfamiliar and foreign fish was introduced simply because this species is, in some respects, easy to raise.

The same constraints did not apply to a French rubber plantation where the management decided in 1955 to try to cut employee food costs by raising fish on the plantation rather than buying them from the distant market. After two years the management of the ponds was well in hand. Each 0.1-ha pond yielded 1600 kg/year on the average, or 16 tons/ha. The new species of fish was readily accepted by the workers on the plantation (Cambodian, Vietnamese, and Cham), although under conditions where they had no choice.

Ironically, the plantation ponds are but a short distance from the government pond in the town of Mimot. The installations were started about a year apart, one has been extremely successful and the other a failure.

Even smaller, subsistence-type aquacultural enterprises may be very successful, as is borne out from the cost accounting for a live-box fish culture operation, also in Cambodia. (Table 6.)

TABLE 6.   COSTS (IN RIELS, 36 RIELS = 1 U.S. DOLLAR) FOR ONE YEAR OF CAGE CULTURE OF *Pangasius* (CAMBODIA)

Box	5,000
Pen in lake, estimated	3,000
Fingerlings	10,000
One coolie	7,200
Fish food	30,000
Total	55,200

NOTE: If these fish are sold at the customary average price of 5 riels/kg and the mean weight is 1.75 kg/fish, then

$$1.75 \text{ kg/fish} \times 10,000 \text{ fish} = 17,500 \text{ kg}$$
$$17,500 \text{ kg} \times 5 \text{ riels/kg} = 87,500 \text{ riels}$$
$$87,500 - 55,300 = 32,300 \text{ riels (923 \$ U.S.),}$$
$$\text{or } 37\% \text{ profit to owner}$$

Comparable examples could be cited from other Asian countries, from Africa, and from Latin America. They suggest that the slow development of aquaculture in many low-income countries is not so much due to inherent shortcomings of the practice—although there are many lacunae in biological knowledge also—but largely to such problems as lack of funds, slow capital formation, lack of credit facilities, overextended administrations, and inadequate infrastructures (roads, markets, etc.). As

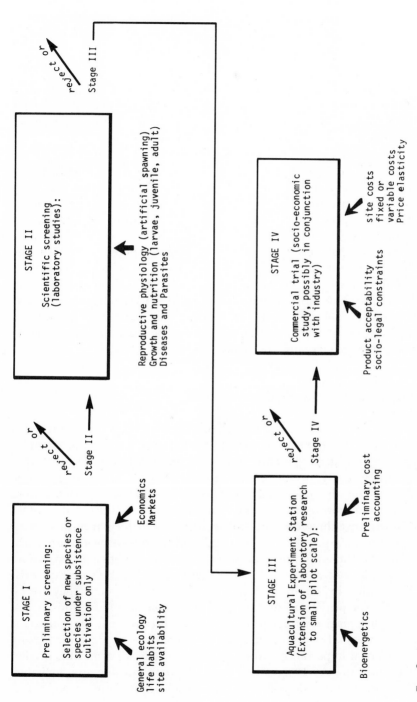

STAGE I

Preliminary screening:

Selection of new species or species under subsistence cultivation only

General ecology
life habits
site availability

Economics
Markets

reject or
Stage II →

STAGE II

Scientific screening (laboratory studies):

Reproductive physiology (artificial spawning)
Growth and nutrition (larvae, juvenile, adult)
Diseases and Parasites

reject or
Stage III

STAGE III

Aquacultural Experiment Station (Extension of laboratory research to small pilot scale):

Bioenergetics

Preliminary cost accounting

reject or
Stage IV →

STAGE IV

Commercial trial (socio-economic study, possibly in conjunction with industry)

Product acceptability
socio-legal constraints

site costs
fixed or
variable costs
Price elasticity

Fig. 2.

23

these are alleviated it behooves governments and private entrepreneurs alike to make the best assessment of the market potential of any species before embarking on large-scale developments. Aquacultural development in emerging nations might also be improved if some foreign technical advisers were less insistent on doing things their own way and more attentive to local preferences and practices. The interplay of many (or some) biological, social, economic, and other factors on the development of a species for commercial aquaculture is schematized in Fig. 2.

Necessary steps in the further development of aquaculture are (1) the establishment of central and regional aquaculture research stations, (2) demonstrations of techniques and facilities, (3) instruction for local officials, teachers, and farmers, (4) adequate extension services, and (5) provision of special credit arrangements. All of these depend on the adoption of a clear policy to promote aquaculture in each country.

In the so-called developed world, the considerations involved in the promotion of aquaculture are much more likely to be purely economic. Government assistance should be but is not necessarily effectively aimed at making the enterprise financially competitive with other food-producing industries. Official policies to improve the diet of the people are hardly seen as incentives except, perhaps, in the socialist planned economies, where, incidentally, cost accounting is also relevant.

A PROGNOSIS FOR AQUACULTURE

The total world tonnage annually produced through aquaculture, as defined by us, has recently been established at over 4 million metric tons. This tonnage is derived mainly from fresh and brackish water, with true mariculture barely in its infancy. Provided there were no economic constraints on the upgrading of culture techniques, the yields from existing aquaculture installations could well be multiplied tenfold within the next three decades. Aquaculture would then furnish us with animal proteins equivalent to more than half the present world fish catch.

In theory, there are opportunities for considerably greater expansion. For example, vast regions of presently unutilized brackish swamps exist in the tropics; in the archipelagoes of southeast Asia alone they are estimated to comprise 6 to 7 million ha. Their wholesale use for aquaculture or any other purpose is not to be advocated, because they are important nursery grounds for a great number of animal species, including many of commercial importance. However, judicious development of a portion of these swamps, say 10% or less, would be both possible and advisable. Within two to three decades these areas could contribute an additional

10 million metric tons of fish and shellfish—more than the current fishery production of the world's leading fish-producing nation.

Mariculture proper is also likely to make great strides in the coming decades. Already, Japanese culturists are producing nearly 30 kg of yellowtail/(m²)(year) in floating net cages. If they succeed in breeding these fish and/or in developing a pelleted feed for them, substantial improvement may be made.

To further illustrate the potential of true mariculture, we quote Shelbourne (1964), who discusses, albeit speculatively, the rearing of marine flatfishes in a very limited space:

The natural diet of the plaice includes small mollusks and marine worms. Mussel (*Mytilus edulis*) culture on a large scale should present no special difficulties; it is not beyond the realms of possibility that a cheap manufactured food based on fish offal or agricultural by-products, with balanced additives, would be acceptable to marine fish in fattening ponds.

Whereas "reaping" the natural stock is the expensive aspect of fishing, as we understand the term, "reaping" a pond stock is the cheapest imaginable process—simply pull out the plug and drain the pond. In this way, really fresh fish would become available to the national market with minimum preservation and processing costs. Processing plant is an expensive investment at the moment, inefficiently operated due to the fickle nature of fish supply. Rotational pond cropping, in a systematic manner, would enable the industry to trim its processing investment to the scale of continuous, guaranteed supply.

As a matter of interest, roughly 200 million North Sea plaice reach marketable size (25 cm) each year. About 75 percent of fish entering the fishery are caught by trawlers, the British effort accounting for 25 percent of the total take. In 1961, for instance, 35 million North Sea plaice were caught by British vessels. If each fish be given a hypothetical allowance of 1 ft² of bottom, then the annual British catch could be housed in shallow ponds covering 1¼ square miles in extent.

Among the imaginative schemes being tested for application in mariculture are various designs to increase the productivity of the surface waters of the oceans by pumping up nutrient-rich water from the depths. It is questionable that such pumping schemes could be economical for the purpose of aquaculture alone but if they were coupled with deep water (which is both cold and rich in nutrients) for use in the condenser cooling systems of power plants, for air conditioning, or perhaps for the production of power by means of vapor pressure differentials of the water from the deep and the tropical shallows, aquaculture could become a viable secondary industry, made possible by the costs of pumping being apportioned to several enterprises. It must be stressed though that such schemes can be potentially disastrous in certain environmental contexts since they may lead to the rapid eutrophication of lagoons within coral atolls. It would be important, therefore, to include in the feasibility studies of such maricultural schemes tests for methods that would use, or remove completely, the nutrients of the injected deep water. If deep water pumping were still envisaged for aquaculture alone, there would, in many locations, be no particular advantage to using deep-sea nutrients as opposed to nutrients of terrestrial origin such as sewage or runoff from fertilized land, provided the latter two were not contaminated with man-made, potentially toxic, inorganic or organic chemicals or human pathogens.

The total theoretical potential of aquaculture is, without question, very high. Whether it can be reached and sustained is problematical. In addition to technical and economic difficulties to be overcome there is the phenomenon of global pollution to be dealt with. Water pollution not only threatens the very survival of aquatic animals, it may also render them unfit for human consumption, as has already happened in a number of instances. Clearly, the future of large-scale aquaculture will be bright only if man thoroughly reappraises his policies of land and water management.

Some forecasters have also discerned a threat to the future of aquaculture in the actual and potential increases in the productivity of terrestrial agriculture. Such speculations underestimate the present and future need for food, as well as global problems of food distribution, which aquaculture can alleviate, to some extent, by being capable of producing moderate volumes of meat in many different places. Yet unless and until man achieves population stabilization, food needs will not be fully met, and regional inequalities in diets will persist. In fact, man's ultimate survival depends on his realizing the urgent need for curbing his numbers to the replacement level. As we proceed down what is hopefully the road to attaining this state, we will strain not only the food-producing capacity of the land, but that of the waters as well; indeed, we have already begun to do so. In the long run then, to discuss the competition of aquaculture with agriculture or any other means of food production seems irrelevant. What is important is to work to maximize sustained production of high-quality human food—along with population stabilization.

It is certain that maximization of the production of aquatic foods can be achieved only if the importance of aquaculture is realized, and its development aided by government and industry-sponsored research and development programs. Then it is not improbable that its development may parallel that of terrestrial agriculture. In a more distant future, the harvest of cultured aquatic stock may even come to rival the traditional, barely managed harvest of natural populations.

REFERENCES

ANON. 1970. A Program of Research for the Catfish Farming Industry. U.S. Bureau of Commercial Fisheries, Ann Arbor, Michigan. September 1970.

BARDACH, J. 1968. Harvest of the Sea. Harper and Row, New York.

BROWN, M. E. 1957. The Physiology of Fishes, Vol. I. Experimental Studies in Growth, 361–398 (M. Brown, Ed.), Acad. Press.

BROWN, L. R., and G. W. FINSTERBUSCH. 1972. Man and his Environment; Food. Harper and Row, New York.

DAVIS, H. S. 1953. Culture and Diseases of Game Fishes. University of California Press, Berkeley. 332 pp.

Esso Agroservice Bull. No. 8, Jan-Feb, 1967.

FAO Fish Cult. Bull. 2(3): Editorial.

GALTSOFF, P. S. 1964. The American Oyster. U.S. Department of the Interior Fish. Bull. **64**.

GINNELLY, G. 1962. The Role of Fish Culture in Cambodian Fisheries. M.S. Thesis, University of Michigan.

GUNDERSEN, K., and P. BIENFANG. 1970. Potential Productivity Value of Deep, Nutrient-Rich Pacific Ocean Water in a Combined System of Power Plant Cooling Water and Aquaculture in the State of Hawaii, U.S.A. Paper XXX, FAO Technical Conference on Marine Pollution. Rome, Italy, December 9–18, 1970.

HOLT, S. 1967. The Contribution of Freshwater Fish Production to Human Nutrition and Well-Being. *In* Symposium on the Biological Basis of Fishery Production. Blackwell, London.

NEESS, J. C. 1946. Development and Status of Pond Fertilization in Central Europe. Trans. Am. Fish Soc. **76**:335–358.

NICHOLSON, H. P. 1967. Pesticide Pollution Control. Science **158**:871–876.

OVCHYNNYK, M. 1968. *In* J. E. Bardach and J. H. Ryther (Eds.), Fish Culture. Part II.

PICKFORD, G. E., and J. W. ATZ. 1957. The Physiology of the Pituitary Gland of Fishes. New York Zoological Society, New York. 613 pp.

RADCLIFFE, W. 1926. Fishing from the Earliest Times, 2nd ed. E. P. Dutton and Co., New York. 493 pp.

ROELS, O. A., and R. D. GERARD. 1970. Artificial Upwelling. Marine Technology Society, Proceedings Conference on Food-Drugs from the Sea, Univ. Rhode Island, Kingston, R.I. pp. 102–122.

RYTHER, J. H. 1969. Photosynthesis and Fish Production in the Sea. Science **166**(3901): 72–76.

SCHÄPERCLAUS, W. 1954. Fischkrankheiten, 3rd ed. Akademie Verlag, Berlin. 708 pp.

SENGBUSCH, R. V., B. LUHR, C. MESKE, and W. STABLEWSKI. 1966. Aufzucht von Karpfen-brut in Aquarien. Arch. Fischereiwiss. 17(2):89–94.

SHELBOURNE, J. E. 1964. The Artificial Propagation of Marine Fish. *In* F. S. Russell (Ed.), Advances in Marine Biology, Vol. 2. Academic Press, London; New York. pp. 1–82.

SPOTTE, S. H. 1970. Fish and Invertebrate Culture. Wiley-Interscience, New York. 145 pp.

THOMAZI, A. 1947. Histoire de la peche. Payot, Paris. 645 pp.

T'UNG-K'AO, WEI-SHU. 1939 and 1954. An Examination of Apocryphal Books. Shanghai. (Chang Hsin-ch'eng, Ed.)

2

Culture of the Common Carp
(*Cyprinus carpio*)

HISTORY AND STATUS OF COMMON CARP IN FISH CULTURE

Of all the species of fish utilized by man the common carp (*Cyprinus carpio*) has the longest history of culture. As early as 475 B.C. spawning of captive carp in China was described and advocated as a profitable business by Fan Li in the first known treatise on aquaculture. Some authors believe the practice dates back as far as 2000 B.C. Aristotle mentions carp and it is likely that both the Greeks and Romans fattened carp in ponds. Further introductions in Europe may have taken place around 1150. The history of carp culture in Austria goes back to 1227, and by 1860 the species was raised in most, if not all, the countries of Europe.

Carp were first introduced into North America in the mid-nineteenth century and subsequently became widespread in streams and lakes there, although carp culture remained unimportant. The countries of southeast Asia have many similar native cyprinids which are used in aquaculture, but common carp were introduced to every southeast Asian country between 1914 and 1957 and are now cultured throughout the region. They were also introduced to Australia at an unknown date. In recent times carp have been widely introduced in Africa and Latin America for aquacultural purposes, but to date they do not play an important role in fish culture in Africa and among the Latin American countries only Guatemala and Haiti support significant carp-raising enterprises.

SUITABILITY OF COMMON CARP FOR CULTURE

The culture of carp has been a remarkably successful and widespread method of producing protein for human consumption. In 1965, carp were estimated to have contributed 210,000 tons to the world fish supply. This estimate did not include the carp production of mainland China, which exceeds that of all other countries combined; 1.5 million tons of common and Chinese carp were grown there in 1965. Furthermore, these figures are based on market statistics which are notoriously inaccurate in developing countries where 50% or more of the production is not offered for sale and therefore not counted. The total world production of carp and similar cyprinids may well approach 2 million metric tons with perhaps half of this figure derived from waters under intensive culture. If one assumes a per hectare yield of carp of 500 kg, which corresponds to the average in Israel for unfed fish in unfertilized ponds (2000 kg or more per hectare are attained with fertilization and feeding), 2 million hectares of water surface would be necessary to produce the estimated tonnage of cultured carp. This area is less than 1% of the total estimated freshwater area of the globe, including brackish lagoons, and certainly a small fraction of the water areas that eventually could be made usable for this type of fish culture.

The success of carp culture is due largely to the relative ease with which carp can be made to spawn in captivity and the hardiness of the species at all life stages from egg to adult. Carp adapt themselves to both acid and alkaline waters and easily tolerate salinities of up to 20‰. In Israel, carp are raised at salinities of up to 30‰. Although production at such high salinities is low, selective breeding is being carried out with the aim of developing a strain of carp which will thrive under such conditions. Carp are naturally tolerant of a wide range of temperatures, and selective breeding has enhanced this advantage by producing strains adapted to a wide variety of temperature regimes. Thus carp are now profitably raised from the tropics to the northern limits of the north temperate zone. Unlike most fish species, carp do well under conditions of high turbidity.

Complementing the general hardiness of carp are their catholic food habits. The natural food of young carp is zooplankton. Later in life they feed chiefly on bottom invertebrates. Both of these animal groups respond by an increase in their biomass to fertilization of the water, which considerably simplifies the aquaculturists' feeding chores. Other foods consumed in nature include algae, small fish, earthworms and other terrestrial invertebrates, and various kinds of detritus, particularly decaying plant matter. As might be expected, in captivity carp quickly learn to accept a wide variety of live and prepared foods.

The significance to the aquaculturist of the carp's remarkable hardiness is that, according to S. Tal, Director of the Inland Fisheries, Ministry of Agriculture, Tel Aviv, Israel, no other fish has yet been found that can be as easily managed for high yields per unit surface or volume of water, nor are there many other species that are as economical to raise. The carp's adaptability is expressed not only in its wide distribution and long history as a cultured fish and in the enormous production of carp flesh throughout the world, but also by the wide variety of techniques employed in carp culture.

COLLECTION OF WILD CARP FOR USE IN CULTURE

Although the carp is notable for the ease with which it is bred in captivity, in some localities low-intensity methods, which do not involve reproduction in captivity, are still employed. In the Soviet Union many carp of various ages as well as other fishes are stranded in shallow pools when spring flood waters of the larger rivers recede. It is common practice to rescue such fish for stocking in other waters. In some years as many as 1½ to 2 million carp are thus rescued.

PLATE 1. Transfer of breeder carp into Indonesian pond, showing tarred bamboo containers for transport of fish. (Courtesy S. Bunnag, FAO.)

Collection of stocks of naturally spawned fish in mainland China is a somewhat more sophisticated operation. Eggs and fry, rather than adult fish, are collected from the larger rivers to be raised in ponds. Common carp are not as highly valued in China as the various Chinese carps, but some are inevitably collected due to the nonselective methods of obtaining eggs and fry. (For a description of these methods see the following section on Chinese carp culture.) Some of these fish are raised, either alone or in polyculture with the Chinese carps, so that common carp accounts for about 5% of the weight of cyprinid fishes cultured in mainland China and Taiwan.

BREEDING

TROPICAL AND SEMITROPICAL WATERS—SEMINATURAL BREEDING

Since the carp is so easily spawned in captivity, most cultured carp are many generations removed from wild stock. Breeding in captivity has the advantage of stabilizing the supply of carp available for culture. More important, it permits selection for various desirable traits.

The classical methods of breeding carp are many, but all are adaptations of the spawning habits of carp in nature. In nature, carp spawn seasonally in temperate climates and year round in the tropics. The stimulus for spawning is a rise in water temperature, often accompanied by flooding. The adhesive eggs are laid near the surface on rooted aquatic plants, on floating plants, or on submerged terrestrial vegetation. The aquaculturist simulates these conditions by bringing spawners together in freshwater slightly warmer than that in his holding areas and providing real or artificial plants for attachment of the eggs. Perhaps the oldest variation of this technique is the Dubisch method, which is still practiced in some parts of Europe and in Indonesia. In this method, grass is grown on the bottom of a dry spawning pond to a height of 40 cm and the pond is filled until the water just reaches the top of the grass. Eggs are deposited on the grass. A trench 0.75 m deep may be dug around the perimeter of the pond to prevent spawning on overhanging terrestrial plants. When spawning is completed the carp are removed and the eggs allowed to hatch in the spawning pond. The classical Chinese method is identical except that filamentous floating plants such as *Ceratophyllum, Myriophyllum,* or water hyacinth are used as egg collectors. These should be thoroughly washed to remove potential egg predators. In Europe, piles of brush are used in place of aquatic plants in the similar Hofer method of carp culture.

Today most carp culturists rely on transporting the eggs to hatching ponds rather than removing the spawners. If the eggs are carefully handled this results in less disturbance than would be caused by netting out the adult carp. The simplest way of accomplishing this is to introduce floating plants confined within a floating frame into carp stock ponds at spawning time and let the carp spawn naturally. When spawning has been completed, the egg-laden plants are removed. Greater ease of handling may be effected by attaching the plants to bamboo poles fixed at regular intervals in the pond. Spawning carp in stock ponds has the advantage of minimizing handling operations and cutting down on the number of ponds required; however, it precludes selection of individual spawners, thus careful primary selection of spawning stock is imperative.

Space, labor, and water supply permitting, it is better to provide separate enclosures for each phase of culture. Spawning enclosures, particularly those used for small breeders, need not be nearly as large as ponds used in other carp culture operations. In India cement cisterns 10 m × 9 m × 1 m are used in commercial breeding. When the fish are ready for spawning, water from a pond heavily populated by carp is pumped into a depth of 0.5m; 15 or 20 kg of thoroughly washed aquatic plants such as *Hydrilla* or *Naias* are used to collect the eggs. In the evening 5 or 6 selected ripe females weighing about 20 kg each and 10 to 15 oozing males of the same weight are introduced. By morning, all or most of the fish will have spawned and the eggs can be removed to hatching tanks.

Containers placed in shallow areas of stock ponds may also be used for spawning. The best known device of this sort is the Indian hapa, a rectangular cloth tank about 1 m in depth, stretched and fixed by bamboo poles. The sizes of male and female carp and the amount of plants used are limited by the size of the hapa. Table 1 is a guide to the approximate amounts of fish and plants used.

TABLE 1.   AMOUNTS OF PLANTS AND FISH TO BE STOCKED IN HAPAS FOR SPAWNING COMMON CARP IN INDIA

DIMENSIONS OF HAPA (M)	NO. ♀♀	WEIGHT OF ♀ (KG)	NO. ♂♂	TOTAL WEIGHT OF ♂♂ (KG)	WEIGHT OF PLANTS (KG)
2 × 1	1	1 or less	2–3	1 or less	2
3 × 1½	1	3–4	2–3	3–4	5
4 × 2	1	5–6	2–3	5–6	7

In the evening the spawners are introduced and the hapa is covered to prevent the fish jumping out. Most of the fish will spawn by the next morning, but to insure complete spawning 30 hours may be allowed before the plants and eggs are removed and transferred to the hatching area.

After spawning, the carp are returned to the stock pond. The females are weighed before and after spawning. The difference in weight in grams multiplied by the average number of ovarian eggs per gram weight of ovary is used as an estimate of the total number of eggs laid.

In Indonesia, carp are spawned in ponds as small as 5 m². In such small ponds freshly cut grass or bunches of the dark, horsehairlike fibers of the indjuk plant (*Arenga pinnata* and *A. saccharifera*) are floated on the water surface to serve as egg collectors. However, most carp spawning in Indonesia is carried out in ponds 20 to 30 m². For such ponds a more easily handled egg collecting device known as a kakaban or egg mat has been developed. Kakabans are made of indjuk fibers which have been strengthened by soaking them in water for about 5 days. A thin layer of fibers 1.2 to 1.5 m long is pressed longitudinally between two bamboo lathes 4 to 5 cm wide. The margins are trimmed to produce an even end. The resulting structure is shaped like a two-sided comb with a width of 40 to 70 cm. Properly used and cared for, kakabans will last 1 to 2 years.

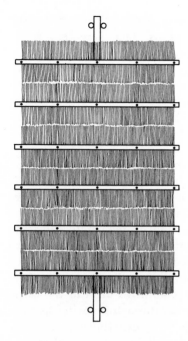

Fig. 1. Arrangement of Kakabans in an Indonesian carp spawning pond. (After Alikunhi, 1966.)

Kakabans are placed in a spawning pond with a bottom free of silt and mud but not hard enough to bruise the breeders. If the bottom tends toward muddiness, the pond may be dried for some days before use. The bottom may also be cemented or covered with sand. To prevent eggs being deposited elsewhere than on the kakabans the margins of the spawning pond should be free of aquatic and terrestrial plants.

Kakabans are laid transversely on a long bamboo pole held in place between two pairs of shorter poles driven into the bottom at either end of the pond. They are spaced so that the fibers of adjacent kakabans just touch. The bamboo pole floats, thus the whole structure moves freely with changes in water level, but the weight of the kakabans keeps it slightly submerged (Fig. 1). The number of kakabans required is calculated on the basis of 5 to 8 kg of female spawners.

While spawning is taking place a gentle flow of water is maintained. As in other methods spawning occurs mostly at night. When the lower surface of a kakaban is covered with eggs it is turned over; if both sides are full, it is replaced with a fresh one. When spawning is completed, the kakabans with eggs attached are placed in hatching ponds.

The spawning methods thus far described, all similar to natural spawning, are quite satisfactory in tropical and subtropical countries where carp are perennial spawners. In warm climates carp of both sexes may become ripe again within three months after spawning or, under exceptionally good conditions, within two months. There are records of individual carp in India which have spawned five times in one year. Thus a commercial carp breeder in the tropics can count on three spawnings a year by merely placing ripe males and females together in warm water and providing an egg-collecting device.

TEMPERATE WATERS-INDUCED BREEDING

In the temperate zones, carp are annual spawners. The carp culturist in a temperate climate is at a disadvantage in comparison to his tropical counterpart, not only because his fish spawn less often but because their spawning is less predictable. Unforeseen changes in the weather, an occasional problem in the tropics, are almost usual in temperate climates. For example, a cold snap during the spawning period may interrupt or delay spawning at a considerable loss in efficiency and total production. Total control of the carp's environment would undoubtedly prove effective in solving this problem, but costs are prohibitive. Instead the carp culturist may regulate the time of reproduction through inducing spawning by pituitary injection. This technique is being experimented with as a means of increasing frequency of carp spawning in temperate climates, but at present its chief advantages are in permitting the fish culturist to schedule his work quite rigidly and in allowing strict genetic control. (A short discussion of the history and significance of induced spawning, using pituitary materials, along with a list of references on the subject, may be found in Chapter 3.)

Ripe female common carp are injected intraperitoneally at the axil of the pelvic fin with fresh or fresh frozen pituitary extract from another common carp of either sex and equal weight or with 2 to 3 mg per kg body weight of acetone-dried carp pituitary in 1 cm³ distilled water. The importance of using common carp pituitaries should be stressed, for although most fish which have been tested with pituitary extracts respond to materials from unrelated species, even animals belonging to different classes, the common carp has thus far been found to respond only to pituitary extracts of its own species.

For best results fish to be injected should be conditioned and spawned in well-oxygenated water at about 20°C. The simplest spawning procedure is to place injected females in small ponds with an equal number of males and a device to collect the eggs and let spawning occur as in the methods already described. The males may also be injected, but if they are very ripe it is not necessary.

More efficient fertilization and a high degree of genetic control can be achieved by hand stripping the fish. This method is best for very large breeders which would normally require excessive space to spawn. After injection, females are kept in holding tanks separated from males. Within 12 to 20 hours they are usually ready to strip. Male breeders are injected in exactly the same way as the females. This has the effect of increasing the amount of seminal plasma, thus providing more milt. Eggs are stripped first, then milt from a male is stripped directly onto the eggs. (For a detailed description of the techniques of egg and milt stripping, see Chapter 20.) The eggs and milt should be mixed thoroughly with a nylon bristle paint brush in a plastic container, to which the eggs will not stick as they do to glass or enamel. Care should be taken to prevent any water from coming into contact with the eggs or milt during stripping. In this connection it is helpful to wipe the fish dry with cheesecloth and work in dry gloves.

Shortly after the eggs and milt are mixed, small amounts of water are dripped onto them; fertilization will occur only in the presence of water. Once fertilized, the eggs will begin to clump and adhere to each other. Before clumping progresses very far the eggs are lifted with the brush and shaken into a holding vessel containing 10 to 15 cm water and a mat of Spanish moss or some other egg collector, or the eggs can be poured onto the egg collector from the plastic container. In either case the water in the holding vessel should be vigorously agitated during transfer of the eggs in order to disperse the eggs in a manner similar to that attained by carp in nature. Dispersion is important to prevent fungus, which will inevitably form on any dead eggs, from smothering adjoining live eggs. The egg collectors with eggs attached are tranferred from the holding vessel to ponds or hatchery troughs for hatching.

It is worth noting in connection with induced spawning of common carp by pituitary injection that while common carp are unusual among fish in responding only to pituitary extracts from their own species, common carp pituitary is by far the most widely used in inducing spawning of other fish species. It is relatively large and easy to remove, retains its potency when dried for up to 2 years or, in some instances, as long as 10 years, and common carp are readily available and easily kept nearly everywhere that fish culture is carried out. Thus many state, commercial, and experimental fish culture stations maintain small populations of common carp, often culls from stocks raised for food, as a source of pituitary glands.

SEXING AND SELECTION OF SPAWNERS

Two common needs in all types of intensive carp culture are sexing and selection of spawners. Often it is not easy to determine the sex of common carp. When ripe, the female usually exhibits a fuller profile than the male. Old males usually develop a few nuptial tubercles on the sides of the head and on the pectoral and ventral fins. The only sure way to sex young breeders is by extrusion of the genital product. To avert the need for examination of each fish, carp culturists in India and elsewhere have developed the practice of spawning each female with two or more presumptive males of such a size that the total weight of the "males" approximately equals that of the female. Thus though a few immature males or females may be included among the brood stock, there is a very low probability of any female's eggs not being fertilized.

Male and female carp to be used as spawners are usually kept separately from each other and from other stock. It is often recommended that ponds used for this purpose be in a sheltered location, for it is believed that exposure to a cold wind with resultant chilling of the water may retard spawning.

In Indonesia spawners are fed a special diet of rice bran, porridge, and corn for 3 days prior to spawning. Immediately after introduction to the spawning pond they may receive a special ration of porridge about 1/20 the total weight of breeders. In general, though, breeders will reach optimum spawning condition under the same dietary regime which produces healthy commercial stock. However, breeders should not be overfed or encouraged to grow too rapidly, for excess fat hampers gonad development. Rice bran may be fed to spent breeders during the period of recovery.

Over the centuries culturists have developed and maintained a number of strains of carp considered to be especially desirable breeders. Fecundity is of course the primary consideration. But fecundity cannot be empirically determined without considerable expenditure of time and effort and the sacrifice of a number of fish. Thus external indicators of fecundity have been sought. Presumed characteristics of good breeders have been summarized as follows:

1. Body moderately soft.
2. Lower side of the belly broad and flattened so that the fish will stand on its belly.
3. Relatively great body depth.
4. Caudal peduncle relatively broad but supple.
5. Small head and pointed snout.
6. Rather large and regularly inserted scales.
7. Genital opening nearer to the caudal peduncle than in the average carp.

According to Hora and Pillay (1962), "Some farmers believe that the best mark of a good spawner relates to the insertion of the last scale before the genital opening; if a line is drawn from the head along the body to the center of the genital opening it should cross this scale and divide it into two equal parts."

There is also at least one behavioral indicator used in selecting spawners. Females which release large numbers of eggs at one time, so that they are bunched on the collector, are considered poor brood stock.

Discriminatory use of spawners displaying the preceding characteristics of course amounts to selective breeding, a subject which will be covered in detail later.

Age and size of spawners is also a factor to be taken into consideration. Age at maturity varies greatly with climate, as does growth. As a general rule in temperate countries males mature by their second or third year and no later than the fourth; females in their third or fourth year. In very cold climates, some individuals may not mature until the fifth or sixth year. In the tropics both sexes usually reach maturity within one year, sometimes in as little as six months.

Carp follow the general rule for fish in that the largest females produce the most eggs. Fecundity of course varies with genetic and environmental factors but Table 2 illustrates the general relationship between size and number of eggs. It should be pointed out that the spawn of very old fish may be low in viability.

TABLE 2.   FECUNDITY OF FEMALE COMMON CARP

SIZE (CM)	NUMBER OF EGGS
15–20	13,512
20–25	29,923
25–30	54,180
30–35	128,434
35–40	141,000
40–45	249,000
45–50	310,000
50–55	488,000
55–60	405,000
60–65	1,507,000
Over 65	2,945,000

It might seem more efficient to spawn the largest and most productive females, but under the conditions of close confinement characteristic of most of the classical carp spawning methods it may be difficult or impossible to breed very large females. Small males are preferred for the same reason and because they are more ardent courters. Most Asian culturists select females weighing 1 to 2 kg and males of the same size or slightly smaller. If induced spawning by pituitary injection and stripping of eggs and milt is employed, however, it does make sense to take advantage of the high fecundity of large spawners.

HATCHING

The time required for carp eggs to hatch in nature varies widely with temperature. Hatching times from 46 to 144 hours have been reported. Prehatching mortality is nearly always high, and may be as great as 80%. The chief causes of mortality are predators, including the parent fish, low rates of fertilization, low temperatures, and fungus brought on by the presence of dead or unfertilized eggs. The newly hatched larvae also suffer heavy mortality due to predation and, in relatively sterile environments, to poor food supply.

Of these causes of mortality, only poor fertilization need be of little concern to the fish culturist. Even if stripping of eggs and milt and artificial fertilization are not practiced, cultured carp are usually spawned in close enough confinement that each egg is almost certain to be reached by a sperm.

Predation by the parent fish is controlled by removing the eggs from the spawners or vice versa. If a hatching pond separate from the spawning area is used, the commonest predators, among them fish, crayfish, copepods, and aquatic insects, may be eliminated by leaving the pond dry until just before use. If it is not practical to dry the hatching pond it may be treated with Lexone at 2.5 ppm for 2 or 3 days before stocking. This will kill most predators and some parasites but will not affect the fertility of the eggs. Other poisons, including quicklime, Camellia seed cake, powdered croton seed, derris root, or commercially available rotenone, may also be used in conjunction with partial draining. Quicklime is especially effective against bottom organisms. Proper dosages are 60 kg/ha applied to a nearly dry pond or 100 kg/ha of quicklime and 150 kg/ha of tea seed cake if there is considerable water. Camellia seed cake or powdered croton seed are applied at from 50 to 200 kg/ha depending on the amount of water in the pond. Raw derris root must be soaked in water for a few hours before use. It is then crushed and the juice containing the rotenone wrung out into a bucket of water and diluted for use. One kilogram of derris root will provide enough rotenone to treat one hectare of pond surface. Rotenone powder should be used at about 5% of this concentration. Predation may also be averted by keeping the eggs indoors in hatching troughs with flowing water until they are eyed. They are then transferred to hatching ponds, however, for it is difficult to provide an adequate amount of zooplankton to feed the newly hatched fry in an indoor environment, although recent experiments in West Germany indicate the potential feasibility of rearing carp fry on a diet of brine shrimp (*Artemia*) in flowing water aquaria.

The temperature for hatching should be the same as that at which the eggs were spawned. The optimum temperature is about 20°C in temperate climates and 25°C in the tropics. Carp are quite temperature tolerant, but eggs and larvae should not be chilled. For this reason, it is best that hatching ponds be sheltered from the wind. During abnormally cold seasons, Indonesian carp culturists hatch eggs in wooden tubs about 12 cm deep which are placed in the sunshine during the day and kept in a warm building at night.

Fungus as a source of mortality is perhaps more prevalent in culture than in nature. The fungus *Saprolegnia* gains a foothold on eggs which are unfertilized or have been killed by physical shock. The white, fuzzy, foul-smelling masses of mycelia which form on such eggs may spread and smother adjacent eggs, killing them as well. In this manner *Saprolegnia* can spread throughout a batch of eggs with disastrous results. Carp eggs are large and remarkably resistant to physical abuse as fish eggs go; nevertheless, they should be transferred from spawning to hatching enclosures with the utmost care. The incidence of fungus can be further reduced by seeing to it that eggs are not allowed to bunch together too closely. If, despite all precautions, numerous dead eggs are seen, growth of fungus may be inhibited when there is a current over the eggs by flushing with malachite green at about 2 ppm.

In primitive methods of carp culture, such as the Dubisch method, the eggs are allowed to hatch in the spawning pond. In more advanced methods the egg collectors with eggs attached are transferred to separate hatching enclosures. Indonesian carp culturists support loaded kakabans in the same manner as for spawning. Before transfer to the hatching pond they are gently washed to insure that none of the eggs are coated with mud. In the hatching pond the fiber margins of the kakabans are not allowed to touch but are separated by 2 to 8 cm. Bamboo poles are placed across the ends of the kakabans parallel to the center pole and held in place by a board at each end of the pond parallel to the kakabans (Fig. 2). The weight of this device is adjusted so that it will compensate for changes in water level in keeping the kakabans about 8 cm below the surface. Hatching ponds are stocked at rates of approximately 1 kakaban per 30 to 50 m² of water surface.

Masses of plants and other egg collectors may be placed directly into hatching ponds or kept in smaller enclosures within ponds. In Indonesia, carp are sometimes hatched in a sump in the center of the pond; thus

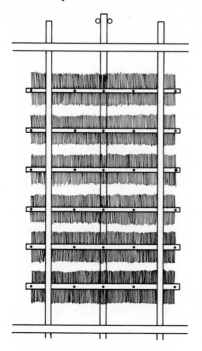

FIG. 2.   Arrangement of Kakabans in an Indonesian carp hatching pond. (After Alikunhi, 1966.)

the pond proper may be kept dry until the eggs hatch. This is especially advantageous where predators are plentiful or water is scarce. In India cloth hapas similar to those sometimes used in spawning are used as hatching enclosures. Hatching hapas are usually 2 m × 1 m × 1 m and made of cloth fine enough to prevent the fry from escaping but coarse enough to permit zooplankton to enter.

In parts of Indonesia the spawning enclosure is a compartment in a larger pond, separated from the pond proper by a temporary earthen dam. When spawning is completed the dam is opened and the spawners allowed to escape. Then the dam is sealed. After a week the dam is opened again and the larvae are allowed access to the pond.

Whether the eggs are hatched in a pond or an enclosure within a pond, it is important that the pond receive ample sunshine to insure the growth of phytoplankton to feed zooplankton which will in turn nourish the young carp. For the same reason the pond should be no more than 80 cm deep at its deepest point and naturally fertile or artificially enriched (see pp. 43–45). A soft bottom is best; moderate turbidity will do no harm. It is convenient to have the pond so arranged that it can be easily drained. A catching device may be set up at the lower end to trap the fry during draining for transfer to rearing ponds.

FRY REARING AND GROWING FOR MARKET

As previously mentioned, the time to hatching varies, but visual detection of larvae is no problem. When first hatched, the larvae do not swim about but attach themselves to plants or the walls of their container by means of a cement gland. Even so, with their elongate shape and beating tails they do not look at all like eggs. A little practical experience will enable the culturist to gauge the hatching rate of his fish under normal conditions. Complications in this regard will be diminished and growth of fry will be more uniform if all the eggs from a spawning are placed in hatching enclosures within one day.

The yolk sac is absorbed and the larvae become free swimming and capable of taking nourishment in 2 to 6 days, again depending on water temperature. The young carp may be transferred to nursery ponds as early as 3 days after spawning or they may remain in the hatching pond for up to 3 weeks. At the time of transfer they may be counted volumetrically. Survival to this point varies greatly; rates as high as 86% of egg production and as low as 20% have been reported, depending on weather, availability of zooplankton, and the success or failure of precautions against fungus and predators.

Once the fry have been transferred to nursery ponds, the culturist's main concern is to raise them to marketable size or for use as breeding stock. In temperate climates, where the carp is an annual spawner, it is often necessary to hold fry at a convenient size so as to have a supply of fish on hand for stocking throughout the year. Carp may be kept at a relatively small size for 2 years or more by judicious crowding and feeding just enough for maintenance but not enough for growth. This amount has been calculated as 30 mg of protein per 100 mg of carp per day at 22 to 24°C, or a daily intake of about 1/1000 of the total protein content of a carp. In general, however, the culturist's aim is to maximize production and profits by growing fish as rapidly as possible consistent with economic considerations. Maximum production is attained by fertilization of the water to increase production of natural carp foods, supplementary feeding, regulation of population density in the rearing areas, periodic culling of inferior fish, selection of breeding stock, and control of various environmental parameters.

POND FERTILIZATION

Pond fertilization may be thought of as an indirect method of feeding. Its purpose is to provide nutrients for phytoplankton, the lowest link in the food chain. An increase in phytoplankton will eventually be reflected in an increase in production at all levels of the food chain on up to the top, in this case the carp. Although it is possible to raise carp with no actual feeding whatsoever if the water used is sufficiently fertile, most successful carp culturists do make some use of direct feeding. But most carp culturists also find it advantageous to let their fish take part of their nourishment from natural foods. There is no economically feasible diet fed to carp which cannot be improved by the presence of naturally occurring bottom-dwelling invertebrates. In most cases the water used for carp culture will not naturally be so rich in nutrients that the numbers of these animals cannot be substantially increased by proper application of organic or inorganic fertilizers. While this may require some investment, the amount of money will usually be less than would be spent in adding the same amount of nutrient directly in the form of food.

In certain cases, carp may be raised in waters which are exceptionally rich in organic nutrients or, to put it another way, heavily polluted. The carp perform an additional service by retarding further organic enrichment of the environment. For example, in Java carp are confined in bamboo cages in rapidly flowing polluted streams. They graze on the carpet of small worms and insect larvae in the highly enriched water and yields of 50 to 75 kg of fish flesh per square meter of surface per year are not rare. Yields of pond fish are usually expressed per hectare or per acre; straight multiplication of the production figures just given would result in a weight of over 500,000 kg/ha. Even with allowances made for the fact that a large portion of a running hectare of such a stream might not be suitable for the placement of cages, this practice clearly represents an extremely efficient and ecologically sound use of sewage. It is, however, possible only in fairly rapidly running warm water and is not consistent with public health considerations.

A more sanitary method of utilizing carp as sewage converters is to grow them in conjunction with conventional sewage treatment operations. In Munich, Germany, the settled and/or partially treated sewage of the city is diluted and led through a 7-km series of 4- to 5-ha ponds, each containing about 5000 2-year-old carp which are fattened over the summer on the abundant invertebrate fauna. The annual increment in fish flesh is about 500 kg/ha without additional feeding. The net income of this now amortized installation owned by the Bavarian Hydropower Company is 50,000 DM (about $12,500). Similar installations in Berlin, Germany, and Kielce, Poland, have produced yields of 800 to 900 kg/ha, and 1300 kg/ha, respectively.

Certain industrial and agricultural wastes may be similarly used. In Czechoslovakia effluents from dairies, sugar mills, slaughterhouses, and

PLATE 2.   Combined duck-carp farming in Malaysia.

starch mills have all been utilized in carp culture. When treated with 5 tons/ha of lime, such wastes have produced yields of carp averaging 500 to 600 kg/ha. In several European and Asian countries ducks are reared in the same ponds as carp and natural fertilization of the ponds by duck manure results in fish yields of up to 500 kg/ha.

These exceptions notwithstanding, most carp culturists will find it highly advantageous to artificially enrich the water. But it should not be assumed that no harm can be done by fertilizers. Whatever combination of decaying plants, human or animal manure, or chemical fertilizers is used, an overdose may actually reduce production or even have a lethal effect on the fish. The correct types and amounts of fertilizer to be used vary from locality to locality and even from pond to pond in the same area. For specific advice it is best to determine the chemical and biological characteristics of the body of water to be used or of the soil on which it is to be constructed and to seek advice from local aquaculturists or from agricultural or fishery personnel.

SUPPLEMENTARY FEEDING

Carp culture usually is economic only when the population density of fish in rearing areas exceeds that which could be supported by the natural or augmented fertility of the water. For this reason, supplementary artificial feeding is necessary except for newly hatched fry.

In most culture methods food for very young fry is produced within the pond by fertilization rather than being introduced by the culturist. Notable exceptions occur in Japan, where water fleas (*Daphnia*) are cultured in ponds prior to the introduction of newly hatched fry, and in India, where young fry are commonly fed on various types of oil cakes mixed with rice bran. Recent research in India tested the effectiveness of 19 carp fry foods fed singly and in combinations of 2, 3, and 4 components. The various food combinations were compared in terms of percentage survival of fry and average growth in millimeters over a 15-day period. Of all combinations tested, a mixture of backswimmers (aquatic insects), freshwater prawns, and cowpeas in a 5:3:2 ratio was judged best. It produced significantly greater growth than any of the other experimental foods, and while a number of other blends, notably a 1:1:1 mixture of prawns, cowpeas and wheat bran, produced higher survival, this was offset by the difference in growth rate. Research on fry foods is in its early stages, but some sort of food for newly hatched fry may eventually be adopted by carp culturists throughout the world.

Sometime within the first three weeks of life carp fry are transferred to nursery or rearing enclosures. From then on supplementary artificial feeding is carried out in all types of intensive culture. The sheer variety of feeds used is bewildering, but selection of a feed or feeds will be less confusing if the culturist bears two principles in mind:

1. Artificial feeding is a supplement to naturally available food and food organisms produced by fertilization of the water. Thus food items selected should not duplicate the contributions of natural food but should compensate for nutrients which are in short supply.

2. The goal of feeding is to achieve maximum growth consistent with economic considerations. Total costs to the culturist will, of course, be higher with feeding than without, but cost per ton may be lower with feeding, thus enabling the culturist to realize a greater profit.

A third consideration which is beginning to be taken more seriously, at least in countries where the rationale for carp culture is not to compensate for inadequate protein supplies, is the effect of a particular feed on the quality of carp flesh produced.

A general index of the efficiency of conversion of a feed into fish flesh is a ratio variously called the growth coefficient, food quotient, or nutritive ratio:

$$\text{food quotient} = \frac{\text{weight of food}}{\text{increase in weight of fish}}.$$

Table 3 lists the approximate food quotients of 23 types of carp feed.

TABLE 3. FOOD QUOTIENTS OF CERTAIN FEEDS IN CULTURE OF COMMON CARP

FEED ITEM	FOOD QUOTIENT
Fresh silkworm pupae	5.0 – 5.5
Dried silkworm pupae	1.3 – 2.1
Silkworm pupae, pressed dry	1.4
Mysis dry	2.0
Chironomids	2.3 – 4.4
Clam meat	1.3
Meat powder	2.0
Dehydrated blood	1.5 – 1.7
Fish meal	1.5 – 3.0
Soybean cake	2.22
Barley	2.60
Oats	2.60
Wheat flour	7.2
Peas	2.7 – 2.8
Potato	20.0 –33.9
Rice bran	5.08
Wheat bran	4.22
Peanut cake	2.13– 2.7
Lupin seeds	3.0 – 5.0
Soybeans	3.0 – 5.0
Maize	4.0 – 6.0
Cottonseed	2.3
Cottonseed cake	3.0

Merely to feed the food with the highest growth coefficient does not guarantee that production of carp flesh will be maximized. Carp, like most other animals, do best on a balanced diet. The proportions of protein, fat, and carbohydrate in an ideal diet vary according to the age of the carp and availability of nutrients in the form of natural foods.

In general, artificial feeding should concentrate on foods which are high in carbohydrates. The natural food of carp is very high in protein, reaching 60% of dry weight in Chironomid larvae, one of the most abundant food organisms in most carp ponds. In nature, part of this protein is converted into energy. Feeds rich in carbohydrates provide a more efficient source of energy and free all of the protein to provide for growth.

It may be necessary to step up the protein intake in growing large carp since the need for protein increases with size. Care should be taken not to feed too much protein, however, for excess amounts will be wasted. Ten to fifteen percent by weight of protein in the diet should be adequate for large carp. Protein intake of carp may be increased by using feeds rich in protein or by fertilizing the water to increase availability of natural foods. The latter method is ofen economically preferable, particularly in countries such as India where all sorts of protein-rich materials command prohibitive prices. Whichever method of adding protein is chosen, it is worth remembering the general nutritional principle that a diet is likely to be better balanced in amino acids if it contains a variety of protein sources. Table 4 lists the percent composition and nutritive value of 37 commonly used carp feeds. Other foods sometimes fed to carp, but for which no data are available, include cottonseed, cottonseed cake, sunflower seed meal, soybeans, sorghum, duckweed, fish roe, fish entrails, fish meal, clam meat, meat powder, and brine shrimp (*Artemia*).

Maximum food consumption by carp is prodigious; at 25 to 27°C a carp is capable of consuming more than its weight in food daily. This does not result in great growth, for gorging leads to poor digestion and inefficient conversion of food. The optimum rate of feeding is difficult to determine, for it depends on a number of variables, among them the type of feed used, the amount of natural food available, and especially the water temperature. At summer temperatures carp may require 30 times as much food as during the winter. Indeed, during the coldest months, long periods of quiescence may occur during which no food whatever is taken. Thus in temperate climates feeding is intensified in summer and may be reduced in winter to occasional very light feeding on warm days. In the tropics the variation in feeding rate is less, but feeding schedules should still take into account water temperature. Tables 5 and 6 are examples of feeding schedules for the temperate climate of Japan and the somewhat warmer climate of South China's Kwangtung Delta.

Feeding schedules are usually made out on a per day basis, but greater efficiency can be achieved through several light daily feedings than by one massive application of feed. On the other hand, frequent feeding has the disadvantage of increasing labor costs. This problem may eventually be circumvented by use of automatic feeders which can be programmed to broadcast specified amounts of feed at set intervals. An improvement on this device is a feeder being tested at Wielenbach, Bavaria, Germany,

TABLE 4. COMPOSITION AND NUTRITIVE VALUE OF PRINCIPAL COMMON CARP FEEDS (% DRY WEIGHT)[a]

NAME OF FEED	CRUDE PROTEIN	FAT	CARBO-HYDRATE	FIBER	ASH
Vegetable products					
Groundnuts	27	45	18	3	2
Groundnut cake	36	10	32	—	19
Coconut oil cake	17–21	7–16	43–44	8–11	5– 6
Mustard oil cake	31	10	29	—	30
Lupin, sweet, yellow	42	6	25	10	2
Cowpea	22	4	71	—	4
Soybean cake	44–50	8–11	33	6	6
Bean meal	51	9	31	5	5
Barley	12–27	2	77	7	3
Rice (hulled)	8	2	77	7	3
Rice, broken, white	8	1	80	<1	1
Rice bran	13–16	4–18	43–47	6– 9	13–15
Wheat	15	2	75	4	4
Wheat flour, white	12	1	87	<1	<1
Wheat bran	15	5	62	12	6
Oats	14	3	74	6	—
Maize (corn)	7–12	5– 6	81	2	2
Maize (fresh)	10	5	70	2	1
Rye	12	2	70	2	2
Italian millet	12	7	75	—	6
Potato	8	<1	84	3	—
Sweet potato	6	2	85	4	—
Ragi	9	9	68	—	14
Animal products					
Silkworm pupae	49	26	7	—	3
Defatted silkworm pupae	83	2	9	—	6
Sardine	71	22	—	—	—
Crustacean (*Mysis*)	74	15	—	—	—
Crustacean (*Daphnia*)	42	7	31	—	—
Small shrimps	66	7	5	—	—
Snail (*Vivipara*)	83	2	—	—	—
Worm (*Limoudrilus*)	48	24	—	—	—
Worm (*Tubifex*)	65	15	14	—	6
Insect (*Notonecta*)	56	4	24	—	16
Insect (*Chironomus*)	60	2	6	—	—
Mixed zooplankton	46	6	23	—	25

[a] Dash = no determination.

TABLE 5.   FEEDING SCHEDULE FOR COMMON CARP IN JAPAN

MONTH	MONTHLY FEEDING RATE IN PERCENTAGE OF TOTAL ANNUAL QUANTITY	ESSENTIAL FOODS	DAILY FEEDING FREQUENCY	TIME REQUIRED TO FINISH FOOD AT EACH FEEDING
January	0			
February	0			
March	0			
April	1	Wheat, snail pupae, soya sauce waste, earthworms, rice bran	1	Within 1 hour
May	4	Mixed foods	1–3	30 min
June	15	Pupae as staple food, also mixed food	3–6	15–30 min
July	20	Silkworm pupae	3–7	15–30 min
August	30	Silkworm pupae	5–9	15–30 min
September	20	Silkworm pupae	5–7	15–30 min
October	9	Pupae, wheat, bean meal, vegetable and fish meal	2–5	15–30 min
November	1		Once to once every 2 days	30 min
December	0		Once, or every other day	

SOURCE: Hora and Pillay (1962) [adapted by Lin (1966) from Shih (1937)].

which permits carp to learn to release food at any time by pressing against an underwater plate. Using this device, carp soon adjust to taking only as much food as they need.

The central questions in determining what, how much, and how often to feed are, of course, questions of economics. Is it more economic to feed or not to feed? What feeding regime is most conducive to profitable

TABLE 6. FEEDING SCHEDULE FOR COMMON CARP AND GRASS CARP IN THE KWANGTUNG DELTA, CHINA

MONTH	APPROXIMATE TEMPERATURE (°C)	ESTIMATED WEIGHT OF FISH IN POND		QUANTITY OF FOOD REQUIRED PER MONTH (KG)			
		INCREMENT PER MONTH (KG)	TOTAL (KG)	GRASS	SILKWORM WASTE	FRESH SILKWORM PUPAE	RICE BRAN
February	15	—	70.00	—	—	—	—
March	16	127.80	197.80	1,534	1,022	639	639
April	20	149.10	346.90	1,739	1,193	745	745
May	24	191.70	538.60	2,301	1,534	959	959
June	28	255.60	794.20	3,067	2,054	1,278	1,278
July	30	319.50	1,113.70	3,834	2,556	1,598	1,598
Quantity harvested at the end of July			800.00		—		
New stocking material added			200.00		—		
Stock			513.70				
August	31	426.00	937.70	5,112	3,408	2,130	2,130
September	28	319.50	1,259.20	3,834	2,556	1,597	1,597
October	27	149.10	1,408.30	1,789	1,193	745	745
November	23	106.50	1,514.80	1,278	852	533	533
December	16	85.20	1,600.00	1,022	682	426	426

SOURCE: Hora and Pillay (1962).

operation? It has already been mentioned that, except in rare cases of extremely fertile water, it is economic to feed carp. The relative merits of feeding versus not feeding can be compared or an economic appraisal of various methods of feeding can be made if the cost of feed and feeding labor and the production of carp per hectare is known. Table 7 gives a sample comparison of costs and production of fed and unfed carp in Israel. This type of analysis may be applied to the comparison of any two feeding schedules simply by plugging in the appropriate data.

TABLE 7.   COSTS PER HECTARE AND PER TON OF COMMON CARP CULTURE IN ISRAEL WITH AND WITHOUT FEEDING

	YIELD WITH FEEDING		YIELD WITHOUT FEEDING	
	PER HA	PER TON	PER HA	PER TON
COSTS	2,100 KG —	— 0.47 HA	1,000 KG —	— 1.0 HA
Charges for capital invested in ponds and fishing gear	$ 360	$169	$360	$360
Water	210	99	210	210
Fertilizers	87	41	87	87
Maintenance	106	50	106	106
Feed	370	174	—	—
Labor	286	116	150	150
Marketing costs	32	15	12	12
Interest on working capital	5	2	2	2
General and overhead expenses	10	5	8	8
Total	$1,466	$671	$935	$935

SOURCE: Tal and Hepher (1966).

One other subject that should be considered in a discussion of carp nutrition is food additives. The use of vitamins and other additives in carp culture rests on the same assumption as their use in human nutrition: optimum amounts of these substances are not present in normal diets. Among the classes of additives which have been given to carp are vitamins, antibiotics, minerals, and tissue preparations. Their use is largely in the experimental stage, but results are encouraging.

Addition to carp diets of hydrolyzed yeast, rich in vitamins of the B and D groups, has resulted in experimental yield increases of 16 to 56%, depending on the other components of the diet, while reducing feed expenditure per unit gain in weight by as much as 15%.

The antibiotic terramycin, applied at a dosage of 5000 to 10,000 units/ kg of feed, has been shown to increase growth by 5 to 25%, with a 10.5% saving in feed costs and a higher survival rate of stock. Terramycin is particularly effective when the feed has a high vegetable content. However, fish culturists may eventually experience a problem encountered by farmers of cattle and poultry who use antibiotics prophylactically. The eventual evolution of strains of disease microbes resistant to antibiotics has in some cases made disease treatment very difficult and may lead to a net decrease in growth and survival. In all likelihood, those fish culturists who refrain from using antibiotics, except perhaps as a therapeutic measure against specific diseases, will do best in the long run.

Cobalt, a component of vitamin B_{12}, when added to carp diets at a rate of 0.08 mg of cobalt chloride/kg of fish/24 hours, or 3.0 g/ton of feed, resulted in an increase of vitamin B_{12} in the liver with an accompanying rise in growth rate of 30% in fingerlings and 15 to 20% in 2-year-old carp. Cost of feed per unit gain of weight decreased by 20%. Cobalt chloride may also be added to ponds as a fertilizer with similar effects.

Commercial tissue preparations, made from the viscera of slaughtered animals and used as a growth stimulant in warm-blooded animals, may also be added to carp feed. Seven kilograms of tissue preparation per ton of feed when added to the rations of 2-year-old carp increased growth by 12.0 to 13.3%.

STOCKING RATES

The amount of space allotted to carp in ponds varies with the characteristics of the pond, the type and amount of supplementary food given, and the size of the carp. Warm, shallow, naturally fertile ponds are best for all sizes and ages of carp, but fertilizers may be added. Once fertilizer dosages and feed rations have been worked out, the carp culturist's chief concern becomes the regulation of population density. Growth will be greatest at low densities, but space and labor considerations limit the extent to which this principle can be applied.

A suitable population density for newly hatched fry would of course amount to gross overcrowding in adult fish. So it is customary to maintain a series of ponds for raising fish of different ages. Segregation by size not only aids in regulating population density; it equalizes competition for food and assures that the somewhat different food requirements of carp at different ages can be met. Ponds may roughly be divided into three categories: nursery ponds, rearing ponds, and production ponds.

Nursery ponds are the first stop for the young carp after they leave the hatching pond and are usually the smallest of the three types of pond. Small ponds facilitate ecological control and recapture of the fry. Depth is usually less than 1.5 m, with some ponds as shallow as 0.5 m, to take full advantage of the warming effect of the sun. Rearing ponds are slightly larger in all dimensions but still less than 2 m deep. Production ponds may be of almost any size consistent with efficient feeding and harvest of fish.

In southeast Asia and the Mediterranean area, rice fields are used as rearing or production ponds. This practice is becoming less prevalent as the use of heavy machinery necessitates periodic draining of the fields and as herbicides and insecticides are increasingly used in doses lethal to fish. A further limiting factor in temperate climates is that the maximum size of carp which can be produced in so shallow an enclosure as a rice field is about 500 g. In most European and some Asian countries this is well below the accepted minimum marketable size.

The size, age, and population density of carp stocked in ponds varies greatly. Table 8 gives samples of still water pond or rice field stocking rates for seven countries. Prospective carp culturists are best advised to follow local custom, at least at first.

The factor limiting the number of carp which can be stocked in a pond is not the amount of space available to each fish, but the volume of water per fish. In still water ponds, which account for the majority of carp culture facilities, available space and water volume are virtually identical, but if water is circulated through an enclosure containing fish, the volume of water per fish is effectively increased with no change in the space allotment.

If conditions are such that a flow of water through a pond can be maintained, stocking rates may be far in excess of those employed in still water. For example, in the Philippines running water ponds are stocked with fry at rates of 280,000 to 850,000/ha as compared to 50,000/ha in still water, with comparable yields.

Often conditions do not permit construction of a flowing water pond. A more frequently applicable method of increasing circulation of water in carp culture involves the use of floating cages as rearing or growing enclosures. Small cages submerged in streams have long been successfully used in growing carp in Java and Cambodia. More recently, Japanese and Russian fish culturists have investigated the feasibility of rearing and growing carp in floating cages in lakes (see p. 559 for a discussion of the advantages of cage culture).

In Japan, carp fingerlings are stocked in rectangular bamboo framed nets, 2 m deep and varying in area from 7 to 81 m². These nets are floated by means of empty oil drums and anchored to wooden stakes driven into the lake bottom, usually in about 3 m of water. When stocked with fingerlings at rates of 10 to 80/m² and heavily fed, carp production may reach 4000 kg/ha.

Russian experiments in growing adult carp at high densities (50 to 250/m²) in floats have not been as successful, with growth generally less than that obtained at lower densities in still water ponds. Nevertheless, experiments in the use of floating cages in all phases of carp culture are continuing.

RECIRCULATING WATER SYSTEMS

A more sophisticated approach to the problem of water circulation in carp culture involves the construction of closed or semiclosed recirculating systems. This approach to fish culture has the advantages of minimizing the amounts of both space and water needed and of allowing nearly complete control over the fish's environment. On the debit side, elaborate filtration and aeration systems are required to compensate for the heavy oxygen demand and the large amounts of waste products generated by the extremely dense populations of carp. Moreover, although the small size of such systems cuts down on the total amount of labor required, a certain amount of specialized technical aid is necessary.

The first recirculating water system used in commercial carp culture was put into operation in 1951 by I. Motokawa of Maebashi City, Japan. In collaboration with Dr. A. Saeki of the Fisheries Faculty, Tokyo University, a pioneer researcher in the use of recirculating systems in fish culture, he converted a concrete fish pond into a 1-ton tank with a closed recirculating system. Dimensions and working capacity are given in Table 9.

Water is pumped from the fish tank to an adjoining concrete tank for settling out of sediment, then through pipes to one of a pair of filtration tanks. It is filtered through 60 cm of 1.5-cm diameter gravel spread on perforated plastic plate placed 20 cm above the bottom of the filtration tank. The filtration tanks are used alternately; the one not in use is washed periodically by compressed air passed through the pipes. Motokawa later built a 5-ton tank along the same lines as the original. Specifications and results of both systems are given in Table 10. When stocked with fingerlings at 30 to 70% of total fish holding capacity, production of carp per unit of water utilized reached 400 kg/m², the highest level ever achieved in Japan by any method.

TABLE 8. STOCKING RATES USED IN CULTURE OF COMMON CARP IN SEVEN COUNTRIES

COUNTRY	TYPE OF POND	AGE OR SIZE OF CARP STOCKED	STOCKING RATE	GROWTH	MORTALITY
India	Nursery; (stagnant; heavily fertilized and fed)	2 days old	1.25 to 2.5 million/ha	to 25 mm in 15 days	—
Indonesia	Nursery (or rice field)	3 weeks old	60,000/ha	to 30–50 mm in 3 weeks	40–60%
Nigeria	Rearing	6 weeks old, 30–50 mm	25,000–30,000/ha	to 50–80 mm in 3 weeks	20%
Philippines	Nursery	8– 10 mm	50,000/ha	to 50–60 mm in 1 month; may be grown up to 180 mm in nursery ponds	10–15%
	Rearing	60–180 mm	50,000/ha	to 20–50 g	—
	Rearing (running water)	Fry	280,000–850,000/ha	—	10–20%
	Production (stagnant)	20– 50 g	5,000/ha	—	—

Japan	Rice field	Fry	3,000–15,000/ha	—
	Rearing 1	1 month old	300–1,500/m²	—
	Rearing 2	2 months old	30–100/m²	—
	Rearing 3	3 months old	10–30/m²	—
	Rearing 4	4 months old	1–3/m²	—
	Production	2 years old	0.8/m²	—
U.S.S.R. (Ukraine)	Rice field (nursery)	Fingerling	10–80/m²	—
U.S.S.R.	Production (floating cage)	40 g	500,000– 2½ million/ha	—
U.S.A. (Alabama)	Nursery (experimental)	3–4 weeks	250,000/ha	—

TABLE 9.   DIMENSIONS AND CAPACITY OF A ONE-TON CLOSED RECIRCULATING
WATER SYSTEM USED IN CULTURE OF COMMON CARP

Dimensions	
Fish tank	76.5 m², 1.3 m deep
Filtration tank	24.8 m², 1.9 m deep
Volume of filter	9.2 m³ (gravel of diameter 1.5 cm)
Head of the two tanks	1.7 m
Total volume of water	205 m³
Working capacity	
Working area of filtration	15 m²
Filtration velocity	102 m/day
Pumps	one 2-KW centrifugal and one 2-KW vertical; one 5-HP gasoline engine (for emergency)
Circulation of water	75 m²/hour
Working oxygen intake	220 liters/day
Filtration	830 g/day as N
Carp reared	up to 1 ton
Carp kept	up to 6 tons

SOURCE: Deguchi (1965), quoted in Kuronuma (1966).

Even more spectacular production of carp with water circulation has been achieved experimentally at the Max Planck Institute in Hamburg, West Germany. There carp have been reared and grown in aquaria at truly incredible population densities. As many as 10 carp have been grown in a 40-liter aquarium. With rapid water circulation, filtration by activated mud supplemented by a constant inflow of freshwater, temperature control and a daily food ration of 3.5% of the fish's weight, growth rates in aquaria were 500 to 600 times higher than for camparable fish kept in ponds. No ill effects due to crowding were observed. Comparable results were achieved in the Soviet Union by use of heated water in a similar recirculating system. It is believed that if it had been financially feasible to build a larger filtration complex, the German system could have been operated as a closed system with similar results. Figure 3 is a diagram of a closed system of the same sort as the semiclosed system used at the Max Planck Institute.

SELECTIVE BREEDING AND HYBRIDIZATION

We have thus far covered the techniques of spawning, hatching, and rearing carp used by fish culturists in maximizing growth, production,

TABLE 10.   DETAILS OF COMMON CARP CULTURE IN ONE-TON AND FIVE-TON
WATER RECIRCULATING SYSTEMS

	ONE TON	FIVE TON
Total volume of water (m³)	205	286.3
Volume of water circulation (m³/hour)	75	135
Oxygen intake (liters/hour)	2.4	3.5
Fish tank		
Surface area (m²)	76.5	125.1
Depth (m)	1.3	1.5
Filtration tank		
Surface area (m²)	24.8	47.6
Volume of filter (m³)	9.2	28.6
Fish harvested (tons)	0.85	4.12
Fish harvested in unit area (kg/m²)	11.1	30.3
Period (days)	55	57
Temperature of water (°C)	16.3 –27.0	17–27.0
Number of fingerlings stocked	1,216	4,370
Weight of fingerlings stocked (kg)	543	2,250
Food given (kg)	625	1,890
Number of fish harvested	1,182	4,308
Weight of fish harvested (kg)	852.2	4,121
Number of fish died	34	62
Weight increase (%)	57	83
Food conversion	2.02	1.01
Average increase of weight (g)	273	442
Same per day (g)	5.35	7.75

SOURCE: Deguchi (1965), quoted in Kuronuma (1966).

and, ultimately, profit. Another factor which has bearing on carp production is selection of stock. Historically this function has also been
the province of the culturist. However, the professional carp culturist
rarely has the time or the facilities to carry out large-scale experiments
on selective breeding. Such work is most appropriately done by government agencies or other large operators. Further, it is questionable whether
any but the largest producers of carp for the market should spawn their
own fish, experimentally or for production. It has been shown that
mating of sibs, half sibs, or even cousins produces a marked inbreeding
depression of growth rate and viability. Given the small amount of brood
stock carried by most culturists, inbreeding is virtually unavoidable.
Since 1964 this rationale has been put into practice in Israel by the Carp

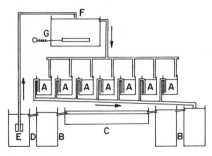

A   Aquaria
B   Clearing Basin
C   Activated Mud Filter
D   Lower Reservoir
E   Pump
F   Upper Reservoir
G   Heater

Fig. 3. Closed recirculating system used in culturing common carp at the Max Planck Institute, Hamburg, Germany. (After Meske, 1968.)

Plate 3. Intensive carp culture. Ten fish (mean weight, 913 g) reared in 40-liter aquarium. (Courtesy Dr. Christoph Meske and Bamidgeh, Israel.)

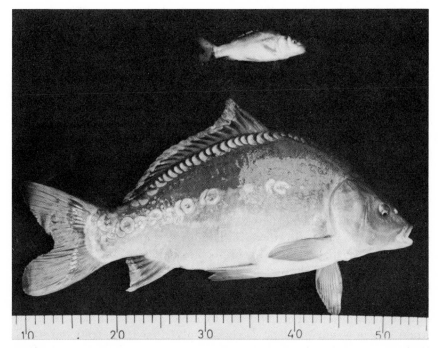

PLATE 4. Two sibling one-year-old carp. The upper fish, raised in a "wild" pond, weighs 40 g. The lower, intensively cultured in an aquarium, weighs 1750 g. (Courtesy Dr. Christoph Meske.)

Breeder's Union, who carry out research aimed at developing improved strains of carp and supply fry of such strains to production farms, thus allowing the commercial culturist to devote his total effort to the quantitative aspects of production.

By virtue of the ease with which common carp are spawned, their hardiness, and their great fecundity the possibility of selective improvement of the species was recognized very early in the history of fish culture. Selective breeding of carp has gone on for centuries, resulting in the diversity of strains available today. Nevertheless, great differences may still be observed between the progeny of different sets of parents of the same strain, so that selective breeding on this basis is still relevant. Further improvements may be made by crossing established strains to combine desirable traits and to take advantage of heterosis (hybrid vigor). Offspring of such crosses may in turn be bred selectively.

Whatever the means and ends of selective carp breeding, extreme care should be taken to maintain the purity of selected stocks. The very reproductive potential which makes the possibility of selective breeding so inviting enables one unwanted fish to destroy the results of years of careful breeding.

Hybridization among strains and subspecies has long played a part in genetic improvement of carp. With the development of artificial methods of spawning, interspecies hybridization is beginning to enter the picture. Most interspecific hybrids of the common carp are sterile, but this is not so disadvantageous as it might at first appear. If adult carp are grown in a pond for a number of years, natural spawning may take place. Spawning retards growth and may result in the introduction of further generations of fish to compete for food with the original stock. Stocking with sterile fish of course eliminates this problem.

It should not be surprising to learn that the qualities most frequently selected for are those directly advantageous to the culturist, among them high fecundity, high viability, good food assimilation, and, in particular, rapid growth. Mass selection for rapid growth has gone on for so many centuries that there is probably little potential for further improvement and such improvement as might occur could only be expressed under optimum conditions. The tendency to rapid growth is linked with the strain known as mirror carp, which has only a few large, scattered scales. In Europe and Israel this is the most commonly cultured strain. It is relatively difficult to spawn, however. Recently young of a strain of blue carp (*Cyprinus carpio,* var. *cerulea*) developed in Poland have also shown more rapid growth than is normal for cultured carp.

One might expect heterosis to be expressed in higher growth rates of hybrids, and this is usually the case when geographically remote varieties and subspecies are crossed. Such hybrids are also commonly hardier and more viable than are pure strains. Some interspecific hybrids, among them mud carp (*Cirrhinus molitorella*) × *Cyprinus carpio* may grow more rapidly than either parent. Carp × goldfish (*Carassius auratus*) is the only common carp hybrid which regularly occurs in nature, and is an exception to the general rule of sterility in hybrids. Backcrosses of carp × goldfish with either parent show somewhat better growth than the parent fish. The hybrid of male common carp with female Prussian carp, *Carassius auratus gibelio* is known as the "Savinsk silver crucian" and is reported not only to grow faster than either parent but to mature one year earlier and to be highly disease resistant. This hybrid has the added advantage of feeding almost exclusively on plankton.

Hybrid carp are in general more disease resistant than are pure strains. Hybridization and selective breeding have also been used to increase

PLATE 5. Regular or wild variety of common carp (above) and mirror carp (below). (Courtesy Dr. Christoph Meske.)

63

hardiness relative to environmental factors. Mention has already been made of Israeli efforts to develop a more salinity-tolerant carp. In the Soviet Union, cold resistance is of equal importance. Crossing European cultured carp (*Cyprinus carpio carpio*) with the wild Amur carp (*Cyprinus carpio haematopterus*) of East Asia produced a cold-resistant variety known as the Kursk carp, which has made possible the extension of carp culture in the Soviet Union as far north as latitude 60°N.

Breeding for quality from the consumer's standpoint has lagged behind selection for characteristics advantageous to the producer. The height-length ratio has long been accepted as an index of quality applicable by both consumer and producer. Carp with a high height-length ratio supposedly grow faster and have better quality flesh. This belief has sparked the production of such breeds as the German *Aischgrund*, which exhibits height-length ratios as high as 1:2, compared to the 1:3 to 1:4 ratios typical of wild carp. Recently, however, it has been shown that body conformation has nothing to do with growth rate and little if any connection with quality. Thus the only justification for continued selection on the basis of body conformation is on economic grounds: If a particular shape is more salable than others it is worth breeding for. The efficacy of even this sort of breeding for shape is questionable, since environmental conditions exert such a profound influence on body conformation.

Recent experiments carried out at the Max Planck Institute with the aim of selectively breeding carp without intermuscular bones (the small, often forked bones which make many fish, carp included, so annoying to eat) might indeed do the consumer a service as well as increasing the market value and salability of carp.

The mirror carp has already been mentioned. Two other varieties with reduced numbers of scales have been developed: the line carp, which has one row of scales along the lateral line, and the leather carp, which is almost devoid of scales. Selection for few scales has been rationalized on the basis that nearly scaleless carp are easier to prepare for the table, but this advantage is doubtful in a fish which has such large scales in any event. Be that as it may, in Europe the "scaleless" varieties of carp are considered superior as food. All three varieties bring better prices than scaled carp, but most culturists concentrate on the mirror carp, which grows faster and is more viable than line or leather carp and, unlike them, breeds true. Scaleless carp were not introduced to Asia until this century. Although they have not done well at low altitudes in the tropics, they have proven superior to the scaled varieties at high altitudes. The Asian consumer, however, is as prejudiced against mirror carp as his European counterpart is against scaled carp.

Before leaving the subject of selective breeding mention should be made of the fancy and colored varieties of carp which have been developed in China and Japan. Among the colors which have been produced are gold, lemon yellow, orange, rose pink, blue, dark green, and gray. Colored carp are used mainly for decorative purposes, but in recent years some of these colors have been used as genetic markers in experimental carp breeding.

Selection and hybridization for resistance to cold are not the only methods available for increasing carp production in temperate climates. In the Soviet Union carp are reared in floating cages placed in cooling reservoirs of power stations. These reservoirs receive effluents 10 to 15°C warmer than neighboring natural waters. When mixed with river water, the resulting temperatures are often within the optimum range for growth of carp. Even in midwinter the temperature of these reservoirs may be above 20°C. Before this practice was adopted, carp culture in floating cages was limited to the extreme south of the Soviet Union. Now it is practiced wherever thermal waters are available in central Russia. Yields compare favorably with, and in cold summers surpass, those achieved in ponds.

POLYCULTURE

Dramatic increases in yield of carp ponds frequently can be obtained through polyculture (rearing of several species together) to make more efficient use of the total pond environment. Chinese fish culturists, who use common carp as one of a complex of species, have developed this method to an art. (For details see Chapter 3 on Chinese carp culture.) Even when monoculture of common carp is intended, the culturist may find it advantageous to stock one of the Chinese species, the grass carp (*Ctenopharyngodon idellus*), if dense growth of weeds, which has been shown to adversely affect growth of common carp, becomes a problem.

Polyculture of carp is not as well developed in other parts of the world, but encouraging results have been obtained in the Soviet Union from raising common carp with *Cyprinus carpio* × crucian carp (*Carassius carassius*) hybrids, goldfish, bream (*Abramis brama*), and sterlet (*Acipenser ruthenus*). In Yugoslavia carp are reared together with tench (*Tinca tinca*). Tench compete for food with carp, grow more slowly, and reach a much smaller size, thus although they bring a high price, the economic feasibility of culturing carp and tench together is doubtful. Carp are commonly reared with roach (*Rutilus rutilus*) in France and experimental polyculture of carp with mullet and *Tilapia* is going on in Israel.

An aspect of polyculture that has developed chiefly in Europe is the use of predatory fish in conjunction with carp raised in ponds which support populations of trash fish or where "wild" spawning of carp takes place. The predators control the population of potential carp competitors and are eventually harvested with the carp. The traditional fish for this purpose is the pike (*Esox lucius*). The rainbow trout (*Salmo gairdneri*), imported from America, has two advantages over the pike. Unlike the pike it inhabits the open waters of ponds as well as shorelines and weed beds, and it brings a better price than pike. Use of rainbow trout in carp ponds is limited by oxygen supply; they require at least 5.6 ppm of dissolved O_2. Russian fish culturists report 25% increases in carp yield through use of another American predator, the largemouth bass (*Micropterus salmoides*), which like the pike is a shoreline dweller. Among the native European fish that have been more or less successfully used as predators in carp ponds are common whitefish (*Coregonus lavaretus*), peled (*Coregonus peled*), brown trout (*Salmo trutta*), English perch (*Perca fluviatilis*), and pike-perch (*Lucioperca lucioperca*).

YIELDS OF COMMON CARP CULTURE

Most of the techniques thus far discussed have been developed for the purpose of realizing the carp culturist's primary goal, a high yield of carp per unit area of water. Just what constitutes a high yield varies with locality and type of culture. Table 11 lists actual yields achieved by various culture methods in various countries.

HARVESTING

Carp are harvested by draining the growing pond or by use of a seine or cast net. Cast nets are preferred for harvesting small quantities of fish, since they can be operated by one man. Seining requires additional labor and, if the pond to be seined is large, the use of a small boat. Size at harvest varies according to local custom. In Indonesia, carp as small as 0.08 to 0.15 kg are not only accepted but preferred for table use. On the other hand, in most European countries 1.2 kg is considered near the minimum marketable size. In general, Asian consumers will accept smaller fish than Europeans, but the carp culturist should ascertain local preference before scheduling for harvest.

It is good practice to periodically harvest the largest, fastest growing fish in production ponds. This not only gives the culturist a constant source of income but serves to thin out the stock, thus improving the

TABLE 11.   YIELDS OF COMMON CARP CULTURE IN VARIOUS COUNTRIES

COUNTRY OR AREA	CULTURE METHOD	YIELD (KG/HA)	
Czechoslovakia	Growth in ponds with ducks	500	
Europe	Natural growth in ponds	25–	400
	Growth in ponds with feeding	100–	400
	Intensive culture in ponds	1,500	
West Germany	Growth in sewage treatment ponds, without feeding	500–	900
Guatemala	Intensive culture in ponds	4,000	
Haiti	Government hatcheries	2,300+	
	Farm ponds (subsistence culture), usually without feeding	550	
India	Natural growth in ponds	400	
	Growth in ponds with management	1,500	
Indonesia	Intensive culture in ponds	1,500	
	Growth in cages in polluted streams without feeding	500,000–	750,000
Israel	Intensive pond culture of mirror carp	1,500	
Japan	Growth in irrigation ponds (low fertility)	76	
	Growth in irrigation ponds (medium fertility)	180	
	Growth in irrigation ponds (high fertility)	490	
	Growth in irrigation ponds with very heavy feeding	5,000	
	Intensive culture in ponds	5,000	
	Growth in running water ponds or streams	400,000–2,000,000 (depends on velocity of water)	
	Growth in rice fields	700–	1,200
	Rearing in floating cages	4,000	
	Rearing in closed recirculating systems	4,000,000	
Nigeria	Commercial culture with fertilization and feeding	371–	1,834

TABLE 11.   (continued)

COUNTRY OR AREA	CULTURE METHOD	YIELD (KG/HA)	
Philippines	Intensive culture in stagnant ponds	5,500	
	Intensive culture in running water ponds, with heavy feeding	80,000	
Poland	Growth in sewage treatment ponds, without feeding	1,300	
U.A.R.	Experimental culture in ponds, with feeding	2,500	
United States (Alabama)	Intensive pond culture with inorganic fertilization, without feeding	314	
	Intensive pond culture with inorganic fertilization and feeding (not economically feasible)	784–	1,930
U.S.S.R. (Ukraine)	Growth in rice fields	150–	200
U.S.S.R.	Experimental culture in ponds in peat hags (unfertilized)	50	
	Experimental culture in ponds in peat hags (fertilized)	250–	300
	Pond culture on collective farms	200–	500
Yugoslavia	Culture in ponds, without fertilization	780	
	Culture in ponds, with manure added	1,500–	2,000

conditions of growth for stock remaining in the pond. Seines are preferred to cast nets for this purpose, since it is easier to sort fish in a seine.

TRANSPORT AND MARKETING

In Asia carp usually are sold live and may pass directly from producer to consumer. Iced or dried carp also are sold, but live fish bring a higher price, particularly in tropical areas, where preservation is a problem. Marketing may consist simply of displaying the fish in cages placed in the growing pond or in small, shallow ponds nearby.

Where it is necessary to transport the fish to market tanks mounted on vehicles are used. Although carp are not as resistant to crowding as certain Asian fish which have accessory breathing organs, they are hardier than most fish and may be transported in closed containers under semicrowded conditions. In Israel it is recommended that adult carp be shipped in tank trucks in a 1:1 ratio of carp to water. Up to four times as much water may be required at high temperatures. If crowding in transport is necessary, aeration by means of a pump or oxygen cylinder is advisable. In regions which have extensive inland waterway systems, oxygen problems are solved by transporting carp in streamline shaped cages suspended over the side of boats, thus providing a constant exchange of water as long as the boat is in motion.

At the market, carp are usually kept in cement cisterns about 1 m deep, which may be supplied with a crude running water system. A 3-m × 2-m cistern with running water is satisfactory for up to 300 kg of carp. If the fish are to remain in the cistern for more than one day, they should be fed. Thus maintained in clear water and fed regularly, they soon lose any "muddy" taste acquired in pond life.

Some carp are sold live wherever carp are caught or cultured for human consumption, but in Europe there is a greater market for processed carp. The oldest method of preservation is to salt the fish and dry them in air. Dried fish is often sold as a snack, to be eaten without cooking. Carp is also sold as fresh or frozen whole fish or fillets or in cans. In the case of canned fish, the producer usually sells to a processor rather than to a retailer or consumer. Carp destined for canning may be cut up and processed as is, in which case it requires cooking before eating, or it may be fried or smoked prior to canning. A common canned fish product is gefilte fish, favored by Jewish people throughout the world. The preferred fish for this dish is whitefish (*Coregonus* spp.), but as stocks of whitefish have become depleted in many countries, carp has increasingly been used for this purpose.

PROBLEMS OF COMMON CARP CULTURE

DISEASES AND PARASITES

We would be remiss in our discussion of culture of the common carp if we did not mention some of the problems the culturist may encounter. Some problems have already been discussed, including methods of compensating for a sterile environment and means of controlling or eliminating predators, competitors, excess plants, and *Saprolegnia* on eggs.

The carp culturist should also be prepared to cope with a number of diseases and parasites of carp. Among the disease and parasites afflicting carp are *Argulus,* ascites disease, bothriocephalosis, caryophyllosis, chilodonellosis, coccidiosis and coccidiosal enteritis, dactylogyrosis, *Ichthyophthirius* (sometimes known as "Ich"), infectious air bladder disease, infectious dropsy or red spot disease, infectious gill necrosis, philometrosis, *Prymnesium,* and sanguinicolosis. Some of these are curable and most are susceptible to treatment to prevent their spread. (For specifics of diagnosis and treatment, the reader is referred to *Culture and Diseases of Game Fishes* by H. S. Davis.)

The carp culturist wishing to provide an ounce of prevention will do well to use disease-resistant strains of fish if they are available, drain ponds which are not in use, maintain good water circulation, handle all carp carefully so as to avoid injury, avoid unnecessary mixing of stock or introduction of wild stock, feed an adequate diet, and to do whatever else is conducive to keeping the stock in good condition. In some instances eradication of specific disease hosts may also be undertaken as a preventive measure.

SOCIOECONOMIC PROBLEMS

The feasibility of carp culture in some regions, particularly the United States and Canada, is reduced by the presence of a complex of attitudes only partially attributable to the carp. Carp were introduced to the United States in 1877 attended by a great brouhaha of publicity anent their value as food fish. About that time, high-ranking fishery officials went so far as to suggest that the carp was so desirable that it might be advisable to eradicate some "undesirable" native species, for example, the largemouth bass and the northern pike. The apostles of the carp seem not to have considered the differences between American and European conditions. In North America human population density was low compared to Europe, land was plentiful, and farming of mammalian and avian stock, together with freshwater and saltwater fisheries, provided for a supply of protein far more abundant than that of any European country. There was simply no incentive for the intensive culture of carp. The enthusiasm accorded the first introductions soon waned, and in 1896 the U.S. Fish Commission ceased to import, distribute, and stock carp.

But the carp had gained a foothold. Escapees from private ponds populated those waters where carp had not been deliberately introduced. Today *Cyprinus carpio* inhabits 46 of the 50 states of the United States as well as the more southerly portions of Canada. Left to fend for themselves, they quickly reverted to the wild strain without a thought for the labors of the European fish culturists who had painstakingly developed the breed. Carp caught in North America are usually scaly, extremely bony, coarse-textured fish with a poor height-length ratio. Sport fishermen, who comprise a much larger segment of the population in the United States and Canada than in Europe, found these fish a far less attractive quarry than the native game fish. Neither did they appreciate the table qualities of the wild carp.

As the human population grew, attendant pollution, siltation, and overfishing decimated game fish stocks while the carp thrived. Anglers had noted the carp's habit of raising small clouds of mud while industriously rooting in the botton and in short order the nuisance became a villain, accused of increasing turbidity, destroying aquatic vegetation, and preying on the spawn of game fish. With the designation of the carp as the scapegoat for the sins of man in the decline of sport fisheries its rejection was complete.

Price and prejudice went hand in hand, so that today in North America carp bring only a third the price of such comparable "rough" fish as buffalo fish. Some carp are marketed by commercial fishermen, but sales are virtually limited to the poorest members of society.

With the continued eutrophication of American waters and the predicted narrowing of the gap between protein supply and demand, sport and commercial fishermen, fish culturists, and housewives may all have to reconsider the carp. For the present, however, carp are scarcely considered edible in North America and their culture is not economically feasible.

PREREQUISITES FOR SUCCESSFUL CULTURE OF COMMON CARP

The common carp remains the easiest of all fish to culture intensively. For this and other reasons it is also one of the best fishes for culture in many countries. Commercial culture of carp is likely to be feasible if the following conditions are met:

1. There must be a market for carp. This may be a mass market, as in countries where carp are traditionally eaten, or a specialty market, as in Guatemala where a small European population sustains a successful carp culture enterprise.

2. There must be an economical means of getting live or iced carp to market or a suitable means of preservation for shipping.

3. The carp culturist must begin with stock which is both adapted to

local conditions and suited to regional tastes in such matters as size and scaliness.

4. There must be adequate space for ponds. While it is possible to raise some carp in a single pond, the competitive advantage goes to culturists who use a number of ponds, each designed for a particular purpose. Ideally there should be a separate pond or ponds for spawning, hatching, nursing, rearing, growing, and holding male and female brood stock.

5. There must be an adequate supply of reasonably warm water. If water is scarce, a recirculating system may solve the problem, but such systems require a very large initial investment of money.

6. Ponds used in carp culture either must be naturally fertile or there must be an economical means of fertilizing them.

7. Unless the waters to be used are extremely fertile, carp feed must be available at a reasonable price.

8. There must be an adequate supply of labor available at moderate cost. If induced spawning or a recirculating system is to be used, employees must have some technical training. Otherwise unskilled labor will do.

9. Sufficient capital must be available to meet the initial expenses for stock, feed, equipment, and so on.

COMMON CARP IN SUBSISTENCE AQUACULTURE

Most of the foregoing conditions assume that the carp are to be sold. Carp also play a role in subsistence aquaculture, for which all that is absolutely necessary is a source of stock and a fertile body of water where they may be held and harvested. Today subsistence aquaculture is being promoted chiefly in Africa and Latin America. *Tilapia* spp. are the fish most frequently advocated, but in Haiti carp is used. Government hatcheries do the actual culturing, attaining yields of 2300 kg/ha and more. Young carp are distributed to small farmers who stock them in ponds, usually less than 100 m² and 50 cm to 1 m deep, then simply wait until they reach harvestable size. Without feeding, this takes 7 to 9 months. It would appear that a considerable local enhancement of protein supply is achieved in this manner.

PROSPECTUS

Culture of the common carp for subsistence and profit may be expected to increase in importance and efficiency for some time as researchers and culturists all over the world, particularly in the Soviet Union, Poland, Hungary, West Germany, Yugoslavia, Israel, Japan, mainland China, and southeast Asia, strive to improve their techniques. For the foreseeable future the common carp will continue to be one of the chief suppliers of fish protein to man and an increasing proportion of that protein will come from intensively cultured carp.

REFERENCES

ALIKUNHI, K. H. 1966. Synopsis of biological data on common carp, *Cyprinus carpio* (Linnaeus) 1758, Asia and the Far East. FAO Fisheries Synopsis 31.1.

ANDRIASHEVA, M. A. 1966. Some results obtained by the hybridization of cyprinids. FAO World Symposium on Warm Water Pond Fish Culture. FR: IV/E-10.

BARDACH, J. E. 1957. Marine fisheries and fish culture in the Caribbean. Proceedings of the Gulf and Caribbean Fisheries Institute, 10th Annual Session, pp. 132–137.

CLEMENS, H. P., and K. E. SNEED. 1962. Bioassay and use of pituitary materials to spawn warm-water fishes. U.S. Bureau of Sport Fisheries and Wildlife, Research Rep. 61.

DAVIS, H. S. 1953. Culture and diseases of game fishes. U. California Press, Berkeley, 332 pp.

GRIBANOV, L. V., A. W. KORNEEV, and L. A. KORNEEVA. 1966. Use of thermal waters for commercial production of carps in floats in the U.S.S.R. FAO World Symposium on Warm Water Pond Fish Culture. FR: VIII/E-7.

HORA, S. L., and T. V. R. PILLAY. 1962. Handbook on fish culture in the Indo-Pacific region. FAO Fisheries Biology Technical Paper, No. 14.

KURONUMA, K. 1966. New systems and new fishes for culture in the Far East. FAO World Symposium on Warm Water Pond Fish Culture. FR: VIII-IV/R-1.

LAKSHMANAN, M. A. V., D. S. MURTY, K. K. PILLAI, and S. C. BANERJEE. 1966. On a new artificial feed for carp fry. FAO World Symposium on Warm Water Pond Fish Culture. FR: II/E-5.

LAVREVSKY, V. V. 1966. Raising of Rainbow trout (*Salmo gairdneri* Rich.) together with carp (*Cyprinus carpio* L.) and other fishes. FAO World Symposium on Warm Water Pond Fish Culture. FR: VIII/E-3.

LIN, S. Y. 1966. General aspect of fish culture in Southeast Asia. Paper given at Fish Culture Symposium of the 11th Pacific Science Congress.

LING, S. W. 1966. Feeds and feeding of warm water fishes in ponds in Asia and the Far East. FAO World Symposium on Warm Water Pond Fish Culture. FR: III–VIII/ R-2.

MESKE, C. 1968. Breeding carp for reduced number of intermuscular bones, and growth of carp in aquaria. Bamidgeh 20(4):105–119.

MINTS, A. G., and E. N. KHAIRULINE. 1966. Intensive forms of rearing fish in ponds in peat hags. FAO World Symposium on Warm Water Pond Fish Culture. FR: VIII/E-6.

MOAV, R., and G. W. WOHLFARTH. 1966. Genetic improvement of yield in carp. FAO World Symposium on Warm Water Pond Fish Culture. FR: IV/R-2.

SARIG, S. 1966. Synopsis of biological data on common carp, *Cyprinus carpio* (Linnaeus) 1785, Near East and Europe. FAO Fisheries Synopsis 31.2.

SCHÄPERCLAUS, W. 1965. Lehrbuch der Karpfenerträge in Teichen durch Stickstoff-düngung Neue Düngungsversuche mit Karpfen, 1965. Deut. Fisch. Z. 13(1):6–14.

SCKHOVERKHOV, F. M. 1966. The effect of cobalt, vitamins, tissue preparations, and antibiotics on carp production. FAO World Symposium on Warm Water Pond Fish Culture. FR: III/E-7.

STEFFENS, W. 1969. Der Karpfen. A. Ziemsen Verlag, Wittenberg Lutherstadt. 156 pp.

TAL, S., and B. HEPHER. 1966. Economic aspects of fish feeding in the Near East. FAO World Symposium on Warm Water Pond Fish Culture. FR: III/R-1.

WOLNY, P. 1966. Fertilization of warm-water fish ponds in Europe. FAO World Symposium on Warm Water Pond Fish Culture. FR: II/R-7.

YASHOUV, A. 1966. Mixed fish culture—an ecological approach to increase pond productivity. FAO World Symposium on Warm Water Pond Fish Culture. FR: V/R-2.

Interviews and Personal Communication

PILLAY, T. V. R. Chief, Fish Culture Section, Inland Fishery Branch, FAO.

TAL, S. Director of the Inland Fisheries, Ministry of Agriculture, Tel Aviv, Israel.

WOHLFARTH, G. W. Fish Culture Research Station, Dor, Israel.

YASHOUV, A. Fish Culture Research Station, Dor, Israel.

3

Chinese Carp Culture

75

PLATE 1. Polyculture fish ponds in Kwangtung Province in southern China. (Courtesy F. Anderegg.)

BASIC PRINCIPLES OF POLYCULTURE

The People's Republic of China leads the world in the production, through culture, of freshwater fish (Plate 1). In 1965 estimates of the harvest ranged from 1.5 million to 3 million metric tons. As early as 1959 Chinese authorities claimed an average individual yield from pond culture of 7500 kg/ha. These estimates may be inflated, but yields of that magnitude are achieved in more than a few ponds. In the last 10 years production per hectare has almost certainly increased greatly, for the innovation of artificially induced spawning, which has advanced fish culture throughout the world, was particularly beneficial in the Chinese situation. Thus the great productivity of Chinese freshwater fish culture is in part due to modern technological innovation as well as to the tremendous effort put forth in recent years by the Chinese people. But of equal importance is the application some 1000 or more years ago of two seemingly obvious principles of fish culture. During the Tang Dynasty (A.D. 618–904) the Chinese, who had been in the fish culture business for some 10 or 20 centuries, recognized that:

1. A body of water is a three-dimensional growing space. To treat it like a field, by planting only one kind of crop, is likely to result in wasting the majority of that space.

2. Any fertile pond will produce a number of different fish food organisms. However, most fish are not omnivorous, but rather selective in their diet. Thus stocking single species wastes not only space but food.

In other words, the Chinese greatly increased the efficiency of fish culture by applying to the pond microcosm the science of ecology, for which the Western world is now acquiring a belated concern. To assure efficient use of the pond environment, Chinese fish culturists usually stock four types of fish; two midwater dwellers, one of which prefers phytoplankton as food, the other zooplankton; a fish which feeds mainly on higher aquatic plants; and a bottom-dwelling omnivore. Often a carnivorous bottom dweller will be added to this basic community. If trash fish or excess young fish inhabit the pond, a predator may be included to control them. This latter practice is not general in China, but is quite common in Cambodia and Vietnam.

HISTORY OF THE CHINESE CARPS IN FISH CULTURE

One might suppose that the practice of polyculture, as it is called, would have spread from China or developed independently elsewhere, but this has rarely been the case. The spread of the Chinese carps themselves was effectively blocked by the inability of fish culturists to spawn them. In nature they spawn in rivers and it was long believed that, under the conditions of still water and high population density prevalent in freshwater pond fish culture, their gonads would not mature. It has since been shown that they will mature in ponds, but in captivity natural spawning almost never occurs. Thus culturists were long limited to wild fish as a source of stock and the spread of the Chinese carps was limited by the available means for transporting live eggs and fry. Nevertheless, Chinese farmers who settled on Taiwan 300 to 400 years ago brought with them the practice of pond culture and the Chinese carps, even though there were no rivers large enough to support natural populations, thus fry had to be imported annually, at considerable risk, from the Chinese mainland. Chinese carps were also transplanted throughout southeast Asia, where local adaptations of the Chinese method developed. Some of the Chinese carps were imported to Japan, but monoculture remained the dominant practice there.

Neither Chinese fish nor the concept of polyculture found their way to the West until recent years. Perhaps improvements in communications have had something to do with the recent Western interest in polyculture, but two other technological advancements have had great effect.

Improved means of transporting fish over long distances were largely responsible for the successful introduction in 1949 of grass carp (*Ctenopharyngodon idellus*), silver carp (*Hypophthalmichthys molitrix*), and big head (*Aristichthys nobilis*) to the Soviet Union and subsequent successful transplantations.

In the 1960s development of techniques of inducing spawning of Chinese carps by hypophysation eliminated dependence on wild stocks. Within the decade various members of the Chinese carp complex became the subjects of experimentation and acclimatization in a number of western countries, including Bulgaria, Czechoslovakia, France, Hungary, Iraq, Israel, Poland, Rumania, the United Arab Republic, the United States, West Germany, and Yugoslavia. More important, the ancient Chinese concept of polyculture is being considered wherever fish are cultured.

The distinctions as to ecological niche among the Chinese carps are not hard and fast. For example, examination of the stomach contents of those species which are considered zooplankton feeders often reveals a preponderance of phytoplankton. Nevertheless, there is enough variation in zone of habitation and/or feeding habits to justify the accepted classification of pond fish into groups, even if centuries of experience had not proven it a useful tool.

SPECIES USED

PLANKTON FEEDERS

The most commonly stocked phytoplankton feeder is the silver carp, which will also accept such artificial feeds as bean meal, rice bran, and flour. In Vietnam it is said to be replaced by the ca duong (*Hypophthalmichthys harmandi*) which may be identical. Another phytoplankton feeder sometimes grown in Malaysia is the sandkhol carp (*Thynnichthys sandkhol*). The classical zooplankton feeder for pond polyculture is the big head, which, although perhaps a more efficient feeder on zooplankton than the silver carp, may take the bulk of its nourishment from phytoplankton. Another cyprinid reputed to be a zooplankton feeder is the ma lang yu (*Squaliobarbus curriculus*), sometimes used in polyculture in South China and the Indochinese Peninsula. A less specialized midwater feeder, which consumes both phytoplankton and zooplankton, is the ca choi (*Labeo collaris*).

There has been some speculation as to how a fish might select a particular type of plankton. The most likely way is by means of the gill rakers, but visual cues or vibration patterns received by the lateral line could also be involved. All plankton feeders have extremely long, fine gill rakers, but these are especially well developed in the silver carp, suggesting that it is capable of more selective filtering than other plankton feeders. Some observers, however, are of the opinion that silver carp do not select for phytoplankton but that zooplankton are crushed by the pharyngeal teeth and digested, so that no trace of them is found when the digestive tract is opened. Whatever the degree of selectivity among plankton feeders, the practice of stocking silver carp and big heads or some similar combination is firmly entrenched in Chinese fish culture and yields good results.

HERBIVORES

Silver carp may also consume some higher plants, and in South China and the Indochinese Peninsula the ca ven (*Megalobrama bramula*) is stocked as an herbivore, but by far the best known and most widely cultivated macrophyte feeder is the grass carp. The grass carp is not an obligate herbivore; indeed it will eat almost anything including decaying clothing, but it does prefer vegetable food. (The U.S. Warm-Water Fish-Cultural Laboratory at Stuttgart, Arkansas, disputes this general assumption, and classifies the grass carp as an "opportunistic" feeder.) Higher plants are usually absent from ponds used in intensive fish culture, but so voracious is the grass carp that even if they are present supplementary feeding is necessary. Offsetting this expenditure somewhat is the nourishment that omnivorous bottom-feeding fishes derive from the partially digested plant remains in grass carp feces.

Despite its extreme voracity, sensitivity to noises (which may lead to such behavior as leaping onto the bank), poor growth in water below 14°C, and bad reputation as a spreader of parasites and diseases, the grass carp is an extremely popular pond fish and the only one of the Chinese carps to find use outside the traditional framework of polyculture. The predilection of the grass carp for vegetable food has led to its use in weed control in ponds. In cold climates, as in Germany, the feeding rate of grass carp is so low that their effect on aquatic plants is negligible. In warmer climates, however, they have been found quite effective in controlling weeds while at the same time adding to the productivity of

TABLE 1.   STOCKING RATES OF GRASS CARP FOR WEED CONTROL

COUNTRY	SIZE OR AGE OF GRASS CARP STOCKED	STOCKING RATIO	EFFECTIVENESS
Japan	Yearling	19/ha	None
	Yearling	55/ha	70% eradication of extremely heavy weed growth in 1 year
	Yearling	50/ha	100% eradication of submerged and floating plants in 1 year; no effect on marginal plants
Rumania	Yearling	500–800/ha followed by 200–500/ha in one year if not 100% eradication	100% eradication in 1–2 years
	3 years	80–150/ha	Control
	4 years	30– 50/ha	Control
	3– 4 years	160–240/ha	100% eradication
United States (Arkansas)	30–40 cm	50–100/ha	Control of *Chara* and pondweeds
	30–40 cm	1,700/ha	100% eradication of submerged plants in 6 weeks

ponds containing other species of fish and invertebrates. Table 1 lists sample stocking ratios of grass carp for weed control.

It can be seen that in warm climates grass carp, if stocked heavily enough, are very effective in controlling submerged and floating vegetation but are of doubtful value against marginal plants. The Arkansas experiments involved 12 species of plants, all of which were eventually eaten, but there was a definite preference for the softer varieties.

The use of grass carp in weed control entails the risk that they may escape into natural waterways and destroy valuable plants. However, since there are few comparable herbivores among the world's fish fauna, and since grass carp are a good deal cheaper than chemical or mechanical methods of weed control, their use will probably become more prevalent. Another fish that may find favor in this regard is the tawes (*Barbus gonionotus*), which, though considered a second-rate food fish, is used in polyculture as well as weed control in southeast Asia.

BOTTOM DWELLERS

Plankton feeders and herbivores are virtually universal in pond polyculture, but bottom-dwelling carnivores and omnivores are not always used. Many Asian ponds are fertilized so heavily that the resulting plankton bloom effectively shades the bottom, so reducing productivity in that zone as to limit greatly fish populations there. Where conditions permit the stocking of benthic fishes the native Chinese omnivore is the mud carp (*Cirrhinus molitorella*), but it is often kept in conjunction with, or supplanted by, the common carp, particularly in northern China. In addition to being more resistant to cold, the common carp is easier to spawn and reaches a larger size. On the other hand, the common carp is a rather energetic feeder and when densely stocked in ponds may substantially increase the turbidity by its actions. High turbidity is never desirable in fish ponds and is particularly unfavorable for silver carp. The mud carp, despite its name, does not root in the bottom like the common carp, but picks its food from the substrate rather daintily. It is also considered a superior table fish to common carp. A third omnivore, the crucian carp (*Carassius carassius*), imported from Europe, is sometimes used. Its principal advantage lies in its extreme hardiness with respect to low temperatures and dissolved oxygen concentrations; it can be bred in ponds where it is impossible to grow other fish. However, in addition to feeding on the bottom it may compete for plankton with silver carp and big head. Since its flesh is of poor quality and the market price correspondingly low, it should not be used if more desirable species are available. Other omnivorous cyprinids cultured in southeast Asia but neither native nor introduced to China include the belinka (*Barbus belinka*), the lampai (*Barbus schwanenfeldii*), the mata merah (*Barbus orphoides*), and the tambra (*Labeobarbus tambroides*).

The usual carnivore for pond culture is the black carp or snail carp (*Mylopharyngodon piceus*). The ca cham (*Mylopharyngodon aethiops*) is also mentioned in the literature in this regard but there is some doubt as to whether or not it is identical with *M. piceus*. Despite the fact that the black carp is the largest of the Chinese carps, reaching lengths of up to 180 cm, it is not highly esteemed as a food fish outside of China. Its inclusion in polyculture in Taiwan, Malaysia, and other nearby countries, is the result of accidental introduction with fry of other species imported from China. The favored food of black carp is mollusks, and Chinese fish culturists sometimes collect snails and clams to feed them, but where conditions are right they require little supplemental feeding. In fact, it has been suggested that black carp could be used to control snail populations in ponds where snails compete for food with herbivorous fishes.

A more generalized carnivore, the bream (*Parabramis pekinensis*), is grown in some ponds. Some authorities consider the bream an omnivore.

PISCIVOROUS FISH

Of the 20 or so species of cyprinids commonly used in polyculture in China and southeast Asia, only the bream, common carp, crucian carp, and lampai (*Barbus schwanenfeldii*) will spawn in ponds without human intervention. If any of these species are present, undesired spawning may occur in growing ponds, in which case a predatory species may be introduced. The same applies to unwanted species such as goldfish (*Carassius auratus*), which often find their way into carp ponds. When predators are used they are usually readily available native fish, such as bass (Serranidae), catfishes (Clariidae and Siluridae) or snakeheads (Channidae). Other predators used in culture of common carp are listed on p. 66.

COLLECTION OF WILD STOCK

COLLECTION OF EGGS AND FRY

The classical method of obtaining stock for pond culture in China involves capture of eggs or fry from rivers, and some wild stock is still distributed. The Chinese carps spawn in the swift waters of large rivers, usually in the spring, although grass carp and mud carp may spawn through the summer. Considerable quantities of eggs and fry are often found drifting with the current, particularly during periods of high water, and fry dealers on the Yangtse and West rivers take advantage of this fact. A fry collecting station consists of 10 to 20 rows of wooden frames, 10 m high by 30 to 50 m wide, to each of which are attached 10 to 15 fine mesh conical bag nets about 5 m long, tapering from 4 to 5 m wide at the mouth to 1 to 1.5 m at the cod end, where a receptacle of finer mesh, about 120 cm long, is attached. The entire net is floated by means of a wooden framework around the cod end and a wooden pole inserted into the bag. A semicircular wooden trough may be used in place of the bag net, with the same sort of receptacle at the cod end. Captured eggs and fry are transferred by dip net into live wells suspended from the frame, then taken ashore. Similar methods are used in collecting fry of *Hypophthalmichthys harmandi, Labeo collaris, Mylopharyngodon aethiops,* and common carp from the Red River of Vietnam.

HATCHING EGGS

If the catch consists of eggs, they are transferred to shallow wood framed cloth hatching boxes, which are anchored in some sheltered spot in the river, often near the mouth of a small tributary, where the water is clearer than in the main channel. Hatching occurs in 2 or 3 days in favorable weather. The newly hatched larvae are kept in the boxes for 3 days, then sorted for distribution or stocking.

SORTING FRY

If the catch consists of fry, they are sorted immediately. The greatest demand in China is for big head and grass carp, but silver carp, black carp, and mud carp are also highly valued. Common carp and bream are less valuable but are usually retained for pond stocking. Goldfish are also kept, not intentionally, but because as fry they can scarcely be distinguished from common carp. Fry of predators and other extraneous fish usually are destroyed.

Sorting the fry is a complex operation carried out by skilled fry experts. A preliminary check of species composition is made at the river. A sample of the catch is placed in a white enamel dish 4 to 6 cm deep and the percentage of each species present is estimated visually. According to Lin (cited in Hora and Pillay, 1962) the characteristics listed in Table 2 are used by fry experts.

Further visual identification occurs at the nursery. Small samples of the catch are placed in bowls of clear water, each type of fry is counted and the percentage composition of the catch calculated. The mixed catch may be sold directly to fish culturists or the fry may be sorted.

The first step in sorting involves pouring the catch through a large net suspended and partially submerged in a pond. There dead fry and debris are picked out by hand. Preliminary sorting is done by passing the catch through a set of 20 or more sieves. At a given time of year fry of a particular species have a characteristic size, so that the sieves have the effect of roughly sorting the catch as to species. The fry of most undesirable species are larger than those of cultivated species and may thus be eliminated in the process.

Next the fry are transferred to a tall sorting basket. Within 10 minutes the fry sort themselves out into layers: silver carp and big heads near the surface; grass carp, black carp, and bream in midwater; and mud carp and common carp near the bottom. There may also be tiny fry of *Elopichthys* or other unwanted fish present. These will usually gather at the sur-

TABLE 2.   DISTINGUISHING CHARACTERISTICS OF SEVEN SPECIES OF CHINESE
CARP FRY, AND THE PREDATOR *Siniperca chuatsi*

Silver carp (*Hypophthalmichthys molitrix*): always swimming in the upper layer
  of water with occasional stops while swimming; head small, interorbital space
  narrow; body short with small roundish air bladder; color dark gray, not as
  transparent as the big head

Big head (*Aristichthys nobilis*): swimming in a slow and steady manner; air
  bladder roundish; body long and robust, head and eyes large

Grass carp (*Ctenopharyngodon idellus*): swimming in the middle layer of water
  with occasional pauses; color dark, eyes medium, air bladder small, roundish,
  body short, tail pointed

Black carp (*Mylopharyngodon piceus*): swimming like the grass carp; air bladder
  long; body long, eyes large; head depressed, triangular-shaped

Mud carp (*Cirrhinus molitorella*): swimming in a sluggish manner near the bottom or along the edge of the dish, eyes small, roundish, body small, slender,
  pinkish in color, a spot on the tail

Mandarin fish (*Siniperca chuatsi*) (a predator): swimming at the bottom, usually
  10 to 20 mm in length; mouth large; color dark

Common carp (*Cyprinus carpio*) and goldfish (*Carassius auratus*): color dark,
  body short with long dorsal fin, fry 6 or 7 days after hatching 5 mm long with
  a black speck at the base of the tail fin

Bream (*Parabramis pekinensis*): swimming in the middle layer of water; color
  dark, a dark speck on front part of dorsal fin; active fry

---

face and can be skimmed off separately. The reasons for this natural
sorting are not clear but apparently involve differences in oxygen requirements. Once the fish have sorted themselves out they must be separated
rapidly to avoid death of fry due to lack of oxygen.

Separation of species which occupy the same layer is difficult at the fry
stage, but silver carp and big head are usually sorted out as much as
possible. This is done by placing them in one part of a partitioned pond
for a day. Then a seine is dragged over the top 5 cm of water. Seining
in this manner will produce only silver carp, which are placed in another
section of the pond. This operation is repeated the next day and so on,
until catches in the surface layer are negligible, then repeated at a depth
of 15 cm, where the catch will consist of equal proportions of silver carp
and big head. The remaining fry will be virtually 100% big head. The
mixed lot of fry taken at 15 cm are reared together and sorted visually
when they are larger.

ARTIFICIAL PROPAGATION

HISTORY

In the 1930s silver carp and grass carp were spawned in China by means of artificial insemination, using ripe wild fish and hand stripping of eggs and milt. (See p. 400 for a description of this technique.) By the early 1950s fry specialists in some parts of China were supplying millions of artificially produced fry of these species and common carp to culturists. But the supply did not approach the demand and until induced spawning using pituitary extracts became common practice Chinese carp culture was largely dependent on collection of wild fry from rivers, despite the great amounts of labor required, the heavy mortality of fry, the high cost of fry attendant on these two factors, and the impossibility of obtaining pure, one-species stock free from predators, parasites, and trash fish.

The practice of hormonally induced spawning of fish was originated in 1934 by Brazilian biologists, but it has found only minor application in South America. The Brazilian innovation has, however, been widely adopted in Asia, Europe, and North America, and ranks among the foremost contributions to the art of fish culture. It has subsequently been found that exogenous hormones (hormones from a donor species other than the one being spawned) may be successfully used in many, if not a majority of, cases. Often even mammalian hormones have been effective. As induced spawning becomes more common and widespread, the techniques used become increasingly diverse. Readers interested in a general discussion of various aspects of induced spawning are referred to Das and Khan (1962), Pickford and Atz (1957), and Sneed and Clemens (1959, 1962). In this text, specific methods will be discussed for each group of fish covered.

The history of induced spawning of Chinese carps begins in 1954 with the spawning ahead of schedule of ripe wild black carp and big head by investigators from the Institute of Hydrobiology, Academia Sinica. Of more significance was the spawning in 1958 by the Fresh Water Section of the South Sea Fisheries Institute of China's Ministry of Aquatic Production of pond-reared silver carp and big head, followed by similar success in 1960 with grass carp. Mass production of artificially produced fry of these three species began in 1961, and in 1962 1200 million fry were produced in China. In 1963 black carp were added to the list of mass-produced species, and as early as 1964 in Chekiang and Kwangtung provinces 100% of the demand for Chinese carp fry was filled by artificially produced fish.

Mud carp are also spawned by induction in Taiwan, where the process was introduced in 1963. In 1964, 5 million fry of silver carp, grass carp, big head, and mud carp were produced there. The first hatchery for grass carp, silver carp, and big head in the Soviet Union was built in 1963 and in 1964–1965 about 90 million larvae were distributed throughout the Soviet Union and in Poland, Czechoslovakia, Hungary, and Bulgaria. Although polyculture has not caught on in Japan, fish culturists there are producing fry of grass carp and silver carp for export.

As one might expect from the short history of induced spawning of Chinese carps, the techniques are by no means standardized. One generalization that can be made is that the Chinese carps, particularly grass carp and silver carp, are more delicate animals than related species of European or Central Asian origin. Therefore for optimum results they should be handled as seldom and as gently as possible.

CONDITIONING

Spawners may be specially reared or selected from stock at large. There are no hard and fast criteria for selection of brood stock, but in general they should be fast growing, lively fish, among the largest and strongest members of their age group, and free from parasites and diseases.

There is no unanimity as to whether brood stock should be segregated by species or kept in mixed aggregations. The common practice in Taiwan is to keep brood stock at a 1:1 sex ratio in ponds of 1000 to 4000 m² and 1.5 to 2 m deep. A normal stocking ratio consists of five pairs each of grass, silver, and mud carp and three pairs of big head per 1000 m². This sort of stocking results in the same efficient utilization of food achieved by polyculture of fish destined for market. It may also be that large ponds are conducive to proper development of the gonads. Yet, since hypophysation and, if the eggs are artificially inseminated, fertilization are quite time-consuming, many of the fish in such large ponds may have to be handled several times, a factor which can cause a decline in potency. Large ponds are also harder to manage, and for these reasons some breeders prefer to segregate their brood stocks in small ponds, despite the loss in feeding efficiency. In China, large ponds are used for mixed brood stock but stocking ratios are weighted in favor of a particular species. Ratios cited include 15:10:1, 10:1:10, and 5:10:1 of grass carp, silver carp, and big head, respectively. Whatever system is used, two rules obtain: Do not stock too heavily, and keep fish of approximately the same size and age together.

Ponds used for holding brood stock should, if possible, be drained prior to use and disinfected with quicklime or teaseed cake. If silver carp or big head are to be raised it is good practice to fertilize the pond and expose it to sunlight for 2 to 3 days before filling, to promote rapid growth of plankton. Water in brood stock ponds should be close to spawning temperature (23 to 29°C) and well oxygenated.

Grass carp spawners should receive heavy feedings of fresh-cut grass or other vegetation—up to 40% of the weight of grass carp per day—to reach prime spawning condition. Other vegetable foods, such as bean meal and rice bran, may be fed, but diets high in fat are to be avoided, since excess fat retards development of the gonads. Other species may be fed peanut cake and soybean cake in conjunction with the light application of organic and inorganic fertilizers.

ATTAINMENT OF MATURITY AND NATURAL SPAWNING CYCLES

Size and age of Chinese carps at maturity varies greatly with climate and environmental conditions, as seen in Table 3.

The spawning cycle is also profoundly affected by climate. In their natural habitat and in other temperate climates the principal species of Chinese carps are all annual spawners. Rising water may play some role in triggering seasonal spawning, but temperature appears to be the most important factor. As is the case with the common carp the spawning cycle breaks down in the tropics, but it cannot be said with certainty that the Chinese carps become perennial spawners, since attempts to spawn them in tropical countries have until recently been largely unsuccessful, no matter what the season. Such information as is available for the tropical counterparts of the Chinese carps suggests that they are seasonal spawners, with rain or rising water replacing temperature change as the triggering stimulus. Grass carp, at least, retain some vestige of their spawning cycle in the tropics. In Malaysia workers at the Tropical Fish Culture Research Institute, although unable to spawn grass carp, were able to obtain infertile eggs in the "summer"; however, there was no reaction whatever to pituitary injections during "winter."

More recently, culturists at the Institute have succeeded in isolating the pituitary hormone and it is now possible, by use of the pure hormone, to spawn all the Chinese carps in Malaysia. Similar techniques have been applied in Thailand, and in 1969 1½ million fry were produced there, along with 500,000 fry of the native *Barbus gonionotus*.

Table 4 lists months and water temperatures associated with sexual maturity and spawning of Chinese carps in eight countries.

TABLE 3.   SIZE AND AGE AT MATURITY OF CHINESE CARPS IN DIFFERENT CLIMATES

COUNTRY	SPECIES	♀♀		♂♂	
		AGE AT MATURITY (YEARS)	WEIGHT (KG)	AGE AT MATURITY (YEARS)	WEIGHT (KG)
China					
South	Big head	3–4	5 –10	2–3	—
	Grass carp	4–5	6 – 8	3–4	—
	Silver carp	2–3	2 – 6	1–2	—
Central	Big head	4–5	5 –10	3–4	—
	Black carp	5	10 –15	4	—
	Grass carp	4–5	6 – 8	3–4	—
	Silver carp	4–5	2 – 6	3–4	—
Northeast	Big head	6–7	5 –10	5–6	—
	Grass carp	6–7	6 – 8	5–6	—
	Silver carp	5–6	2 – 6	4–5	—
India					
Wild	Grass carp	3	4 – 8	3	4.5
Pond-bred	Grass carp	2	1.5	0	0.9
Malaysia	Grass carp	1–2	2.3– 3.2	1–2	1.2–2.0
Taiwan	Big head	3–4	5 or more	2–3	—
	Black carp	5	10 or more	4	—
	Grass carp	4–5	3 or more	3–4	—
	Mud carp	3–4	1 or more	2–3	—
	Silver carp	2–3	2 or more	1–2	—
U.S.S.R.					
Krasnodar area	Big head, grass carp, silver carp	5	—	4	—
Turkmenia	Grass carp	3–4	—	2–3	—
Ukraine	Grass carp	8–9	2.7– 3.8	7–8	2.7–3.8
Siberia	Grass carp	8–9	6.5– 7.0	8–9	6.5–7.0
Moscow area	Grass carp	10	—	9	—

SEXING

The first requirement for artificial spawning of any fish is of course to distinguish the sex of the spawners. This presents no serious difficulties with the Chinese carps, at least not when they are ripe. Males of all species betray both their sex and degree of ripeness by releasing milt when handled. Males may also be distinguished by the presence of finely

TABLE 4. MONTHS AND WATER TEMPERATURES ASSOCIATED WITH SEXUAL
MATURITY AND SPAWNING OF CHINESE CARPS IN EIGHT COUNTRIES

COUNTRY	SPECIES	TYPE OF SPAWNING	MONTHS	WATER TEMPER-ATURE (°C)
China				
Central	Grass carp	Natural	April-August	26 –30
Central	Mud carp	Natural	April-May	—
South and central	Big head	Natural	April-June	—
Yangtse River	Silver carp	Natural	April-June	—
India	Grass carp	No spawning, gonads mature	March-August	—
Japan	Grass carp	Natural	June-July	20 –22
Malaysia	Grass carp	Gonads mature	January-October	—
Taiwan	Big head	Induced	March-September (poor results in July and August)	20 –30
	Grass carp	Induced	May-July	20 –30
	Grass carp	Natural	May-September	24 –28
	Mud carp	Induced	June-September	—
	Silver carp	Induced	March-September (poor results in July and August)	—
Thailand	Tambra (*Labeobarbus tambroides*)	Natural	July-August (rainy season)	—
U.S.S.R.				
Siberia	Grass carp	Natural	June-July	17.5–22.5
Ukraine	Grass carp	Natural	June	22 –24
Vietnam	Ca cham (*Labeo collaris*)	Natural	June	—

serrated ridges on the pectoral fin rays. Grass carp males may be further distinguished by a thickened first pectoral ray. Male grass carp and black carp are reputed to develop nuptial tubercles or "pearl organs" on the head at spawning time, but some workers have been unable to discern these.

Females are distinguished not by the presence of secondary sexual characters, but by the absence of the male characters. Ripeness in females may be indicated by a large, soft abdomen, but this is not a sure indication, as fish with excess intestinal fat present the same appearance. Lin (1965) cites a "swelling cloaca, pinkish in front" as an indication of ripeness, but this is not always observable.

ESTRUALIZATION

In the process of injection with pituitary materials, or estrualization as it is sometimes termed, as in other operations involving handling of spawners the incidence of injury to the fish may be reduced by transporting each fish individually in a carrying cradle about 15 cm deep made of cloth and attached to a wooden frame of suitable size. Another piece of cloth attached to one side of the frame only is laid over the top of the fish. During transportation this cloth and the cradle itself should be kept wet.

Most of the Chinese carps respond to pituitary preparations from many different kinds of animals, but the most commonly used are common carp pituitary and human chorionic gonadotropin. Grass carp do not respond to human gonadotropin, but small amounts of it added to fish pituitary preparations reinforce their effect. Any cyprinid pituitary is likely to be effective on all the Chinese carps, so if the culturist's stock includes goldfish, crucian carp, or other cyprinids of low market value, some economy may be effected by setting these fish aside for use as pituitary donors. The generally accepted dosages for big head, black carp, mud carp, and silver carp are 2 to 3 mg of dried cyprinid pituitary, 3 fresh pituitary glands, or 700 to 1000 IU of human chorionic gonadotropin per kilogram of spawner. Grass carp require a slightly higher dosage (3 to 4 mg of dried pituitary/kg) but overdoses should be avoided for all species since they may cause partial or complete infertility. Fractional injection is preferred, with $\frac{1}{8}$ to $\frac{1}{10}$ the total dosage injected on the first dose and the remainder 6 to 24 hours later. If fractional injection is used, females only should be injected at first, else the males will ripen too soon. Indeed, it may not be necessary to inject the males at all, although injection facilitates the flow of milt.

SPAWNING

Spawners may be placed in the spawning pond as early as 12 hours before injection, or stocking the spawning pond may be deferred until immediately after injection. The spawning pond should be 100 to 140 m² in area and 1.5 to 2.0 m deep with a smooth bottom to avert possible injury to the spawners. The same end may be achieved by placing the spawners in floating cloth cages anchored in the center of the pond. Spawning cages should be about 1 m deep with the bottom extended by weights tied to the corners of the cloth. A net placed over the cage will prevent the fish from jumping out. If the bottom of the pond slopes, it will facilitate draining after spawning is completed. It is helpful, particularly when spawning grass carp, if a current of 30 to 45 liters/sec can be maintained. Water for spawning should be clear, clean, preferably slightly alkaline, and have a dissolved oxygen content of 4 ppm or better. Although natural spawning of grass carp has been observed at temperatures as low as 17.5°C, 20°C is generally considered the minimum for spawning Chinese carps artificially, and 23 to 29°C is regarded as the optimum range.

If, after injection, the fish are to be allowed to spawn naturally, spawning ponds may be stocked at roughly the same density as holding ponds for spawners. Stocking rate is relatively unimportant if artificial insemination is to be practiced. The proper ratio of males to females is 2:1 for natural spawning or 1:1 with artificial insemination. Although natural spawning requires twice as many males, it is usually preferable, since the fish are more likely than the culturist to time their spawning so as to produce maximum fertilization of eggs.

Within 6 to 20 hours after introduction to the spawning pond, and usually within 10 to 14 hours, the males will be seen chasing the females. This time interval should be taken into account by the aquaculturist in deciding whether he prefers to have spawning occur at night or during the day. Daytime spawning is of course more convenient from a human standpoint and may result in lower labor costs. On the other hand, if eggs are laid at night, they will hatch during the day and thus be warmed by the sun and protected from sudden chilling during the most critical hours. If artificial insemination is to be practiced, the spawners should be removed for stripping within 15 min to 1 hour after the start of chasing. Grass carp should be allowed to chase a bit longer than the other species.

Natural spawning takes 30 to 90 min, after which time the spawners are removed. If a separate hatching pond is used, 1 to 2 hours are allowed for the eggs to harden, then the pond is drained and the eggs collected in a floating net placed at the outlet. If a spawning cage is used, a fine piece of netting is attached under and around the coarse net which holds the spawners to prevent eggs from being scattered about the pond.

After spawning, the adult fish may be returned to the spawning pond or a similar pond for recuperation. This is especially important when the fish have been stripped for artificial insemination. Recuperating fish should receive plenty of food; grass carp in particular require an abundance of fresh vegetable matter at this time.

HATCHING

Hatching of wild fry in boxes has already been described. Primitive methods of hatching artificial fry employ similar but shallower boxes stocked with eggs at a density of 4 to 5/cm² and anchored in a small stream of moderate flow. Hatching boxes may also be placed in standing water provided there is sufficient oxygen available, but under such circumstances the eggs must be gently stirred every 20 to 30 min. In either case the empty egg membranes should be removed after hatching.

Well-equipped fry farms use various indoor hatching devices, which permit control of temperature, light, and dissolved oxygen concentration as well as exclusion of predators. The original method of indoor hatching involved placing the eggs in flat trays or raceways and maintaining a flow of water over them, but better results are now obtained through adaptations of the jar method used in culture of many game fish in the United States. A wide variety of apparatus is used, but all are designed so that water enters the hatching container from below. Thus water is passed among the eggs rather than over the top of them, assuring a steady circulation of water around each egg and substantially increasing the rate of gas transport by the eggs.

One of the most successful versions of this method makes use of a series of hatching baskets constructed of nylon netting. A series of these are placed in a trough provided with baffle plates to produce an upward flow of water through the baskets (Fig. 1). Baskets near the trough are made of coarser net than those at the tail. Thus unhatched eggs are retained at the upper end of the trough, but fry escape and are washed down to be collected in the finer-mesh baskets. Alternate techniques involve various jars and funnels equipped with hoses at the bottom. Fry are either netted out or allowed to escape over the top, in which case a device to catch them is attached. Density of eggs in the apparatus and current velocity would seem to be important factors in this type of hatching, but a survey of successful hatching operations shows a fairly wide range of stocking rates and great variations in the rates of flow used. Russian hatcheries stock eggs of grass carp and silver carp in jars at 5000–10,000/liter and maintain a current of 0.6–0.7 liter /min. But the Saitama Fisheries Experiment Station in Japan maintains a hatching rate of 90% in nylon baskets where flow is maintained at 20 liters/min and eggs are stocked at approximately 3000 to 6000/liter. The variation in current as measured is probably much greater than the

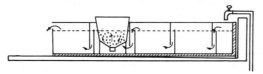

Fig. 1.   Hatching trough used in raising Chinese carps in Taiwan. (After Lin, 1965.)

differences in force applied against an individual egg in these devices. Measurement of current takes into account only the volume of water passing a point in a given time. To maintain a certain current through a small space such as a glass hatching jar or its inlet hose requires a great deal more pressure than is necessary to maintain the same current through a large trough. Thus the eggs in a trough receiving a current of 20 liters/min are probably no more liable to physical damage than eggs in a jar with a current of 0.6–0.7 liter/min. It may also be that, in systems where the eggs are tumbled about, the nature of the hatching container has some relation to the optimum water velocity. Certainly, an egg is more likely to sustain damage as a result of collision with the rigid wall of a glass jar than through contact with a resilient nylon net. For the present, it is up to the culturist to determine a rate of flow which will provide adequate circulation without damaging the egg membranes.

Time to hatching and absorption of the yolk sac varies with temperature. In Taiwan, Chinese carps hatch in 24 hours at 28°C and in 30 to 32 hours at 25°C. Within 3 to 6 days after hatching the yolk sac is absorbed and the larvae commence to swim freely and feed actively. Sac fry may be left in the hatching apparatus until this time or they may be placed in shallow net cages or other containers with perforations less than 0.2 mm in diameter. Such cages may be floated indoors in troughs or outdoors in nursing ponds. Ponds used for this purpose must be entirely free of small predatory fish for, although they may not be able to gain access to the cage, some species become adept at sucking larvae through the netting. Wherever the larvae are placed, they require a dissolved oxygen concentration of at least 4 ppm.

FRY REARING

PREPARATION OF PONDS

As soon as the yolk sac is absorbed, the fry are transferred to the first of a series of nursing and rearing enclosures. There is little uniformity in the size and number of such enclosures or in the techniques used in nursing and rearing, but the techniques of preparing nursing and rearing ponds are similar wherever Chinese carps are cultured. First the pond is drained as completely as possible, then it is poisoned to remove any potential fry predators. After a waiting period of up to a month, the pond is levelled, filled, and fertilized. If predatory aquatic insects reestablish themselves in the pond after filling, a small population of 3-year-old common carp may be stocked to feed on them.

The classical Chinese prescription calls for an initial application of fertilizer which is identical for all species, followed by periodic addition of specific types and quantities of manure according to the species being cultured. The initial manuring uses odoriferous plants, such as goatweed (*Agertium conyzoides*), with or without an admixture of cow dung, at a rate of 1200 to 2000 kg/ha. If adequate amounts of goatweed or similar plants are not available, a 7:1 mixture of cow dung with ammonium nitrate or ammonium sulfate may be applied at 800 kg/ha. Within a week of fertilization the water should turn greenish-brown, indicating an abundance of phytoplankton. At this time undecayed bits of fertilizer may be removed and the pond stocked with fry. Thereafter, periodic fertilization is carried out as outlined in Table 5.

TABLE 5.   FERTILIZATION SCHEDULE FOR REARING FRY OF FOUR SPECIES OF CHINESE CARP

SPECIES	FERTILIZER	DOSAGE
Big head Silver carp	Goatweed	100 kg/ha every 3 days
Grass carp	Goatweed and cow dung	200 kg/ha every 4 days
Mud carp	Cow dung	1,200 kg/ha every day

In Malaysia, nursery and rearing ponds are fertilized while dry with 12,000 kg/ha of cow dung, which is dug into the bottom. As knowledge of the action of fertilizers increases and synthetic fertilizers become more readily available in Asia, more sophisticated methods may become prevalent, but the traditional methods remain effective.

STOCKING

The size of nursery ponds in China is determined primarily by convenience. The range of sizes in use is from 100 m² to 2 ha. In Malaysia smaller ponds of about 50 m² are usual. No matter what the surface area, nursery ponds should be 0.5 to 1.0 m deep. In regions where ponds are scarce or hard to build, paddy fields may be adapted for nursing fry by levelling them and raising the dikes to 50 to 70 cm. In Taiwan, Chinese carp fry are sometimes kept in aquaria for the first few days of their life. Floating net cages made of nylon cloth with 12 to 16 meshes/cm are in use in Japan. Cages 50 to 70 cm deep are anchored in ponds, and when equipped with a hose to provide a slight current through the cage, they can be used to hold fry at densities of 15,000 to 62,000/m², as compared to the usual nursery pond stocking rate of 1400 to 1800/m². Whatever the rate at which fry are to be stocked, if they are kept indoors through the yolk sac stage it may be helpful to acclimate them to the nursery pond by occasionally introducing small amounts of pond water into their containers.

SUPPLEMENTARY FEEDING

Feeding very young Chinese carp fry is secondary to fertilization, as the principal food of all species at this stage is plankton. Nevertheless, supplementary feeding is often undertaken. The considerable diversity of feeding schedules and stocking rates at the nursing stage becomes nothing if not bewildering as the fry approach salable size. The most that can be done by way of explanation is to present data from various countries in tabular form (Table 6). Table 7, which lists the principal natural foods of Chinese carp fry at different sizes, may also be helpful.

The growth rates suggested by Table 6 could easily be improved upon by lowering population density or feeding more protein. However, the fry specialist is not interested in raising fish for food but in supplying fry for other culturists to stock in growing ponds. Except in Singapore and Malaysia, where it is customary to rear Chinese carp up to 8 months of age and 150 mm, the most popular size of fry for pond stocking is about 30 mm. However, there is some demand for fish as small as 10 mm and as large as 120 mm. Under normal conditions fry of the principal species reach 30 mm in 2 to 4 weeks, but the enterprising fry farmer attempts to keep at least a few fish of all salable sizes on hand at all times. Size control is accomplished by stunting and using carefully controlled diets and stocking ratios. This is a delicate business, because underfeeding can cause malnutrition and a small excess of food in a crowded pond can cause severe oxygen depletion. If fry are being kept in a stunted condition, the culturist maintains a careful watch over them and every 2 weeks or so removes the largest fish for sale or transfer to another pond.

TABLE 6. SAMPLE STOCKING AND FEEDING RATES FOR CHINESE CARP FRY

COUNTRY	SPECIES	AREA OF POND	DEPTH OF POND	SIZE OF FISH	AGE OF FISH	STOCKING RATE	FEED	FEEDING RATE
China Rearing for growth	Black carp, big head, grass carp, mud carp, silver carp	—	0.5–1 m	up to 20 mm	—	1 mila/ha	Egg yolk paste or soybean milk, plus peanut cake after 10 days	1 egg/2,500–7,500 fry/day or milk from 300–500 g beans/50,000 fry/day
	All species	—	0.5–1 m	20–100 mm	up to 1 month	—	Soybean meal	100 lb/5,000 fry/month
	Black carp	—	2.5 m	15–30 cm	1 mo.–1 year	100–2,400/ha	Barley, bean cake, small snails	Little food needed
Stunting	Big head, black carp, silver carp	1 ha	1.5 m	40–100 mm	—	1 mil/ha	Rice bran, soybean milk or peanut cake oil	0.5–2.0% of wt. of fish/day
	Mud carp	1 ha	1.5 m	—	—	2 mil/ha	Rice bran, soybean milk or peanut cake oil	0.5–2.0% of wt. of fish/day
	Grass carp	1 ha	1.5 m	40–100 mm	—	1 mil/ha	Duckweed	3–5% of wt. of fish/day
Hong Kong	All species	1,000 m²	0.8 m	0–8 cm–3 cm; 3 mg–1 g	up to 25–30 days	150/m²	Soybean milk and peanut cake meal	100 kg soybean milk or 200 kg peanut cake meal/month
	All species	1,400 m²	1 m	3–12 cm; 1–15 g	30–70 days	35/m²	Peanut cake, rice bran or soybean cake	Start at 1.5 kg/day, build up to 5 kg/day

Japan	Grass carp, silver carp	—	50–70 cm	10–20 mm	up to 17 days	15,000–62,000/m² (net cage) 1,400–1,800/m² (pond)	52% egg yolk, 18% soybean cake, 14% liver, 9% soybean milk, 5% flour	—
	Grass carp, silver carp	—	50–70 cm	20–30 mm	approx. 17 days–approx. 20–30 days	as above, minus mortality	38% fish meal, 38% flour, 16% rice bran, 8% silkworm pupae meal	—
Malaysia	All species	50 m²	45–60 cm	up to 5–7.5 cm	—	40/m²	Wheat flour and *Wolffia*	Enough to cover surface, twice daily
	All species	15 m²	60 cm	5–7.5 cm–12–15 cm	—	all survivors from above	Duckweed and *Wolffia*	—
	All species except grass carp	450 m²	1 m	12–15 cm and up	up to 8 months	500–600/pond	Peanut cake	—
	Grass carp	450 m²	1 m	12–15 cm and up	up to 8 months	500–600/pond	Grass	—
U.S.S.R.	Grass carp	—	—	up to 5 g	—	40,000–50,000/ha	Pond fertilization only	—
	Grass carp	—	—	5 g and up	—	—	Algae, duckweed, terrestrial plants, animal protein	—

a mil = million

97

TABLE 7. PRINCIPAL NATURAL FOODS OF CHINESE CARP FRY

SIZE OF FRY (MM)	GRASS CARP	BLACK CARP	SILVER CARP	BIG HEAD	COMMON CARP	MUD CARP
7–9	Protozoa, rotifer, nauplii	Protozoa, rotifer, nauplii	Protozoa, rotifer, nauplii	Protozoa, rotifer, nauplii	Protozoa, rotifer, nauplii, *Cladocera*	Protozoa, rotifer, nauplii, *Cladocera*
10–12	Same as above, plus small daphnids and *Cyclops*	Same as above, plus small daphnids and *Cyclops*	Same as above, plus small *Cladocera*	Same as above, plus small *Cladocera*	Same as above, plus copepods	Same as above, plus diatoms, minute organic detritus
13–17	Daphnids, copepods, minute benthic animals	Large daphnids, minute benthic animals	Same as above	Same as above	Same as above	Minute organic detritus, diatoms, phytoplankton
18–23	Same as above, plus organic detritus	Same as above, organic detritus	Same as above, plus phytoplankton	Same as above, plus phytoplankton, zooplankton	*Cladocera*, copepods, minute benthic animals	Same as above
24–30	Same as above, plus phytoplankton, minute algae	Same as above, minute benthic animals, insect larvae	Phytoplankton mostly	Zooplankton, phytoplankton	Same as above, plus insect larvae, organic detritus	Same as above, small blue-green algae, minute benthic flora

98

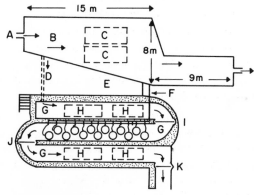

A   Main Water Supply
B   Spawning Pond
C   Spawning Boxes
D   Water Supply Pipe
E   Injection Platform
F   Water Supply Pipe to the Hatching Funnel
G   Hatching and Rearing Tank
    with Hatching Funnels (Circles)
H   Rearing Troughs
I   Water Flows Over this First Gate
J   Water Flows Under this Second Gate
K   Water Flows Over this Third Gate

Fig. 2. Schematic plan of the spawning pond and hatching-rearing tanks of the Wu-Shan-Tou Hatchery (Taiwan). (After Lin, 1965.)

DESIGN OF FRY HATCHERIES

The layout of hatcheries for Chinese carps follows no set pattern, but operations should center around the facilities for spawning, hatching, nursing, and early rearing, since these are the most critical activities in terms of fry production. Figure 2 is a diagram of the spawning, hatching, and rearing facilities of Wu-Shan-Tou Hatchery, a small and unelaborate but efficient operation in Taiwan.

The source of water for the hatchery is Wu-Shan-Tou Reservoir, where temperature averages 16°C in January, the coldest month, and never falls below 10°C. Summer water temperatures average 28 to 30°C, so that there is not only an optimum spawning temperature during the summer but enough of a seasonal temperature variation to ensure proper maturation of Chinese carp gonads. The water is slightly alkaline, a condition particularly favorable for grass carp and mud carp. The suitability of the water supply is perhaps best demonstrated by the fact that Wu-Shan-

Tou Reservoir is one of the few places in Taiwan where natural spawning of big head, grass carp, and mud carp occurs.

Water passes directly from the reservoir to the spawning pond (B), which is 1.5 m deep at the head and 2 m deep at the tail, thus providing a good slope for drainage. During use, a flow of 30 to 45 liters/sec is maintained. A platform (E) is provided for injection of breeders, which are spawned in anchored floating boxes (C).

Water from the spawning pond is fed into the S-shaped concrete hatching and rearing pond (G) and the hatching facilities, both situated at a slightly lower level than the spawning pond. Good circulation in the tank is maintained by having the water pass alternately over and under a series of three gates (I, J, and K). Nylon mesh hatching funnels (designated by circles), each with its own water supply hose attached at the bottom, are suspended in the middle section of the hatching and rearing tank. The surface of each funnel is kept about 10 cm above the water surface. The shape of the hatching and rearing tank makes it easy to net out the newly hatched fry and transfer them immediately to floating rearing troughs (H) anchored in the outer portions of the tank.

Other facilities not shown in the diagram include ponds for holding spawners, recuperation ponds, a pond for culture of common carp pituitary donors, and a number of ponds for rearing various sizes and species of fry.

The facilities needed to insure economically feasible production of fry vary according to the scale of operation and the number of species to be cultured. Lin (1965) suggests, in addition to spawning and hatching facilities, the minimum number and area of ponds as listed in Table 8. He calculates that with these facilities 18 pairs each of big head, grass carp, mud carp, and silver carp can be raised to full maturity and injected. With an estimated 50% responses to injection and 67% hatching rate, 7,200,000 fry can be produced annually.

TABLE 8.   MINIMUM NUMBER AND SIZE OF PONDS FOR A CHINESE CARP FRY HATCHERY

Spawner holding ponds	5 × 1,000 m²	5,000 m²
Common carp donor pond	1 × 1,000 m²	1,000 m²
Fry nursery ponds	10 × 1,000 m²	10,000 m²
Fry nursery ponds	10 × 2,000 m²	20,000 m²
Total	26 ponds	36,000 m² or 3.6 ha

SALE AND TRANSPORT OF FRY

Within the traditional framework of Chinese carp culture, fry are usually sold to culturists in the vicinity of the fry hatchery. In China today, fry are commonly grown to edible size by People's Cooperatives but some are stocked in large reservoirs where they are not cultured further but contribute to fisheries. Transport of fry over long distances is sometimes necessary, but this incurs the risk of heavy mortality. Modern techniques have made it possible for a fry hatchery to supply culturists hundreds of miles away. One of the best methods of fry transport, developed in the Soviet Union, employs hermetically sealed plastic bags filled with water and oxygen in equal proportions. Using such containers it is possible to ship fry at densities of 2000 to 4000/liter.

Some fry hatcheries find it useful to raise their own food supply. For example, the Tsingpu Experimental Freshwater Fish Farm near Shanghai, China, with 12,000 ha of fish ponds, maintains 25,500 ha of fields planted to soybeans and rice, the rice being exchanged for soybeans.

Duckweed for feeding fry may be grown in separate ponds, and in Malaysia *Wolffia* is often cultured in ponds 50 m² in area and 40 cm deep. *Wolffia* ponds are initially fertilized while dry with 0.6 kg/m² of cow dung and 1/10 that amount of prawn dust, dug into the bottom. After filling, a few *Wolffia* are introduced and the pond is treated with 1/10 the original dose of cow dung and prawn dust every week thereafter.

SPECIALIZED PROPAGATION TECHNIQUES FOR INDONESIAN CARPS

NILEM

In Indonesia, specialized techniques, not involving pituitary treatment, are used in spawning and rearing some of the native cyprinids. In West Java, a series of ponds is used in breeding nilem (*Osteochilus hasselti*). First, breeders are kept for a week in 1-m² flowing water conditioning ponds and fed on rice bran, beaten maize, or dried manihot. Males used for spawning are generally 1 year old, about 20 cm long, and 100 g in weight. Females as young as 8 months, and as small as 18 cm and 50 to 60 g may be used, but the best results are obtained with 1½- to 2-year-old, 25-cm, 150-g specimens. At the rate of one spawning every 3 to 6 months, female breeders may be used up to eight times.

In the spawning method used at the Galunggung hatchery, a sediment tank is built above the spawning pond to supply the silt-free water required by nilem. The spawning pond itself measures 2.5 m × 2 m and slopes from 40 cm deep at the inlet to 70 cm at the outlet. Grass is grown on the bottom of the shallow end of the pond, which is the actual spawning area. At the Mage hatchery, Indjuk palm fibers, held in place by a bamboo latticework, are substituted for the grass. Three or four pairs of spawners are placed in the pond, along with floating leaves or branches of plants to provide shade.

Spawning is induced by creating a strong current through the pond, beginning about 2 hours after the introduction of the fish. Breeding commences within about 2 hours and is finished in 1 to 2 days.

The spawning pond is not provided with a screen at the outlet; rather the eggs are washed down into a 3- to 25-m² hatching pond. Like the spawning pond, the hatching pond has a sloping bottom, but in this case the depth increases from 10 to 20 cm at the inlet to 20 to 40 cm at the outlet. The bottom is paved with flat stones, covered with 1 cm of gravel, over which the eggs are spread evenly to prevent them from sticking together. When spawning is completed, the spawning pond is drained, so that clean water from the sedimentation tank can be supplied directly to the hatching pond.

Hatching begins 3 days after spawning, and the first larvae can be cropped, by partial draining, 2 days later. Usually cropping of the 50,000 or so fry is completed within a week of spawning, at which time the hatching pond is completely drained and the gravel washed for reuse.

Variations on the Galunggung method exist, but all rely on the creation of a current and/or a rise in water level to induce spawning. It is believed that tree leaves trailing in the water also have a favorable effect. Rearing ponds vary in size from 0.5 to 5 ha and may or may not be directly linked with the hatching pond.

TAWES

Tawes (*Barbus gonionotus*), like most tropical cyprinids, are potentially year-round spawners, but hatchery operations are usually confined to the beginning of the rainy season, when there is adequate water to supply the strong current necessary. Tawes spawning ponds, which are 200 to 500 m² in area, 35 to 50 cm deep, and located no more than 700 m above sea level, are prepared by drying for at least 5 days, or until there is a crust 4 to 5 cm thick on the bottom. In drying, care must be exercised that cracks are not formed, because eggs may drop into them and be lost. A sand bottom, ideal from this standpoint, is, however, infertile, so a mixture of sand and silt is considered best. If sun drying is not adequate to prepare the pond properly, paddy straw may be burned on the bottom.

Water is introduced to the spawning pond, beginning in the morning, and 2-year-old breeders, about 300 g in weight and 25 cm in length, are introduced when it is half full. During the 3 to 4 months between spawnings, brood stock are kept in special ponds and fed on rice bran and soft plant leaves, but they should not be overfed or they will become undesirably fat. For the last 3 to 5 days spawners may be conditioned in running water. Brood stock may be used a maximum of five times. One pair of spawners/50 to 70 m² of surface area is a good stocking rate. A strong current of slightly turbid, well-oxygenated water is maintained until spawning is completed.

Mating generally occurs at night and is accompanied by an audible humming noise. Sometimes the fish are reluctant to spawn, in which case the water is beaten with bamboo slats. A similar excitement may be created by tying a stick to one of the fish. If such tricks fail, partial draining and exposure of the pond to sunlight may be effective.

When spawning is over, the current is shut off and the eggs spread evenly on the pond bottom. The spent spawners are left in the pond and fed once or twice a day with as much rice bran as they will consume in 2 to 3 hours. After a week, leaves of such soft plants as *Carica, Colocasia,* and *Manihot* are added to the diet and the quantity of both food items gradually increased.

Hatching occurs within 2 to 3 days; after about 20 days the fry are strong enough that the flow of water through the pond may be restored. Depending on the fertility of the pond, the fry may be reared for 25 or 50 days. In fertile ponds they reach lengths of 2 to 3 cm in about 50 days. Common carp fry, about 10 cm long, may be stocked with the tawes fry, as the agitation of the bottom caused by common carp feeding is believed to be beneficial. Small, 25-day-old fry are collected in the conventional manner, but larger fry have a tendency to swim against the current and are thus not well suited to collection by drainage. Such fry are collected by partially draining the pond, upon which the fry gather in a central ditch, then restoring the flow so that they move upstream and may be collected at the inlet. Average production of tawes fry is 10,000 to 20,000 25-day-old fish or 5000 50-day-old fish/female.

In parts of East Java where a large head of water is not readily available, a different technique of spawning tawes has been developed. A 0.1- to 0.2-ha spawning pond, 0.6 m deep, is constructed near a small stream which is known to swell after rains. In the bottom of the pond is a square pit, 4 m on a side and 0.5 m deep, in which 20 pairs of breeders are stocked at the start of the rainy season. Prior to a storm, the pond is kept dry except for the pit. Following heavy rainfall, the turbid flood water is let in to fill the spawning pond. Once the fish have dispersed, the pond is beaten with sticks to induce spawning. The fry obtained are transferred to rearing ponds a week after spawning, at which time the pond may be prepared for the next spawning. Some losses of eggs are experienced as a result of siltation, but yields of 100,000 fry/spawning are not unusual.

Tawes were introduced to Malaysia in 1953, and the Fisheries Division of the Ministry of Agriculture and Co-operation maintains stocks at its fry hatcheries. A research program has been initiated with the intentions of improving methods of fry production and selecting a fast-growing strain.

MATA MERAH

Artificial propagation of mata merah (*Barbus orphoides*) is a more recent development. Mata merah may be spawned in ponds or rice fields; in either case the water should be 40 to 60 cm deep. Current, turbidity, and dissolved oxygen concentration should be as described for tawes. The pond bottom, which is dried for 2 to 4 days before use, should be covered with grass or rice stalks; if these are not available, artificial substitutes may be provided.

Breeders, which should be at least 8 months old, 13 to 17 cm long, and weigh 60 to 85 g, are stocked in the morning at 1 pair/24 m² in ponds, or about half that density in rice fields. Sometimes combined spawning of mata merah and kissing gourami (*Helostoma temmincki*) is practiced, in which case 1 pair of mata merah/48 m² and 1 pair of kissing gourami/ 70 m² are stocked.

The spawners are fed with rice bran in the afternoon, and nocturnal spawning follows. The eggs, which are attached to the grass or rice stalks but never found on mud, hatch in about 2 days. Rice stalks, if present, are cut off near the bottom 3 days later. The fry are fed on fine rice bran after the first week. Manuring may also be helpful in their nourishment.

Fry are normally reared for 30 to 40 days, but it may be necessary to thin the stock after 20 days. Cropping is done as described for tawes. If kissing gourami are present, the species are separated by essentially the same method used after combined rearing of kissing gourami with common carp or nilem (see p. 224). From 240 to 400 2- to 3-cm fry/pair are obtained in pond culture and 100 to 170 fry/pair in rice fields or in combination with kissing gourami.

Breeders are removed and reconditioned on rice bran in separate fertile ponds. Under good conditions, they can be spawned every 3 months.

FIG. 3. Habitat and feeding niches of the principal species in classical Chinese carp culture. (1) Grass carp (*Ctenopharyngodon idellus*) feeding on vegetable tops. (2) Big head (*Aristichtys nobilis*) feeding on zooplankton in midwater. (3) Silver carp (*Hypophthalmichtys molitrix*) feeding on phytoplankton in midwater. (4) Mud carp (*Cirrhinus molitorella*) feeding on benthic animals and detritus, including grass carp feces. (5) Common carp (*Cyprinus carpio*) feeding on benthic animals and detritus, including grass carp feces. (6) Black carp (*Mylopharyngodon piceus*) feeding on mollusks.

GROWING CHINESE CARPS FOR MARKET

Although the techniques used by fry producers are interesting and sophisticated, it is the final stage in Chinese carp culture which is unique. The principles of full utilization of growing space and fish food resources, first worked out long ago in China, are only beginning to find application elsewhere. In practice, polyculture of fish also employs a number of other principles of stocking.

PRINCIPLES OF POND STOCKING

Plankton feeders should usually make up the largest portion of the stock, since plankton is by far the most abundant source of food in most ponds. Phytoplankton feeders such as the silver carp are particularly good basic fish for polyculture since they operate at the very bottom of the food chain.

Unless there is an abundance of higher plants in the pond, herbivores should be stocked in small numbers or provision should be made for supplementary feeding.

Use of bottom-feeding omnivores and carnivores depends on the depth and clarity of the pond. In a very deep pond or one with a heavy plankton bloom the productivity of the bottom water may be too low to support a good number of these fish. On the other hand, in clear, shallow ponds, where benthic insect larvae and the like are abundant, it may be a waste of resources not to stock such fish. Bottom feeders may also be stocked if it is economically feasible to provide supplementary food which sinks.

Piscivorous fish should be stocked only where uncontrolled spawning occurs in growing ponds, or where populations of small fish with little or no market value exist.

A suitable population density must be maintained in the pond as a whole and within each ecological niche. Understocking may result in underutilization of food. For this reason, and because at high population densities development of the gonads of many species is retarded, thus enhancing growth and eliminating the possibility of undesired spawning, heavy stocking is preferred. However, if a pond is overpopulated at any trophic level, interspecific or intraspecific competition will occur and productivity will decline. The precise population density for maximum production in a given pond must be determined empirically through experience.

The species cultured must be suited to the environment; one would not attempt to raise mud carp in northern China knowing that they cannot tolerate prolonged exposure to temperatures below 12°C or to place silver carp in a very turbid pond where silt particles would clog their extremely fine gill rakers.

Economic factors should also be considered in choice of species for culture. For example, goldfish utilize plankton well, but one would be foolish to stock goldfish in a pond which will support silver carp or big head, which have many times the market value. Regional preferences also enter into the picture, as in the case of the black carp, a very logical choice as a pond carnivore in China, where it is highly esteemed as a food fish, but a poor choice for stocking in Taiwan, where there is little demand for black carp.

PLATE 2.   Feeding grass carp in Taiwan. Greens are placed in one corner of pond behind floating bamboo rod. (Courtesy Ziad Shehadeh, Oceanic Institute, Hawaii.)

SUPPLEMENTARY FEEDING

In pond polyculture (Plate 2), the key word in feeding is "supplementary." A well-managed pond is capable of producing substantial quantities of edible fish without the addition of food. Supplementary feeding should be undertaken only if the resulting increase will more than offset the cost of feed and feeding. The general principles of feeding, as discussed for common carp in Chapter 2, are applicable to Chinese carps, as are the data on nutritive value of various feeds.

STOCKING AND FEEDING SYSTEMS

These principles find application in many systems of stocking. A number of the systems used in the Far East and the rationale for each are outlined here, along with comments on feeding.

Kiangsu Province is in the north of China, where temperatures preclude the mud carp, so common carp and bream are substituted as bottom-feeding omnivores. Plant growth in ponds is sparse, hence grass carp may not be stocked. Snails are common inhabitants of ponds and streams in the area, thus culture of black carp, with or without supplementary feeding, is emphasized (Table 9).

TABLE 9.   STOCKING RATES OF CHINESE CARPS IN PONDS TWO TO THREE METERS
DEEP IN KIANGSU PROVINCE, CHINA

	LENGTH OF FISH STOCKED (MM)	NUMBER OF FRY PER HA		
SPECIES		SYSTEM I	SYSTEM II	SYSTEM III
Black carp	250–350	1,000	2,000	2,400
Black carp	150–180	2,000	200	—
Grass carp	200–300	—	2,000	—
Silver carp	150–200	3,000	3,000	2,400
Big head	150–200	600	1,600	—
Common carp	150–200	200	200	1,200
Bream	120–150	500	500	—
Total		7,300	9,500	6,000

In deeper ponds in Kiangsu the productivity of the bottom water is
substantially reduced and higher plants are virtually absent. Therefore,
in stocking such ponds herbivores and bottom dwellers are reduced in
numbers or eliminated and great emphasis is placed on plankton feeders
(Table 10).

In most localities Chinese carps are stocked in growing ponds as fry
or yearlings, left there for 3 years, and harvested. But in the Tinghai and
Shaoshing regions of Chekiang, another northern Chinese province, a
3-year rotation is used (Table 11). Fingerlings are stocked in ponds 1.3 to
1.5 m deep, with a preponderance of black carp to take advantage of the
abundant bottom fauna. The second year the fish are moved to ponds
about 2 m deep and the balance of numbers shifted toward plankton
feeders. In the final year ponds 2.5 to 3 m deep are used and plankton

TABLE 10.   STOCKING RATES OF CHINESE CARPS IN PONDS THREE TO SEVEN
METERS DEEP IN KIANGSU PROVINCE, CHINA

	WEIGHT OF YEARLINGS (G)	NUMBER OF YEARLINGS PER HA			
SPECIES		SYSTEM I	SYSTEM II	SYSTEM III	SYSTEM IV
Big head and silver carp	500	4,500	4,500	9,000	9,000
Grass carp	500	600	—	3,000	—
Black carp	500	—	450	—	3,000
Common carp	200	200	200	200	200
Total		5,300	5,150	12,200	12,200

TABLE 11.   STOCKING RATES OF CHINESE CARPS IN PONDS IN CHEKIANG PROVINCE, CHINA

| SPECIES | FIRST YEAR | | SECOND YEAR | | THIRD YEAR | |
	INITIAL WEIGHT (G)	NO. OF FINGERLINGS PER HA (AGE 3 TO 9 MONTHS)	INITIAL WEIGHT (G)	NO. PER HA (AGE 15 TO 21 MONTHS)	INITIAL WEIGHT (G)	NO. PER HA
Black carp	35	6,000	500	1,300	2,750	240
Silver carp	40	400	600	1,000	600	400
Big head	40	400	600	80	600	300
Common carp	30	400	350	50	350	200
Total		7,200		2,430		1,140

feeders predominate. By this time the black carp are much larger than the other species and some must be removed for marketing. Where this system is used mortality is about 10% during the first 2 years and negligible in the third year.

Further south, in the lower Yangtse River valley, black carp are the principal food fish raised, accounting for 50% of the stock at harvest time. Black carp and grass carp are given supplementary feedings of snails and plants, respectively. Grass carp feces act as a source of fertilizer for plankton to nourish silver carp and big head.

At the same latitude, in Hangchow, a simpler stocking system is employed. Ponds in that region are usually very rich in plankton due to the alluvial soil and ample run-off from urban areas and heavily fertilized fields. Culturists there stock 65% silver carp, 30% big head, and 5% crucian carp. Such a community feeds almost entirely on plankton, but the crucian carp will also feed on the bottom and thus probably makes good use of the meager benthic resources of plankton-rich ponds. Since the crucian carp will reproduce in ponds, it might be advisable to include a predator in this association or stock only one sex of crucian carp.

In other coastal regions of China a more complex association is stocked, with heavily fed grass carp as the principal species. Next in numbers are the mud carp and striped mullet (*Mugil cephalus*), which make use of algae and detritus on the bottom and in midwater, respectively. Also quite heavily stocked are the plankton-eating silver carp and big head. Small numbers of the omnivorous common carp and the carnivorous sea perch (*Lateolabrax japonicus*) complete the community. The sea perch feed on the prolific gambusia (*Gambusia affinis*), a small American livebearer which has established itself as a pest in coastal ponds in southeast

TABLE 12.   FEEDING SCHEDULE FOR CHINESE CARP-MULLET-MILKFISH PONDS
IN TAIWAN

MONTHS	AMOUNTS OF FEEDS (KG)		
	GREEN FODDER	AGRICULTURAL PRODUCTS	ANIMAL PRODUCTS
March	50 (*Wolffia* or duck weed)	400	40
April	150 fodder	600	100
May	300 fodder	800	200
June	500 fodder	1,000	300
July	1,000 fodder	1,200	500
August	2,000 fodder	1,200	500
September	3,000 fodder	1,000	300
October	4,000 fodder	600	100
November	2,000 fodder	400	—
	13,000	7,200	2,040

Asia, as well as shrimp, which commonly invade such ponds, and common carp fry.

Milkfish (*Chanos chanos*), which feed on fresh and decaying algae, are cultured along with striped mullet and Chinese carps in slightly brackish ponds in Taiwan. Plankton feeders predominate in such ponds and grass carp are few but heavily fed. Table 12 is a feeding schedule for such a community, consisting of 3000 striped mullet, 2000 silver carp and big head, 2000 milkfish, 1000 mud carp, 500 common carp, and 250 grass carp, for a total of 8750 fish in a 1-ha pond. Total production of such a pond is about 3500 kg/ha of fish.

Striped mullet are also stocked in slightly brackish ponds in Hong Kong, where they are readily available. When they have reached a length of about 12 cm the stock is thinned by fishing. Mud carp are stocked in relatively large numbers since they do not attain as large a size as the other species. Bream are sometimes stocked as well.

The customary method of growing carp in Hong Kong is similar to that practiced in Malaysia (discussed earlier) with animal feces used as fertilizer. Such culture is still carried on to some extent, but it has run up against the types of difficulties one might expect would be created by such practices in an urbanized area. As a result many fish culturists in Hong Kong are switching to fry culture, which does not require the use of animal manures. Stocking systems used in growing ponds 1 to 2.5 m deep in Hong Kong are shown in Table 13.

TABLE 13.   STOCKING RATES OF CHINESE CARPS AND MULLET IN PONDS IN
HONG KONG

SPECIES	LENGTH OF FRY (CM)	NUMBER OF FINGERLINGS PER HA		
		I	II	III
Striped mullet	2.5– 4.5	12,000	27,360	15,000
Grass carp	5.0–10.0	1,524	300	700
Silver carp	7.5	2,208	2,568	600
Big head	7.5	1,536	524	600
Mud carp	5.0	9,600	3,120	9,000
Common carp	5.0	—	—	4,800
Total		26,868	33,872	30,700

In the West River drainage of South China and North Vietnam, the
mud carp, which withstands moderately high temperatures well, is emphasized. The numbers of mud carp and grass carp stocked are dictated
by the quantity of supplementary food available. Although snails are
available, few black carp are stocked, since they do not grow well in the
tropical climate. Silver carp and big head are stocked in greater numbers
in the deeper ponds (Table 14).

Some very shallow ponds (about 60 cm) in the West River area are
also stocked, although they are not very productive, as shown by the low
total number of fish stocked (Table 15). Black carp and mud carp may
suffer mortality due to extremely high temperatures and are thus not
stocked in such ponds. Shallow ponds are often used to hold surplus fish

TABLE 14.   STOCKING RATES OF CHINESE CARPS IN DEEP PONDS IN THE WEST
RIVER VALLEY OF CHINA AND NORTH VIETNAM

SPECIES	LENGTH OF FISH (CM)	NUMBER OF FISH STOCKED PER HA					
		PONDS WITH AVERAGE DEPTH OF 1½ M			PONDS WITH AVERAGE DEPTH OF 2 M		
		I	II	III	IV	V	VI
Grass carp	6.0–15.0	1,200	1,200	600	480	1,200	2,400
Silver carp	10.0	300	240	600	1,200	1,800	400
Big head	10.0	300	240	1,200	510	1,800	400
Mud carp	7.5	1,800	4,800	—	2,400	3,600	7,200
Common carp	5.0	1,200	360	3,600	—	1,200	300
Black carp	5.0–10.0	100	24	120	60	—	150
Bream	7.5	—	—	—	600	—	—
Total		4,900	6,864	6,120	5,250	9,600	10,850

TABLE 15.   STOCKING RATES OF CHINESE CARPS IN SHALLOW PONDS IN THE
WEST RIVER VALLEY OF CHINA AND NORTH VIETNAM

| | INITIAL WEIGHT OF | NUMBER OF YEARLINGS PER HA | |
SPECIES	EACH FISH (G)	I	II
Grass carp	50	240	480
Silver carp	20 – 40	1,200	972
Big head	100 –150	120	240
Common carp	0.4– 0.5	1,680	972
Total		3,240	2,664

in the winter. In the summer the stock is removed, those individuals
which weigh more than 500 g are marketed, and the smaller fish are used
to stock deeper ponds.

Pond culture practices in Malaysia and Thailand are based on the
Chinese method, but more use is made of supplementary feeding. Grass
carp in particular are heavily fed, since there is an abundance of suitable
vegetable feed in the area. With heavy feeding and the help of the warm
tropical climate, fish grow rapidly and the culturist may realize two crops
per year. Most ponds in Malaysia and Thailand are 1.2 to 1.5 m deep
and would thus be suitable for culture of mud carp despite the high
surface temperatures. However, since Chinese carps have only begun to
be bred in the tropics, stocks are imported from China. It is not economically feasible to import fry of a species which attains such a small maximum size as the mud carp, so it is largely excluded from culture in
Malaysia and Thailand.

Fish culture in this region might benefit from a more thorough investigation of some of the cyprinids native to the region, many of which can
easily be bred locally. For example, the mata merah is a good table fish,
accepts a variety of foods, and can be bred in ponds in the tropics, but it
has only recently begun to see wide use in pond culture.

In addition to the stocking ratios shown in Table 16, Gopinath (1950)
suggests a 2:1:1:3 ratio of grass carp, big head, silver carp, and common
carp for Malaysia.

In Singapore, the nursing and rearing period is extended to 8 months
for Chinese carps (although not for common carp), at which time the
fish are transferred to fattening ponds for 6 months. Stocking ratios used
in both periods are shown in Table 17. The native tawes, bream, or
Java tilapia (*Tilapia mossambica*) may supplement or replace the grass

TABLE 16. STOCKING RATES OF CHINESE CARPS IN PONDS IN MALAYSIA

SPECIES	INITIAL WEIGHT OF EACH FISH (G)	NO. OF FISH STOCKED PER HA			
		I	II	III	IV
Grass carp	350–600	300	320	375	500
Big head	350–600	100	120	75	175
Silver carp	350–600	100	125	75	200
Common carp	30– 60	145	150	120	250
Total		645	715	645	1,125

carp as a herbivore and mud carp or catla (*Catla catla*) may totally or partially fill the omnivore niche usually given over to common carp.

The only significant monoculture of cyprinid fishes in the Orient outside of Japan occurs in Cambodia, where *Barbus altus, Barbus bramoides,* and *Barbus gonionotus* are sometimes grown in floating cages (see p. 209–210).

POND FERTILIZATION

Pond fertilization is always practiced in Chinese carp culture. The type and amount of fertilizer used depend on the characteristics of the soil and water supply and the quantity of plankton-feeding fish to be grown. An excess of fertilizer may cause oxygen depletion, but short of that point the role of fertilizer in production of plankton-eating fish may be stated succinctly: The more fertile the water, the more plankton is available. The more plankton in the pond, the more plankton-eating fish that can be grown. However, plankton feeders, although the principal constituents of most pond fish communities, are not the only species

TABLE 17. STOCKING RATES OF CHINESE CARPS IN PONDS IN SINGAPORE

SPECIES	WEIGHT OF FRY PUT INTO THE NURSERY (G)	NO. STOCKED PER HA	WEIGHT OF FISH STOCKED IN FATTENING POND (G)	NO. STOCKED PER HA
Grass carp	30	1,250	1,500	450
Big head	40	250	2,000	150
Silver carp	40	250	1,800	175
Common carp	30	—	—	225
Total		1,750		1,000

cultured. The general effect of fertilizers on other groups of fish may be summarized as follows:

Midwater carnivores benefit from fertilization, but not as greatly as plankton feeders, for they are several links removed from plankton in the food chain.

Herbivores that feed on higher plants are usually dependent on supplementary feeding in pond culture, so they neither benefit nor suffer from fertilization.

Bottom feeders are usually adversely affected by the type of fertilization used in polyculture. Fertilization may have some beneficial effect on benthos, but this frequently is cancelled by the shading effect of a heavy plankton bloom, which reduces productivity in the bottom layer of the pond.

Trash fish are of several varieties, but many of the common species are plankton feeders, in which case they benefit greatly from fertilization.

The traditional methods of fertilizing growing ponds in the Far East involve organic fertilizers, usually animal manures. However, there is experimental evidence suggesting that inorganic fertilizers are superior. Research in Malaysia comparing the two types of fertilizers showed that, once a pond has been treated with the proper amount of lime to establish a neutral pH, superphosphate is the only fertilizer needed to enhance fish production. As compared to organic fertilizers, superphosphate, applied at 333 kg/ha, produced better growth of blue-green algae, the principal agents of nitrogen fixation in ponds; reduced production of superfluous algae, which limits benthic productivity; lessened the danger of oxygen deficiency due to overfertilization or a sudden phytoplankton die-off; and diminished the need for supplementary feeding, which in itself may contribute excess fertilizer.

Continued use of organic fertilizers has been justified on the basis of cost, but considering the small amount of labor required to apply superphosphate, the slight expense, and the smaller amount of feed required when it is used, inorganic fertilization may actually be cheaper. The use of superphosphate is gradually being adopted by commercial fish culturists, though not by subsistence farmers, in Malaysia, but it will take some time before substantial amounts of inorganic fertilizers can be diverted from land use in much of the Far East. The long-term effects of inorganic fertilizers on ecosystems have not been investigated, but experience in North America suggests that caution is in order.

There are many local methods of preparing and applying organic fish pond fertilizers, but the one universally used method, fertilization by the fish themselves, is often not thought of as such. Fertilization by fish is

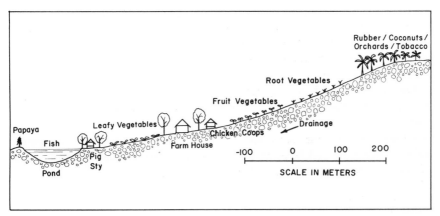

Fig. 4. Profile of integrated aquaculture-agriculture system used in Singapore. (After Ho, 1961.)

particularly effective where grass carp are stocked, because their feces contain partially digested plant remains, which decay to the benefit of plankton.

Compost for pond fertilization may be prepared similarly to that used in terrestrial agriculture. In the Kiangsu and Hunan districts of China, grass, sheep manure, and human manure in a 4:2:1 ratio are placed in a hole in the ground with 1% of quicklime. The hole is then filled with water and sealed with clay, and the contents are allowed to ferment until needed. Compost may also be created in the pond; in Kwangtung and Kwangsi provinces, manure is mixed with rice, legumes, or plants of the chrysanthemum family, placed in the water around the margin of a pond, and turned over daily until consumed. A quicker method of achieving a similar result is to mix manure in a 1:2 ratio with water, strain the resulting soup, retain the liquid portion, and pour it around the edge of a pond daily.

The most interesting and thorough use of organic fertilization in Chinese carp culture occurs in Singapore, where fish, grown for profit by the Chinese or as a subsistence crop by Malays, are an integral part of an operation which also includes fruits, vegetables, and livestock as well as such cash crops as rubber and tobacco. Figure 4 is a diagram of such a farm. Excess fertilizer, compost, and soil minerals from all the terrestrial crops eventually find their way downhill to the fish pond. Chickens and the human inhabitants also contribute their share of organic nutrient, but it is the pigs (or cattle or buffalo in the case of the Malays, who, as Muslims, are forbidden to raise pigs) which are the key to the operation.

Manure from the pig sty is periodically washed into the pond, which usually receives no other direct fertilization. In time the pond becomes a dilute solution of fertilizer, which may be applied to land crops. Periodically the sludge at the bottom of the pond is dug out and incorporated into vegetable beds.

Pigs are also used as a source of fertilizer for Chinese carp ponds in Hong Kong, Thailand, and China's Kwangtung province. In Hong Kong, 100 pigs weighing about 30 kg each or 2500 1-kg ducks kept in pens overhanging the water are considered adequate to fertilize a 1-ha pond. Human waste is used in some areas, and in South China and Indonesia, latrines are built over fish ponds for this purpose.

YIELDS

Reliable data on yields achieved through pond polyculture in the Far East are scarce, but it is known that under favorable circumstances 7500 to 8000 kg/ha may be attained. The average yield in intensive operations is estimated at nearly 4000 kg/ha in Malaysia and 3000 kg/ha in China and Taiwan. Despite the scarcity of hard data it is certain that yields of freshwater pond fish culture in the Far East are high compared to those normally achieved elsewhere, and that the practice of polyculture has a great deal to do with this success.

That it is the technique of polyculture and not the species of fish cultured, the type of feed or fertilizer, or some accidental property of the environment which is primarily responsible for high yields has been demonstrated by Russian experiments. Polyculture, without feeding of grass carp, silver carp, and big head with common carp in ponds formerly used for monoculture of common carp has resulted in production increases of 400 to 600 kg/ha in Central Russia and 600 to 1000 kg/ha in the South. With feeding and intensive pond fertilization, increases of 3000 to 4000 kg/ha have been achieved. Even in the rather sterile environment of a common carp pond in a peat hag, addition of grass carp and silver carp increased total fish production by 127 kg/ha, with a 15% reduction in food expenditures.

HARVEST AND MARKETING

A large percentage, perhaps as much as 50% of the pondfish production of Asia, is used as a means of subsistence. Harvest and marketing of commercially raised Chinese carps generally proceed in the manner described earlier for common carp. However, the Chinese carps, particularly silver carp and grass carp, are much more high-strung than common carp, and great care must be exercised in catching and transporting them or they will injure themselves, often fatally. Containers used in transport should be lined with fine mesh or in some way padded. Fishing and transport should be carried out in well-oxygenated water at the lowest possible temperature; at temperatures of 1 to 6°C Chinese carps are semidormant, but above 10°C they become very excitable. If fish must be transported at high temperatures, anesthesia may be used. A 6.7 to 7.7 μg/liter solution of sodium barbital or a 1 to 4 g/liter solution of urethane is effective in transporting Chinese carps at temperatures of 25.5 to 32°C.

PARASITES AND DISEASES

In addition to injuries sustained as a result of their nervous temperament, Chinese carps are susceptible to the usual array of parasites and diseases. Most of the parasites found on Chinese carps have not been recorded as causing disease in Asian fish, but some of them, notably the tapeworm *Bothriocephalus gowkongensis,* introduced to Russia with grass carp from China, have become serious problems with European cyprinids. At present Chinese carps, or at least the grass carp, have a bad reputation in Europe as spreaders of disease, but they probably present no greater problems in this regard than are occasioned by most introductions of exotic species. Nor are European fish the only ones to suffer, for Chinese carps have been infested with diseases of European fish. The parasitic diseases coccidiosis, lernaeosis, postodiplostomatosis, synergasilosis, and tetracotylosis as well as infectious dropsy have thus far been recorded for big head, grass carp, and silver carp. For information on prevention and cure of these and other diseases the reader is referred to Davis (1953).

PROSPECTUS

RESEARCH IN POND FERTILIZATION

Problems of an esthetic or sanitary nature arise from the traditional practice of using animal manures, and particularly human wastes, in pond fertilization. Research in the use of inorganic fertilizers may eventually result in the elimination of these methods, though the long-range effects on production may be unfavorable and equally undesirable side effects may be incurred. There is a definite need for scientific evaluation of

Asian fish pond fertilization techniques, not only in the hope of increasing fish production, but from a more holistic viewpoint as well.

SELECTIVE BREEDING AND HYBRIDIZATION

Another promising area for improvement, through research, of Chinese carp culture comprises selective breeding and hybridization. The history of spawning Chinese carps in captivity is too short for selective breeding to have accomplished significant improvement of fish stocks, but work is proceeding in several countries. Hybridization has already yielded some interesting results. Male silver carp × female big head and mud carp × common carp hybrids (sexes of the parents unknown) produced in Taiwan both show indications of growing faster than either parent. The latter cross is especially interesting, for a bottom-dwelling omnivore combining the resistance to cold and large size of the common carp with the superior table qualities and nondestructive feeding habits of the mud carp would be a most desirable fish for pond culture. All possible hybrids of big head, grass carp, and silver carp have been produced at the Tzimljanskoje Hatchery in the Soviet Union, and some of these also show superior growth characteristics as well as high viability. The gill rakers of the hybrids are intermediate in size and structure to those of the parents, suggesting intermediate feeding habits, which might prove useful in ponds where the characteristics of the plankton population are known.

USE OF NEW SPECIES

New species as well as new hybrids will surely be brought into polyculture. Many of the native southeast Asian species, a number of which have been discussed here, may be superior to the true Chinese carps for culture in tropical climates. Recently the Department of Fisheries of Thailand has begun to study the pla kaho (*Catlacarpio siamensis*), a river fish native to that country, for possible use in pond culture. The Chinese government is also interested in new species for culture, particularly in the northern part of the country. Among the cyprinid species currently under consideration are *Elopichthys bambusa, Erythroculter ilishaeformis, Hemiculter levisculus,* and *Xenocypris argentea.*

SPREAD OF POLYCULTURE

Of course Chinese carps, particularly the grass carp, which has no phytophagous counterpart in many countries, will continue to be introduced about the world. But in many countries the adaptation of native fishes to polyculture seems at least as promising. For instance in South America, where fish culture is scarcely developed, one can visualize an analog of Chinese carp culture using various members of the family Characidae, which rivals the Asian Cyprinidae in number of species and diversity of life habits. In fact, while Chinese carp culture in the People's Republic will continue to be vital to the peoples of Asia, the greatest contribution of the Chinese fish culturists may be not to their own people, but to the world, as the contingencies of population excess and protein shortage force us all to apply the principles of ecology as they did centuries ago.

REFERENCES

ANDRIASHEVA, M. A. 1966. Some results obtained by the hybridization of Cyprinids. FAO World Symposium on Warm Water Pond Fish Culture. FR: IV/E-10.

AVAULT, J. W. 1965. Biological weed control with herbivorous fish. Proc. 5th Weed Control Conf. 18:590–591.

DAS, S. M., and H. A. KHAN. 1962. The pituitary and pisciculture in India, with an account of the pituitary of some Indian fishes and a review of techniques and literature on the subject. Ichthyologica 1(1):43–58.

DAVIS, H. S. 1953. Culture and diseases of game fishes. U. of California Press, Berkeley, 332 pp.

GOPINATH, K. 1950. Freshwater fish farming in the Malay Archipelago. J. Zool. Soc. India 2(2):101–108.

HEPHER, B. 1967. Some biological aspects of warm-water fish pond management. Symposium on Biological Basis of Fish Production. Blackwell, London.

HICKLING, C. F. 1967. On the biology of a herbivorous fish, the white amur or grass carp Ctenopharyngodon idella Val. Proceedings of the Royal Society of Edinburgh. Section B (Biology), Vol. LXX, Part 1 (No. 4).

HO, R. 1961. Mixed farming and multiple cropping in Malaya. Proceedings of the Symposium on Land Use and Mineral Deposits in Hong Kong, Southern China, and Southeast Asia. Paper No. 11, pp. 88–104.

HORA, S. L., and T. V. R. PILLAY. 1962. Handbook on fish culture in the Indo-Pacific Region. FAO Fisheries Biology Technical Paper 14. 204 pp.

HSIEN-WEN, W., and C. LING. 1964. Progress and achievements in the artificial propagation of four farm fishes in China. Contributions at the 1964 Peking Symposium, pp. 203–218.

KONRADT, A. G. 1966. Methods of breeding the grass carp Ctenopharyngodon idella (Val.) and the silver carp Hypophthalmichthys molitrix. FAO World Symposium on Warm Water Pond Fish Culture. FR: IV/E-9.

KURONUMA, K. 1966. New systems and new fishes for culture in the Far East. FAO World Symposium on Warm Water Pond Fish Culture. FR: VIII-IV/R-1.

KURONUMA, K., and K. NAKAMURA. 1957. Weed control in farm pond and experiment by stocking grass carp. Proc. Indo-Pacific Fish. Council 7(II):35–42.

LIN, S. Y. 1954. Chinese systems of pond stocking. Proc. Indo-Pacific Fish. Council, 5th Meeting. Technical Paper 1 (mimeo).

LIN, S. Y. 1965. Induced spawning of Chinese carps by pituitary injection in Taiwan. Chinese-American Joint Commission on Rural Reconstruction, Fisheries Series, No. 5. 31 pp.

LING, S. W. 1966. Feeds and feeding of warm-water fishes in ponds in Asia and the Far East. FAO World Symposium on Warm Water Pond Fish Culture. FR:III-VIII/R-2.

PICKFORD, G. E., and J. W. ATZ. 1957. The physiology of the pituitary gland of fishes New York Zoological Society, New York. 613 pp.

PROWSE, G. A. 1966. A review of the methods of fertilizing warm-water fish ponds in Asia and the Far East. FAO World Symposium on Warm Water Pond Fish Culture. FR:II/R-2.

SNEED, K. E., and H. P. CLEMENS. 1959. The use of human chorionic gonadotrophin to spawn warm-water fishes. Prog. Fish Culturist 21(3):117–120.

SNEED, K. E., and H. P. CLEMENS. 1962. Bioassay and use of pituitary materials to spawn warm-water fishes. U.S. Bureau of Sport Fisheries and Wild Life, Research Report 61. 30 pp.

SHU, T. 1966. Description of Chinese fish culture, Vol. 1.

YASHOUV, V. 1966. Mixed fish culture—an ecological approach to increase pond production. FAO World Symposium on Warm Water Pond Fish Culture. FR:V/R-2.

Interviews and Personal Communication

BAILEY, R. M. University of Michigan Museum of Zoology, Ann Arbor, Michigan.

HICKLING, C. F. 95 Greenway, London N20, England.

LING, S. W. FAO Regional Office, Bangkok, Thailand.

4

Culture of the Indian Carps

It has become a cliche that the people of India need protein. Even the most socially unaware persons have some grasp of the magnitude of India's problem. One can cite statistics *ad infinitum* to demonstrate that India is in a steadily worsening state of nutritional crisis, but however

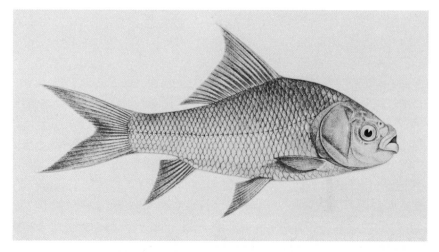

PLATE 1. Catla (*Catla catla*). (Courtesy V. G. Jhingran, Central Inland Fisheries Research Institute, Barrackpore, West Bengal, India.)

one defines the problem, it is clear that the solution, if there is ever to be one, will not come as any single panacea, but from effective population control coupled with increased and more efficient protein production by many means, including fish culture.

Fish culture shows promise in India, for there is a considerable demand for fish (total landings of fish in 1965 amounted to 1331 thousand metric tons), a number of native fish well suited to culture, a tradition of fish culture, and a great abundance of cultivable waters (an estimated 7,307,642 ha of fresh and brackish water). The potential for fish culture in India has not gone unnoticed, and according to R. V. Pantulu, formerly of the Indian Fresh Water Fisheries Research Institute at Barrackpore, fish culture operations in India have increased sevenfold to eightfold during the last decade. Yet large areas of potentially rich waters still lie fallow and, technologically, Indian fish culture lags behind much of the world.

Though brackish water culture is practiced in some areas, notably the states of West Bengal and Kerala, freshwater culture is better developed and will probably remain more important. Freshwater fish culture is favored not only by the availability of suitable waters and fish, but by the strong preference for fresh fish in most parts of the country, which permits locally raised freshwater fish to compete favorably with saltwater fish in inland areas.

As elsewhere in Asia, the most commonly cultured fish in India and

PLATE 2.   Rohu (*Labeo rohita*). (Courtesy V. G. Jhingran, Central Inland Fisheries Research Institute, Barrackpore, West Bengal, India.)

Pakistan are members of the carp family (Cyprinidae). Indian cyprinids used in fish culture are popularly separated into two groups: the most desirable species are referred to as major carps; the smaller, less desirable species are called minor carps. The minor carps persist in Indian fish culture largely because as fry they are extremely difficult to distinguish from the major carps, and are thus unintentionally stocked. Minor carps may also be deliberately stocked when stocks of major carps are scarce. In the south of India, in the states of Madras and Mysore, a third group, the Cauvery carps, largely supplant the major carps.

Indian carps have not yet become as popular for introduction and culture outside their native habitat as the common carp (*Cyprinus carpio*) or the Chinese carps, but the catla (*Catla catla*) (Plate 1) has been introduced to Ceylon, experimentally cultured in Israel and the United States, and commercially raised in Malaysia. All the Indian major carps have been recommended for use in the Philippines.

ECOLOGICAL NICHES OF THE INDIAN CARPS

THE MAJOR CARPS

The major carps comprise the catla, the rohu (*Labeo rohita*), and the mrigal (*Cirrhinus mrigala*) (Plates 2 and 3). Some authorities also include the calbasu (*Labeo calbasu*). These three or four species are often grown in polyculture, though their ecological niches are by no means as distinct

PLATE 3.   Mrigal (*Cirrhinus mrigala*). (Courtesy V. G. Jhingran, Central Inland Fisheries Research Institute, Barrackpore, West Bengal, India.)

and well defined as those of the Chinese carps, nor is Indian polyculture nearly as sophisticated as the ancient Chinese practice. In general, the catla feeds on plankton and decayed macrovegetation on the surface and throughout the water column, the rohu is a column feeder on decayed vegetation with a taste for higher plants, the mrigal a bottom-feeding herbivore, and the calbasu a benthic omnivore. Certain of the Chinese carps have been experimentally included in Indian carp polyculture systems and in the future further combinations of Indian carps, Chinese carps, and other fish may be expected in both experimental and practical culture.

THE CAUVERY CARPS

The niches of the Cauvery carps overlap; thus polyculture of these three species could probably be enhanced by introduction of other species. The species association for fish culture in southern India has already been diversified by the introduction of catla. The catla joins the fringe-lipped carp (*Labeo fimbriatus*), which consumes mostly filamentous algae and some zooplankton; the white carp (*Cirrhinus cirrhosa*), a plankton feeder with a preference for zooplankton, and the Cauvery carp (*Labeo kontius*), which competes with the fringe-lipped carp for filamentous algae, but also eats pieces of plants and detritus.

THE MINOR CARPS

Minor carps present in fish ponds vary from region to region. Among those commonly found in one or another part of India are the reba (*Cirrhinus reba*), which feeds largely on phytoplankton and decayed plants; the nagendram fish (*Osteochilus thomassi*), a benthic and marginal feeder on filamentous algae and diatoms; and the sandkhol carp (*Thynnichthys sandkhol*), which consumes mostly phytoplankton plus some zooplankton; also the bata (*Labeo bata*), the carnatic carp (*Barbus carnaticus*), *Barbus chola, Labeo boga, Labeo dyoechilus, Labeo gonius, Labeo nandina,* and *Barbus sarana.*

One more Indian cyprinid should be mentioned: the omnivorous copper mahseer (*Barbus hexagonolepis*). Not exactly a "minor" carp, the copper mahseer attains a maximum length of 90 cm. It is cultured in some parts of India but is not included in the traditional Indian polycultural scheme.

SPAWNING THE INDIAN CARPS

For many years a major obstacle to the development of culture of the Indian carps has been the inability of culturists to consistently breed them in captivity. All species spawn naturally in rivers and will not reproduce in standing water, although the major carps may be spawned in specially constructed reservoirs, or bunds, where there is enough current to approximate river conditions. The copper mahseer will also spawn in running water ponds provided the temperature is about 20°C, or ripe fish may be hand stripped and the eggs fertilized artificially. Little or no effort has been expended on hand stripping or reservoir spawning of the Cauvery carps or minor carps.

BUND SPAWNING

The technique of bund spawning can be applied only where the drainage from an extensive catchment area can be accumulated in a natural depression. The low end of such a depression is blocked off by a strong embankment, so that during the rainy season intermittent streams will inundate the depression. Or a dam may be built in the uplands to concentrate runoff in one place, from which it is channeled into the bund. During the dry season a minimum of 1.5 m of water is maintained in a pool perhaps 3000 m² in area, which is stocked with catla, rohu, mrigal, and calbasu in a 4:2:1:1 ratio at a sex ratio of 2 males per female; males may be distinguished by the rough dorsal surface of the pectoral fin. The exact number of fish stocked depends on the extent of the spawning area; about 1 fish per 15 m² is appropriate. The pool is usually fished just before the start of the monsoon, and the numbers of each species and sex estimated, so that losses can be rectified by stocking.

The spawning area is a flat piece of land on which grass is grown during the dry season so that the water will not be easily muddied when it is flooded. It may comprise the entire bund other than the permanent pool, or it may be just one corner of the bund.

When the rains come and flood the spawning ground an outlet channel in the embankment is opened up so that water circulates through the bund. The breeders then move out of the pool and begin chasing in the shallow areas. Spawning usually occurs at night during the full moon or new moon.

When spawning is completed, the eggs are collected by dragging seines, about 4 m long × 1.5 m deep, made of mosquito netting, over the spawning ground. Unfortunately a large percentage of the eggs is often destroyed by being trampled by the fishermen. The eggs are placed in spawning pits dug as near as possible to the bund and supplied with bund water. These pits may be 1 to 1.3 m long, 0.6 m wide, and 0.3 m deep and accommodate from 100,000 to 300,000 eggs. When the inlet to the pit is closed, the water temperature rises rapidly, accelerating hatching time to 12 to 18 hours, instead of the 36 hours that might be required at lower temperatures.

The rate of hatching in these pits is low due to bacterial decay and lack of aeration in the stagnant water. Some improvement may be effected by suspending a hatching net or "hapa" in a hatching pit. The hapa is rectangular in shape and constructed as a net within a net. The eggs are placed in the inner net, which is made of material just fine enough to hold them. The newly hatched larvae pass through into the outer net of fine cloth and are nursed there until they are 4 to 5 days old, by which time they have become fry.

ARTIFICIALLY INDUCED SPAWNING

As mentioned, the technique of bund spawning may be used only in areas where the topography permits. Induced spawning with the aid of pituitary injections is more generally applicable. The first attempts to apply this technique to the Indian carps were made in 1956, but it is only in the last five or six years that consistent success has been achieved. (For a general discussion of induced spawning with pituitary materials, see p. 85.) Today all four major carps and a number of other Indian cyprinids are bred in this manner. Success has been found to hinge on the quality of pituitary glands used and the correct dosage. Pituitaries used in induction should be taken from fully mature, ripe, or freshly spawned fish of the same species as the fish being spawned, or a very closely related species. Dosage is variable, depending on the stage of maturity of the breeders. Since best results are obtained only with fully mature fish, only dosages for such fish will be described.

Breeders are usually selected from 1.5- to 5.0-kg, 2- to 4-year-old fish stocked in spawning ponds a few months prior to the breeding season, which usually coincides with the southwest monsoon, then segregated at maturity. Maturity of males is easily determined; fully mature specimens ooze milt when the abdomen is gently pressed. Selection of females is more difficult, but fish with soft, rounded, bulging abdomens and swollen, reddish vents are preferred. A catheter may also be used to assess ripeness.

Intramuscular injections are made on the caudal peduncle or near the shoulder region. Females are injected two or three times, while males receive only one injection. The first injection, of females only, consists of 2 to 3 mg of pituitary extract/kg of body weight, after which the sexes remain separated for 6 more hours. Then males are given a dose equal to the first injection administered to females, while the females receive a second dose of 5 to 8 mg/kg.

Following this injection, the fish are placed together in groups of three (two males and one female) in covered breeding hapas, 1.6 to 6.5 m² in surface area and 0.9 m deep, fixed on bamboo poles in the marginal waters of ponds. Spawning ordinarily takes place within 3 to 6 hours, but if after 10 to 12 hours no spawning occurs, females only are given a third, slightly higher dosage of pituitary extract.

If the water temperature is near optimum (about 27°C for most species), 60 to 100% success may be expected by this method. Once the eggs have hardened, 8 to 10 hours after spawning, they are transferred, in batches of 75,000 to 1,000,000, to 1.6-m², 0.9-m deep hatching hapas set in marginal waters of ponds, where they hatch in 15 to 18 hours at 27 to 31°C. Spawners may be sacrificed to obtain their pituitary glands.

As induced spawning of cyprinids becomes more prevalent in India, selective breeding and hybridization of Indian carps can assume greater importance. Experimental hybridization of Indian carps with each other and with Chinese carps has already been done, but the resulting offspring have for the most part been unpromising if not incapable of survival. One exception is the hybrid ♂ catla × ♀ rohu, which combines the wide body of the catla with the small head of the rohu. These traits would give it an advantage at the market wherever consumers do not eat the head of the fish, while at the same time effectively giving the consumer more nourishment for his money. The catla × rohu hybrid is fertile and has produced a healthy F_2 generation.

Despite recent improvements in induced spawning methods, the main source of stock for culture in India and Pakistan continues to be collections of eggs, larvae, fry, and fingerlings taken from rivers.

COLLECTION OF EGGS, LARVAE, FRY, AND FINGERLINGS

COLLECTION AND HATCHING OF EGGS

Eggs are collected only from the Halda River in the Chittagong area of Bangladesh. Drifting fertilized eggs are captured 12 to 14 hours after spawning, which generally takes place within 3 weeks before or after each full moon from April through July. The precise time of peak spawning is determined by test collections at the start of the season. The number of eggs caught in these collections enables the collectors to forecast the peak of spawning so that they can fish about that time or shortly after, when the maximum number of eggs is available.

The net used in egg collecting is simple, consisting of no more than a rectangular piece of mosquito netting 11 to 12 m long and 2.7 m wide with a bamboo pole attached at each end. Such nets are usually operated by two men in a boat stationed at right angles to the current, but they may also be moored in the river or used by men wading.

Captured eggs may be placed in a compartment of the collecting boat for some time, even up to hatching, but this results in high mortality due to congestion and inadequate aeration. Better methods of holding and hatching eggs involve nets suspended from a bamboo framework, bamboo baskets lined with fine cloth and suspended in the river (Plate 4), or hatching pits dug in the river bank and supplied with water through a system of pipes. Suspended nets are more often used as a temporary device to accommodate the eggs immediately upon collection, after which the eggs are transferred to pits or baskets, which are believed to produce healthier larvae. A typical hatching pit was 448 cm long × 244 cm wide × 46 cm deep and could accommodate 120 to 300 kg of eggs (900,000 to 2,200,000 eggs).

The hatching rate in all these traditional devices is usually 25 to 50%. Somewhat better results may be obtained using hatching pits in conjunction with the double net device already described in connection with bund spawning.

PLATE 4.   Hatching enclosure or hapa for eggs and freshly hatched larvae of Indian carps. (Courtesy V. G. Jhingran, Central Inland Fisheries Research Institute, Barrackpore, West Bengal, India.)

COLLECTION AND TRANSPORT OF LARVAE AND FRY

The most commonly collected life stages of the Indian carps are the larvae and fry, which are taken at various times of year, depending on the locality. Favored locations for fry collection are along gently sloping banks of rivers where the current is not too strong and at the mouths of small creeks. In such locations special fry-collecting nets are fixed in 1 to 3 m deep water (Plate 5). These nets are funnel shaped, tapering from 2.5 to 3.5 m wide at the mouth to 20 to 25 cm at the cod end, which is kept open by means of a ring. Two lateral wings may be attached at the mouth so as to cover a wide area. A detachable tail piece, or "gamcha," shaped like a monk's hood, is attached at the cod end. The gamcha may be 1 to 2 m long and 40 to 100 cm wide at the rear end. These and other dimensions of fry nets vary widely according to local need. Equally variable is the mesh of the net, which may be of mosquito netting or nearly any available cloth. Experiments conducted by the Allahabad Substation of the Central Inland Fisheries Research Institute have shown that 0.3 cm. netting is much more effective than the materials in general use. Finer netting may be called for where there is a fast current and low turbidity.

PLATE 5. Net for collecting cary fry in Indian rivers. (Courtesy V. G. Jhingran, Central Inland Fisheries Research Institute, Barrackpore, West Bengal, India.)

Fry nets are anchored in the stream bottom by bamboo poles, two each at the mouth and middle of the net, and one at the tail. Usually a battery of 15 to 25 nets, owned and operated by a group of 7 to 12 fishermen, is placed at each collection site. While the nets are in use two fishermen in a boat move from net to net emptying the gamchas every ½ to 1 hour or as often as is necessary to prevent crowding of the catch. Captured fry and larvae are passed through a wire or bamboo sieve to separate them from debris, then transported in large earthen basins to rectangular cloth live wells about 2 m long × 1.25 m wide × 0.7 m deep anchored in the river so that the top is just above the water level. The average daily catch from one fry net may be from 300,000 to 750,000 fry and larvae.

The next step is transportation of the fry to market at one of the larger cities where they are sold to owners of nurseries for raising to fingerling size. The classical container for transport is an open earthen vessel of variable size, called a "hundi." The density of fry in a hundi varies according to the time to be spent in transit (Table 1).

However many fry are to be transported, it is advisable that the water in the hundies be kept constantly agitated. Manual agitation is sometimes necessary, as when fry are shipped by rail and the train is stopped in a station. Oxygen depletion is further prevented by the addition of 50 to

TABLE 1.   NUMBERS OF INDIAN CARP FRY RECOMMENDED FOR TRANSPORT IN
EARTHEN HUNDIES OF 27.3-LITER CAPACITY

LENGTH OF FRY (MM)	NO. PER HUNDI	MAXIMUM DURATION OF TRANSPORT (HOURS)	PERCENTAGE OF MORTALITY
12–19	1,500	24	2 –5
	1,200	36	2 –5
19–25	1,000	20	2 –5
	800	30	2 –5
25–51	500–800	24	10.0
51–77	200	8	10.0

SOURCE: Alikunhi (1957).

100 g of colloidal earth to each hundi. It has been suggested that the colloidal earth serves as a buffer to maintain proper pH, but its principal function is to attract and concentrate in the bottom sludge dead fry and other potential sources of organic pollutants.

In recent years a variety of more sophisticated transport containers made of metal or plastic and equipped with circulating pumps, aerating devices, or a supply of oxygen have started to supplant the earthen hundi, but many fry are still shipped with no means of oxygenation other than agitation of the water by hand or by the motion of the transport conveyance. Even without supplementary oxygenation, metal containers are sometimes preferred to hundis since there is no danger of breakage. The number of fry which can be transported in closed metal containers with and without oxygen is shown in Table 2. As a more general rule of thumb in transporting Indian carp fry, the minimum volume of water per fish has been estimated for different sizes of fry (Table 3).

On arrival at the market the fry are transferred to 70- to 90-liter earthen pots, or "galmas," each containing 150 to 200 g of colloidal earth. After 10 to 15 min, the colloidal earth is removed and with it the dead larvae and fry.

Young fry sold at market are usually a mixture of major and minor carps, for no method of fry sorting comparable to that used for Chinese carps (see Chapter 3) has been developed for Indian carps, although numerous attempts have been made. Catla can be separated from other species with fair success by placing the mixed fry in a tall, narrow container and allowing the oxygen supply to become severely depleted, at which point the catla come to the surface and most of them can be skimmed off, but the other species do not sort themselves out as the

TABLE 2.   NUMBERS OF INDIAN CARP FRY RECOMMENDED FOR TRANSPORT IN
CLOSED 22.7-LITER METAL CONTAINERS WITH AND WITHOUT OXYGENATION

INITIAL DISSOLVED OXYGEN CONTENT (PPM)	SIZE OF FRY TO BE TRANSPORTED (MM)	NO. OF FRY TO BE PUT IN EACH CONTAINER	APPROXIMATE SAFE PERIOD DURING WHICH TRANSPORT CAN BE EFFECTED (MIN)
4	6–7	50,000	19
4	6–7	30,000	31
4	6–7	20,000	47
4	15–20	1,000	40
4	15–20	500	80
4	30	300	120
4	30	150	240
5	6–7	50,000	25
5	6–7	30,000	42
5	6–7	20,000	62
5	15–20	1,000	60
5	15–20	500	120
5	30	300	165
5	30	150	330
6	6–7	50,000	31
6	6–7	30,000	51
6	6–7	20,000	77
6	15–20	1,000	75
6	15–20	500	150
6	30	300	207
6	30	150	414
oxygenated	12–20	1,000–1,250	12 hours
oxygenated	39–51	500	24 hours
oxygenated	25–32	400	16 hours

SOURCE: Alikunhi (1957).

Chinese carps do. About all that can be done otherwise is to separate and
discard young predatory fish, most of which are slightly larger than the
carp fry, by sieving. Visual identification of advanced fry is possible but
excessively tedious.

COLLECTION AND CONDITIONING OF FINGERLINGS

Fingerlings as well as fry of catla, carnatic carp, and Cauvery carps are
collected in Madras and certain parts of East Bengal, Delhi, and Uttar

TABLE 3. MINIMUM VOLUME OF WATER REQUIRED DURING TRANSPORT BY INDIAN CARP FRY OF DIFFERENT SIZE GROUPS

LENGTH OF FRY (CM)	AV. WT. OF FRY (G)	MINIMUM WATER VOLUME REQUIRED PER FRY (CC)
4–7	1.91	25
3–5	0.92	15
2–4	0.35	8
2–4	0.25	7
1–2	0.076	2

SOURCE: Saha et al. (1956).

Pradesh. Fingerlings are collected from back waters of rivers, paddy fields, irrigation channels, tanks, and so on, which they enter from large rivers. In such confined waters they may be captured with seines, dip nets, cast nets, or, in Madras, with special rectangular basket traps made of palmyra roots, called "mavulu." These traps, 1.2 m long, 30.5 cm wide, and 0.9 to 1.2 m high, are placed in gaps in dams constructed across channels for the express purpose of obstructing the movements of young fish. A number of holes about 15 cm in diameter are located along the side of the mavulu near the bottom. Fish enter the holes and swim into the trap but find it very difficult to leave due to inwardly narrowing funnel-shaped structures connected to each hole. In parts of Andhra Pradesh state, similar traps are used to collect fingerlings of catla, mrigal, and fringe-lipped carp from irrigated paddy fields. Fry of these species enter paddy fields naturally in June and July and are allowed to grow there until September or October, when they are trapped.

In the Cauvery River delta Indian carp fingerlings are captured by regulating the flow of water through irrigation sluices. Fingerlings congregate below the sluices and are stranded when the sluice gates are closed. This operation is carried out at intervals of about 4 hours and the stranded fingerlings easily captured by seining. Since by the fingerling stage Indian carps have acquired adult specific characters, they are easily sorted visually.

If the fingerlings are to be stocked nearby, they may be transported in jars and stocked immediately, but if the rearing ponds are distant, it is considered advisable to "condition" them so as to eliminate all food and excreta in the gut. If fingerlings are transported in crowded containers without conditioning, feces and vomited remnants of food may severely pollute the water. Conditioning basically involves nothing more than starvation for 48 to 72 hours, although it may also help accustom the fish to crowded conditions. Rapid elimination of food and feces may be achieved by placing the fingerlings in a net basket fixed in a pond, then splashing water on them from all sides, which frightens them so that they pass excreta and vomit immediately. No analysis has been made of the stress factor induced by this practice. It has been suggested that some feeding with animalcules such as cladocerans may be preferable to total starvation in both conditioning and transport.

Practically any container with mesh or perforated sides which can be placed in a pond or stream may be used for conditioning. Conditioning containers should be kept in a shaded area to protect the fingerlings from sudden changes in temperature; an optimum temperature for conditioning is considered to be between 26 and 29°C. Conditioning containers should not be placed where the water will be muddied by the activities of fishermen.

NURSERY PONDS

PREPARATION AND STOCKING

Captured and conditioned fingerlings may be placed directly into growing ponds to fatten them for consumption or stocked in rearing ponds for further intensive care. Fry, however, must be nursed for 12 to 15 days to insure their health and satisfactory growth. Both seasonal and perennial ponds, ranging from 7 m² to 0.5 ha in area and 0.9 to 3.6 m deep, are used as nurseries. Sample stocking rates for nursing ponds are shown in Table 4.

Perennial ponds have the serious disadvantage of harboring a host of predators, parasites, and competitors of carp fry, which seldom can be entirely eliminated. For this reason, in stocking perennial ponds it is best to scatter the fish, rather than dumping them all in at once. Stocking at night is also advisable, since most fry predators feed visually and the fry are especially vulnerable to predation during the period of acclimatization to a new environment.

Temporary ponds are somewhat better, but the ideal is a pond which can be drained at the discretion of the culturist. In the relatively few modernized nursery units in India, the ponds are drained and sun-dried before use, the bottom is plowed, and a short-season crop of leguminous plants is grown as a source of nitrogen for the soil. After harvesting the legume crop, the plants are plowed under and the bottom is levelled. Further fertilization may be carried out using various kinds of manure or refuse, mixed with oil cake, applied at 200 to 325 kg/ha before the pond is filled. Vegetable manures are usually applied in heaps, weighted

TABLE 4.   SAMPLE STOCKING RATES OF INDIAN CARPS IN NURSERY PONDS

TYPE AND/OR AREA OF POND	STAGE AND/OR SIZE	APPROXIMATE NUMBER PER HA OF WATER SURFACE	DURATION OF REARING (DAYS)
Shallow, seasonal nurseries, 91–152 cm deep, 1,524 cm × 1,524 cm or 1,828.8 cm × 914 cm in area, and paddy fields with a depth of over 30 cm	Fry up to 8.5 mm	741,300–1,235,500, depending on density of plankton available as food	15–30
Seasonal nurseries 1,524–1,828 cm × 914–1,219 cm × 91–122 cm	Fry up to 8.5 mm	1,235,500–1,976,800	15
Seasonal, shallow nurseries 1,524 cm × 1,524 cm in size and 91–122 cm depth	Fry few hours to 3 days old	222,390 without feeding, 1,235,500 with artificial feeding	15
0.62 ha in size	—	—	30
Seasonal nurseries, 0.9–1.5 m deep, 2.4 × 3 m in size, or perennial ponds 1.8–3.6 m deep and 0.5 ha in area	Fry	7,812,500	12–15
Cement cisterns	Fry up to 19–25 mm	8,401,400 without artificial feeding (must be thinned after 10 days)	—

SOURCE: Jhingran (1966).

down with stones rather than scattered about the pond, so as to reduce the danger of widespread deoxygenation. Animal manure is placed in bags or baskets for the same reason. Dilute sewage is used as a fertilizer in some areas, notably in the Bidyadhuri Spill in Calcutta. In using sewage, care should be taken that the dissolved oxygen concentration in the pond does not drop below 3 ppm. Inorganic fertilizers, though too expensive for use by many culturists, have been used on an experimental basis. They would appear to be especially promising for use in arid areas, where organic manures are relatively scarce and more appropriately used as a source of humus for the soil. Unfortunately, most of the research on inorganic pond fertilization in India has involved use of the N-P-K mixtures, with little testing of the individual elements. Use of N-P-K mixtures may be wasteful, as limnological data from many parts of India show an abundance of potassium. The nitrogen-fixing blue-green algae so prevalent in Indian fish ponds, along with the occasional practice of growing legumes in dry ponds, may make the addition of nitrogen superfluous as well. There is some experimental evidence that phosphates alone are as effective as or better than N-P-K mixtures in many situations, but the lack of adequate controls in this research leaves room for doubt. In parts of Bangladesh a 3:1 mixture of cow dung and superphosphate is applied to ponds at 555 kg/(ha)(year). Clearly, pond fertilization practices in India and Pakistan can be improved, through research and by greater appreciation of the fact that the waters of the region vary widely in chemical characteristics, so that each body should be treated individually rather than according to some custom or general formula.

If the soil is acid, it may also be necessary before filling to treat the pond with lime until a pH of 8 or 9 is reached. Large perennial ponds are generally acid, but 300 to 500 kg/ha of lime is usually sufficient to correct this situation.

In ponds which cannot be drained and dried, pest control is a major preoccupation of the culturist. Predators include fish such as the snake-heads (*Ophicephalus* spp.), frogs, several varieties of insect, and aquatic birds. Competitors of the carp fry include many species of small, commercially undesirable fish as well as tadpoles. Fishing is usually inadequate to control predator and competitor fish, and poisoning must often be relied on. A number of pesticides of plant origin are used, but the most common is derris root powder. Applied at 4 to 6 ppm, it eliminates virtually all fish as well as killing some aquatic insects and tadpoles. Larger doses may be more effective, but the prescribed dosage is preferred since it initially only stuns the fish, and edible predators can be salvaged by quickly transferring them to clear water. Further control of insects can be achieved by applying an emulsion of 56 kg of mustard or coconut oil and 18 kg of washing soap per hectare. These poisons lose their effectiveness within 2 to 12 days of application, at which time manuring or stocking may be initiated. Ducks and geese must be fenced out of nursery ponds.

Weeds may also cause a problem by competing for nutrients with plankton and by providing hiding places for predators. Manual or mechanical removal is preferable to chemical control, but the possibility of ecological control, using shading or herbivorous fishes, should be considered. A lengthier treatment of weed control methods follows in the discussion of growing ponds.

FERTILIZATION OF NURSERY PONDS AND SUPPLEMENTARY FEEDING OF FRY

The fry remain in the nursery 12 to 15 days and are not ordinarily fed, since plankton produced by fertilization is adequate for growth. It has been customary to stock nurseries when the water turns bottle green, indicating a heavy growth of phytoplankton. Inorganic fertilizers are particularly effective in producing this condition. Recent research, however, has shown that early fry of the major carps prefer zooplankton. Application of fresh or partially dried cow dung at about 11,000 to 16,000 kg/ha will, after about 10 days, produce a vigorous growth of zooplankton lasting 7 to 10 days. Heavier doses are actually less effective in producing zooplankton. This sort of manuring should be carried out before stocking. If the initial dose does not produce an adequate amount of zooplankton, treatment may have to be repeated at 2500 kg/ha or less every 4 to 5 days until production is satisfactory. Fertilizers applied after the pond is filled should be handled in much the same manner as fertilizers placed in dry ponds: they should be placed in a heap at one corner of the pond. Or a separate culture of zooplankton may be grown in a small, heavily manured pond.

The presence of adequate quantities of zooplankton can be tested for very simply. If 50 liters of pond water are filtered through a plankton net into a test tube and a few drops of formalin or a pinch of common salt is added to kill the plankton, a layer of sediment about 1 cm deep should develop. The color of the sediment, brown or green, indicates the relative proportions of zooplankton and phytoplankton, respectively.

If a pond simply does not produce enough food organisms, or if other problems such as deoxygenation or reinvasion of predators develop, the fry may be transferred to another pond. This is not necessarily an emergency measure, as a change of environment may accelerate growth in any case.

If a heavy bloom of phytoplankton develops, it may be controlled by dissolving cow dung or dye in the surface water to block the penetration of light into deeper water. An abundance of duckweed or other small floating plants serves the same purpose.

Some Indian carp culturists make use of artificial fry feeding. Various types of dried and powdered oil cakes mixed with rice bran are the customary food, although water fleas (*Daphnia*) have also been recommended. The normal feeding schedule for oil cakes and rice bran is given in Table 5.

There has been speculation that oil cakes and rice bran placed in ponds

TABLE 5. FEEDING SCHEDULE FOR INDIAN CARP FRY

	ARTIFICIAL FOOD TOTALLING:
First five days after stocking	One to two times the weight of the fry at stocking daily
Second five days after stocking	Two to three times the weight of the fry at stocking daily
Third five days after stocking	Three to four times the weight of the fry at stocking daily

function more as fertilizer than as food. Extensive feeding experiments carried out at the Central Inland Fisheries Research Substation, Cuttack, India, compared the effects on common carp and rohu fry of 19 food items, fed singly or in combinations of two, three, and four items, with the effects of the oil cake-rice bran diet. In laboratory tests many of the diets studied produced better growth and/or survival than did the normal diet. The best of the diets tested, a mixture of notonectids (aquatic insects), prawns, and cowpeas, was then field tested on nursery fry of catla, rohu, and silver carp. Although fry of these species are plankton feeders in nature, all showed better growth and survival on the mixture than on conventional diets. Mrigal fry also grew well on the mixture but were not compared to mrigal fry fed oil cakes and rice bran.

The Biometry Research Institute of the Indian Statistical Institute has been the site of other research on diets for Indian carp fry, in this case involving micronutrients. Addition of yeast, vitamin B complex, and ruminant stomach extract with cobalt nitrate to ponds containing 3- to 26-day-old, Daphnia-fed fry of catla and rohu resulted in higher survival and, especially in the case of yeast, better growth. Yeast and B complex also reduced density effects on survival. Yeast, in particular, may find application in commercial culture of the Indian carps. However, while addition of yeast to fry diets enhances growth, it also decreases the total protein per gram dry weight of fry.

REARING PONDS

When the nursing period is over, at which time the fry have reached a length of 20 to 50 cm, they should be left in the nursery pond for about 2 days, during which time they are not fed, then captured in a fine-mesh seine and transferred to rearing ponds. The nursery pond may then be prepared for the next lot of fry.

Rearing ponds are poisoned, fertilized, and if necessary limed in the

TABLE 6.   SAMPLE STOCKING RATES OF INDIAN CARPS IN REARING PONDS

TYPE AND/OR AREA OF POND	STAGE AND/OR SIZE (MM)	APPROX. NO. PER HA WATER SURFACE	DURATION OF REARING (DAYS)
Perennial or seasonal rearing ponds, 122–183 cm deep, 0.62–1.24 ha in area, and paddy fields with a depth of 46 cm or more	25–51	49,420–74,130 without feeding; 148,260–197,680 with regular feeding	30–60
Rearing ponds slightly larger than nursery ponds	25–38	98,840–123,550 without feeding, 148,260–197,680 with artificial feeding	60
Perennial or seasonal rearing ponds retaining water for a long period, but not deeper than 183 cm, long and narrow in shape for easy, inexpensive fishing operations, and paddy fields 45–61 cm deep	19–25	24,710 without feeding; 197,680 with artificial feeding	60–90
Rearing ponds	25–51	4,000–5,000	60
Rearing ponds	35	250,000–500,000	—

SOURCE: Jhingran (1966).

same manner as nursery ponds. Stocking rates for rearing ponds are shown in Table 6.

From this time on, artificial feeding is only rarely practiced in the culture of the Indian carps, the plankton production of a fertilized pond providing adequate food for growth.

PRODUCTION PONDS

PREPARATION

Production ponds are prepared similarly to nursing and rearing ponds. First all bottom deposits are removed, either manually after draining or, if draining is impossible, by use of nets or long-handled scoops. Lime is added not only to adjust the pH of acidic ponds but also as a disinfectant. If a pond has not previously been limed, a heavy dose, perhaps as much as 10,000 kg/ha, may be necessary, but in regularly limed ponds 100 to 200 kg/ha is sufficient, except where the soil is very acidic or poor in carbonates.

Manuring practices vary greatly; one recommended dosage for production ponds is 1000 kg or more of cow dung, 560 to 1200 kg of poultry manure, and 5000 kg of green compost/ha. Other fertilizers such as oil cakes or commercial inorganic fertilizers are also used. Water containing sewage may be used in fattening Indian carps for consumption. This practice might seem questionable from a public health standpoint, but at concentrations of sewage great enough that human pathogens are abundant, dissolved oxygen levels are too low for high survival of fishes.

STOCKING

Stocking practices in Indian carp production ponds seem haphazard when compared with polyculture practices in China and elsewhere. This is partly due to the extreme difficulty of distinguishing the various species as fry, but also because the whole question of suitable polycultural techniques for India has not been adequately explored. Not only are the numbers and proportions of the different species of fish stocked not standardized, neither is the size at introduction to the growing pond. All sizes from fry to juveniles 300 mm long are used. Table 7 includes a sample of various stocking policies which have proven at least partially satisfactory.

Since the characteristics of the bodies of water in which these stocking practices are employed are unknown, it is not possible here to make

TABLE 7.   SAMPLE STOCKING RATES OF INDIAN CARPS IN PRODUCTION PONDS

SIZE STOCKED	FISH/HA	SPECIES RATIO
50–100 mm	4,000	catla:rohu:mrigal, 6:3:1
75–130 mm	ca. 11,000	catla:rohu:mrigal, 2:3:4
juveniles	—	catla:rohu:mrigal, 3:3:4
80–130 mm	6,250	catla:rohu:mrigal, 3:6:1 or catla:rohu:mrigal:calbasu, 3:5:1:1
—	—	catla:rohu:mrigal, 3:3:4 or catla:rohu:mrigal:calbasu, 3:3:3:1

recommendations for pond stocking. In general, the numbers of the major carps stocked should be determined by the relative availability of preferred foods for each. Calbasu are usually added to the basic three species when there are mollusks available in the growing pond, since these are not utilized by the other major carps.

GROWTH AND YIELDS OF INDIAN CARPS

In stocking production ponds, allowance is usually made for an annual mortality of 30% or more. Very large fish bring poor prices, so Indian carps are seldom left in growing ponds longer than 3 years. Most fish are sold after the first year, at which time catla, rohu, mrigal, and calbasu may have attained weights of 900 to 4100, 675 to 900, 675 to 1800, and 450 g, respectively. The Cauvery carps are smaller fish, attaining first-year weights of up to 450 g for the fringe-lipped carp, 330 g for the white carp, and 300 g for the Cauvery carp. In very fertile waters, comparable weights may be reached in as little as 6 to 8 months. Few data are available on yields of Indian carp culture, but they vary widely. In semiwild waters, where the fish are merely stocked and forgotten until harvest, yields seldom exceed 110 kg/ha. With cultivation this expectation may be increased to 300 to 900 kg/ha. With artificial feeding yields as high as 2802 kg/ha have been achieved. In recent experiments at the Central Indian Fisheries Research Station at Cuttack, the unprecedented yield of 3564 kg/ha was obtained by stocking catla, rohu, mrigal, silver carp (*Hypophthalmichthys molitrix*), grass carp (*Ctenopharyngodon idellus*), common carp, and calbasu in the ratio of 5:10:5:10:4:5:1 at 5000/ha, with fertilization and supplementary feeding.

MARKETING

Indian carps usually are sold fresh, in accordance with the regional preference. Most fish are not sold directly to the consumer, but through various middlemen; sometimes as many as five are involved in handling one lot of fish. Obviously, this greatly increases the price to the consumer and reduces the amount of protein food available to those who most need it. In some states of India, notably Ahmedabad and Maharashtra, cooperatives market a substantial proportion of fishery products at considerable saving to the consumer, a practice which may be extended to aquacultural products.

PROBLEMS OF INDIAN CARP CULTURE

Among the routine problems facing Indian fish culturists is the presence of unwanted plants in growing ponds. Depending on the plant species to be dealt with, a diversity of methods, including poisoning, is used in weed control. Many of the most effective poisons are chlorinated hydrocarbons such as 2,4-D, which are scarcely to be recommended for introduction into ecosystems or human food supplies. Other poisons, such as copper sulfate, sodium arsenite, and anhydrous ammonia gas, lack the long-term cumulative effects and the dangers to man presented by the chlorinated hydrocarbons, but they may be toxic to fish. In all but the most desperate cases therefore, it is safer (and usually cheaper) to use mechanical or ecological methods of plant control. For floating weeds, outright manual removal is the best treatment. Emergent plants may be controlled by cutting of leaves at weekly intervals for about 6 to 8 weeks before fruiting. Marginal weeds may be controlled by many methods, including plowing under, grazing by livestock, burning during the dry season, or deepening the margin of the pond. Rooted submerged weeds may be removed, albeit with considerable labor, by netting with strong nets, by dragging chains, or with the aid of various types of fork and rake.

The commonest agents of ecological weed control are herbivorous fish. The most commonly used species for this purpose is the grass carp, which in addition to destroying or reducing most submerged and emergent plants and adding to pond fish production, contributes to the nourishment of other fish by dropping partially digested plant remains in its feces. All four of the major carps and a number of the minor carps, although unable to utilize fresh macrophytes as food, consume the partially decayed plant material in grass carp feces. However, grass carp are not a panacea for weed problems, since there are some plants, such as *Eichornia* and *Salvinia*, which they eat only with reluctance if at all. There is also the possibility that if grass carp find their way into natural waterways, they may be destructive of the native flora. Another herbivore, already introduced to Ceylon for use in weed control, is the tawes (*Barbus gonionotus*). A number of species of tilapia, including *Tilapia melanopleura, Tilapia mossambica, Tilapia nilotica,* and *Tilapia zillii,* may effectively control certain types of weed. However, *T. mossambica* has on at least one occasion been found to depress total yield in polyculture involving catla (for details on use of tilapia in weed control, see p. 375–376).

Another ecological method of weed control is shading. Trees around the border of a pond may provide sufficient shade to discourage marginal weeds, but to control offshore weeds in most ponds the culturist must rely either on floating plants such as duckweed (*Lemna*), individuals of which are small enough not to interfere with netting and routine management operations, or on creation of an algal bloom. A heavy bloom can be induced by repeated application of N-P-K fertilizers, but this may be prohibitively expensive. An unusual combination of effects results from application of superphosphate or urea at 50 ppm or more. These substances at that concentration are toxic to most submerged plants, but they act as fertilizers and produce an algal bloom as well. The desirability of an algal bloom must of course be evaluated in terms of the characteristics of the individual pond and the feeding habits of the fish present.

Other pond management practices used in India and Pakistan include the treatment of foul water by doses of 1.5 ppm or less of potassium permanganate, raking the pond bottom to release accumulated gases, and addition of minute amounts of alum to settle suspended or colloidal matter and reduce turbidity.

Among the diseases reported in Indian carps are gill rot, *Saprolegnia* infection, dye disease, fin rot, Ichthyophthiriasis, costiasis, argulosis, ligulosis, gyrodactylosis, and dropsy. For a detailed discussion of some of these diseases and methods of treatment, the reader is referred to Davis (1953).

PROSPECTUS AND RECOMMENDATIONS

The prospectus for culture of the Indian carps is one of growth, as it must be if India's protein crisis is to be abated or merely kept from worsening. Significant growth can be achieved if a greater proportion of India's available waters are brought into fish production. But to even approach the type of production increase that is needed, great scientific, technological, and social strides must be made. Among the steps which must be taken are the following:

1. The ecological niches of the various Indian carps must be better understood, so that pond stocking and management can be carried out on a more rational basis. If research discloses that not all available niches are being filled, additional species, for example some of the Chinese carps, could be introduced to Indian polyculture. More precise knowledge of the ecological role played by the species cultured would also eliminate such unfortunate practices as encouraging a phytoplankton bloom in a pond full of zooplankton feeders.

2. More effective means of spawning Indian carps must be developed. This almost certainly means perfecting existing techniques of induced spawning by hormone injection. Only when Indian fish culturists can

consistently and selectively spawn their stock will Indian fish culture approach its full potential.

3. When successful breeding is assured, it will be possible and desirable to proceed with experiments in selective breeding and perhaps hybridization.

4. Far better methods of hatching eggs must be adopted. A hatching rate of 25 to 50% is simply not satisfactory for a major fish culture enterprise.

5. Mortality in transportation of fry must be reduced and transport made more efficient. Again the technology exists, but economic considerations restrict its usage. Some progress is being made in this area, however.

6. As long as river-caught fry sustain a major portion of the market, there will be a need for a workable method of sorting fry to species. The present inability of culturists to determine what species they are stocking renders current knowledge of suitable stocking ratios useless and results in inadvertent stocking of inefficient protein producers, that is, the minor carps.

7. More and more efficient hatcheries and culture stations must be built, using artificial ponds that can be drained and otherwise ecologically controlled. Losses to predators and other natural hazards which could be controlled undoubtedly significantly reduce the total yield of Indian fish culture.

8. Much more research needs to be done on pond fertilization. In conjunction with this research, surveys should be made of the chemical characteristics of Indian soil and waters so that the culturist may know precisely what the effect of a certain dosage of a substance in a given pond will be. This would eliminate loss of fish due to fouled water, unutilized phytoplankton blooms, and so on. For the present at least, emphasis should continue to be placed on organic fertilizers, for ecological as well as economic reasons, but inorganic fertilizers should not be ignored.

9. Ecological means of weed control, particularly those involving herbivorous fish, should be encouraged and use of chlorinated hydrocarbon herbicides discouraged.

10. Artificial feeding of all ages of Indian carps should be studied, and if useful and economically feasible foods are found, such feeding should be encouraged.

11. Indian fishery and fish culture research should be directed toward potential application of its findings. The Indian scientific literature should not be further burdened with masses of needlessly precise data

detailing, for example, the lengths of carp fry in hundredths of a millimeter.

12. Cooperatives and other schemes to eliminate middlemen and supply protein food efficiently and cheaply to those who need it should be encouraged and aided.

Though research undoubtedly holds the key to many of the problems of Indian fish culture, given the urgency of the situation, development must be given preference. Implementation of presently known feasible techniques for increasing protein production in India is of utmost importance and must succeed if India is to feed her people.

REFERENCES

AHMAD, N. 1948. Methods of collection and hatching of carp ova in Chittagong with some suggestions for their improvement. J. Bombay Nat. Hist. Soc. 47:586–602.

ALIKUNHI, K. H. 1956. Fish culture techniques in India in progress of fisheries development in India. Central Inland Fisheries Research Station and Department of Fisheries, Orissa. 96 pp.

ALIKUNHI, K. H. 1957. Fish culture in India. Indian Council of Agricultural Research, New Delhi. Farm Bulletin No. 20. 143 pp.

ALIKUNHI, K. H., H. CHAUDHURI, and V. RAMCHANDRAN. 1951. A possible method for segregation of species from a mixed collection of carp fry. Science and Culture 17:220.

BHASKARAN, T. R. 1952. Effect of organic manures on the oxygen budget in fish ponds. Proc. Nat. Inst. Sci., India (B) 18(4):257–259.

CHAUDHURI, H. 1966. Breeding and selection of cultivated warm-water fishes in Asia and the Far East—a review. FAO World Symposium on Warm Water Pond Fish Culture. FR: IV/R-3.

DAS, B. C. 1966. Effects of micro-nutrients on the survival and growth of Indian carp fry. FAO World Symposium on Warm Water Pond Fish Culture. FR: II/E-13.

DAVIS, H. S. 1951. Culture and diseases of game fishes. U. of California Press, Berkeley, 332 pp.

HICKLING, C. F. 1966. Fish-hybridization. FAO World Symposium on Warm Water Pond Fish Culture. FR: IV/R-1.

HORA, S. L., and T. V. R. PILLAY. 1962. Handbook on fish culture in the Indo-Pacific region. FAO Fisheries Technical Paper 14. 204 pp.

JAGANNADHAN, N. 1947. A note on the collection, conditioning and transport of fingerlings of Catla in Madras Presidency. J. Bombay Nat. Hist. Soc. 47:315–319.

JHINGRAN, Y. G. 1966. Synopsis of biological data on Catla, Catla catla (Hamilton) 1822. FAO Fisheries Synopsis No. 32.

LAKSHMANAN, M. A. V., D. S. MURTY, K. K. PILLAI, and S. C. BANERJEE. 1966. On a new artificial feed for carp fry. FAO World Symposium on Warm Water Pond Fish Culture. FR: III/E-5.

PAKRASI, B. 1953. Preliminary observations on the control of aquatic insects in nursery ponds. Proc. Indian Acad. Sci. (B) **38**:211–213.

PROWSE, G. A. 1966. A review of the methods of fertilizing warm-water fish ponds in Asia and the Far East. FAO World Symposium on Warm Water Pond Fish Culture. FR: II/R-2.

SAHA, K. C., D. P. SEN, and P. MAZUMDAR. 1956. Studies on the mortality in spawn and fry of Indian major carps during transport. Part II. Effect of oxygen pressure, free surface area, water volume and number of fry in the medium of transport. Indian J. Fish **3**(1):127–134.

Interviews and Personal Communication

PANTULU, R. V. Economic Commission for Asia and the Far East.

5

Early Attempts at Fish Farming in the South Central United States Using Buffalofish and Paddlefish

BUFFALOFISH CULTURE

RATIONALE FOR SELECTION OF BUFFALOFISH AS A CROP

Fish farming in the United States is centered in the south central states, which are also among the largest producers of a number of agricultural crops, including rice. The history of commercial fish culture in the region is short, and stems from the institution, in the 1950s, of federal restrictions on the amount of land that could be planted to terrestrial crops. Fish farming was thus initiated as a profitable and legal alternative use for farm land. Rice farmers, who must rotate their crops in order to maintain soil fertility, found the prospect of fish farming particularly inviting.

Among the most important commercial fishery products in the area are catfish (*Ictalurus* spp.) and a group of large suckers (family Catastomidae) known as buffalofish (*Ictiobus* spp.). Both catfish and buffalofish had been experimentally cultured in the early part of the twentieth century, when American fish culturists were still attempting to propagate virtually everything that swam, but little had been done with them since that time. Although buffalofish bring only half the price of catfish, they were selected for the first attempts at fish-rice rotation and other forms of food fish culture. From a biologist's point of view buffalofish seemed the logical choice, since they are low on the food chain, consuming principally plankton, benthos, and detritus. Theoretically, therefore, more buffalofish than catfish can be produced in a given amount of water. Farmers were also more enthusiastic about buffalofish because, whereas the catfish fishery was fairly stable, the availability of buffalofish fluctuated seasonally. A very crude form of buffalofish culture, consisting of no more than holding fishery-caught specimens in small ponds for sale during the off season, had been tried periodically, but very high mortality due to handling prevented it from being a success. It was thought that intensive culture of buffalofish would enable the farmer to more profitably supply the off-season demand.

SPECIES USED AND HYBRIDS

There are three species of buffalofish, the bigmouth buffalo (*Ictiobus cyprinellus*), the black buffalo (*Ictiobus niger*), and the smallmouth buffalo (*Ictiobus bubalus*), all of which are fished commercially. All were cultured experimentally, but the bigmouth buffalo was found to be superior on all counts, since it grows faster, matures earlier, and is more prolific than the other species.

Ictiobus spp. hybridize in nature, and in 1963 to 1964 all possible hybrids of the three species were produced, under hatchery conditions, at the U.S. Fish Farming Experimental Station at Stuttgart, Arkansas. Most of the hybrids were of academic interest only, but one, ♀ *I. niger* × ♂ *I. cyprinellus,* exhibited 33% better growth in weight than either parent in experimental culture. This hybrid was subsequently recommended for practical culture, but it has not been well evaluated, since its introduction coincided with the decline of buffalofish culture.

HISTORY OF BUFFALOFISH CULTURE

Within a decade of the start of fish farming in the south central states, bigmouth and hybrid buffalo were supplanted as the chief fish crop by the channel catfish (*Ictalurus punctatus*) (for details on channel catfish culture in the region see Chapter 6). Table 1 illustrates the change in relative importance of the two types of fish in Arkansas.

TABLE 1. AREA DEVOTED TO CULTURE OF BUFFALOFISH AND CATFISH IN ARKANSAS, 1958–1966

YEAR	AREA OF WATER PRIMARILY DEVOTED TO BUFFALOFISH CULTURE (HA)	AREA OF WATER PRIMARILY DEVOTED TO CHANNEL CATFISH CULTURE (HA)
1958	1,378	0 (some catfish grown as a supplemental crop with buffalo-fish)
1960	1,434	100
1963	297	428
1966	100	5,800

Today, monoculture of buffalofish is virtually nonexistent. The bulk of the buffalofish on the market is contributed by fisheries, with the rest coming from low-intensity polyculture, or buffalofish stocked as supplemental fish in catfish ponds.

The reason for the wholesale switch to catfish culture was that, as long as the culturist's goal is to make a profit, southeast Asian methods, usually cited as a model for fish culture, are not advantageous in the United States. Under the unusual, and perhaps temporary, economic conditions which prevail in the United States, the farmer can usually afford to feed fish heavily on a diet rich in animal protein. Under such circumstances he is economically justified in producing the fish that fetches the highest price, regardless of ecological niche. This is demonstrated by the success in catfish farming of a number of culturists who realized low profits, or even net losses, from buffalofish farming.

BREEDING AND REARING THE YOUNG

SELECTION AND CONDITIONING OF BROOD FISH

One of the real advantages for culture of buffalofish is the ease with which they can be spawned. The first step in breeding buffalofish starts well before the spawning season, in February or earlier, when brood stock are selected and stocked in wintering ponds. Brood fish should be 1.3 to 3.4 kg in weight, and among the fastest growing fish of their age group. One-year-old fish, of whatever size, should not be used, because a large proportion of the females are likely to be immature.

Spawners should be free of disease and injuries, but buffalofish are rather delicate, and injuries may occur any time they are handled. This can be guarded against by handling potential spawners a few at a time and covering vessels containing buffalofish to keep the interior dark. If, despite all precautions, some fish are injured, the wounds may be swabbed with a piece of cotton dipped in 20% potassium permanganate. Some culturists routinely disinfect breeders in 10 ppm potassium permanganate for 1 hour, followed by 15 ppm formalin for 12 hours, before stocking wintering ponds.

Satisfactory dimensions for a wintering pond are 0.3 ha in area, and 1 m in mean depth, with some water up to 2 m deep. Buffalofish are among those fish that emit a substance which, in high concentration, inhibits spawning. Culturists take advantage of this fact by stocking wintering ponds as heavily as is consistent with the health of the fish; stocking densities cited in the literature vary from 400 to 2000 kg/ha. Although spawning of even ripe fish can be retarded for up to a month by such heavy stocking, the culturist should nevertheless be prepared to draw the pond down if heavy rainfall occurs, since this may trigger spawning.

Whenever the water temperature rises above 13°C, fish in wintering ponds are fed 1% of their total body weight daily; just prior to spawning, this is increased to 4%. Pellets of Auburn No. 1 fish feed, which contains 35% soybean meal, 35% peanut meal, 15% fish meal, and 15% distiller's dried solubles, have been found satisfactory for buffalofish breeders. A more complex and variable formulation, developed at the U.S. Fish Farming Experimental Station, Stuttgart, Arkansas, has also produced good results (Table 2). Little is known about feeding buffalofish, and these feeds undoubtedly can be improved.

TABLE 2. FEED FORMULA FOR YOUNG BUFFALOFISH, DEVELOPED AT U.S. FISH FARMING EXPERIMENTAL STATION, STUTTGART, ARKANSAS

Rice bran (not hulls)	31.6%
Cottonseed meal	31.6%
Dehulled soybean meal (fine ground)	15.8%
Poultry by-product meal	10.5%
Wheat shorts (fine grade)	2.6%
Ground yellow cornmeal	2.6%
Dry whey or other dried milk product	5.3%
Pure vitamin B-12	25 mg/ton
Pure vitamin A	22,000 units/kg (finished feed)
Pure vitamin D	9,900 units/kg (finished feed)
TM-10	1.8–9.0 kg/ton (finished feed)[a]
Salt	18.0 kg/ton
Di-calcium phosphate	18.0 kg/ton

[a] Sometimes fed only when disease outbreak is feared.

SEXING AND SPAWNING IN COMMERCIAL CULTURE

Spawning may be carried out in the spring anytime after water temperatures rise to 18 to 21°C, but hatching and survival of the young will be best if it is delayed until the likelihood of a cold snap is negligible. Spawning ponds, which are similar in all respects to wintering ponds, should be kept dry until just before stocking, to prevent the establishment of predatory insects.

It may be necessary, before stocking, to disinfect the spawners in 10 ppm potassium permanganate for 1 hour. If possible, stocking and filling should be carried out simultaneously. As the pond is filled with water drawn from a well or other source which does not contain buffalofish, 3 to 20 pairs of spawners, of one species only, are added.

Sexing presents no problem; as spawning time approaches, the vent of the female becomes enlarged and inflamed and begins to protrude. Ripe males, if touched near the vent, will produce a small amount of milt. Males also develop nuptial tubercles, which give the body the feeling of sandpaper, as opposed to the smooth texture of the female.

The combined stimuli of freshly drawn water and greatly reduced population density should be sufficient to induce spawning within 24 hours of stocking, but it may be delayed if the fish are not fully mature or if a cold snap occurs. If spawning does not occur within a week, the pond may be drained nearly dry and refilled to trigger mating behavior.

HATCHING

Hatching occurs in 5 days at 18 to 21°C, and the fry are free swimming in another 2 days. The newly hatched fry are very delicate and should be left in the spawning pond for some time; the parents are removed to reduce the danger of transmission of diseases and parasites. If the fry are not to starve, an abundant supply of plankton must be available soon after they become free swimming, so spawning ponds are fertilized after the eggs are laid. Commercial 8-8-2 fertilizer applied at 100 kg/ha has been found to produce an adequate bloom.

REARING EARLY FRY

Light feeding with suitable sized particles of the same feeds described above should commence about 10 to 15 days after spawning. The rate of feeding is determined empirically, based on the appearance, growth rate, and behavior of the fry. Underfed buffalofish fry may sometimes be observed swimming in schools around the edge of the pond; at such times they are extremely susceptible to disease.

The only other management measure commonly undertaken is the periodic application of a mixture of kerosene and motor oil to the pond surface to control predatory insects. Details of this technique and its pros and cons will be discussed later.

Fry may be left in the spawning ponds until fall, when they are ready for stocking in growing ponds, but it is preferable to stock them in special nursery ponds when they attain lengths of 12 to 40 mm. Yields of up to 1 million such fingerlings/ha have been achieved, using the methods described, at the U.S. Fish Farming Experimental Station, but commercial culturists have more commonly produced 50,000 to 250,000 fingerlings/ha.

HATCHERY PROPAGATION

Buffalofish have been artificially spawned and the eggs hatched under hatchery conditions, but commercial buffalofish culture never became sufficiently important for these techniques to be adopted in practical culture. In the event of a large demand for cultured buffalofish, hatchery culture would have been advantageous in that eggs can thus be hatched more efficiently, in larger numbers, and, most important, synchronously, so that fry of a uniform size can be produced.

Buffalofish used in the previously described hybridization experiments were artificially spawned as follows. Spawners, usually 1.3- to 2.7-kg fish, were isolated in 20-gallon aquaria at 24 to 25°C, injected with pituitary extracts, and, when the eggs were free flowing, stripped of eggs and milt. (See p. 85 for a general discussion of artificially induced spawning.) Females were injected with 315 units of human chorionic gonadotropin or 0.9 mg of acetone-dried buffalofish pituitary per kilogram of body weight; males received half these dosages. Some females spawned within 22 to 24 hours of injection, but others could not be made to spawn with repeated injections.

Eggs were hatched in a closed recirculating water system, using the jar method. (See pp. 92–93 for description of hatching eggs in jars.) Hatching rates of 95 to 98% were obtained in 24 hours at 24°C, but uniformly poor rates were observed at 19°C.

NURSERY PONDS AND THEIR MANAGEMENT

Ponds similar to wintering and spawning ponds may serve as nurseries, but ponds as small as 0.04 ha have also been used effectively. The most important consideration in selecting a nursery pond is that it be absolutely free of predatory fish, since buffalofish fry are unusually susceptible to predation. Before transfer to nurseries, disinfection for 2 to 4 hours in 15 ppm formalin plus 1 ppm acriflavine is advisable.

Stocking rates in nurseries vary from 800 to 4000/ha. The precise rate is determined by the size of fingerling which is desired; the amount of buffalofish of any size which can be produced from 1 ha of water is usually 200 to 400 kg. Fingerlings weighing about 0.1 kg or slightly more each were considered adequate for the intensive rice-fish rotation originally envisioned, but for the less intensive types of culture currently practiced, a larger fingerling is best, because the growing areas commonly contain large predatory fish. Another factor that must be taken into account in nursery stocking is mortality, which is highly variable, but may exceed 50%.

Fingerlings accept the same artificial feeds as fry and breeders, but feeding is optional at this stage. Fertilization is essential, however. The dosage described above should be applied monthly.

RICE-BUFFALOFISH ROTATION

As mentioned previously, the type of rice-fish rotation originally instituted in the south central states was an economic failure. However, it
will be described in some detail because it was fairly successful in terms
of producing fish, and could be greatly improved in that regard. In the
future, if the abundant supply of protein for human consumption in the
United States is not maintained, the economics of American aquaculture
may be profoundly altered, and rice-fish culture and/or buffalofish culture may again be attempted.

Fields used in rice-fish rotation were divided by inside levees to create
ponds of suitable size. Some compromise always had to be reached in
this regard since fields 16 ha or more in area are best for rice farming,
whereas buffalofish culture proved to be most efficient in ponds smaller
than 16 ha. In practice, ponds of all sizes from 8 to 80 ha were used.
Whatever their surface dimensions, productive ponds were designed to
be no less than 0.5 m deep at the shallowest point, and mostly 1 to 2 m
deep, to discourage rooted aquatic plants, provide lower bottom temperatures, and reduce predation by birds.

Production ponds were stocked at various times from late summer to
spring. One of the most efficient stocking methods took advantage of
the heavy fall rains which usually occur in the south central states.
Prior to the rains, ponds were partially filled, preferably with well water.
Fingerlings were then stocked in the borrow ditches leading into the
catch basin (see the Appendix for a generalized description of such pond
construction features) and automatically released into the rest of the pond
when the rains filled it.

Recommended stocking rates varied from 25 to 250 fingerlings/ha in
unfertilized ponds. If potential competitors such as minnows were suspected of being present, largemouth bass (*Micropterus salmoides*) fingerlings were sometimes stocked at 60 to 125/ha. Direct feeding of buffalofish
in growing ponds was never done, and fertilization was only occasionally
carried out. Not only did effective dosages of fertilizer vary greatly from
pond to pond, but the farmer had to consider the effect on the forth-
coming rice crop as well as on the fish.

The growing period ordinarily extended for 2 years. Midway through
the second year of growth, 25 to 33% of the crop was sometimes harvested
with nets or traps. At the end of the 2-year period, the pond was drained
to harvest the remainder of the fish, then planted to rice for 2 years.
Survival of fingerlings was usually high (about 90%), but production was
limited by the preference of consumers for fish weighing about 2.5 kg.

Yields of fish this size commonly ran 200 to 1000 kg/ha for the 2-year growing period, whereas yields of more than 1000 kg/ha of 1-kg fish are not difficult to obtain in 18 months. It is also possible that substantially higher yields could have been obtained with fertilization. Fertilized experimental ponds at Auburn University, stocked with bigmouth buffalo fingerlings at 1080/ha, yielded 656.5 kg/ha in a 6-month growing period. These fish, averaging about 1.3 kg, were too small to be profitably marketed, but the implications of the experiment are clear: If consumer habits change and/or if economically feasible methods of pond fertilization consistent with good rice farming practice are developed, practical intensive culture of buffalofish may still be a possibility.

PRESENT USES OF BUFFALOFISH IN FISH CULTURE

At present, the principal aquacultural uses of buffalofish are in very low intensity polyculture and as supplemental fish in catfish culture. The former method is practiced in natural sloughs and backwaters which abound in the lower Mississippi Valley. The principal species stocked is usually channel catfish, and details of stocking are given in Chapter 6.

Buffalofish are sometimes stocked in small numbers with intensively cultured channel catfish, since they will eat smaller particles of food than the catfish and thus may benefit from partially disintegrated food pellets which would be wasted by the catfish. In addition to providing a supplemental fish crop and preventing waste, buffalofish so stocked may actually enhance the production of catfish by reducing pollution of the water by excess food. There are no data available to support this assertion for buffalofish, but Java tilapia (*Tilapia mossambica*) and *Tilapia nilotica* have been shown to have such an effect on channel catfish in experimental ponds at Auburn University.

PROBLEMS

DISEASES AND PARASITES

A number of diseases and parasites, including fungus infections, bacterial infections, and the anchor parasite (*Lernaea*), have been observed in cultured buffalofish. Research in disease control was only beginning when buffalofish culture in the United States began to be deemphasized, thus specific remedies have not been developed. Culturists can probably handle most situations by adapting treatments used on catfish and other commonly cultured fishes. One problem peculiar to buffalofish is ulcerations on the gills at spawning time. These respond well to the treatment recommended earlier for wounds.

ECONOMIC PROBLEMS

It has been standard practice among buffalofish farmers to harvest their crop in summer and winter, when buffalofish fisheries are least productive, but even so they have not been able to compete with fishermen. Among the economic problems besetting buffalofish culturists is the large weight loss in dressing these fish. While there are local outlets for live or iced whole buffalofish, the bulk of the crop must be cleaned, dressed, and iced or frozen. The total dressed weight of bigmouth buffalo so processed averages 53% of their live weight.

It has been mentioned that small buffalofish are not favored by American consumers. Their poor acceptance is due largely to the many small, v-shaped bones characteristic of Catastomid fishes, which are particularly irritating in small specimens. Workers at the U.S. Bureau of Commercial Fisheries Laboratory in Ann Arbor, Michigan, developed a method of processing small (under 1.2 kg) buffalofish that was hoped would enhance their commercial value but, perhaps because buffalofish culture was already close to nonexistent in 1965, when the method was introduced, it has had no effect. The processing technique involved smoking the rib section and grinding the bony loin section to produce a variety of frozen, reconstituted products. A disadvantage of these products was their tendency to rapidly develop rancid off-flavors and odors.

PROSPECTUS OF BUFFALOFISH CULTURE

Unless and until the status of buffalofish culture in the United States improves, there is little likelihood of *Ictiobus* spp. being cultured elsewhere. In the late 1940s and 1950s, when Israeli fish culturists were searching for supplemental species to stock with common carp (*Cyprinus carpio*), bigmouth buffalo were tested but proved even less suitable for culture than in the United States. Among the objections raised were that they grew too slowly, did not convert artificial feed well, were very susceptible to infection with *Lernaea,* competed with carp, and brought a low price on the Israeli market.

Buffalofish will certainly continue in their relatively unimportant role in low-intensity fish culture in the South Central United States. Interest in polyculture in that country seems to be rising, so it is likely that their use as a supplemental crop in ponds primarily devoted to other fish will increase. But cultured buffalofish will not assume an importance near that which was predicted for them in the 1950s without a major technical breakthrough in culture methods or a drastic alteration of the economy. For the present, American fish culturists would be ill advised to invest heavily in buffalofish.

EXPERIMENTAL CULTURE OF PADDLEFISH

Another fish suggested for culture during the early years of fish culture in the lower Mississippi Valley was the paddlefish (*Polyodon spathula*). Paddlefish, although they reach lengths of up to 150 cm and weights of up to 85 kg, are well suited to traditional pond fish culture in that they feed primarily on zooplankton. Not only is the quality of the flesh excellent, but the roe can be made into a good grade of caviar. At least until recently, paddlefish supported a sizable commercial fishery in the Mississippi-Missouri Basin.

Though sometimes referred to as "spoonbill catfish," *P. spathula,* far from being a catfish, is a member of the primitive fish family Polyodontidae, which contains only one other living species, the huge *Psephurus gladius* of China's Yangtze River. *P. gladius* is reportedly important as a food fish, but it is not known if its culture has been attempted.

The principal disadvantage for culture of paddlefish is that they do not spawn in standing water. In 1962, attempts were made at the United States Fish Farming Experimental Station to hormonally induce spawning. Though fertilization of eggs was not achieved, it was demonstrated that adult paddlefish of both sexes will respond to chorionic gonadotropin. Only large paddlefish responded to the treatment, and sexual dimorphism was observed to be slight. In view of these findings, plus the fact that adult paddlefish do not spawn each year, it was concluded that it would be necessary for practical paddlefish culturists to maintain considerable numbers of large brood fish.

Further experiments were not pursued as first buffalofish, then catfish took over the spotlight in fish culture in the lower Mississippi Valley.

In the last decade, most paddlefish populations have declined rapidly, due to pollution, dam construction, and overfishing. Experimental culture may eventually be resumed as a conservation measure, but the outcome seems dubious, and attempts at commercial culture are most unlikely in the foreseeable future.

REFERENCES

BRADY, L., and A. H. HULSEY. 1959. Propagation of buffalo fishes. Report of the South East Association of Game and Fish Commissioners, 13th Annual Conference, pp. 80–89.

MEYER, F. P., and J. H. STEVENSON. 1962. Studies on the artificial production of the paddlefish. Prog. Fish. Cult. 24(2):65–67.

OLDEN, J. H. 1959. Fish-farming industry. U.S. Bureau of Commercial Fisheries. TL 19.

STEVENSON, J. H. 1957. Report on rearing buffalofishes and catfishes. Trans. Am. Fish. Soc. 87.

SWINGLE, H. S. 1957. Revised procedure for commercial production of bigmouth buffalofish in ponds in the southeast. Proceedings of the 10th Annual Conference of the South East Association of Game and Fish Commissioners, pp. 162–165.

6

Catfish Culture in the United States

159

INTRODUCTION AND HISTORY

One of the most highly publicized aquacultural developments of recent years is the growth of the catfish industry in the south central United States. Although Ictalurid catfish have been experimentally cultured for 50 years or more and some small catfish farms were operating as early as the 1950s, commercial culture of catfish on a significant scale goes back no further than 1963 when a few thousand kilograms were produced, chiefly in Arkansas. By 1966, U.S. catfish production was up to 9 to 11 million kilograms and by 1969 it was about 30 million. Arkansas, Mississippi, and Louisiana account for a large share of the production, but commercial catfish culture on some scale occurs in at least 18 states. Though catfish are important contributors to sport and commercial fisheries as far north as Iowa, and markets for catfish exist in all the major northern cities, the southern states will probably continue to dominate catfish culture in the United States, due largely to the longer growing season in that part of the country. Some successful operations may eventually be carried out in the north using cooling water from factories and power plants, but the only serious threat to the preeminence of the southern states is likely to come from Central and South America.

Catfish farming has been an important source of revenue for the south central states, which constitute the poorest section of the United States. As an illustration of its value as an industry, in 1965 in Lonoke County, Arkansas, 44,000 ha were planted to soybeans and yielded a total income of $6 million. The following year 3800 ha wholly or partially devoted to culture of catfish produced over $5 million in revenue. Although such revenues may have no direct effect on the poorest people of the area and the fish produced make no contribution to their substandard protein intake, the economic effect of such a new industry must eventually be felt at all levels of the local economy.

The mushrooming development of catfish farming has been paralleled, and in fact preceded, by a strong supporting program of research, beginning in 1957 with the pioneering work on spawning of channel catfish

(*Ictalurus punctatus*) carried out by Kermit Sneed of the U.S. Bureau of Sport Fisheries and Wildlife and H. P. Clemens of the University of Oklahoma. Applied research on a large scale began with the opening in 1961 of the U.S. Fish Farming Experimental Station at Stuttgart, Arkansas. In 1965 this laboratory merged with similar facilities at Kelso, Arkansas, and Marion, Alabama, to become the U.S. Warm-Water Fish Cultural Laboratories, thus providing a complex of facilities for conducting basic biological investigations and applied research as well as follow-up studies of all subjects of concern to catfish culturists. The economic aspects of catfish farming have also been evaluated in detail many times so that, despite its virtually overnight development, catfish culture in the United States is one of the most thoroughly documented forms of fish culture and serves as an excellent example of the potentials and pitfalls of intensive fish farming.

The real impetus for catfish culture in the United States was the economic failure in the late 1950s and early 1960s of buffalofish (*Ictiobus* spp.) as an aquatic crop in rice fields and sloughs (see Chapter 5). In 1960 there were, in Arkansas alone, 1400 ha of water devoted to buffalofish culture, with only 100 ha in catfish culture. By 1966, 1800 ha were devoted to monoculture of catfish, with an additional 2000 ha in polyculture with catfish as a principal crop; buffalofish accounted for a very small portion of fish production in the state. Cultured catfish received a further boost as consumers discovered that their flavor was better, or at least more consistent, than that of wild fish.

SPECIES USED

CHANNEL CATFISH

At the same time, intensive pond culture began to supplant rice field and slough culture as the most important form of food fish culture in the area. One of the first problems in conversion to a new form of fish culture was selection of the most suitable species. Catfish, which vie with buffalofish for importance in freshwater commercial fisheries in the southern United States, were the likeliest choice, but which catfish? All seven of the widely distributed, edible sized North American catfishes (family Ictaluridae) have been tested, if not applied, in fish culture, but by far the most widely used is the channel catfish, which is generally considered the most desirable Ictalurid for table use. It is not clear whether the channel catfish really possesses superior table qualities or whether this is an assumption based on the more esthetically pleasing appearance of the channel catfish and the fact that it is more highly esteemed than other Ictalurids by sport fishermen. Nor is it clear that it is the most practical catfish for culture in all situations. However, channel catfish have often yielded excellent results and far more is known about their culture than that of any of the other species. Among the demonstrated virtues of channel catfish are their ready adaptation to artificial feeds and their resistance to crowding. On the other hand, they have a fairly nervous temperament, which may cause problems, particularly when they must be netted or handled.

BLUE CATFISH

The second most frequently used species is the blue catfish (*Ictalurus furcatus*). Although blue catfish at the U.S. Fish Farming Experimental Station have exhibited much poorer growth than channel catfish, some culturists find that they grow more uniformly and produce fewer "giants" and "runts." Blue catfish of all sizes dress out better than channel catfish, weight of dressed fish averaging 60 to 62% of live weight as compared to 56 to 58% for channel catfish. Unlike channel catfish, blue catfish readily learn to feed at the surface, which enables the culturist to inspect his stock for health daily and to determine if they are accepting feed well. They are less nervous and easier to seine than channel catfish, and the males are less prone to fight at breeding time, but they exhibit poor survival when shipped live. Other disadvantages are poorer conversion of most artificial feeds and greater age at maturity.

FLATHEAD CATFISH

The largest of the Ictalurids, the flathead catfish (*Pylodictis olivaris*), differs from the rest of the family in being highly piscivorous and cannibalistic. It has seldom been used in conventional catfish culture but occasionally finds use as a predator in culture of various other species. Rather high mortality has often been experienced in rearing flathead catfish fry, but were as much attention paid to this species as to the channel catfish, fry rearing would doubtless prove comparably feasible. In some areas the flesh of the flathead catfish is considered, at least by fishermen, to be inferior to that of other Ictalurids.

WHITE CATFISH

White catfish (*Ictalurus catus*) are native not to the south central states but to streams emptying into the Atlantic Ocean along the east coast of the United States. Perhaps this explains their infrequent use in fish culture, for they have much to recommend them. Smaller than channel, blue, or flathead catfish, they occupy a place between these species and the bullheads. Unlike the bullheads, white catfish can be easily grown to a size acceptable to all consumers, although they do not grow nearly as rapidly as channel catfish, and due to the rather large head do not dress out as well as channel or blue catfish. They are fully comparable to other Ictalurids in terms of quality of flesh and are superior in converting food. They are among the hardier Ictalurids, withstanding crowding, low dissolved oxygen concentrations, turbidity, and high temperatures much better than channel catfish. Corollary to the temperature tolerance of white catfish is their tendency to feed voraciously throughout the summer. Finally, white catfish are easier to spawn than any of the larger Ictalurids. It seems likely that in the near future white catfish will see wide use in fish culture.

BULLHEADS

The smallest Ictalurids commonly used as human food are the bullheads. The most frequently used bullhead in experimental aquaculture is the brown bullhead (*Ictalurus nebulosus*), but the yellow bullhead (*Ictalurus natalis*) and the smaller black bullhead (*Ictalurus melas*) have also been tested on occasion. In addition to their small size, bullheads readily reproduce in ponds without specific measures being taken, thus over-population and stunting frequently occur. There is a small, steady market for bullheads in some areas, but prices are usually low, and this demand can probably better be filled by fisheries than by fish culture. In other areas, retailers and consumers make no distinction between bullheads and other catfishes, so that although some dealers may profit from passing off bullheads as channel catfish, the culturist would do well to concentrate on larger, faster growing species.

The only real superiority bullheads have over other Ictalurids is their extreme hardiness with regard to the physical environment. On the other hand, they are more susceptible to disease than are channel catfish. Disease is much more of a danger in well-managed intensive fish culture operations than extremes of temperature, dissolved oxygen concentration, and so on, so that this sort of hardiness is no real advantage to the culturist. After an initial flurry of interest among fish culture researchers in the late 1950s, bullheads were overshadowed by the larger Ictalurids, and they are seldom used in fish culture today.

In the accounts of catfish culture techniques that follow, the species referred to is the channel catfish unless otherwise specified. In attempting to raise other Ictalurid species methods similar, if not identical, to those used in culture of channel catfish are likely to be effective.

CATFISH FARM SITE SELECTION

The first question a prospective catfish farmer must ask himself is whether or not he has a suitable site for pond construction. His decision must be made on the basis of topography, soil quality, and quality and quantity of water available. The general principles of fish pond location and construction are discussed in the Appendix. Here we shall limit ourselves to factors of special concern to Ictalurid catfish farmers.

The temperatures and chemistry of the water supply must be suitable for the species to be cultured, or be capable of being inexpensively modified to meet its needs. Although all the Ictalurids can survive temperatures down to the freezing point, virtually no growth occurs at low temperatures. Channel catfish grow slowly at 60 to 70°F, but they do better at 70 to 80°F; above 85°F, feeding falls off and growth slows.

In addition to temperature, pH and alkalinity should be known. Here the culturist may wish to call upon the Soil Conservation Service for technical assistance, but it is advantageous for the serious catfish farmer to have his own inexpensive pH meter and keep daily records, since the pH of a pond is not constant. Channel catfish suffer no ill effects in the pH range 5 to 8.5, and 6.3 to 7.5 is considered optimal; a pH over 9.5 is likely to be lethal. Alkalinity may be measured as total hardness or total alkalinity. Values of these indexes should fall between 20 and 150 ppm or 30 and 200 ppm, respectively. Alkalinity and pH may be adjusted artificially to some extent; methods are given in the Appendix.

An aspect of soil quality which is of particular concern to American catfish farmers is pesticide contamination. The lower Mississippi Valley, the Imperial Valley of California, and some other producing regions are among the most heavily sprayed agricultural regions in the world, and the prospective aquaculturist would be reckless and remiss in his duty if he did not consider the possible effects of these compounds on his stock and their consumers. For a more detailed discussion of the dangers of chlorinated hydrocarbons in fish farming see the Appendix or McLarney (1970).

GROWING CATFISH FOR MARKET

DESIGN OF PONDS

There is money to be made from the production and sale of fingerling catfish as well as edible sized fish, but most catfish farmers purchase fingerlings from specialists and concentrate on growing adults for mar-

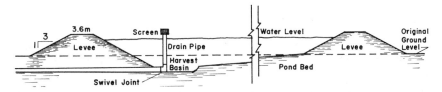

FIG. 1. Typical catfish pond cross section. (After Mitchell and Usry, 1967.)

ket. Accordingly, the most important facilities of a catfish farm are the production ponds. Channel catfish have been effectively reared in all sizes of pond from less than 0.4 ha to more than 40 ha. There is little unanimity on the proper size, but most commercial operations are conducted in ponds of 0.4 to 5.0 ha, although there is a trend toward larger ponds, up to 16 ha. In general, smaller ponds are easier to manage and thus preferable for the inexperienced catfish farmer, but there is no "best" size. In deciding on pond size, the culturist should consider the respective advantages of small and large ponds as outlined in the Appendix.

There is somewhat more of a consensus as to what constitutes the proper depth for a catfish rearing pond than there is as to surface area. Most authorities cite depths of 0.9 to 1.8 m for use in the South and 1.8 to 3.0 m in the North, where winterkill is a possibility in shallow ponds. Deep water in a catfish pond is superfluous in the South as it would probably be virtually devoid of oxygen for much of the year.

The preferred type of drainage system for catfish production ponds incorporates a harvest basin containing an L-shaped adjustable drainpipe. These and other features of a well-designed catfish pond are illustrated in Fig. 1. Many of the features of Fig. 1 are discussed in more detail in the Appendix.

An alternative to the use of a harvest basin is to harvest fish by opening the drainpipe and draining them directly into holding basins. Such basins may be permanent concrete structures, or made of wood for mechanized loading onto trucks, or they may take the form of small ponds. Holding basins must be sufficiently large that the fish do not suffer mortality from overcrowding while waiting to be loaded on trucks. When large ponds are used, this means that rather large holding ponds are required. For example a 20-ha pond would require at least five 0.4-ha holding ponds. There is a trend toward highly mechanized havesting, which may eventually eliminate the need for harvest basins and holding ponds.

OBTAINING AND STOCKING FINGERLINGS

Catfish farmers who spawn their own stock will need additional facilities. It is nearly universal practice to stock growing ponds with fingerlings rather than fry. Fingerlings are available from commercial hatcheries in most parts of the South or they may be shipped to culturists in other parts of the country. Most catfish farmers prefer to stock fingerlings at least 13 to 15 cm long, since they are less subject to mortality and usually result in larger fish at harvest time. However, fingerlings as small as 5 cm are sometimes available at substantially lower prices.

A commercial catfish farming operation can be wiped out before it is fairly started if fingerlings contract disease. The best precaution against this is to purchase stock from an established, reputable hatchery. Most hatcheries treat fingerlings for external parasites and diseases before delivering them to the pond. Untreated stock should be treated by the culturist. Two effective prophylactic treatments are:

1. Place fingerlings in 10 ppm potassium permanganate for 1 hour, then wash with freshwater. Follow this with 15 ppm formalin for 5 to 12 hours, then 1 ppm acriflavine for 5 to 12 hours.

2. Place fingerlings in 15 ppm formalin for 24 hours, then treat them briefly in 0.001% acriflavine. The second step can be carried out in transit.

It is customary to stock growing ponds in the spring, when the water temperature is 13°C or higher, when mortality is lower than is the case for fall stocking and the fingerlings will start to grow immediately. In the early days of catfish culture, it was necessary to harvest in the fall, but culturists found that this produced a substantial percentage of fish too small for the market. It is now customary to grow catfish for $1\frac{1}{2}$ to 2 years, at the end of which time virtually all the stock should be marketable.

Channel catfish and white catfish, at least, can withstand any reasonable population density. The more of these species stocked per acre, the greater the production of fish will be. However, the catfish farmer does not aim so much for maximum meat production as for maximum dollar production. Therefore his desire to maximize total yield is tempered by a knowledge of the size of fish favored by the prospective consumer of his product. In most cases, the optimum market size for catfish is about 0.45 kg live weight, but in the southeastern United States smaller fish may be preferred. The best stocking rate to produce 0.45 kg fish in one growing season in ponds supplied with well water will almost always be between 2200 and 6600 fingerlings/ha. The precise optimum stocking rate will vary with the amount of food and dissolved oxygen available; 3300 to 4400 fingerlings/ha is a good starting rate. Beginners at catfish culture should err on the low side in stocking since overstocking can lead to disease and oxygen depletion. While either of these conditions can usually be corrected with slight loss of fish, the neophyte is less likely than the experienced operator to be sensitive to the early symptoms.

Stocking rates may be increased if it is possible to aerate or circulate the water, as in raceway culture. Research on such techniques is far from complete, but some commercial catfish growers already employ raceways. Mary Porter of Fayetteville, Arkansas, grows channel catfish in a series of 25 to 30 m × 3 m ponds through which water is pumped at 2800 liters/min by a 7½ HP pump. Each pond is separated from its neighbors by a concrete dam with a narrow sluiceway at the top. Just below the sluiceway is a baffle board to reoxygenate the water.

Mrs. Porter stocks each segment with 2000 fingerlings (223,221 fingerlings/ha). Production of 450 kg of catfish per week is claimed. Over a 150-day growing season, this would amount to more than 40,000 kg/ha. It should be noted that experimental raceway culture to date has suggested stocking rates far lower than that used by Mrs. Porter—6600 to 110,000 fingerlings/ha.

Ponds supplied by surface water should be stocked at only 1650 to 2200 fingerlings/ha, since if oxygen depletion occurs during hot dry weather, there may be no way of adding fresh, well-oxygenated water rapidly enough.

In some areas there is a market for large catfish up to 1.8 kg. To produce fish of this size requires at least two growing seasons. It may also be necessary during the second season to reduce the population density to 1540 to 3300 fish/ha, each fish weighing 0.2 to 0.45 kg.

ROLE OF POLYCULTURE

American catfish culture, as contrasted to most forms of freshwater pond fish culture, is largely based on monoculture. While nothing comparable to Asian polyculture is foreseen for the south central United States, it may be that mixed culture of Ictalurids and/or admixture of other species with Ictalurids will become more common. Already some culturists, particularly in Louisiana, find it advantageous to include 10% blue catfish in their channel catfish ponds. Where more or less intensive culture is carried out in natural ponds, largemouth bass (*Micropterus salmoides*) are usually added to control trash fish populations. Experiments at Auburn University have demonstrated that addition of tilapia to channel catfish ponds may prove beneficial. Channel catfish stocked at 4400/ha yielded 1400 kg/ha to monoculture with supplemental feeding. When Java tilapia (*Tilapia mossambica*) were added at 1250/ha, catfish production increased to 1568 kg/ha. Adding to this the 266 kg/ha of potentially marketable tilapia produced gives a total production of 1834 kg of fish/ha, for an increase of 434 kg/ha or 27.3%. Efficiency of food conversion by the catfish was the same with or without tilapia. Apparently the tilapia fed on plankton, wastes, and excess food intended for the catfish. Similar results were obtained when ponds where channel catfish were stocked at 7500/ha were supplemented by Nile tilapia (*Tilapia nilotica*) at 2500/ha.

POND FERTILIZATION

Although in catfish farming, as in most systems of fish culture, the fish derive a substantial portion of their nourishment from food organisms produced within the pond, pond fertilization is not emphasized. Use of organic fertilizers is especially discouraged, because they are reputed to adversely affect the flavor of catfish.

Where catfish ponds are fertilized, it is often done not to increase pond productivity per se, but to produce an algal bloom to shade out rooted plants. This may not be necessary in many ponds; as a general rule if a Secchi disk is not visible below 46 cm, fertilization is not necessary.

Fertilization is generally done in the winter, well before stocking; once fish are present in a pond their wastes, plus unused food fragments, are adequate to maintain productivity. Commercial fertilizers high in nitrogen and phosphorus but containing little or no potassium are best, at least for the south central states. An initial application of 73 to 90 kg/ha should be adequate, but if Secchi disk visibility remains greater than 46 cm, the process may be repeated.

Attendant on fertilization is the danger of a heavy algal bloom developing, then dying off, resulting in severe pollution. This can happen virtually overnight, and the culturist must be prepared to replace a considerable proportion of polluted water with freshwater as rapidly as possible. For this reason and because pond fertilization is by no means vital to successful catfish culture, the beginning catfish farmer is advised not to fertilize his ponds unless it becomes apparent that to do so will materially enhance fish production.

SUPPLEMENTARY FEEDING

One of the most extensively investigated aspects of catfish culture is nutrition, and well it might be, for proper nutrition is crucial to the success of any commercial catfish farm. Although some of the food consumed by commercially raised catfish is produced within the ponds they inhabit, the bulk of their nourishment is derived from supplementary feeding.

Research in nutrition is far from comprehensive, but in general a channel catfish diet should satisfy the requirements outlined in Table 1. A few comments are in order on some of the items listed in the table.

TABLE 1.   NUTRITIONAL REQUIREMENTS FOR A DIET FOR CHANNEL CATFISH

Protein	Minimum	32%
Crude fat	Minimum	4%
	Maximum	8%
Crude fiber	Minimum	8%
	Maximum	20%
Fish meal	Minimum	8%
Calories	Minimum	540/kg
Protein calories	Minimum	243/kg
Calcium	Minimum	1%
Phosphorus	Minimum	1%
Vitamins and minerals	(see text)	

Protein.   At least 50% should be animal protein. Research at Kansas State University's Agricultural Experiment Station has shown that protein in excess of 25% may be used not for growth but as a source of energy, and could thus economically be replaced by carbohydrates. Authorities elsewhere are in disagreement and generally cite 32 to 33% as the minimum protein requirement. The role of carbohydrates in protein-sparing has not been adequately studied, but indications are that inclusion of up to 18.6% carbohydrates in catfish diets is beneficial in this regard.

Fish Meal.   This seems to be the one absolutely essential food item. Most of the ingredients of standard catfish feed formulas can be substituted for, but whenever fish meal has been left out of catfish diets poorer growth and food conversion have resulted.

Vitamins and Minerals.   Such additives as methionine, Lyamine-50, and vitamin B_{12} may be individually added to catfish diets, but most culturists use a commercially available vitamin premix originally developed for inclusion in poultry diets. Ingredients of the vitamin premix are listed in Table 2.

TABLE 2.   INGREDIENTS OF THE VITAMIN PREMIX COMMONLY INCORPORATED
IN DIETS FOR CULTURED CHANNEL CATFISH

Vitamin A	450,000 USP units
Vitamin D_3	200,000 IC
Riboflavin	300 mg
Pantothenic acid	600 mg
Niacin	3,500 mg
Choline chloride	40,000 mg
Vitamin B_{12}	1 mg
Vitamin E	150 IU
Vitamin K (menadione sodium bisulfite)	100 mg
Ethoxyquin (antioxidant)	6.5 g
Folic acid	40 mg

Use of the vitamin premix and an otherwise balanced diet should eliminate problems associated with vitamin deficiencies. If deficiencies do occur, they may be recognized by the symptoms listed in Table 3.

Mineralized salt is the most common mineral additive in catfish feeds, but dicalcium phosphate and limestone have also been included.

Other ingredients sometimes used include antibiotics such as aureomycin and binders to prevent rapid disintegration of the food in water.

The small farmer may find it more convenient and economical to buy prepared catfish feed, but better results are obtained using individually formulated feed. If a commercial feed is used, the ingredients should be carefully checked, since many of the products on the market are far too high in carbohydrates.

If the culturist prepares his own feed, the precise formula will be determined not only by the nutritional requirements outlined above, but by the availability and cost of various ingredients. The four feed formulas listed in Tables 4 to 7 are included only as samples and general guidelines.

General feed formulation specifications at the U.S. Fish Farming Experimental Station are listed in Table 4. Another very high protein diet developed at Auburn University and successfully used in Texas has the composition given in Table 5. Table 6 gives a computer-derived formula based on December, 1966, prices in St. Louis, Missouri, and meeting or exceeding most of the requirements for a catfish diet, though not as high in protein as the above two diets. This formula was modified to simplify production and inventory control with no appreciable loss in efficiency. The new formula is indicated in Table 7. Similar formulas could of course be derived for any time and place.

Among the substances which have been successfully fed to catfish but

TABLE 3.   VITAMIN DEFICIENCIES OF CHANNEL CATFISH AND THEIR SYMPTOMS

VITAMIN DEFICIENCY	EFFECTS
Pyridoxine	Erratic swimming, tetany, gyrations and muscular spasms when stressed, reduced weight gain, and mortality
Pantothenic acid	"Flabby" body tissues, "mummy" textured skin, excessive mucus on gills, clubbed gill filaments, and eroded gill membranes, lower jaw, fins, and barbels, lethargy, reduced weight gain, and mortality
Riboflavin	Opaque lens of one or both eyes, mortality
Thiamine	Reduced weight gain, lethargy, and difficulty in maintaining equilibrium, convulsive spasms, partial paralysis, and curvature of the spine
Folic acid	Lethargy, reduced food consumption, mortality
Nicotinic acid	Tetany and eventual death brought about by stress, lethargy, reduced coordination
B-12	Reduced weight gain
Choline	Hemorrhagic areas in the kidneys and enlarged livers, reduced weight gain
A	"Pop-eye," fluid in body cavity, hemorrhagic kidneys, and edema of the body cavity
K	Hemorrhages on body surface

SOURCE: Dupree (1966).

which are not included in the formulas just given, are meat meal, blood meal, beef heart, beef liver, animal fat, chicken entrails, chopped fish, cottonseed meal, peanut meal, sesame meal, ground corn, ground sorghum, wheat bran, and vegetable oil.

All prepared catfish foods must be in the form of pellets for best results. Although catfish are perfectly capable of engulfing large chunks

TABLE 4.   GENERAL FEED FORMULATION SPECIFICATIONS FOR CHANNEL CATFISH AT THE U.S. FISH FARMING EXPERIMENTAL STATION, STUTTGART, ARKANSAS

INGREDIENT	COMPOSITION (%)
Grain by-products	45
Protein concentrates	45
Dehydrated alfalfa	4
Distiller's dry solubles	5
Mineralized salt	1

TABLE 5.   FORMULA FOR PELLETED CHANNEL CATFISH FEED DEVELOPED AT AUBURN UNIVERSITY

INGREDIENT	BY WEIGHT (%)
Soy bean oil meal (44% protein)	35
Peanut cake (53% protein)	35
Fish meal (60% protein)	15
Distiller's dry solubles (24% protein)	14
Bentonite clay (binding material)	1

SOURCE: Hastings (1964).

TABLE 6.   COMPUTER-DERIVED FEED FORMULA FOR ICTALURID CATFISH, BASED ON NUTRITIONAL REQUIREMENTS AND ECONOMIC CONSIDERATIONS

INGREDIENT	COMPOSITION (%)
Fish meal (menhaden)	13.7
Soybean meal (solvent, dehulled)	22.0
Meat scraps	0.6
Feather meal	6.4
Blood meal	1.3
Alfalfa meal (required)	5.0
Rice bran	42.1
Rice hull fractions	7.9
Vitamin premix (required)	1.0

SOURCE: Report by W. H. Hastings, U.S. Fish Farming Experimental Station.

TABLE 7.   SIMPLIFIED VERSION OF TABLE 6

INGREDIENTS	COMPOSITION (%)
Fish meal	12.0
Soybean meal	20.0
Feather meal and/or blood meal	10.0
Distiller's solubles	8.0
Rice bran	35.0
Rice hull fractions	10.0
Alfalfa meal	4.0
Vitamin premix	1.0

SOURCE: Report by W. H. Hastings, U.S. Fish Farming Experimental Station.

of food, feeding large particles encourages development of feeding hierarchies, resulting in uneven growth. On the other hand, use of dry meal mixes and other small particle feeds reduces the efficiency of food conversion by 50% and heightens the danger of pollution.

Newly planted fingerlings do best on 60-mm diameter pellets; pellet size can be gradually increased to 1.2 cm as the fish grow. Pellets should not dissolve or disintegrate rapidly in water. Ideally, 90% of a pellet should remain after 10 min in water. Durability of pellets may be enhanced by inclusion of bentonite clay as a binder, by using fairly high amounts of fibrous materials, or by subjecting the feed to high-temperature dry steam before pelleting.

The culturist seeking to formulate his own feed will find that local feed mills have access to most of the ingredients and the machinery to do the job. If possible, the mill proprietors should be cautioned to clean the pelleting heads with vegetable oil rather than the commonly used petroleum products, which may contaminate the first portion of feed to pass through the dies after cleaning.

Feeding rates are dependent on the poundage of fish to be fed and the water temperature. Rates are generally expressed as a percentage of the total weight of fish in the pond. This figure will of course be known at stocking time, but subsequent values should be estimated by periodically seining and weighing a sample of the stock.

Suggested feeding rates for channel catfish at various temperatures are outlined in Table 8. At high temperatures these rates may need to be

TABLE 8.   SUGGESTED FEEDING RATES FOR CHANNEL CATFISH AT DIFFERENT TEMPERATURES

WATER TEMPERATURE (°C)	DAILY WEIGHT OF FEED AS A PERCENTAGE OF TOTAL WEIGHT OF STOCK
More than 32	1.5 or less, depending on dissolved oxygen concentration
21–32	3
16–21	2
7–16	1
Less than 7	0.5 on warm, sunny days only

adjusted upward in very hot climates such as south Texas, where local catfish stocks may be adapted to hot weather. The listed rates would certainly have to be so adjusted in feeding white catfish. Feeding rates may also be revised upward where there is little chance of seriously depleting dissolved oxygen by decay of uneaten food, for example, in raceway culture.

In feeding catfish over 0.45 kg, expenses can be reduced by stocking fathead minnows (*Pimephales promelas*) at 1 to 5 kg/ha. As the catfish grow and become acclimated to eating minnows, the feeding rate of pellets at optimum temperatures can be gradually reduced, from an initial 3% of the total weight of stock, to 2%, and eventually to 1%. Proportional reductions may be made at other temperatures.

The makers of Purina fish chow suggest a sliding scale of feed weight: fish weight for use with their product (Table 9). Adjustments must be made for temperature with this feed as with any other.

TABLE 9.   FEEDING RATES FOR CHANNEL CATFISH AS SUGGESTED BY THE MANUFACTURERS OF PURINA FISH CHOW

WEIGHT (KG)	FEED PER DAY (KG)
4.5	0.45
9.0	0.54
14.4	0.94
27.0	1.44
41.8	2.30
50.4	2.70
81.0	4.14
147.6	5.40
197.5	7.20
229.1	9.00
295.2	11.25
382.5 and over	13.50

Whenever and however catfish are fed, the feed should be scattered so that all fish have a chance to feed, rather than being dumped in one spot. Another rule applicable to catfish culture is the universal rule of fish feeding: too little is better than too much. An underfed catfish represents the loss of a certain percentage of potentially marketable meat. An overfed catfish may well be a dead catfish—a total loss. There are several precautions which can be taken to prevent overfeeding and resultant pollution:

1. An easy way to check whether or not feed is being wasted is to use submerged feeding tables which can be removed some hours after feeding and inspected for leftover fragments.

2. Feeding should not be carried out in water over 1.5 m deep.

3. Feeding should be reduced or discontinued on extremely hot days, cloudy days, or whenever any environmenal factor, such as a heavy blue-green algae bloom, suggests that dissolved oxygen levels may be lower than normal.

4. Most catfish farmers allow themselves a margin of error by feeding only 6 days a week. It has been shown that by feeding every day produc-

tion can be increased by 12 to 15% over the course of a growing season, but more caution must be exercised.

5. The maximum amount which should be given under any circumstances is 30 kg/ha at one time.

Automated feeding has been explored as a labor-saving device in catfish culture, but since there are so many variables in determining the rate of feeding it is doubtful whether conventional time-regulated feeding devices will find much favor. However, a self-feeding device, which permits fish to release a small quantity of food by pressing on an underwater plate, is becoming popular (Fig. 2). In addition to saving labor, the self-feeder prevents food waste by allowing the fish to determine how much they will be fed, eliminates the guesswork inherent in winter feeding, and enables the culturist to determine whether or not his fish are feeding. The latter attribute is particularly important, since cessation of feeding is one of the most common early signs of oxygen deficiency, disease, or other impending trouble. Self-feeders currently available are not suitable for use with fingerlings under 8 cm because these small fish are not heavy enough to operate the mechanism. It is not yet known whether the rates of feeding chosen by catfish are consistent with the economic priorities of the culturist.

Even if all possible precautions are exercised in feeding, the culturist must constantly be on guard for such symptoms of oxygen depletion as foul odors or dark streaks of decayed matter. It is a good idea to periodically check ponds very early in the morning, when dissolved oxygen levels are usually at their lowest. At this time if inadequate amounts of oxygen are available, fish may be seen at the surface gasping for air.

If for any reason oxygen depletion does occur, the only remedy is to replace part of the water with fresh, oxygenated water. Pond aeration would be a good preventive measure, but it is not as yet widely applied.

LOW-INTENSITY GROWING

There exist in the south central states large expanses of standing water where catfishes and other species could be cultured by methods far less intensive than those just described. Although yields in natural waters without supplemental feeding are unlikely to approach those achieved by intensive pond culture, there is nevertheless a certain amount of untapped potential for fish production in sloughs of up to 40 ha, which commonly occur in river bottoms in that region. A suggested stocking ratio for such waters is 220 channel catfish, 165 buffalofish, 165 crappies (*Pomoxis* spp.), and 100 largemouth bass per hectare. After stocking in the spring little

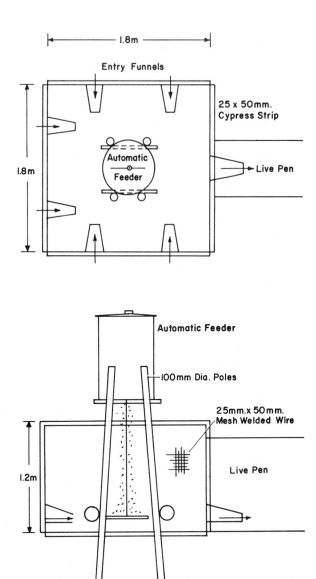

FIG. 2. Automatic feeder—live pen catfish trap.

need be done until fall when the fish may be harvested. One cannot look to fish culture of this sort as a source of livelihood in itself, but if one happens to own a slough or other suitable natural water, stocking and harvesting fish may yield some supplementary revenue.

In the early years of fish culture in the south central states, rice field culture, first using buffalofish, then with increasing emphasis on catfish, was widely touted. Today, although fish can be raised to marketable size in rice fields and may in addition improve rice yields by providing fertilizer, rice field fish culture has dwindled in importance in the United States. Unlike slough culture, rice field culture requires a certain amount of input in terms of labor, feed, and so on. Since catfish production in rice field culture does not compare with that achieved by intensive pond culture, it cannot compete economically. This situation, plus the increased use by rice farmers of pesticides and mechanized methods incompatible with fish culture, has led to the virtual disappearance of rice field fish culture in the United States. Where it is still practiced, fields are usually stocked in late summer, fall, or winter with 22 to 220 buffalofish fingerlings and 11 to 165 channel, blue, or white catfish fingerlings per hectare.

GROWING BULLHEADS IN PONDS

Though the bullheads are usually scorned as fish for pond culture due to their tendency to reproduce at a very high rate, one series of experiments conducted at Auburn University suggested that brown bullheads could be pond-raised in a commercially feasible manner. When 265 10.0-cm fingerlings were stocked at 6600 to 13,200/ha in June, a repressive factor developed so that in 13 months of rearing, with initial fertilization and supplementary feeding whenever the water temperature was above 16°C, reproduction did not occur and most of the fish recovered were of marketable size. However, when brown bullheads were stocked at lower densities, as in channel catfish culture, reproduction did occur, and 59 to 70% of the fish harvested were only 8 to 10 cm long.

HARVESTING

Harvest of catfish crops may be total or partial. Partial harvest is indicated when there are so many fish in a pond that total harvest would result in large-scale mortality or when there is a ready market for a certain fraction of the stock on hand. This market may be composed of anglers. As high as 62% of the population of a growing pond has been removed by angling. If an appropriate fee is charged, the monetary return per surface area of water may be higher than that obtained from conventional harvest and sale of fish. More conventional means of partial harvest involve concentration of fish in shallow water by baiting and capture with seines or basket traps. Nets used in partial harvest are usually treated with tar to prevent snagging of the pectoral and dorsal spines.

Total harvest traditionally involves partial drainage of the pond to concentrate the fish in the harvest basin. This is an inconvenience at best, as even with a large-diameter drainpipe, an 8- to 20-ha pond may take several days to drain. More important, it imposes a number of restrictions on the culturist, the most severe being the necessity to limit harvesting operations to the cold months, since the danger of concentration of large numbers of fish in the relatively small area of the harvest basin, resulting in severe oxygen depletion, is multiplied greatly during hot weather. The problem may be alleviated somewhat by continually pumping fresh, oxygenated water into the harvest basin.

The usual method of capturing fish concentrated in the harvest basin is by seining. The fish may be further concentrated by feeding in one spot just before seining. Baiting is not recommended if fish are to be loaded on trucks immediately upon capture, nor should they be fed the day before harvest, since when live fish are transported the gut should be empty.

Seines used in harvesting catfish should be 2.4 to 3.0 m deep, of 2.5-cm mesh to avoid gilling or otherwise injuring the stock, and of appropriate length to cover the entire harvest basin. In most ponds it will be found advantageous if seines are constructed with a 30-strand sisal twine rope or a heavy jute rope in place of the usual lead line, as the leads tend to dig into soft bottoms, causing the seine to be partially filled with mud.

If the pond to be harvested is regularly shaped and has a smooth bottom, seining could theoretically be employed to harvest the entire crop without drawing down the pond. However, in practice this would require excessive amounts of time and labor. Researchers of the U.S. Bureau of Commercial Fisheries have resolved this problem by employing a modification of the mechanical Lake Erie-type haul seine rope puller. The unit is powered by a $7\frac{1}{2}$ HP gasoline engine, which pulls a manila tow rope connected to the bottom of the seine by short toggle ropes. Seines used with this gear may be 600 m or more long and 3 m deep, with a fish bag 2.5 m wide × 3 m long × 3 m deep in the center. Detachable 120-m wings enable the operator to adjust the seine to various pond sizes. This seine puller has made it possible to harvest ponds up to 20 ha in size with no drawdown and reduced the time required to concentrate a crop of catfish in the harvest basin from a matter of days to a few hours. A three-man crew using this device was able to harvest 85% (16,300 kg) of the fish in a 16-ha reservoir in a single morning.

Catfish may also be concentrated for harvest by inducing migration. Experimental setups of this sort involve a number of ponds laid out along a central canal. At harvest time runways connecting each pond with the canal are opened and the fish are lured or driven into the canal. The same principle may be used in harvesting from conventional ponds, using traps rather than canals. Recently a trap utilizing an automatic feeder has been shown to permit efficient harvest even in sloughs and other waters which are full of obstacles to conventional harvest equipment (Fig. 2). Methods of inducing migration by noise or artificially created currents are also being studied.

With the fish captured, or at least concentrated, the problems of harvest are still not over. In fact, the most difficult part remains—transferring the fish to trucks. Since many catfish are marketed live and almost all the remainder are hauled to processing plants live in tank trucks, they must be rapidly transferred to trucks or to holding ponds, which create the necessity of subsequent smaller harvests, with additional labor costs. Most of the emphasis in research and development has been placed on methods to rapidly transfer the fish to trucks, but the use of holding ponds has the advantage of permitting the culturist to keep supplies of fish on hand for ready delivery to buyers. If fish have been baited or fed immediately before harvest, they may be placed in holding ponds for 24 hours to void the gut, thus eliminating vomit and excreta as sources of pollution in tank trucks.

A practical size for holding ponds is 0.4 ha. They should be at least 1.2 m deep with a bottom sloping at about 1.3% toward a 15-cm diameter drainpipe. A supply of fresh water should be available for flushing. Stocking and holding ponds should be limited to 1000 fish/ha in summer and twice that in winter.

Among the devices proposed to facilitate transfer from harvest basin or holding pond to truck are a hydraulic boom, a conveyor connected to the mechanical seine puller, and a vacuum pump. The vacuum pump appears most promising. It consists of a 5460-liter vacuum tank mounted on the back of a truck and equipped with a compressor driven by a farm tractor power take-off shaft. A 15-cm diameter hose passes from the tank to the bag of the seine. As air is removed from the tank the intake gate is opened and the tank fills with water and fish in about 5 min. When the tank is full the compressor cuts off automatically. This pump has successfully lifted 300 channel catfish weighing 126 kg without injury, and much greater success is deemed likely.

Use of the conveyor entails pulling the seine onto the conveyor apron so the fish can be mechanically loaded onto the truck. By using the conveyor two men have been able to load 1575 kg of fish in 50 min. Its use is limited by the fact that many pond levees are structurally unsuited for it.

The hydraulic boom is a very straightforward device which lifts containers of fish from water level to the truck on which it is mounted. While this eliminates the drudgery of hauling fish up the levee, there is still the time-consuming chore of dipping fish into the container.

Up-to-date information on catfish harvesting gear and techniques may be obtained from the National Marine Fisheries Service, P.O. Box 711, Rohweir, Arkansas 71666.

YIELDS OF AMERICAN CATFISH CULTURE

The efficacy of intensive pond monoculture of catfish in the south central states is attested to by the fact that in the 1950s, when rice field culture of buffalofish and catfish was in vogue, the normal range of fish production was 225 to 1170 kg/ha over a 2-year growing period, whereas today the average *annual* production of catfish ponds in Arkansas is 900 kg/ha (Plate 1). Table 10 lists a few of the higher yields obtained and briefly outlines the methods employed in achieving them.

BREEDING CHANNEL CATFISH

In 1966 in Arkansas alone the demand for catfish fingerlings was an estimated 8 million. Since successful hatcheries can produce 44,000 to 220,000 fingerlings/ha, it can be seen that there is considerable opportunity for profit in large-scale production of catfish fingerlings. Some large catfish farmers breed their own stock and thus operate on a completely self-sustaining basis, but the majority of producers of catfish for the market continue to purchase fingerlings from specialists.

Catfish breeding requires more skill and experience than growing fingerlings to marketable size and should not be undertaken by the neophyte (Plate 2). After he has gained a certain amount of expertise, the culturist may wish to expand into production of fingerlings for his own use or for sale if he has sufficient space to accommodate brood stock ponds, spawning ponds, fry rearing ponds, and so on, without reducing the acreage devoted to growing fish for market. The culturist anticipating breeding his own stock should also determine whether his annual need for fingerlings is large enough to justify the additional expense and la-

PLATE 1. Aerial view of catfish farm in Lonoke County, Arkansas. (Courtesy James White, Ed., *American Fish Farmer.*)

bor. As a general rule, culturists who stock less than 100,000 fingerlings annually will find it more economical to buy them.

PROBLEMS

Spawning of channel catfish in captivity has become commercially feasible only since 1960. The most severe problems of the early catfish culturists were brought on by the nervous and aggressive temperament of the channel catfish. Fighting between males at spawning time was a particularly unfortunate consequence, but the reluctance of some fish to spawn in captivity was also problematical. Unless wild fish could be captured at the peak of sexual ripeness, the pioneers in the field preferred to acclimate stock for two years in captivity before attempting to breed them. This practice has been eliminated by the establishment of hatchery stocks going back several generations. Fighting as a source of injury and mortality among brood stock has been virtually eliminated by the development of sophisticated methods of handling and stocking spawners.

Another problem faced by the early culturists was the frequent confusion of channel and blue catfish. Although channel catfish are generally

TABLE 10. STOCKING, FEEDING, AND PRODUCTION OF ICTALURID CATFISHES IN DIFFERENT LOCATIONS (EXAMPLES OF REPRESENTATIVE YIELDS)

LOCATION	SPECIES STOCKED	NO. STOCKED PER HECTARE	SIZE OF FISH STOCKED	FEEDING RATE	GROWING TIME	SIZE AT HARVEST	PRODUCTION (KG/HA)	TYPE OF CULTURE
Auburn Univ., Alabama	Brown bullhead	1,200	2.5–10.0 cm	Unknown	13 months	0.33 kg or more	2,024 (marketable size fish only)	Experimental culture in fertilized ponds supplied with well water.
Auburn Univ., Alabama	Channel catfish	1,184	6.4 g	9,130 kg total for growing period	188 days	Unknown	2,008	Experimental culture in ponds supplied with well water.
Dumas, Arkansas	Blue catfish; Channel catfish	5,441 60	5–10 cm 10–15 cm	3% of total body weight when water temperature is above 16°C; 5% on warm days only during winter	251 days	Blues, 0.45 kg Channels, 0.67 kg	Blues, 1,189 Channels, 145 Total, 1,334	Commercial culture in a 2.1-ha pond supplied with well water.
Dumas, Arkansas	Blue catfish; Channel catfish	32 736	8–20 cm 8–20 cm	3% of total body weight when water temperature is above 16°C; 5% on warm days only during winter	273 days	Blues, 0.85 kg Channels, 0.54 kg	Blues, 123 Channels, 1,732 Total, 1,855	Commercial culture in a 10-ha pond supplied with well water.

TABLE 10. (*Continued*)

LOCATION	SPECIES STOCKED	NO. STOCKED PER HECTARE	SIZE OF FISH STOCKED	FEEDING RATE	GROWING TIME	SIZE AT HARVEST	PRODUCTION (KG/HA)	TYPE OF CULTURE
Stuttgart, Arkansas	Channel catfish	600	15 cm	3% of total body weight when water temperature is above 16°C (6 days a week); 0.75% 1–2 days a week when temperature is lower	210 days	0.57 kg	1,527	Experimental culture in ponds supplied with well water.
Stuttgart, Arkansas	Channel catfish	2,000	10 g	Unknown	Unknown	0.24 kg	2,206	Experimental culture in 0.04-ha ponds supplied with well water.
Stuttgart, Arkansas	Channel catfish	2,000	10 g	Unknown	Unknown	0.27 kg	2,621	Experimental culture in 0.04-ha ponds supplied with well water and aerated with compressed air.
Stuttgart, Arkansas	Channel catfish	2,152	10 g	Unknown	Unknown	0.29 kg	2,864	Experimental culture under semiraceway conditions.
Ames, Iowa	Channel catfish	about 800	fingerlings	Unknown	Unknown	Unknown	3,977	Experimental culture in tertiary sewage treatment ponds.

183

PLATE 2.   James W. Avault of Louisiana State University holding a breeder channel catfish. (Courtesy James W. Avault.)

lighter in color than blue catfish, some individuals are quite dark and are erroneously called "blue" catfish by fishermen. This source of confusion has been greatly reduced by the establishment of pure hatchery stocks of both species. Should any question arise, examination of the anal fin will reveal the identity of the fish in question. In channel catfish this fin has a rounded margin and 24 to 29 rays; in blue catfish the margin is straight and there are 30 to 36 rays.

Sexing was also once a source of difficulty for catfish farmers. Today, however, experienced workers approach 100% accuracy in determining sex. Males may be superficially distinguished from females by

their darker coloration and shorter, wider head. The female genital opening is slitlike rather than tubular as in males. As spawning time approaches, males develop areas of dark pigmentation under the jaw and
body and the genital papilla becomes enlarged and protrudes. In females,
the vent becomes loose, inflamed, and swollen and may pulsate when
examined. When, as happens on occasion, the sex of a fish is not apparent
on visual examination, a broom straw rubbed longitudinally over the vent
will hang on the vent if the fish is a male.

Prominent, reddish genitalia are one of the signs that a fish is ready
to spawn, but the principal criterion in selecting breeders of either sex
is general condition. Further indications of ripeness in females are soft,
distended ovaries, palpable through the body wall, and a full, rounded
abdomen, extending posteriorly past the pelvis to the genital orifice. To
avoid being misled by a gut distended by food it is best not to feed brood
stock prior to selection of spawners.

CONDITIONING OF SPAWNERS

Channel catfish may mature as early as 2 years of age and as small as 33
cm in length and 0.33 kg in weight, but for reliable spawning it is best
to use brood stock 3 or more years old. Fish of 1 to 4.5 kg are preferred,
since larger fish are hard to handle, require too much space, and are not
as reliable spawners as smaller animals.

From December on, spawners are generally kept segregated by sex in
ponds of about 0.4 ha in area. Recommended stocking density for 1.0- to
1.5-kg fish is 270 to 360 kg/ha if further growth is desired or twice that
if it is not. Feeding of brood stock is one of the most important phases
of catfish culture, because it influences time of spawning, number and
size of eggs produced, incidence of fighting among the males, and general health of both spawners and offspring. Of course when water temperatures are low, brood stock as well as other fish should be fed only on
warm days, but whenever the water temperature exceeds 13°C spawners
may be fed 2 to 3% of their body weight 3 or 4 days a week. However
much is fed, it is important that a substantial percentage, if not all of
the diet be composed of fresh or frozen meat or fish, partly because in
cold weather catfish seem to utilize animal protein better than cereal
feeds and partly because meat and fish diets enhance the attainment of
spawning condition. Some culturists provide brood stock with a ready
source of animal protein by stocking their ponds with fathead minnows.

In the spring, after the breeders are transferred to spawning ponds,
they may be switched to conventional pelleted dry feeds at 4% of body

weight per day, or more if they will eat it. Dry feed should still be supplemented by fresh animal protein, however. All feeding is discontinued at the onset of spawning.

The State Fish Hatchery at Centerton, Arkansas, has had good success feeding brood stock with cut fish, supplemented by a specially prepared pellet feed, according to the schedule given in Table 11.

TABLE 11.   FEEDING SCHEDULE FOR CHANNEL CATFISH BROOD STOCK USED AT THE STATE FISH HATCHERY, CENTERTON, ARKANSAS

December–February	Small portions of cut fish, fed on warm days only
March	Cut fish; gradually increase amount until by April 1 the breeders are fed as much as they will eat
April 1–15	After stocking in spawning ponds, continue cut fish as a supplement to pellets; feed as much as the breeders will eat, twice a week
April 15–onset of spawning	Cut fish; feed as much as the breeders will eat, twice a week

SPAWNING IN NATURE

In nature, channel catfish migrate to the shallows of rivers and lakes to spawn sometime between April and July, depending on latitude. Vicious fighting between the males may precede spawning, with severe injury sometimes resulting. Not infrequently wounds received in fighting may become infected, resulting in the death of the injured fish.

Once a pair is formed, the male chooses a spawning site, usually in a sheltered place, such as under a bank, where he constructs a crude nest by cleaning the bottom, removing as much silt and debris as possible. He will defend this location against any intruder until spawning is completed, the fry hatched, and their yolk sacs absorbed.

Spawning takes place at 20 to 23°C. If the female is not ready to spawn, she may be attacked or driven away. The spawning act consists of the deposition of successive layers of adhesive eggs by the female and individual fertilization of each layer by the male. The whole process may require 4 to 12 hours.

When the female is spent, the male drives her away and commences guarding and caring for the eggs. In addition to defending the eggs and fry until they are free swimming, the male circulates water through the eggs by fanning them with his fins. Fanning is occasionally supplemented by a more vigorous disturbance of the egg mass with the body and pelvic fins.

POND SPAWNING

The three types of spawning practiced in channel catfish culture, in decreasing order of similarity to natural spawning, are pond spawning, pen spawning, and aquarium spawning. Most culturists still use the most primitive method, pond spawning, because it requires minimal facilities as well as demanding the least time, labor, and skill. Pond spawning is recommended for all situations where there is insufficient skilled labor for proper application of more sophisticated methods or wherever it will consistently produce an adequate quantity of high-quality fry for stocking.

Channel catfish have been propagated in ponds as large as 26 ha, but spawning ponds average about 0.4 ha. Spawning ponds should be constructed similarly to growing ponds and be no deeper than 2.1 m at any point. They should be drained in winter, disked, and, if the soil is acid, limed. Spawning ponds should not be filled until 30 to 40 days before spawning is expected, to minimize the chance of establishment of predatory insects. During this 30 to 40 day interval two or three applications of 16-16-4 (N-P-K) fertilizer at 45 kg/ha are recommended.

Equal numbers of males and females are stocked at 50 to 330 fish/ha and pairing is allowed to occur naturally. This entails some fighting among the males but has the advantage of permitting pairs of comparable ripeness to be formed, thus reducing the incidence of intersexual fighting. For this reason, pond spawning is to be preferred when brood stock is of marginal ripeness.

Spawning receptacles are provided in the form of 45-liter milk cans, nail kegs, earthenware crocks, and so on, spaced 9 to 12 m apart with the open end toward the center of the pond. There need not be one spawning receptacle for each pair of fish, since spawning will not be synchronous. Successful catfish breeders provide anywhere from 50 to 90% as many receptacles as spawning pairs. Natural spawning occurs at depths from 15 cm to 1.5 m, but it is best to place spawning receptacles in at least 0.6 m of water to minimize disturbance to the spawners. Receptacles placed deeper than arm's length present difficulties in handling and management. Nearly any suitable sized container may be used, but milk cans or similarly shaped containers are preferable to kegs since the constricted opening reduces the chance of fry getting out. All metal spawning receptacles should be swabbed inside and out with asphalt paint.

Each spawning receptacle should be numbered and checked periodically, and records should be kept of spawning and hatching. Checking may be carried out by visual examination of the inside of the receptacle if, after frightening the male out of the receptacle, it is lifted out of the pond and most of the water gently decanted. Less disturbance is occasioned by gently probing inside the container with a rubber hose or some such device. If a stationary, spongy mass is felt, eggs are present. If instead a wriggling mass is felt, the eggs have hatched. The male may leave when a hose or the like is inserted, or he may bite it. In the latter event he may often be pulled out of the receptacle so that another probe can be made. Hands are not recommended for this purpose, as an angry male catfish can deliver a nasty bite. Some culturists eliminate both the necessity of probing and the danger to hands by constructing doors on the tops of spawning receptacles so that the progress of spawning may be visually inspected without moving the receptacle.

PEN SPAWNING

Pen spawning is an advancement on pond spawning in that it permits delaying of the time of spawning to suit the convenience of the culturist, genetic selection of breeding pairs, protection of the spawn from intruders, the immediate removal of spawned-out fish for reconditioning and the treatment of reluctant spawners with hormones. Spawning pens may be constructed of various materials but are usually in the form of three-sided enclosures of wire mesh, with the shore of the spawning pond constituting the fourth side. Pens range in size from 1.2 × 2.4 m to 1.8 × 3.6 m and may be up to 1 m deep at the deep end. The sides should be embedded 15 cm into the pond bottom and extend 0.3 to 0.6 m above the water surface to prevent the spawners from tunneling or leaping out. Each pen is provided with a receptacle of the same sort used in pond spawning.

As spawning time approaches, the brood stock are sexed and paired in the pens. This is the most crucial phase of pen spawning, since if the female is not ready, she may be killed by the male in only 15 to 20 min time in close confinement. On the other hand, large females may attack small males, thus pairs should be selected so that the male is at least as large as, and preferably larger than, the female. If there is an excess of brood stock, the largest, oldest fish should be spawned first, as they tend to ripen earlier. Excess brood stock are separated by sex and stocked in holding ponds for later use. If sufficient amounts of cold well water can be supplied to such holding ponds so that they remain at 16 to 19°C, the onset of sexual ripeness in fish held therein can be retarded for some weeks. If spawning starts off well, then lags, a 2.5- to 5.0-cm rise in water level has been found to stimulate breeders.

Once spawning has started, the procedure of pen spawning is the same as for pond spawning, except that the female is removed immediately on completion of spawning to prevent her from eating the eggs or being damaged by the male. Where separate hatching facilities are available, the eggs may also be removed and artificially hatched. Artificial hatching reduces the chance of transmission of disease from parents to young and permits the use of good males in a second, or even third, spawning. Males to be so used should receive at least a day's rest and one large portion of feed between spawnings.

AQUARIUM SPAWNING

Injection with pituitary hormones is only occasionally practiced in pen spawning, but it is an integral part of aquarium spawning. Since it requires a considerable amount of skill, the aquarium method is the least commonly practiced technique of obtaining catfish spawn. On the other hand, it is the most efficient in terms of use of space and producing a high rate of successful spawning. Aquarium spawning also permits very accurate timing of spawning and the production of fry of uniform size and age, as well as eliminating the chance of predation by the parents or transmission of disease from parents to offspring.

Channel catfish respond well to a wide variety of fish pituitary extracts as well as to human chorionic gonadotropin. Females are injected intraperitoneally with three doses of 2.2- to 22-mg of acetone-dried fish pituitary material or a single dose of 600 to 2200 IU of human chorionic gonadotropin per kilogram of body weight. (For a general discussion of induced spawning of fishes by hormonal injection, see Chapter 3.) Males are not injected.

After injection the fish are paired in 23- to 240-liter aquaria provided with running water and tarpaper mats to collect the eggs. As in pen spawning, pairing is crucial. If the male severely bites the female, he should be removed and the female kept by herself until she has received two or three more injections.

Most injected fish will spawn within 16 to 24 hours after the last injection. Upon completion of spawning the eggs may be removed for artificial hatching and a new pair of spawners placed in the tank.

HATCHING

If either the pond or pen method of spawning is used, the eggs may be left in the pond to hatch naturally in 5 to 10 days, depending on the water temperature. Once eggs are found, disturbance should be minimal until the culturist suspects they may have hatched. Excessive handling, activity in the water nearby, or even loud noises such as slamming car doors or discharge of firearms in the vicinity may cause some males to relocate or eat the eggs.

PLATE 3.   Channel catfish egg mass, Brawley, California. (Photograph by W. O. Mc-
Larney.)

Allowing hatching to occur naturally is of course the simplest way of
proceeding, but artificial hatching enables the culturist to use his spawning facilities repeatedly, thus attaining higher production. If conditions
in the spawning pond are less than optimal, artificial hatching will
usually also increase the rate of hatching. If spawn is obtained by the
aquarium method, artificial hatching is of course mandatory. The jar
method has been experimentally employed in hatching, but commercial
operators rely on the use of hatching troughs.

Hatching troughs may be of any convenient size, but they should be
about 25 cm deep and supplied with running water. Eggs are placed in
7.6-cm deep wire mesh baskets hung along the sides of the trough (Plates
3 and 4). Alongside each basket is a paddlelike agitator extending slightly
deeper than the bottom of the basket and driven by an electric motor or
water wheel. The agitation thus provided must be sufficient to move the
entire contents of the basket but not to throw the eggs out.

Where possible, water should be supplied to hatching troughs by
gravity flow, since this system is less liable to failure than pumps and
the like. If well water is used, it may need to be aerated and heated with
a gas or electric heater or by letting it stand in a pond for a few days.

When large numbers of eggs are placed close together in hatching

PLATE 4. Catfish hatching trough with paddle wheels to agitate egg mass, Brawley, California. (Photograph by W. O. McLarney.)

troughs small amounts of a 2 ppm solution of malachite green should be added daily at the head of the trough to control fungus. This treatment should be discontinued as hatching time approaches, as malachite green is toxic to fry.

When the eggs hatch, the fry pass through the sides of the hatching baskets into the trough. Hatching time is about the same as at comparable temperatures in nature, but up to 98% success may be expected.

REARING CHANNEL CATFISH FRY

EARLY REARING IN TROUGHS

It is possible to rear fry to fingerling size in the spawning pond if the adults are removed, but usually the fry are stocked in separate nursery facilities. Fry which have been hatched in troughs can simply be siphoned into tubs for transport to the nursery area. If hatching has been allowed to occur in the spawning pond, the male must be driven out of the spawning receptacle, which is then lifted out of the pond and partially emptied, and the remaining fry are poured into a tub. It is good practice when emptying a spawning receptacle to splash water back into the receptacle while pouring, so as to wash all the fry out.

Lower mortality of fry occurs if they are started in nursery troughs rather than ponds. Troughs used for this purpose may be of wood, aluminum, or fiberglass, 2.4 to 3.0 m long, 20 to 50 cm wide, and about 30 cm deep. Running water should be supplied at about 23 liters/min. One large spawn or two small to medium spawns are placed in each trough. Usually the fry are set free in the trough, but some culturists prefer to transfer fry in the spawning receptacle, which is then laid on its side in the trough with the mouth facing into the current.

Fry in troughs will start to feed shortly after becoming pigmented and free swimming, or about 3 to 5 days after hatching. For the first 4 to 5 days they should be fed sparingly with a good prepared feed, ground suitably fine. As time goes on the amount fed can be increased, but as long as the fry are in troughs any food uneaten after 2 hours should be siphoned off. Some culturists treat fry in troughs with 5 ppm acriflavine for 4 hours twice a week as a prophylactic measure.

PREPARATION OF REARING PONDS

Fry may be brought to fingerling size in troughs, but normally they are transferred to ponds within a few weeks of hatching. Fry-rearing ponds vary in size from 0.04 ha to 2 ha or more, but ponds of 0.4 ha or less are preferred. If possible they should be left dry until just before stocking to prevent establishment of predatory insects. Many culturists use various insecticides in permanent ponds or even in ponds where water has been standing for a few days before stocking, but the dangers presented by these substances outweigh their benefits. Air-breathing species, which account for the greater number of piscivorous insects, may be eliminated by treating fry ponds with a 1:20 mixture of SAE-30 motor oil or cottonseed oil and kerosene at 17.5 liters/ha twice a week on days when there is just enough wind to distribute the mixture across the surface of the pond. Use of oil on fish culture ponds may, however, contribute to oil pollution of natural waters receiving drainage from the hatchery, for which the culturist could be held liable.

STOCKING AND FEEDING

Fry may be placed directly into the pond, but as insurance against predators it is better to stock fry, particularly small ones, in some sort of cage. If they are transported to the pond in tubs, cages may easily be improvised by placing each tub on its side in the pond inside a retaining frame of 60-mm mesh hardware cloth, or floating cages may be constructed. The fry should be so confined for 1 to 2 weeks or longer, while being fed 4 to 5% of their body weight 6 days a week. The feed already described for conditioning breeders (Table 12) has been found effective in rearing fry when the proper sized particles are provided. Even after the fry have been permitted access to the entire pond, the culturist should continue to feed in the same places.

TABLE 12.   COMPOSITION OF THE FEED USED FOR CHANNEL CATFISH BROOD STOCK AT THE STATE FISH HATCHERY, CENTERTON, ARKANSAS

INGREDIENT	PERCENT
Dried milk	10.00
Wheat shorts (best grade)	14.00
Soybean meal (fine grind)	14.00
Cotton seed meal	14.00
Yellow corn (fine grind)	14.00
Meat scraps	16.00
Fish meal	15.75
Vitamin A feeding oil (15,000 units per gram)	0.25
Iodized salt	1.00
Brewers dried yeast	1.00

TM-10 Terramycin is added to the feed at the rate of 0.6% of the total weight.

The Texas Agricultural Extension Service recommends a different feeding regime. About 1 kg of meat scraps or tankage per spawn (5000 to 20,000 fry) is to be fed 6 days a week until the fry begin "topping" or coming to the surface to feed. Once topping begins, it will be possible to determine how much feed is being taken and the fry can then be fed as much as they will eat. When they are 5 to 8 cm long the feed is changed to a 1:1 mixture of meat scraps or tankage and pellets of the same sort used in growing ponds. This feed is continued until spring, when the fingerlings are ready for stocking or sale.

The Texas Agricultural Extension Service also recommends that no more than 1 spawn be stocked per 0.1 ha of pond surface and that no more than 3 spawns be placed in a 0.4-ha pond, but fry are sometimes stocked as heavily as 550,000/ha. Of course the more fry are stocked, within reason, the more fingerlings will be harvested, but since large fingerlings bring better prices than small ones the culturist may do better to stock fairly sparsely. Yields of as high as 176,000 10-cm fingerlings/ha have been achieved, with 35% mortality, but the U.S. Soil Conservation

Service is considerably more conservative in its recommendations. The SCS suggests stocking rates of 2.5 to 5.0 cm fish to reach various sizes in one growing season as given in Table 13.

TABLE 13.   RECOMMENDED STOCKING RATES FOR CHANNEL CATFISH FINGER-
LINGS AT DIFFERENT SIZES

			ESTIMATE AT END OF SEASON		
NO. STOCKED (PER HA)	GROWING PERIOD (NO. OF DAYS)	NUMBERS[a]	TOTAL WEIGHT (KG)	AVERAGE WEIGHT (G)	AVERAGE LENGTH (CM)
88,000	180	30,000	270	9	10
66,000	180	22,500	324	14	13
44,000	180	15,000	405	27	15
33,000	180	11,250	471	42	18
22,000	180	7,500	378	50	20
11,000	180	3,750	338	81	25

[a] Assuming a 25% loss.
SOURCE: Grizzell et al. (1968).

Some culturists have had success in growing very large (23 to 25 cm), robust "fingerlings" using a 2-year rearing schedule. The first year fry are crowded in ponds at 110,000/ha, then they are thinned the following year to 22,000 to 26,400/ha.

BREEDING AND REARING FRY OF OTHER ICTALURIDS

BLUE CATFISH

Blue catfish have been experimentally spawned using all of the methods described but, since male blue catfish are less prone to fighting among themselves than are male channel catfish, most commercial producers use the pond method. Brood stock is wintered in ponds at 450 kg/ha and fed pelleted feed supplemented with cut fish, beef liver, and small live fish, then stocked in spawning ponds at 44 pairs/ha in the spring. Spawning ponds are prepared in the same manner as described for channel catfish. Spawning occurs soon after the water reaches 22°C.

Blue catfish fry may be reared in the spawning pond with or without the parent fish, but better results are obtained by stocking 2.5- to 5.0-cm fry in separate nursery ponds at 88,000/ha. In either case fry are fed 3 to 5% of their total body weight in prepared dry feed, 6 days a week.

FLATHEAD CATFISH

Production of flathead catfish fingerlings has proven far more difficult than is the case for other Ictalurids. Flathead catfish have been successfully spawned using the pond and aquarium methods, but often captive flathead catfish fail to achieve full sexual maturity. This failure is often due to improper feeding; flathead catfish are very piscivorous and may not adjust well to feeds, particularly dry feeds, preferred by the culturist. Flathead catfish also require considerably more space than other Ictalurids.

Yet another problem in spawning flathead catfish is the difficulty of accurately sexing them. Unlike channel catfish, in flathead catfish it is the female which becomes darker as spawning time approaches, but this is not a completely reliable character. The best indicators in mature fish are the genital papillae. In the mature female these are reddish and slightly raised. The genital opening may also be slightly dilated.

Female flathead catfish respond as readily as channel catfish to injections of fish pituitary preparation or human chorionic gonadotropin, and if sexing is accurate, no problems are to be anticipated in spawning them in aquaria of appropriate size, say 225 liters or larger.

Hatching has been successfully carried out in ponds, and in hatching troughs, as described for channel catfish, it occurs at about the same rate and presents no special problems.

Most of the difficulties in attempting to produce flathead catfish fingerlings for the trade have occurred in rearing the fry. Reported survival rates from fry to fingerling stage have been as low as 4 to 6%. Among the problems encountered have been predation, cannibalism, diseases, and parasites, but remedial and/or preventive measures are known for all these causes of mortality. A more fundamental difficulty has been the failure of very young fry to accept food. One successful technique for starting fry on feed is to spread a paste of liver, powdered milk, and egg yolk on a board floated in the nursery trough. The fry, which tend to hide in the shade of the board, are thus able to immediately detect the odor of the food. Feeding should commence when the fry first exhibit food-seeking behavior by circling the trough.

Egg yolk alone is successfully used as a first food in a feeding schedule employed to rear flathead catfish fry up to the age of 7 weeks in troughs at the U.S. Fish Farming Experimental Station (Table 14). The eggs are first boiled for 15 min, then the yolks are pulverized in a small amount of water.

Originally fresh carp flesh was not supplemented by beef liver, but fry fed trout chow and carp flesh without liver developed an apparent

TABLE 14.   FEEDING SCHEDULE FOR FLATHEAD CATFISH FRY

AGE OF FRY (DAYS)	COMPOSITION OF FEED	FREQUENCY OF FEEDING
1–7 (or until the yolk sac is absorbed and feeding behavior is first observed)	None	—
8–9	Pulverized boiled egg yolk	Hourly during the day; one night feeding
10–12	Pulverized boiled egg yolk and live *Daphnia*	Hourly during the day; one night feeding
13–17	Frozen shrimp and live *Daphnia*	Hourly during the day; one night feeding
18–49	Equal parts of finely ground commercial trout chow, fresh carp flesh, and beef liver	As above, but discontinue the night feeding; trout chow should be presented 15 min before the carp flesh and beef liver to encourage acceptance of dry food

thiamine deficiency, which may have been responsible for 27% mortality before it was corrected.

WHITE CATFISH AND BULLHEADS

Reproduction of white catfish has not been extensively studied, but they respond to the same general techniques used for channel catfish. Any fish culturist wishing to raise bullheads will find that fry production is the least of his problems, as bullheads placed in a pond will successfully reproduce with or without the culturist's blessings.

SELECTIVE BREEDING AND HYBRIDIZATION

Selective breeding of Ictalurid catfishes is in its infancy. One of the chief goals is the development of fish with smaller heads in proportion to body size. Success in this endeavor would benefit both producer and consumer. Albino channel catfish, which occasionally crop up in the course of ordinary breeding operations, are said to possess this characteristic. The albino fish, which are a light golden color, are also more attractive to consumers when marketed with the skin on and are said to be resistant to Ichthyophthiriasis. They possess additional value as a novelty fish for stocking in pay fishing ponds. The albino strain breeds true and is being cultured in some areas. However, the survival rate of albino channel catfish fry is significantly less than that for normal fish. Other goals pursued by selective breeding include resistance to low dissolved oxygen concentration and more efficient feed conversion.

Hybridization is another means to the same ends sought by selective breeding. Many Ictalurid hybrids have already been produced experimentally, including virtually all the possible crosses involving channel catfish. All hybrids with one channel catfish parent have shown better growth than either parent. The most promising hybrid thus far appears to be ♂ blue catfish × ♀ channel catfish, which has not only shown 11 to 65% better growth than the parents under a variety of stocking and feeding regimes, but also grows more uniformly.

In recent years, the incidence of deformities and other hereditary disorders in cultured channel catfish has increased considerably, and inbreeding has been implicated in this problem. Intraspecies hybridization of different strains may therefore be emphasized in the future.

PROBLEMS

PESTS AND COMPETITORS

Among the pest organisms sometimes found in catfish ponds, the most common are tadpoles of various species, which compete for food with fry. Properly constructed ponds, with no shallow areas along shore, are the best prevention but frogs and tadpoles can never be completely excluded from ponds. If tadpoles are numerous enough to create a problem, partial control may be achieved by shooting or spearing adult frogs and by removing frog egg masses whenever they are found. There is also a commercially available tadpole poison, called Tad-Tox, which can be used in severe cases.

Two crustacean pests, the fairy shrimp (*Streptocephalus texanus*) and the tadpole shrimp (*Apus longicaudatus*), occasionally form dense clouds in fish ponds. Fairy shrimp compete for food with catfish fry, while tadpole shrimp cause excessive turbidity. Both species seriously interfere with visual detection and capture of fry. At present, the only known methods of eradication entail the use of chemicals which are even less desirable than fairy shrimp or tadpole shrimp.

DISEASES AND PARASITES

The catfish culturist is more often called on to deal with diseases and parasites than with pest or competitor organisms. Fry and small fish are particularly susceptible to protozoan diseases which, due to the schooling habit of small catfish, may spread like wildfire and cause losses of epizootic proportion. Adults are less susceptible to protozoan diseases, with the exception of ichthyophthiriasis or "ich," but they suffer from a number of major and minor parasites.

Among the commonest health problems of adult Ictalurids are bacterial infections and fungus. The most important preventive measure against these and all diseases and parasites is to maintain the fish in good condition. This means proper feeding, suitable temperatures and, especially, a good supply of dissolved oxygen at all times. It has been estimated that the latter condition is not met in 90% of all catfish ponds. Even healthy fish may be damaged in handling, though the damage may not be apparent, and the slightest injury may give bacteria or fungus a foothold. Therefore it is a good idea when handling or transporting catfish to add an antibiotic to the water.

Some diseases and parasites may be readily detected and treated by even the inexperienced culturist. Among these are ich, fungus infections, *Pseudomonas* and *Aeromonas* (bacterial infections), and gas bubble disease. Details on diagnosis and treatment of these and other diseases may be found in Davis (1953).

If signs of poor health (reluctance to feed, wasting away, sluggishness, visible parasites, sores, or inflammation on any part of the body, etc.) are seen but none of the foregoing problems are apparent, the culturist is advised to call the local representative of his state conservation department immediately. Most diseases and parasites of catfish can be controlled if diagnosed in time, but diagnosis often requires the services of a trained fishery biologist. Among the important disease-producing or parasitic organisms which fall into this category are the protozoans *Chilodon, Chilodonella, Costia, Scyphidia, Trichodina,* and *Trichophyra;* the copepod crustaceans *Achtheres, Argulus, Ergasilus,* and *Lernaea;* and various monogenetic trematodes (flukes) and acanthocephalans (spinyheaded worms). Digenetic trematodes, tapeworms, roundworms, and leeches are ordinarily less dangerous but may reduce weight gain, produce unsightly fish, or even cause losses in heavily infested fish. The experienced culturist may become able to diagnose and treat many of the "difficult" diseases and parasites.

It has been a source of annoyance to some catfish farmers that treatment of diseases and parasites must often be carried out using "antiquated" methods, due to Food and Drug Control Administration restrictions on the use of a number of chemicals which have been proven effective remedies. Although many of these chemicals are commercially available, their use can result in the culturist's fish being condemned by the FDA for sale for human consumption. It is small consolation to the culturist faced with the loss of part or all of his crop, but these restrictions are not a matter of caprice, and are designed to prevent serious damage to human health and ecosystems by as yet untested chemicals.

OFF-FLAVOR

Another problem occasionally encountered by catfish farmers is off-flavor. Since most causes of off-flavor can be corrected and the flavor improved before marketing, it is best to capture and cook a few sample fish shortly before harvesting. Off-flavor has several possible causes.

Heavy Algal Bloom. If sufficient supplies of water are available, algal blooms can be eliminated by thorough flushing with well water. Otherwise, treatment with copper sulfate is indicated. Consult the Soil Conservation Service for details. Flavor of fish should improve within a few days after treatment.

Benthic Algae. A type of strong-smelling benthic algae commonly known as musk grass may develop and impart a musty taste to fish in ponds which are too shallow or underfertilized. Copper sulfate is the only effective treatment.

Overfeeding. Pollution by spoiled food may affect the flavor of fish. If this occurs, cease feeding and flush the pond, drawing the water off the bottom if possible.

Other Decaying Organic Matter. Fallen trees, leaves, animal manure, and so on, may cause a musty flavor. Such substances should be prevented from entering ponds. If problems occur, it will probably be necessary to hold the fish in clean water for several days before sale.

Chemicals. Various agricultural chemicals, particularly those applied as sprays, may find their way into ponds from nearby agricultural operations. If chemicals are suspected as a source of off-flavor, the first step is to determine the possible sources of contamination. If chlorinated hydrocarbon insecticides or other substances hazardous to human health are suspected, the fish should not be marketed.

ECONOMICS

PRESENT SITUATION

Probably no branch of aquaculture has been subjected to as much economic analysis as the catfish industry in the United States. However,

due to the youthfulness of the industry, a good deal of what was published as recently as 1967 seems irrelevant or inapplicable to the present
and future catfish farmer. About all that can be predicted with certainty
is that the most efficient producers will be rewarded and the least efficient
will fail. This is of course an economic truism, but in catfish culture
the need for efficiency has been obscured by the lack of direct competition
characteristic of an industry in the first stages of expansion and by the
optimistic tendency of culturists and market analysts alike to think not
in terms of average but of superior farm management. Thus there has
been considerable unfortunate speculation in the catfish business and
there will continue to be if prospective catfish culturists are not advised
of the risks as well as the potential for profit. Perhaps the situation can
be brought into focus by pointing out that J. E. Greenfield, an economist
of the U.S. National Marine Fisheries Service, has asserted that, of the
operators who have entered the catfish business to date, less than 25%
have made a profit. Even if it is assumed that the remaining 75% include
a substantial number of recently established farmers who had not counted
on showing a profit in their first year or two, it can be seen that the deck
is stacked against all but the most thoroughly aware and prepared
entrants. Present indications are that the economic prognosis for new
catfish farmers is getting poorer.

Before making any commitment of resources or labor, the prospective
catfish farmer must assess the demand for catfish and the price they will
bring in whatever markets are available, as well as calculate all the costs
he will incur in starting and maintaining his business and thus determine
whether he will make a profit and what sort of income he can expect. He
must also attempt to foresee future trends in costs, prices, and demand,
realizing that cultured catfish is essentially a new product, therefore the
best informed forecasts rest on a shaky foundation of generalities and
assumptions. Finally, he must temper his judgment with the knowledge
that it takes time to acquire the skills of a superior catfish grower, and
that beginners almost always make less money than experienced hands.

Figure 3 illustrates the marketing options currently available to a
catfish farmer in the United States. Readers interested in a detailed
economic analysis of the catfish industry are referred to Greenfield (1969)
and Jones (1969).

PROSPECTUS

A serious threat has recently been posed to the U.S. catfish industry in
the form of competition from various other species of fish. Various catfishes from Mexico and Brazil, as well as marine wolffish (*Anathichus*

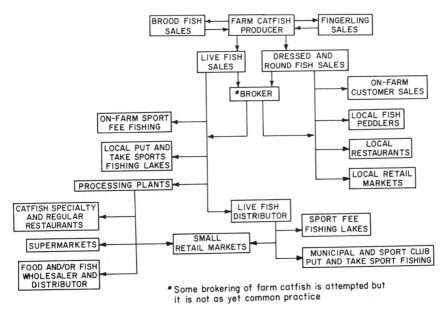

FIG. 3. Farm catfish marketing opportunities and options, 1969. (After Greenfield, 1969.)

spp.) captured by American commercial fishermen have been falsely represented as channel catfish. Most of these species are reputed to be inferior to channel catfish as food, so that in addition to competing with cultured catfish they present a potential threat to the image of cultured catfish as a high-quality food.

Further competition may eventually come from channel catfish cultured in Latin American countries. Attempts have been made in Mexico to rear catfish for the American market, but to date operations there have not been efficient enough to compete with catfish grown in the United States. As the market expands, however, incentive will also increase and Latin American catfish farmers, who already have the competitive advantage of cheap labor, may be expected to become a serious force in the U.S. market.

In brief, the economic prospectus for catfish culture is one of levelling growth, if not decline, after the boom of the 1960s. As the market matures, fewer and fewer culturists will realize more and more of the profits. The successful minority will increasingly be composed of the largest and/or most efficient growers and their economic success will continually enable them to enhance their position by taking advantage of biological and technological advances in the field.

PLATE 5. Cage culture of channel catfish in White Oak Lake, Arkansas. (Courtesy Arkansas Game and Fish Commission, photograph by Dick Lawrence.)

FUTURE OF THE INDUSTRY

INCREASED PRODUCTION

The chief effect of most biological and technical advances will be to increase production per hectare. Average production is expected to level off by 1978 at about 5400 kg/ha, but individual producers will certainly exceed this level. Technological advances in such areas as harvesting and processing, while having no direct effect on production, will increase the amount of catfish reaching the market by making higher per hectare and total production feasible for the culturist. Improved methods of processing which facilitate use of presently discarded portions of the fish could serve to make catfish culture more profitable.

CAGE CULTURE AND OTHER MEANS OF INCREASING STOCKING RATES

Among the areas of research which may yield significant results are nutrition, disease control, hybridization, and selective breeding, but at the moment the greatest potential for increasing production appears to lie in methods of increasing the dissolved oxygen content in waters used for catfish culture, thus raising the number of fish which can safely be stocked in a given area of water. Aeration and raceway culture have already been mentioned, but have yet to find commercial application. More recently, culture of catfish in floating cages (Plate 5), such as are used in marine fish culture in Japan, has been receiving considerable attention. This method has already been applied by such pioneer catfish farmers as Roy Prewitt of Lonoke, Arkansas, who has been able to grow 675 kg of marketable size channel catfish in a cage 3 m × 1.5 m × 1.4 m deep. (See Chapter 29 for a discussion of the unique advantages of floating cages as enclosures for fish culture.)

GEOGRAPHICAL SHIFTS

Temperature control has been considered as a means of increasing catfish production. It has been found that, using present techniques, about 180 growing days are required to grow average size fingerlings to marketable size. However, in the principal catfish growing areas, the effective growing season (total growing season less excessively hot days, cloudy days, etc.) is approximately 150 days. Clearly there would be application for some sort of pond water heater which could add 30 days onto either end of the growing season. To date no economically feasible device of this sort has been developed. Thus there has been a trend toward 1½- or 2-year growing periods. Catfish farmers in southern Mississippi, Louisiana, and southeast Texas, however, are favored by a longer growing season, so that they can produce nearly 100% marketable fish from fingerlings in one summer. The greatest concentration of catfish farms is presently in northeast Arkansas, but the competitive edge enjoyed by culturists further south is causing a southward shift in production acreage. This may be partially offset if efficient techniques of growing catfish in thermal effluent are developed.

Another geographic shift, into the southeastern states, may occur if the catfish fisheries there decline. The decline of wild catfish populations in the south central states has played a positive role in encouraging catfish culture there. Similar fishery declines in the upper Mississippi Valley, the lower Great Lakes, and the Chesapeake Bay region have expanded the market for cultured catfish. The southeastern states presently support the only major catfish fishery which is fully capable of supplying local demands. If overfishing, pollution, and so on, eventually take their toll on wild catfish stocks in that area, there will be considerable impetus for the development of catfish culture in Georgia, Florida, and South Carolina, where it is presently of little or no importance.

ADVERTISING AND MARKETING

An area of research which is seldom considered by most culturists, but which has great effect on their success, is that of advertising and marketing. One signal success in this area has already been scored; in the last decade the image of catfish has been considerably improved. While there are still segments of the population which are repelled by the "whiskery" appearance of catfish or consider them scavengers and thus unfit for human consumption, there is a greatly increased market for catfish advertised as such. A significant factor in the increased appeal of catfish to housewives has certainly been the circulation of recipe booklets by various state and federal agencies. The image of catfish in the restaurant trade has also improved. Not only are there catfish specialty restaurants, but other restaurants no longer find it necessary to use such meaningless terms as "tenderloin of trout" on their menus. In fact, the situation has been reversed; it is now common practice to falsely label various other fish as "catfish."

Current research in marketing, conducted primarily by the U.S. National Marine Fisheries Service, is aimed at getting farm-raised catfish to markets they have yet to reach, for example, the grocery chains and the heat-and-serve food packagers. Progress in this areas depends not only on the success of marketing research, but on the increased ability of processors to economically produce a uniform product and the capability of culturists to meet the demands of processors for a year-round dependable supply of high quality, uniform size catfish. Hopefully, the combined efforts of culturists, biologists, technologists, and economists will succeed in further expanding both the supply and demand for catfish.

REFERENCES

CLEMENS, H. P., and K. E. SNEED. 1962. Bioassay and use of pituitary materials to spawn warm-water fishes. U.S. Bureau of Commercial Fisheries Research Rep. 61.

DAVIS, H. S. 1953. Culture and diseases of game fishes. University of California Press, Berkeley. 332 pp.

DAVIS, J. T., and J. E. HUGHES. No date. Channel catfish farming in Louisiana. Louisiana Wild Life and Fisheries Commission, Wildlife Education Bull. 98.

DUPREE, H. K. 1966. Vitamins essential for growth of channel catfish. Bureau of Sport Fisheries and Wildlife, Tech. Papers, No. 7.

GIUDICE, J. 1966. Growth of a blue × channel catfish hybrid as compared to its parent species. Prog. Fish Culturist 28(2).

GREENFIELD, J. E. 1969. Some economic characteristics of pond-raised catfish enterprises. U.S. Bureau of Commercial Fisheries, Division of Economic Research, Working Paper No. 23.

GREENFIELD, J. E. 1970. Economic and business dimensions of the catfish farming industry. U.S. Bureau of Commercial Fisheries, Ann Arbor, Mich. Offset publication.

GRIZZELL, R. A. 1966. Commercial production of blue catfish in ponds and reservoirs. J. Soil Water Conserv. 21(3):104–106.

GRIZZELL, R. A. 1967. Pond construction and economic conditions in catfish farming. Paper presented at 21st Meeting, Southeastern Association of Game and Fish Commissioners, New Orleans, La. September 25–27, 1967.

GRIZZELL, R. A. 1969. Use of self-feeders in fish farming. Am. Fish Farmer 1(1):13.

GRIZZELL, R. A., E. G. SULLIVAN, and O. W. DILLON. 1968. Catfish farming—an agricultural enterprise. U.S. Department of Agriculture. 4-26433.

HASTINGS, W. H. 1964. Catering for channel catfish. Feedstuffs 36(23).

JONES, W. G. 1969. Market Alternatives and Opportunities for Farm Catfish. Paper given at Fish Farming Conference, Texas A&M University, College Station, Texas. October 7–8, 1969. 13 pp.

McLARNEY, W. O. 1970. Pesticides and aquaculture. Am. Fish Farmer 1(10):6–7, 22–23.

MITCHELL, T. E., and M. J. USRY. 1967. Catfish farming—a profit opportunity for Mississippians. Miss. Research and Development Center, Jackson, Miss. 83 pp.

STRAMEL, G. 1970. Trapping catfish can be successful. Am. Fish Farmer 1(3):17–18.

SWINGLE, H. S. 1954. Experiments on commercial fish production in ponds. Proc. S.E. Ass. Game Fish Commissioners 8:69–74.

SWINGLE, H. S. 1957. Commercial production of red cats (speckled bullheads) in ponds. Proc. S.E. Ass. Game Fish Commissioners 10:156–160.

SWINGLE, H. S. 1966. Biological means of increasing productivity in ponds. FAO World Symposium on Warm Water Pond Fish Culture. FR: V/R-1.

Texas Agricultural Extension Service and Department of Wildlife Science, Texas A&M University. 1967. Proceedings of the Commercial Fish Farming Conference. February 1–2, 1967.

TIEMEIER, O. W., and C. W. DEYOE. 1967. Production of channel catfish. Kansas State University of Agriculture and Applied Science, Agricultural Experiment Station, Bulletin 508.

Interviews and Personal Communication

JONES, W. G. U.S. Bureau of Commercial Fisheries, Ann Arbor, Michigan.

MEYER, F. P., Director, Fish Culture Station, U.S. Bureau of Commercial Fisheries, Stuttgart, Arkansas.

OWENS, C. Little Rock Chamber of Commerce, Little Rock, Arkansas.

PREWITT, R. King-Prewitt Farm, Lonoke, Arkansas.

7

Culture of Catfishes Native to Australasia and Europe

Recent progress in the culture of the channel catfish (*Ictalurus punctatus*) in the United States has served to focus attention on the potential for culture of other catfishes. The suborder Siluroidei is large and varied but, in general, its larger members, which are represented in the inland and coastal waters of every continent, are of high quality as food fish and adapt well to heavy stocking and artificial feeding. Many are also extremely hardy with respect to environmental conditions and can be grown where few other fishes will survive. Their only serious drawback from a fish culturist's point of view is their great voracity, which makes supplemental feeding a must. This expense is compensated for somewhat by their catholic taste.

CULTURE OF PANGASIUS SPP. IN ASIA

CAPTURE OF WILD FRY AND SPECIES USED

Though channel catfish farmers in the south central United States may be the most highly publicized growers of catfish, they were by no means the first. The history of catfish culture probably began on the Indochinese peninsula, where a number of members of the genus *Pangasius* (family Siluridae) have been grown in ponds since ancient times. The most important species is *Pangasius sutchi* (pla swai), which reaches a maximum length of 150 cm. The smaller *Pangasius larnaudi* (pla tepo), which reaches 70 cm, is less important, but it is economically significant. *Pangasius micronemus* (pla sangawad), which grows to only 50 cm, is less esteemed as a species for culture, but it is sometimes mistakenly stocked with the two preferred species. *Pangasius sanitwongsei* (pla thopa), considered a good food fish, is also excluded from culture where possible, because it is too large—up to 250 cm.

As is the case with many forms of Asian fish culture, culture of *Pangasius* spp. was originally dependent on the capture and rearing of naturally produced fry, and a substantial proportion of the total production still originates in this way. Natural spawning occurs in large rivers at various times between June and November, and it is then and there, particularly in the Chao-phya River, the Sakakrong River, and the Kreang Kri Canal of Thailand, and the Mekong River in Cambodia, that 3- to 8-cm fry are captured in 1-cm mesh seines. Juvenile fish, weighing 80 to 150 g each, are also collected, with dip nets, from the Tonle Sap and Grand Lac of Cambodia during February to April.

Inadvertent stocking of *Pangasius micronemus* is due largely to the difficulty of distinguishing between fry of that species and those of *Pangasius sutchi* and *Pangasius larnaudi*. The identifying morphological and behavioral characteristics of these three species have been worked out by officers of Thailand's Department of Fisheries (Table 1). Less is known about the fry of *Pangasius sanitwongsei*, which has a somewhat limited distribution, and is apparently commonly cultured only in Vietnam.

TABLE 1.   DISTINGUISHING CHARACTERISTICS OF FRY OF *Pangasius* SPP.

	P. sutchi	*P. larnaudi*	*P. micronemus*
Mixing of shoals	With or without *P. larnaudi*	With or without *P. sutchi*	Without both species
Movement of caudal fin	Fast movement, splashing water at the surface	Faster movement than *P. sutchi*	Slow movement of caudal fin, no water splashing
Ratio of head width to body length	1:8	1:6	1:7
Angle of insertion of dorsal fin base	Parallel to body axis	Parallel to body axis	At about 30° to body axis
Number of ventral fin rays	8–9	6	6
Mouth cleft	Wide	Wider	Narrow
Number of gill rakers	More than 12 long gill rakers	12 short gill rakers	More than 12 long gill rakers
Dorsal and pectoral fin color	Greyish black	Absolutely black	Colorless
Color of upper and lower lobes of caudal fin	Greyish black	Absolutely black	Colorless

SOURCE: Pongsuwana and Varikul (1962).

SPAWNING AND FRY REARING IN CAPTIVITY

In the 1960s, in Thailand at least, the effects of overfishing, pollution, and destruction of spawning grounds began to be felt in the scarcity, and consequent high price, of *Pangasius* fry. Accordingly, in 1966, Thai fishery authorities began to apply the results of earlier experiments with the induction of spawning in *P. sutchi* by means of pituitary injection. Both sexes receive fractional injections of *P. sutchi* or *Clarias batrachus* pituitary, followed by fertilization by the dry method (see p. 400). Proper dosages and time intervals for injections have yet to be worked out, thus the range of hatching rates is extremely wide, from 0 to 85%.

Immediately following fertilization, the eggs are placed in fine mesh hatching nets containing fine fibrous materials—either aquatic plants or such artificial substitutes as palm and jute fronds—and allowed to attach. Hatching occurs 24 to 33 hours after the nets are placed in clear water at 26.5 to 31.0°C.

The mouths of the larvae open about 36 hours after hatching, and feeding commences about 12 hours later. Zooplankton is the chief natural food of the early fry, but small insect larvae and worms also may be taken. When availability of these foods is poor, cannibalism may occur. Fry are maintained in the hatching tanks for 16 to 21 days, by which time they attain lengths of 2 to 6 cm and may be stocked in rearing ponds. The best survival of very young fry (32.2%) has been achieved by feeding live *Daphnia*. After the fry attain a length of about 2 cm, *Daphnia* may be supplemented with small pieces of mollusks, worms, and cooked fish.

GROWING FOR MARKET IN PONDS IN THAILAND, CAMBODIA, AND VIETNAM

In Thailand, *Pangasius* spp. are stocked alone in ponds, or in combination with tawes (*Barbus gonionotus*) or sepat siam (*Trichogaster pectoralis*). When stocked at about 25/m³, and fed on kitchen waste, bananas, cooked broken rice, rice bran, and soft aquatic and terrestrial plants, *Pangasius larnaudi* fingerlings may attain weights of 0.45 kg at the end of 1 year, and 1 kg in 2 years. *Pangasius sutchi* grows more slowly initially, but may reach 4 kg in 2 years. When fish wastes are added to the diet, as is common in Vietnam, weights of 1.0 to 1.2 kg may be reached in 8 to 10 months.

Data on the yields obtained by pond culture of *Pangasius* spp. in Thailand, Cambodia, and Vietnam are lacking, but there are verbal reports that far higher yields are obtained through culture in floating cages set in series along the edges of rivers, where there is a gentle current. Very high yields are also obtained when *Pangasius* spp. are used as one component of a pig-poultry-fish farming complex similar to that described for Cyprinid fishes in Malaysia (see Chapter 3).

GROWING FOR MARKET IN FLOATING CAGES IN THAILAND AND CAMBODIA

The floating cage technique, which is traditional in Cambodia, has been practiced in Thailand for only a little over 50 years. Fry and fingerlings are initially stocked at 150 to 300/m³ in cages 1 to 9 m² in surface area and no more than 1.5 m deep, made of bamboo poles or wooden planks, covered with mosquito netting. In each cage a food plate is suspended about 10 cm under the water surface. The fry are fed once or twice daily with ground trash fish mixed with boiled rice or rice bran. As soon as they reach 15 cm in length, the young catfish are stocked in larger cages.

Cages used for growing *Pangasius* for market are more commonly built of 25-mm planks than of bamboo, and they may incorporate living quarters for the culturist. A gap of about 25 mm is left between planks to permit passage of water. Growing cages vary from 2 to 16 m long, 1 to 6 m wide, and 1.5 to 2.5 m deep; the most popular size is 4 m × 3 m × 2 m. Normally bamboo poles are used as floats in order that the top of the cage, which is provided with 1 m × 1 m holes so the fish may be fed, remains about 30 cm above the water surface. Though the cages themselves last 10 to 12 years, the bamboo floats must be replaced every 2 years.

High stocking rates (20 to 60 kg/m²) are employed to offset the usual high mortality in cages. Trash fish is the basic feed during times of year when fish are plentiful; the daily feed ration is 10 to 15% of the weight of catfish. When trash fish are scarce, maize, broken rice, rice bran, and aquatic vegetation in large amounts substitute for or supplement the fish diet.

Pangasius spp. in cages may be reared from 0.08 to 1 kg in 8 to 10 months. If the growing period is extended to 2 years, weights of 2 kg may be reached.

CULTURE IN LAOS, MALAYSIA, AND INDIA

Pangasius spp. are not limited in their distribution to Thailand, Cambodia, and Vietnam but range throughout southeast Asia, from East India to the Malay Peninsula, Indonesia, and neighboring islands. Over much of this area they are not cultured, in some regions perhaps because of the lack of suitable grounds for collecting large numbers of fry or fingerlings. This may be the case in Laos, where *P. larnaudi* and *P. sutchi* fingerlings are imported from Thailand in January and stocked at 25/m³ in ponds 0.2 ha in area and 3 m deep. Laotian culture techniques are generally the same as those employed in Thailand and Cambodia, but fish are not used as food. Fertilization with pig and chicken dung is carried out, and it is believed that the fish consume some of the resulting plankton. The growing season is limited to 8 months, and growth is relatively poor, with fish averaging only 200 g in weight when harvested.

The only known instances of culture of *Pangasius* spp. outside Indochina occur in Malaysia, where *Pangasius micronemus* may be stocked in combination with lampai (*Barbus schwanenfeldii*) and sepat siam, and, in the Indian state of West Bengal, where *Pangasius larnaudi* and/or *Pangasius sutchi* are cultured on a limited scale.

CULTURE OF CLARIAS SPP. IN THAILAND

One of the most widely distributed catfish families is the Clariidae, members of which are found throughout southeast Asia, the Indian subcon-

PLATE 1. *Clarias batrachus* (Linnaeus). (Courtesy Royal Thai Department of Fisheries.)

tinent, and Africa, as well as in parts of the Near East. The Clariid catfishes are distinguished by the possession of an accessory air breathing organ, which enables them to exist for hours at a time out of the water, or indefinitely in oxygen-poor waters and even moist mud. They may take further advantage of their versatility by coming ashore at night in search of food. The extreme hardiness of the Clariids renders them well suited to culture in arid regions such as tropical east Africa, where their use has been promoted in the last decade (see Chapter 12).

Interest in culture of Clariid catfishes outside of tropical Africa is even more recent, but already spectacular success has been achieved in Thailand, where *Clarias batrachus* and *Clarias macrocephalus,* two of the most highly valued native fish, support a thriving industry which is becoming progressively more efficient. *C. macrocephalus* is preferred by consumers for its more tender flesh, but culturists favor *C. batrachus* because it grows faster, and it has become the dominant species.

COLLECTION OF WILD FRY

The major source of fry of both species is natural spawning. *C. macrocephalus* spawns during the rainy season in nests constructed on the bottom of paddy fields in 20 to 50 cm of water. *C. batrachus* spawns at a similar depth in a horizontal hole in the bank. Hatching requires 20 hours at 25.0 to 32.2°C.

PLATE 2.   *Clarias macrocephalus* (Gunther). (Courtesy Royal Thai Department of Fisheries.)

Fry are collected in hand nets from May to October; generally 2000 to 15,000 fry are collected from a nest. About 50 million fry are collected annually in this manner.

ARTIFICIAL PROPAGATION

Collection of naturally produced fry, extensive though it is, cannot provide enough stock for commercial culture. Accordingly, Thai fishery officials have undertaken, with considerable success, to spawn *Clarias* spp. in captivity. As early as the mid-1950s it proved possible to spawn *C. batrachus* by methods similar to those used in spawning channel catfish in the United States. In place of the metal or wooden containers used to spawn channel catfish, horizontal holes, 20 to 35 cm in diameter, were dug in the bank a little over 1 m apart, and aquatic plants were provided nearby. About 80% of the pairs stocked near such holes spawned within 7 to 10 days, yielding 2000 to 5500 fry per spawning.

C. macrocephalus was induced to spawn by means of pituitary injection. Single intramuscular injections of 100- to 190-g fish with 13 to 26 mg of pituitary extract per kilogram at 25 to 32°C produced spawning within 14 to 16 hours in 60 to 80% of cases, even when the breeders were placed in small aquaria.

Fertilized eggs are placed in shallow troughs and jars, where hatching takes place within 20 hours at 26 to 33°C. The water should be changed twice daily and fungused eggs removed as soon as they are detected.

FRY REARING

The larvae absorb their yolk sacs after 5 days, at which time they are transferred to earthen ponds about 3 m² in surface area. Ponds up to 1 m deep may be used for fry rearing, but best results are obtained in water 10 to 18 cm deep. Fry may be stocked very heavily—5000 to 6000/m²—for the first few weeks of life. The best early food is zooplankton; after 2 or 3 weeks, boiled fish may be added to the diet. When the fry reach 1.5 to 10.0 cm in length, they are ready to be stocked in production ponds.

GROWING FOR MARKET

Ponds used in growing *Clarias* spp. for market are 100 to 1000 m² in area and 1 to 3 m deep. They are prepared for stocking by firming the banks to discourage climbing or burrowing, and by erecting a fence about 50 cm high around each pond. If a small amount of water is allowed to continually run into the middle of the pond at the surface, the fish may be discouraged from digging into the bottom.

Fry in production ponds are stocked at about 180/m² and fed on ground trash fish, mixed with rice bran at a ratio of about 9:1. For the last 15 days of growth the feed formula is changed to a 17:2:1 mixture of ground trash fish, broken rice, and rice bran. Fertilization, with about 30 kg of farmyard manure per 100 m², is carried out only if growth is poor.

By use of these methods, three crops of *C. batrachus* may be annually grown to marketable size (about 26.5 cm and 145 g). Thai culturists thus achieve an average survival rate of 37%, a 5 to 6:1 food conversion ratio, and the tremendous annual yield of 97,000 kg/ha.

CULTURE OF CLARIID, PLOTOSID, AND SILURID CATFISHES ELSEWHERE IN AUSTRALASIA

CLARIIDS

In Cambodia, *C. batrachus* is grown in cages like *Pangasius* spp., but on a smaller scale. *C. batrachus* is sometimes stocked as a minor component of polyculture systems in ponds and rice fields in India, Pakistan, Malaysia, Vietnam, and the Philippines. Only in the Philippines is it occasionally fed, on trash fish and small shrimp, and in none of these countries is it intensively cultured.

The highly productive culture of *Pangasius* spp. and *Clarias batrachus* in Thailand and Cambodia, along with the recent culture of the channel catfish in the United States, has already sparked interest in culture of catfishes native to other regions. In the United Arab Republic, biologists at the Fish Culture Research Station at El Kanatir, El Khairia, are experimenting with culture of *Clarias lazera*. They have solved the problems of large-scale spawning and have developed an artificial feed, made of dried liver, dried vegetables, yeast, and other ingredients, which has enabled them to rear substantial numbers of fry past the stage where cannibalism is a serious threat. At present they are capable of producing fair amounts of edible sized fish, but only with high rates of stocking and feeding. Experiments conducted at the Serrow Fish Farm indicate that polyculture of *C. lazera* with common carp and *Tilapia* spp. would result in better yields than are presently achieved through monoculture of carp, the prevailing mode of fish culture in the UAR.

Another member of the Clariidae, *Heteropneustes fossilis,* has been cultured for some time in Pakistan. Indian biologists have recently shown an interest in the species, and have had some success with artificially induced spawning.

PLOTOSIDS

Two estuarine species of the Australasian catfish family Plotosidae, *Plotosus anguillaris* and *Plotosus canius,* occur as incidental fishes in Indonesian tambak culture of milkfish (*Chanos chanos*) (see Chapter 17), and another Plotosid, the freshwater *Tandanus tandanus,* is sometimes marketed as a food fish in Australia. There are no established fish culture enterprises of any magnitude in Australia, but when the possibility of freshwater pond culture is suggested, *T. tandanus* always figures in the speculations. Experiments carried out at the Inland Fisheries Research Station, Narrandera, New South Wales, indicate that this catfish can easily be kept and bred in small ponds. It is a nest-building species, and for this purpose gravel should be provided, though it will spawn, with much less success, on a mud bottom. It seems to be necessary that the water temperature be at least 24°C for the fish to engage in reproductive behavior. Spawning, which has been observed in 0.5 to 1.2 m of water, may be accelerated by a slight rise in the water level, but this is not a necessary condition. A subsequent drop in the water level will cause the abandonment of nests. One or both parent fish guard the nest until after hatching, but it is likely that, as is the case with many species exhibiting similar behavior in nature, parental care would prove unnecessary in intensive culture, where predators are excluded. It is said that *T. tandanus* can be sexed by observing the shape of the urogenital papilla, but details of sexing are not known to us.

SILURIDS

Traditional Chinese fish culture methods make no use of catfish, but the native Silurid catfish *Parasilurus asotus* is sometimes included in pond communities in Taiwan, and the possibility of its culture, as well as that of the Bagrid catfish *Pseudobagrus fulvidraco*, is being studied in mainland China.

Another Asian Silurid, *Wallagonia attu,* sometimes called "freshwater shark" because of its high dorsal fin, is sometimes used in low-intensity fish culture in reservoirs and swamps in Pakistan. *W. attu* is not stocked in ponds because of its extremely voracious and piscivorous feeding habits. *Ompok bimaculatus,* which is occasionally stocked in ponds, has been artificially spawned by means of injections of cyprinid pituitary extract.

CULTURE OF SILURIS GLANIS IN EUROPE

A Silurid catfish which has been more thoroughly studied is the sheatfish (*Siluris glanis*), native to many of the large rivers of Europe. Sheatfish have occasionally been used as predators to control excess reproduction in pond culture of various fishes in Europe, but it is only in the last few years that they have been bred on a large scale, first in Hungary, then in Yugoslavia and Czechoslovakia.

ARTIFICIAL PROPAGATION

Sheatfish spawners may be obtained from rivers, or they may be overwintered in running water in 300- to 800-m² storage basins 1.0 to 1.5 m deep. In the latter case, 30 to 40 kg of small fish must be provided monthly as food for each 100 kg of sheatfish. The storage basins may double as spawning ponds, but the fish must be segregated by sex in the spring, before the spawning season, as soon as the water temperature reaches 12 to 14°C. The sexes may be distinguished externally by the shape of the genital papilla, which is narrow, pointed, and nipplelike in males and wide and rounded in females.

When the water temperature reaches 20°C, the basins to be used for spawning are filled to a depth of 1.0 to 1.2 m, and artificial spawning nests are constructed. The first step in constructing a sheatfish spawning nest is to bind three 1.8- to 2.0-m sticks together to form a tripod, whose legs are inserted in the pond bottom. Roots of willow or alder or twigs or pine or thuya are suspended from these frames to collect the adhesive eggs. Mats made of reeds are placed under the collectors to salvage eggs that fail to attach. The number of nests placed in each spawning basin, usually 2 to 4, corresponds to the number of pairs of sheatfish stocked. The spawners are placed in the basin the same day it is filled, and spawning usually occurs within a few days, if not the following night.

The nests are checked daily from a small boat. Those that contain eggs are removed and the twigs and mats, with 'eggs attached, are hung on wires and suspended in covered hatching crates 150 cm × 50 cm × 60 cm deep, made of 0.5- to 0.8-mm mesh perlon. The loaded crates, each containing 30,000 to 50,000 eggs, are affixed next to the inflow pipe in storage basins similar to those used in overwintering and spawning. Hatching occurs in about 60 hours at 20 to 22°C. The population density of the newly hatched fry is adjusted to 50 to 70/m². Basins containing young fry are provided with ample cover in the form of twigs, reeds, and bricks.

REARING FRY AND GROWING FOR MARKET

When, 4 to 5 days after hatching, the fry attain the length of 12 mm, they begin feeding. At this time it is necessary to provide large amounts of plankton. At the age of 10 to 12 days the fry may be switched to a diet of ground fish and scrapings of liver and spleen. These foods are presented by smearing them on a black tray or the side of a flowerpot. In addition to provision of shelter and heavy feeding, periodic removal of the fastest growing fish is necessary to forestall cannibalism.

At the age of 30 to 35 days the 5- to 6-cm sheatfish fingerlings are stocked in ponds at 1500 to 2000/ha. A ready supply of food can be assured by spawning common carp (*Cyprinus carpio*) and tench (*Tinca tinca*) in the pond 15 days before stocking it with sheatfish. At the end of their first summer, the young sheatfish weigh 50 to 100 g and are ready for stocking in carp ponds at 50 to 150/ha. Most sheatfish are harvested along with the carp at the end of the following summer, at which time they weigh 0.5 to 1.4 kg. In some cases enough 2-year-olds are left to populate the pond at 35 to 60/ha. These fish are harvested the following fall as 1.5- to 2.0-kg specimens.

WORLDWIDE PROSPECTUS OF CATFISH CULTURE

As the success of Asian catfish culturists receives more publicity, it is certain that the transplantation of some of the Asian catfish in other areas will be suggested. One accidental introduction has already occurred in Florida, where *Clarias batrachus,* the albino form of which is commonly brought into the United States as an aquarium fish, has become acclimated. Ecologists and sport fishermen tend to view the "walking catfish" as a destructive force in the Florida ecosystem and hope to limit its spread. Local fish culturists, who have on occasion seen their channel catfish undersold by imported catfishes, also have a negative opinion of *C. batrachus.* However, recently Fred Meyer of the United States Fish and Wildlife Service has admitted that "Exotic fish such as *Clarias* and *Pangasius,* African and Asian catfishes, though in disrepute at the moment, could conceivably replace the channel catfish."

The fears of sportsmen and ecologists seem well grounded, and it is recommended that introductions of exotic catfishes, in the United States and elsewhere, be undertaken cautiously, if at all. On the other hand, there is no reason not to follow the example of the Hungarian sheatfish culturists and proceed energetically in exploring the possibilities of culture of native catfishes throughout the world.

REFERENCES

BOONBRAHM, M., W. TARACHALANUKIT, and W. CHUAPHOHAK. 1966. Induced spawning by pituitary injection of pla-sawai, *Pangasius pangasius* (Hamilton) in captivity. Kasetsart J. 6(3, 4):97–110. (In Thai with English summary.)

HORA, S. L., and T. V. R. PILLAY. 1962. Handbook of fish culture in the Indo-Pacific region. FAO Fisheries Biology Technical Paper, No. 14. 204 pp.

INOUE, K., and S. SWEGMAN. 1970. Economic survey on cat fish culture in Suphanburi Province, Thailand. Thai Fisheries Gazette, Dept. of Fisheries, Bangkok, Thailand 23(2):251–269.

LAKE, J. S. 1967. Rearing experiments with 5 species of Australian freshwater fishes. Aust. J. Mar. Fresh Water Res. 18:137–153.

LING, S. W., A. SIDTHIMUNKA, and S. PINYEYING. 1965. On the induced spawning of Pla Sawai, *Pangasius sutchi.* Proc. Indo-Pac. Fish. Comm. 12(II):244–252.

MEYER, F. 1969. Where do we go from here? Am. Fish Farmer, 1(1):25.

PONGSUWANA, U., and V. VARIKUL. 1962. Morphological differences among the fry of 3 species of catfish, *Pangasius sutchi, Pangasius larnaudi,* and *Pangasius macronemus.* Proceedings of the Indo-Pacific Fisheries Council, 10th Session. Section II.

SIDTHIMUNKA, A., J. SANGLERT, and O. PAWAPOOTANEN. 1966. The culture of catfish (*Clarias* spp.) in Thailand. FAO World Symposium on Warm Water Pond Fish Culture. FR: VIII/E-1.

WOYNAROVICH, E. 1966. New systems and new fishes for culture in Europe. FAO World Symposium on Warm Water Pond Fish Culture. FR: VIII-IV/R-3.

Interviews and Personal Communication

ZELDAM, U. T. Economic Commission for Asia and the Far East. Mekong Committee, Bangkok, Thailand.

8

Culture of Labyrinth Fishes (Family Anabantidae)

Several species of labyrinth fishes (family Anabantidae) are widely cultured in Asia, often in conjunction with freshwater shrimp or other fishes. Monoculture of anabantid fishes also occurs, and this practice, along with their unique environmental adaptations and spawning behavior, dictate that they be treated separately.

The Anabantidae are differentiated from all other fish by possession of a unique accessory air-breathing organ, the labyrinth, located on either side of the gill chamber. As is the case with other air-breathing fish, the labyrinth fishes are capable of surviving in waters nearly devoid of dis-

218

solved oxygen. Indeed, so dependent have they become on the labyrinth that if denied access to atmospheric air and forced to rely entirely on their gills for respiration, they will "drown."

The anabantids are also unusual among freshwater fish by virtue of producing floating eggs. Typically, the eggs are deposited in a nest built of air bubbles produced by the male and preserved by a hardened secretion of the mouth. Some labyrinth fishes, including three of the five cultured food species, do not construct bubble nests but indiscriminately scatter their eggs or construct nests solely of plant materials.

THE GOURAMI

NATURAL HISTORY AND DISTRIBUTION

The largest, and probably the most important of the anabantids is the gourami (*Osphronemus goramy*), which reaches a maximum length of about 65 cm. The gourami builds a submerged nest of pieces of plant material in about 30 cm of water. The eggs are of approximately the same specific gravity as water and remain in the nest, which is sealed by the parents after spawning. Hatching time is variously reported in the literature at from 30 hours to 30 days. Certainly the shorter time periods seem more likely for a tropical animal. The larvae float belly up for 5 days before feeding begins. In large bodies of water, gouramis apparently breed mostly during the dry season, but in ponds they may spawn at any time of year.

The gourami is an excellent food fish, which has attracted the attention of European fish culturists and has been introduced unsuccessfully to France. Introductions nearer to its native Indonesia have proven more successful, and *O. goramy* is now cultured throughout southeast Asia and in China, India, Ceylon, and the Philippines.

ARTIFICIAL PROPAGATION

Very simple breeding techniques are used in India and the Philippines. Ripe breeders are simply stocked in large ponds, 1.5 m or more in depth, with a good marginal growth of such plants as *Typha,* and nature is allowed to take its course. Ripe fish may be distinguished by their full, rounded bellies, the reddish color of the fins, and, in males, thickened lips.

More specialized spawning methods are used in Indonesia, particularly in western Java. In many cases, gourami spawning ponds double as rearing ponds for fry of nilem (*Osteochilus hasselti*), tawes (*Barbus gonionotus*), and common carp (*Cyprinus carpio*), which may be stocked before or after spawning. Such ponds are at least 700 m² in surface area and planted with *Hydrilla verticillata* to bind the bottom soil and provide shelter for the fishes. *H. verticillata* is not suitable as a nest-building material, so artificial substitutes for marginal plants are provided by driving branching stems of bamboo into the pond bottom where the depth is 70 cm or more. Bunches of indjuk palm (*Arenga saccharifera*) fibers or bamboo sticks are loosely fastened in the crotches of the stems, about 20 cm below the surface, for use by the fish. Stems are placed about 5 m apart, with one being provided for each female gourami stocked.

Before being introduced to spawning ponds, brood fish are stored in ponds about 50 cm deep and conditioned on soft plant leaves, soft fruit, and rice bran. Leaves of *Carica* and *Colocasia* are believed to be especially good for this purpose.

Female spawners are stocked at 1/100 to 150 m² of water surface. Generally two males are provided for every three females, but this ratio may be altered if significant numbers of very large males are present. Although both sexes of gourami are believed to reach maturity at about 1½ years of age, the preferred age for spawners varies locally from 4 to 8 years.

Nest building, followed by spawning, usually occurs within a few days of stocking. The first indications of spawning are a fishy smell and an oily substance emanating from the nest.

Smaller ponds, about 100 m² in area, stocked only with gourami, may be used for both conditioning and spawning. Nest-building materials are provided as described, and 10 females and 5 males stocked per pond. The breeders, fed on leaves of *Carica, Colocasia,* and other soft plants at 5 kg/(pond)(day), plus 4 liters of rice bran twice weekly, generally begin nest building in about 10 days and spawn 3 days later.

In some parts of Java, the nests are removed from the ponds and the eggs hatched artificially. This has the advantage of permitting more frequent utilization of the spawning pond by fresh brood fish. Nests are opened underwater in 10-liter earthenware jars, filled to 75 to 80% of capacity, which serve as hatching chambers. Each nest contains 3000 to 4000 eggs, which are distributed among 3 or 4 jars. Jars of eggs are kept cool by placing them in the shade or floating them in the pond. At first the water in the jars must be changed very frequently to remove all the oily substance from the eggs; even after this has been eliminated there should be one complete change of water daily. Hatching reportedly takes about 10 days.

Changing the water involves holding down the eggs with a fine sieve and bailing from above. This is exceedingly laborious, particularly when many jars are involved, so labor-saving methods have been sought. Several adaptations of the jar method, involving circulating water in the hatching chambers, have been suggested but, as far as is known, not applied.

Fish farmers in Singaparna, Java, have successfully applied a circulating system which completely dispenses with hatching containers. Nests are opened and the contents, along with the nest fragments, scattered on the surface of small, flowing-water ponds similar to those used in spawning nilem (see pp. 101–102). Straw is also scattered over the pond surface to prevent the eggs from sticking together. About 6 m² of pond surface are required for each nest.

FRY REARING

The young gouramis feed on zooplankton, which develop naturally in the decaying nest, until they are about 10 days old and 1 cm long. At this time, they are captured, usually by draining, and stocked in rearing ponds at rates of 1 nest/40 to 200 m². The best food at this stage is white ants, which are given at the rate of 1 teacup/(nest)(day); twice as much peanut waste is an acceptable substitute. Fry about 3 cm long may be thus obtained in about 3 months. Such fish may be sold for stocking, but if the fry must be transported long distances, lengths of 5 to 8 cm are desirable. For growing to this size, the fry are transferred into larger rearing ponds, allowing about 3200 m² for the progeny of one nest. The primary food at this time is the floating plant *Azolla pinnata,* supplemented after the second month by minced plant leaves. The desired size may thus be reached in about 5 months.

The methods described could be improved. For example, at the Jabasso fish hatchery in West Irian, Indonesia, a floating fry chamber has been designed and tested. The combination of this chamber and intensive fertilization of rearing ponds with cattle manure has enabled workers there to achieve a reported 100% survival of larvae. Methods of induced spawning of gourami are also being perfected.

SEPAT SIAM AND THREE-SPOT GOURAMI

In Indonesia, the word "gurami" is applied only to *Osphronemus goramy,* and properly speaking, the term, in all its variant spellings, should probably be used only to refer to that species. Aquarists, however, designate a number of Asian anabantids, belonging to at least six genera, as "gouramis." Three of these species, the snakeskin gourami (*Trichogaster pectoralis*), the three-spot gourami (*Trichogaster trichopterus*), and the kissing gourami (*Helostoma temmincki*), are also cultured as food fishes. The snakeskin gourami is better known among food fish culturists as Sepat Siam, and we so refer to it; the other two species will be designated by the names just given.

DISTRIBUTION AND IMPORTANCE

Both species of *Trichogaster* are typical bubble nest builders, and thus spawn chiefly during the dry season. Of the two, the Sepat Siam, native to Thailand, South Vietnam, and the Malay Peninsula, is the larger, attaining a maximum length of about 25 cm, and the more important. It has been introduced and cultured to some extent in the Philippines, Indonesia, Burma, Pakistan, and Ceylon, but it is still much more prevalent in its native lands. The natural distribution of the three-spot gourami, which reaches only 15 cm in length, covers roughly the same area as that of the Sepat Siam, plus some of the western islands of Indonesia. A blue subspecies, *Trichogaster trichopterus sumatranus,* native to Sumatra, is popular among aquarists, who refer to it as the "blue gourami," but is not grown as a food fish. Unlike the Sepat Siam, the three-spot gourami is almost never reared as a principal crop; it is a species of secondary importance in Thailand, Vietnam, and Malaysia.

PROPAGATION

Breeding of Sepat Siam is simple; all that is necessary is a well-oxygenated pond containing a rich growth of aquatic vegetation, particularly *Hydrilla verticillata*. The natural spawning season in Thailand is April to October, but it may be bred in ponds at any time as long as the water temperature is 26 to 28°C. Conditioning of the 100-g breeders, in the manner described for *Osphronemus goramy,* is desirable but not essential. Hatching occurs in 24 to 48 hours, and within 3 to 7 days the fry have absorbed the yolk sac. The fry are reared in the spawning pond, where they feed on plankton.

Three-spot gouramis are not intentionally bred by culturists but are allowed to reproduce naturally in ponds and rice fields where other species of fish are cultured. There are reportedly two spawning seasons in western Java, one each at the beginning and end of the rainy season, but some authorities state that it spawns in every month.

KISSING GOURAMI

The kissing gourami (*Helostoma temmincki*), so named because of its habit of puckering its thick lips and "kissing" other individuals of its species as well as other fishes, plants, and inanimate objects, is native to Thailand, Vietnam, the Malay Peninsula, and the western islands of the East Indies. The kissing behavior apparently has no sexual function; some observers believe it to be a threat display. The kissing gourami has been introduced and cultured in Celebes, Ceylon, and the Philippines. It reaches a maximum length of about 30 cm and matures at 20 cm, or 12 to 18 months of age.

ARTIFICIAL PROPAGATION

Spawning this species, which scatters free-floating eggs rather than building a nest, is a bit more difficult than breeding bubble nest builders. Kissing gouramis are bred in special ponds and in modified rice fields. Spawning ponds are 50 to 80 cm deep and 30 to 100 m² in surface area; the larger sizes are preferable. The best results are obtained at elevations of 200 to 700 m above sea level, but kissing gouramis are also spawned at low elevations. Wherever they are located, ponds must be supplied with an abundance of clear water so that dead and decaying plankton can be flushed out. The outlet must be supplied with sieves to prevent eggs being washed out during such operations.

Before introducing the spawners, ponds are drained and dried for a week and the bottom or, in the case of large ponds, selected parts thereof, covered with damp paddy straw. When the pond is filled, the straw floats and protects the eggs and larvae from rain and sun.

While the spawning pond is being prepared, brood fish are held in 4-m² conditioning ponds about 60 cm deep. Each such pond contains about 40 fish, which are held for 30 to 40 days and fed regularly with rice bran.

Kissing gouramis are spawned as soon as they reach sexual maturity and at 6-month intervals thereafter. They are short-lived fish and can generally be bred only about five times. Breeders are stocked, at about 1 pair/30 to 50 m², when the pond is about ⅔ full. If the males are appreciably smaller, more may be stocked, so that the total weights of the sexes are approximately equal. The start of spawning is indicated by a characteristic fishy odor; it usually takes place within 18 hours of stocking, most often in the early morning. Sometimes some of the eggs are eaten by the parents; an abundance of vegetable food in the pond may be of value in reducing the incidence of such behavior. Pond spawning may be carried out in conjunction with spawning of Mata Merah (*Barbus orphoides*) (see p. 104).

Rice field propagation of kissing gouramis is done shortly after the rice is harvested, in fields where the dikes have been reinforced so that the water can be maintained at a depth of 30 cm. The procedures followed are basically the same as for pond spawning, but paddy straw need not be added, since it is present naturally.

(see p. 104)

FRY REARING

The eggs hatch in about 2 days and the larvae float on the surface for 3 to 4 days, after which they move into deep water. Starting 7 to 10 days after hatching, the pond is manured with decaying plants, animal manure, or a mixture thereof. Green manures are applied at about 0.5 to 2.0 kg/ha; animal manures at half this rate. Half of the total quantity is applied on the eighth day after the eggs hatch; the remainder is added in 5- to 10-kg lots every 2 to 3 days. The culturist must be careful not to overfertilize and pollute the pond. If this is inadvertently done, part of the water must be drained off and replaced immediately. The water should be bright green and the fry move vigorously at all times; brown water and sluggish fry are danger signals. Some culturists stock fry of common carp or nilem after the kissing gouramis have hatched as an additional barometer of water quality; the pond then doubles as a cyprinid rearing pond. Abnormal mortalities of cyprinid fry indicate that partial replacement of the water is imperative. Usual stocking rates are 5 to 10 2-week-old common carp or 15 to 50 5-day-old nilem/m^2.

Month-old fry are harvested, by draining, for sale or stocking to grow for market. Where common carp or nilem fry are present, a special technique is employed to separate them from the kissing gouramis. With the pond partially drained, a ditch about 50 cm wide and 10 cm deep is dug from the inlet to a 4-m^2 sump near the outlet. When drainage is resumed, all the fry are collected in the sump. Then a second sump is dug in the center of the pond, and a small amount of water let in. The carp or nilem fry swim upstream and congregate in the second sump, while the kissing gouramis stay in the original sump. The few carp or nilem fry which remain with the kissing gouramis will come to the surface when the water is stirred and may then be skimmed off.

Spawned-out breeders are stocked in large ponds at 1 to 4 fish/m^2 to recuperate. If the population density is relatively low, fertilization with animal manure may be all that is necessary to provide food for the convalescent fish, but if it is high, an amount of rice bran equivalent to 1 kg for each 24 fish should be given twice daily.

CLIMBING PERCH

The climbing perch (*Anabas testudineus*), native to India, all of southeast Asia, Indonesia, the Philippines, and southern China, makes particularly extensive use of its air-breathing apparatus. It may leave the water, particularly during rainstorms, and prowl about on land, aided by the stiff edges of its opercula. If its home pond dries up, it may walk to another body of water or burrow into the mud and remain dormant through the dry season.

As one might expect, fish farmers take no particular pains to care for a species that may walk away. Although climbing perch are of minor importance in fresh and brackish water fish culture in Cambodia, Vietnam, and the Philippines, particularly in areas where low temperatures are a problem, nowhere are they bred by culturists. Natural spawning, which proceeds as described for the kissing gourami, requires temperatures of 25 to 29°C and is reportedly confined to the rainy season, although several spawnings are said to occur each year in Ceylon.

USE OF ANABANTIDS IN FOOD PRODUCTION

MONOCULTURE

As mentioned, the chief uses of labyrinth fishes are in polyculture, but some monoculture occurs, particularly in stagnant waters which will not support other fish. Monoculture of kissing gourami in manured ponds has resulted in average annual yields of 500 kg/ha, that of Sepat Siam, 250 to 350 kg/ha. *Osphronemus goramy* is a very slow-growing fish, often requiring 2 to 3 years to reach marketable size, thus it yields only about 200 kg/(ha)(year). The low yield is, however, partially compensated by its high price. The climbing perch is occasionally stocked alone in cages in Cambodia (see p. 209 for a description of this technique), but yield data are not available.

Recently, higher yields have been effected in rice field fish culture in Smudprakarn province, Thailand, through monoculture of Sepat Siam. Fields are prepared by excavating a ditch 3 m wide and 1 m deep along the dikes and manuring the field with cut grass. Sepat Siam breeders are stocked at 375/ha and fed periodically with cut grass. Breeders and young are harvested 9 to 10 months later, with yields of 600 to 1600 kg/ha.

POLYCULTURE

Polyculture must, of course, be based on a knowledge of the foods of fish. Among the cultured anabantids, the climbing perch is the only habitual carnivore, deriving about 70% of its nourishment from animal matter, mostly invertebrates. It is somewhat of an opportunist, however, and in captivity has been observed to accept a wide variety of food, from lettuce and rice to white ants and dead fish. The gourami browses in nature on floating plants and overhanging leaves of terrestrial plants. In culture, it is fed chiefly on soft leaves of aquatic and terrestrial plants, with occasional helpings of animal food. The Sepat Siam and three-spot gourami feed on both phytoplankton and zooplankton, as well as decayed algae and higher plants; supplemental feeds are exclusively vegetable, consisting chiefly of aquatic plants. The kissing gourami is by nature a surface and midwater plankton feeder, but will accept such artificial feeds as rice bran and starch of cassava roots.

Anabantids are most often stocked as minor components of polyculture systems based on Chinese carps in southeast Asia (see Chapter 3) or Indian carps in India, Pakistan, and Ceylon (see Chapter 4). In Malaysia and Thailand, kissing gouramis, Sepat Siam, and three-spot gouramis are also stocked together with fresh water shrimp (see Chapter 32). The only important polyculture system based largely on an anabantid fish is applied in Malaysia and Singapore, where a combination of Sepat Siam, common carp, Java tilapia (*Tilapia mossambica*), and lampai (*Barbus schwanenfeldii*) or tawes is an alternative to the usual Chinese carp complex.

Experimental polyculture involving labyrinth fishes is proceeding in a number of Asian nations. For instance the College of Fisheries, Kasetsart University, Thailand, has obtained good preliminary results with 1:1 and 2:1 ratios of Java tilapia and Sepat Siam. Such experiments may eventually result in expansion of the role of the labyrinth fishes in Asian aquaculture.

REFERENCES

Esox, P. 1931. Vom "Sepat" (*Trichogaster trichopterus*). Blatt Aquar-Terrarienk. **42**(1): 4–7.

Hofstede, A. E., R. O. Ardiwinata, and F. Botke. 1953. Fish culture in Indonesia. Special Publication of the Indo-Pacific Fisheries Council, No. 2. 129 pp.

Hora, S. L., and T. V. R. Pillay. 1962. Handbook on fish culture in the Indo-Pacific region. FAO Fisheries Biology Technical Paper No. 14. 204 pp.

Sterba, G. 1966. Fresh Water Fishes of the World. Viking Press, New York.

9
Culture of Pikes and Perches

The larger members of the pike family (Esocidae) and the perch family (Percidae), though not related, are similar in being elongate, largely piscivorous fishes of holarctic distribution. From an economic point of view, they are all valued as sources of food and sport, with the perches of somewhat higher quality as food, while the pikes offer superior sport. Representatives of both families are thus cultured in Europe and North America, both for sale and for stocking to augment sport and commercial fisheries.

PIKES

SPECIES CULTURED

The pikes yielded sooner than the perches to the efforts of culturists to artificially propagate and rear them, and thus have a longer history of culture. By far the most commonly cultured pike is the northern pike (*Esox lucius*), which may reach weights of more than 20 kg in nature. Although, as mentioned, the Esocidae and Percidae are holarctic, the northern pike is the only holarctic species in either family, ranging as far south as Iran in the Eastern hemisphere, and Missouri in the Western hemisphere. In North America, it is cultivated almost exclusively as a game fish, but in Europe pike have long been stocked in carp ponds to control excess reproduction.

An even larger pike, the muskellunge (*Esox masquinongy*), which has a rather limited distribution in the eastern United States and Canada, is cultured as a sport fish but is unsuited to pond life and will probably never be important as a food fish.

Another American pike, the chain pickerel (*Esox niger*), though of considerable importance to anglers, is not presently cultured. However, as it is by nature a pond dweller and attains relatively small sizes (maximum 5 kg; average taken by anglers, less than 1 kg), it should be considered by American fish culturists for a role similar to that played by the northern pike in Europe.

Yet another pike, the Amur pike (*Esox reicherti*), is presently being considered for use as a predator in fish ponds in China. *E. reicherti* is presently confined to the Amur River basin and Sakhalin Island, but in 1968 it was imported by the United States Bureau of Sport Fisheries and Wildlife from the Soviet Union for purposes of breeding. Its habits are little known, but it is difficult to see what niche it could occupy, in nature or culture, that is not already satisfactorily filled by one of the three large American species of *Esox*.

ARTIFICIAL PROPAGATION AND REARING OF NORTHERN PIKE

An ancient European method of pike culture involves the artificial flooding of alpine meadows in the spring. Pike enter and spawn naturally, after which the meadows are left flooded until the young reach fingerling stage. Then they are captured and used for stocking. Modern techniques of propagation of the northern pike have been largely standardized and vary little from place to place. Essentially the same methods are used with muskellunge, and presumably they could be adapted to other pikes.

Brood stock are collected, usually from the wild, in the spring when water temperatures reach about 10°C. Eggs and sperm may be extruded without application of hormones, and they may be fertilized using the dry method (see p. 400). Without exception, the jar method (discussed in Chapter 3) is used in hatching; this requires about 2 weeks. As soon as the fry are washed out of the jars they are collected and stocked in ponds.

It is at this stage that the culturist's problems begin. Young pike are carnivorous and voracious from the start, and an abundance of food must be available to them at all times. For the first few days, zooplankton is satisfactory, but once the fry reach a length of 4 to 5 cm, smaller fish must be constantly supplied. Even with a great abundance of food, cannibalism is so prevalent as to be the chief economic limiting factor in pike culture. Under good conditions, the survivorship of fry stocked in the spring at the rather conservative rate of 2000 to 2500/ha may be expected to amount to only 2 to 5% by fall.

STOCKING NORTHERN PIKE IN POLYCULTURE

As indicated, when pike are stocked by culturists it is always as predators in a polyculture system. This function is especially important in such countries as France and Yugoslavia, where such small cyprinids as the roach (*Rutilus rutilus*) and the tench (*Tinca tinca*) are commonly cultured in ponds where common carp (*Cyprinus carpio*) are the chief crop. Stocking rates and yields representative of French practices are given in Table 1.

TABLE 1.   STOCKING RATE OF NORTHERN PIKE AND CYPRINID FISHES IN POLY-CULTURE PONDS IN FRANCE

SPECIES	WEIGHT AT STOCKING (G)	STOCKING RATE PER HECTARE	YIELD (KG/HA)
Common carp	150–240	250	60
Roach and/or tench	100	200	55
Northern pike	100	15	10
Total			125

The roach and tench may be eliminated, but even in carp-pike ponds, pike seldom exceed 10% of the total fish population. Growth of pike in ponds is generally good; in the Ukraine they may reach a weight of 800 g at the end of the first summer.

In most European countries, the pike is gradually being replaced as a pond predator by other fish which are more economical to propagate and rear to fingerling size and, in some cases, bring better prices. Among these are the sheatfish (*Siluris glanis*), the largemouth bass (*Micropterus salmoides*), the rainbow trout (*Salmo gairdneri*), and the common pike-perch or zander (*Lucioperca lucioperca*). Another percid fish, the European perch (*Perca fluviatilis*), has on occasion been stocked, but it is not large enough to be a really effective predator, thus, despite its excellent quality as a food fish, its future in aquaculture is doubtful.

PIKE-PERCH AND WALLEYE

ARTIFICIAL PROPAGATION

Pike-perch and the very similar walleye (*Stizostedion vitreum*) of North America have been artificially propagated since the late nineteenth century, but spawning and rearing were inefficient processes until the 1960s, when methods especially suited to these species were developed. Today, pike-perch fry and fingerlings are readily available in many European countries, and fish farmers are stocking them increasingly in place of northern pike. In addition to being better food fish, pike-perch are somewhat preferable to pike by virtue of having a smaller gullet, thus enabling the culturist to plant them earlier in the season or at a larger size without danger to his main crop. On the other hand, pike-perch are less tolerant of high temperatures than are northern pike. They are also reputed to be very intolerant of turbidity, but fish farming experience in Hungary long ago indicated otherwise.

Advances in the culture of the pike-perch in Europe have been paralleled by improvements in walleye culture in North America. Walleye are not used in pond culture but are stocked in lakes and streams to augment sport and, in a few waters, commercial fisheries. Pike-perch are similarly stocked in East Germany, France, and the Soviet Union.

Spawning of pike-perch in captivity usually entails the use of artificial nests. As originally developed in Hungary, these nests are made of willow or alder roots, collected during winter, and tied in thin layers onto a 50-cm × 50-cm frame. The frame is weighted to sink it, and lines are tied to the four corners so that the nest may be removed from the water in a level position. In recent years, synthetic materials have begun to supplant the willow and alder roots. In recent experiments in Czechoslovakia, the best substrate for spawning was found to be nylon shearings with a diameter of 0.2 mm.

Pike-perch spawning basins are usually 300 to 1000 m² in surface area, and 1.2 to 1.8 m deep. In the spring, the basins, which have previously been supplied with one nest for each pair of spawners, are stocked at the rate of 1 pair/2 to 5 m². Breeders are normally 3 years old or older. Stocking is not usually done until the water temperature has reached 8 to 9°C and a considerable drop is unlikely.

Both lesser and greater amounts of manipulation may be carried out.

Some culturists in East Germany and Yugoslavia are able to produce fingerlings at a profit through natural spawning in ponds. One female and 1 or 2 male pike-perch/ha are stocked in the spring and left untended.

In some instances pituitary injection may be resorted to (see p. 85 for details of this technique), in which case smaller concrete spawning basins may be used. Both sexes are injected, and spawning follows in 1 to 3 days after stocking in the basins.

Pike-perch eggs in nests may be covered with a wet cloth and transported overland for several hours, if necessary. When packed in crates with wet moss and ice, the travel time may be extended to several days.

The old method of hatching involves suspension of the nests in ponds at a depth of 0.5 m with the covered egg surface facing downward. Nests may be so suspended directly into ponds containing one- and two-summer-old carp, in which case they should be covered with a fine mesh wire shield to prevent predation. The number of eggs in a pike-perch nest reportedly varies from 20,000 to 200,000, although natural fecundity ranges from 200,000 to 1 million. Enough nests should be stocked so that there are 1000 to 2000 eggs/ha.

Better hatching rates may be obtained by use of the Woynarovich method in which the nests are placed in chambers and exposed to a continuous foglike spray of water at 1 to 4 atm. From 30 to 60 nests can be placed in each cubic meter of sprayed space; at 10° the eggs hatch in 11 days.

Walleye spawners used in culture are usually captured and stripped during their spawning migrations; pituitary injection is seldom or never used. The jar method of hatching is employed and requires 7 to 8 days at 17 to 20°C, or 14 days at 10°C.

FRY REARING

Pike-perch may be stocked in carp ponds at the egg stage. Some culturists, however, prefer to stock fry or fingerlings. Although young pike-perch are not as prone to cannibalism as pike, in general they present the same problems to the culturist. In addition, they are quite fragile with respect to handling. Rearing is carried out in ponds so fertilized as to produce an abundance of zooplankton. Under optimum conditions fingerlings reach 12 to 25 cm in length and weigh 30 to 100 g at the end of their first summer. Culturists in West Germany aim for more and smaller (6 to 10 cm) fingerlings.

European techniques of pond fertilization for rearing pike-perch fry and fingerlings are far from perfected, and experiments conducted with walleye by the Minnesota Department of Conservation at Waterville,

Minnesota, may have some bearing on the subject. Traditional walleye rearing methods in Minnesota involve fertilization in the spring with organic materials, usually barnyard manure. The bacteria thus produced are more than sufficient to support an abundance of cladocerans, which are the primary food of walleye fry and small fingerlings. Growth and survival are thus very good in May and June. The single dose of fertilizer is not, however, adequate to maintain dense populations of the copepods and benthic organisms required by larger fingerlings. Manuring through the summer is not the answer, as it results in dense growths of algae and other aquatic vegetation. Walleye culturists were thus faced with a choice between tolerating a high percentage of cannibalism and premature stocking of small (10 cm) fingerlings, which are believed to be significantly more susceptible to predation than 15-cm fish.

A large variety of combinations of organic and inorganic fertilizers were tested to see whether pond fertility could be sustained and fingerlings reared through the summer without undesirable side effects or cannibalism. Very good results were obtained with a single spring application of dried sheep manure in a quantity sufficient to supply 1.1 kg of nitrogen per 1000 m³ of water. Higher rates of fertilization were effective but also resulted in an overabundance of phosphorus. Manuring was not repeated, but starting in mid-June, weekly applications of brewer's yeast at 112 kg/ha were made. This treatment supported large populations of copepods and benthic animals through late summer. The Michigan Department of Conservation is currently attempting to rear walleye fry in sewage ponds, which may offer a similar environment.

The methods developed in Minnesota would of course have to be adapted to local conditions wherever they were applied. Further problems would be presented by the excessive amounts of phosphorus released and the high cost of brewer's yeast. The latter problem might be circumvented if it were known which constituents of brewer's yeast are effective.

Minnesota biologists have also been experimenting with alternative means of feeding walleye fry. Attempts to confine fry in tanks and pump through water containing large amounts of zooplankton resulted in no feeding whatsoever. Attempts to induce walleye fry to accept artificial feed were similarly unsuccessful until 1968, when hatchery workers at New London, Minnesota, were able to induce feeding on commercial trout pellets by confining 5- to 10-cm fingerlings in small cribs and starving them for 10 days.

STOCKING PIKE-PERCH IN POLYCULTURE

Pike-perch may be stocked in carp ponds at greater densities than pike; common rates are 50 to 100/ha for one-summer-old fish and 30 to 60/ha for two-summer-olds. Thereafter, management is negligible until harvest time.

There is no difficulty in harvesting pike-perch separately from carps and other fishes in drainable ponds. They are very sensitive to currents and will attempt to escape the pond prior to other species. Three-summer-old specimens, 37 to 55 cm long, weighing 500 to 1500 g, are normally retained for marketing, while smaller individuals and selected breeders are overwintered.

Pike-perch which are to be overwintered must be transported and handled carefully in fresh, clean water. They should not be stored or carried in the same containers as carp, since their oxygen demands are much greater. Food, in the form of small fish, must be provided through the winter so that cannibalism does not occur; 150 to 250 kg of fish will support 100 kg of pike-perch through the winter.

We do not foresee great expansion of the roles of any of the pikes or perches in aquaculture.

REFERENCES

DOBIE, J. 1966. Experiments in the fertilization of Minnesota fish rearing ponds. FAO World Symposium on Warm Water Pond Fish Culture. FR: II/E-16.

STEFFENS, W. 1960. Ernahrung und Wachstum des jungen Zanders (*Lucioperca lucioperca*) in Teichen. Z. Fisch. 9(3/4):161–272.

UNGER, E. 1939. Die Zucht des Zanders in Karpfenteich Wirtschaften und in Freien Gewässern. Handb. Binnenfisch Mitteleurop. 4:723–742.

WOYNAROVICH, E. 1959. Erbrütung von Fischeiern in Sprühraum. Arch. Fischwiss. 10(3):179–189.

10

Black Bass and Sunfishes (Family Centrarchidae) in Fish Culture

THE FARM POND PROGRAM

The black basses and sunfishes (family Centrarchidae), indigenous to North America, are of virtually no value in commercial fisheries but are extremely popular with sport fishermen, so much so that in some areas their sale as food is forbidden. Since most Centrarchids have extremely high reproductive potentials and in fact tend to overpopulate small bodies of water, sport fisheries are sustained mainly on the basis of natural

234

reproduction, and black bass and sunfish are seldom propagated in hatcheries for liberation in streams and lakes. The only important use of Centrarchids in fish culture in North America is in "farm ponds," which are common throughout the United States and in southwestern Ontario, Canada.

The purposes of the farm pond program, which reached a peak of popularity in the 1950s, are to conserve water and wildlife and provide food and recreation for residents of rural areas by encouraging farmers to build small ponds on their property and stock them with fish obtained free of charge from state or federal government agencies.

STOCKING RATES IN FARM PONDS

The traditional farm pond stocking scheme involves only Centrarchids. By far the most commonly stocked species are the largemouth bass (*Micropterus salmoides*) and the bluegill (*Lepomis macrochirus*). Bluegills, which are normally stocked at rates varying from 220 to 2200/ha, feed chiefly on invertebrates, but they may also derive some nutrition from algae. Largemouth bass are higher on the food chain and feed exclusively on other animals, usually larger ones than those eaten by bluegills. Bass are stocked at 100 to 200/ha, as necessary to provide fishing and control the bluegill population. In recent years, fishery biologists in a number of locales have begun to recommend the redear sunfish (*Lepomis microlophas*), which has a very low reproductive potential, as a supplement or substitute for the bluegill. In a few ponds, channel catfish (*Ictalurus punctatus*) are stocked as an additional source of food and sport. A number of other fish species have been experimentally or accidentally stocked in farm ponds, usually with poor results.

MANAGEMENT AND YIELDS

After construction, stocking, and perhaps an initial dose of fertilizer, management of American farm ponds is virtually nil unless and until it is decided to poison and restock a pond. In practice, most ponds are inadequately fished and otherwise neglected and soon become poor fish producers. Even in ponds which are properly fished, recreation, not food production, is usually emphasized, thus farm ponds are sparingly treated in this volume. (In some ponds in the southern United States, where bluegill are intensively fished by women and children, food and recreation are of approximately equal importance.) Readers seeking more detailed information should consult or contact local sources, such as state conservation departments and university agricultural extension services.

The last decade has seen the formation in the United States and Canada of a considerable number of communes, experimental communities, and other groups which attempt to be more or less self-sufficient in terms of food production. The great majority of these groups have not attempted fish culture, but where it has been considered, polyculture of Centrarchids has been the method proposed, if only because stock is available free of charge. It is thus possible that the next decade may see the first attempt to manage largemouth bass-bluegill communities or the like as a food resource. Perhaps if this occurs there will be some yield data available for this crude sort of polyculture. It has been estimated that average annual production of fish in North American farm ponds is 250 to 450 kg/ha, but it is not known how these figures are related to the weight of food fish which could be harvested on a sustained basis. A few commercial fish hatcheries, particularly in Arkansas, raise edible size centrarchids, including largemouth bass, bluegill, white crappie (*Pomoxis annularis*), and black crappie (*Pomoxis nigromaculatus*) for sale to operators of fee fishing ponds; they are able to produce only 44 kg/ha of these fish.

GROWTH OF FISH

Growth data for centrarchids are much more readily available than data on production and yield; in fact there is an overabundance of such information in the sport fishery literature. Table 1 shows the average growth rate for the three principal farm pond species in Illinois. Growth is affected not only by pond management but by the length of the growing season; hence one could generally expect growth to be more rapid south of Illinois and slower to the north.

THE LARGEMOUTH BASS AND CRAPPIES IN COMMERCIAL FISH CULTURE

The centrarchids have not been totally neglected by practical and experimental fish culturists in the United States, but the results achieved have thus far been unspectacular. In the 1950s, when fish culture was just beginning to become a significant industry in the south central states, several centrarchids were considered for commercial culture. On the basis of experiments conducted at Auburn University, the bluegill and the flier (*Centrarchus macropterus*) were rejected as unsuited for culture, but interest was retained in the crappies and the largemouth bass.

During the period when bigmouth buffalo (*Ictiobus cyprinellus*) were the fish most commonly farmed, it was general practice to stock about 125

TABLE 1.   AVERAGE GROWTH OF LARGEMOUTH BASS, BLUEGILL, AND REDEAR
SUNFISH IN FARM PONDS IN ILLINOIS

SPECIES	AGE (YEARS)	LENGTH (CM)	WEIGHT (KG)
Largemouth bass	1	16.0	0.1
	2	22.9	0.2
	3	29.5	0.3
	4	34.3	0.5
	5	40.1	0.9
	6	44.2	1.1
	7	48.0	1.4
	8	50.3	1.8
	9	51.6	2.2
	10	52.6	2.5
Bluegill	1	8.1	0.01
	2	11.7	0.04
	3	14.5	0.07
	4	16.8	0.09
	5	18.8	0.14
	6	21.3	0.21
Redear sunfish	1	13.2	0.06
	2	16.3	0.11
	3	18.0	0.13
	4	19.8	0.19
	5	23.4	0.28
	6	23.6	0.29
	7	24.1	0.30

largemouth bass fingerlings per hectare to control trash fish. With the
development of better methods of excluding trash fish and the almost
total replacement of the bigmouth buffalo by the channel catfish, culture
of largemouth bass became less important, though some marginal operators may still stock them in catfish ponds.

Largemouth bass and crappies are still stocked, along with channel
catfish and bigmouth buffalo, as part of a very low intensity form of
culture practiced in natural sloughs. (See Chapter 6.) This practice,
however, accounts for only a tiny percentage of the annual cultured food
fish production of the south central states.

The only centrarchid which has been widely introduced outside North
America is the largemouth bass. Today it is much more common in commercial fish culture in Europe than in its native land. It has also been
stocked in Latin America and Africa, and in 1965 experimental culture
was initiated in Tunisia.

In Europe, largemouth bass occupy the niche traditionally accorded to the northern pike (*Esox lucius*), as a piscivore to control trash fish and excess young of other cultured species. It is a hardy fish, and generally well suited for the purpose, but it possesses one serious disadvantage for pond culture: it can successfully spawn in most ponds and may produce an abundance of young, so that it may actually contribute to the problem it was intended to alleviate. This situation is further complicated by the fact that it prefers soft-rayed fishes to its own spiny-finned young as prey. Nevertheless, it is widely stocked, particularly in France and the Soviet Union.

POSSIBILITIES OF CULTURE OF SUNFISH AS FOOD FISH

Although largemouth bass spawn readily, their reproduction is difficult to control. Anyone can spawn them by simply placing pairs in ponds in the fall and letting nature take its course in the spring, but to date bass culturists are totally dependent on natural spawning and cannot produce fertile eggs on demand. The normal way of achieving this result would be through treatment with pituitary hormones, but largemouth bass and most other centrarchids do not respond well to this technique. Spawning has, however, been successfully induced in white crappie and rock bass (*Ambloplites rupestris*) at the Southeastern Fish Cultural Laboratory, Marion, Alabama, and it is expected that the methods developed there could eventually be adapted to largemouth bass.

Those American fish culturists who are still interested in centrarchids presently concern themselves less with such largely piscivorous fishes as the black basses and crappies and more with *Lepomis* spp., most of which feed mainly on invertebrates. Not only are such fish more readily trained to take artificial feeds (although there are more than a few reports of crappies eating catfish pellets), they are the finest food fishes among the Centrarchidae. The principal obstacle to their commercial culture in the past has been their high rate of reproduction, which often leads to stunting. Similar problems have been encountered by growers of *Tilapia* spp., and they have found at least two ways to cope with them—monosex culture and cage culture.

MONOSEX CULTURE

Monosex culture, as the name implies, involves growing of only one sex. To visually sort tilapia by sex is not easy; to do so with centrarchids would be even more difficult except during the breeding season, when males of most species display brilliant coloration. Fortunately for tilapia culturists, certain hybrids produce all or nearly all male offspring. Not so much attention has been paid to hybridization of *Lepomis* spp.; nevertheless, many of the possible hybrids of the edible size species have already been produced. Of the twelve crosses made thus far, only one, ♀ redear sunfish × ♂ green sunfish (*Lepomis cyanellus*), has produced young with a near-normal sex ratio. The following crosses have produced F_1 generations highly skewed toward males:

♂ green sunfish × ♀ pumpkinseed (*Lepomis gibbosus*)
♀ green sunfish × ♂ pumpkinseed (*Lepomis gibbosus*)
♂ green sunfish × ♀ bluegill
♀ green sunfish × ♂ bluegill
♂ green sunfish × ♀ longear sunfish (*Lepomis megalotus*)
♀ green sunfish × ♂ longear sunfish (*Lepomis megalotus*)
♀ green sunfish × ♂ redear sunfish
♂ bluegill   × ♀ redear sunfish
♀ bluegill   × ♂ redear sunfish
♂ bluegill   × ♀ pumpkinseed
♀ bluegill   × ♂ pumpkinseed

Many of these hybrids are of low fertility, if not sterile, rendering them even more suitable for intensive pond culture. It is possible to sterilize bluegills, without otherwise harming them, by means of low doses of radiation, and this technique may also find application in practical fish culture.

CULTURE IN FLOATING CAGES

Culture in floating cages does not prevent tilapia from spawning, but, since the eggs drop through the bottom of the cage, it does prevent the fish from exercising the parental care necessary for successful reproduction. Parental care is as necessary for eggs and fry of centrarchids as for tilapia, but to our knowledge cage culture of centrarchids has not been attempted. In the spring of 1970, the editors of Farm Pond Harvest magazine announced their intention to attempt intensive cage culture, with supplemental feeding, of bluegills. Results are not yet available.

Although the small-scale use of crappies and largemouth bass in catfish culture in the United States and the stocking of the latter species as a predator in European policulture ponds are the only current instances of commercial culture of centrarchids as food fish, the members of the family, particularly *Lepomis* spp., present possibilities which have yet to be explored, and fish farmers should not be too quick to write off the Centrarchidae.

REFERENCES

CHILDERS, W. F., and G. W. BENNET. 1961. Hybridization between 3 species of sunfish (*Lepomis*). Illinois Natural History Survey, Biology Notes, No. 46. 15 pp.

HUBBS, C. L., and L. C. HUBBS. 1931. Increased growth in hybrid sunfishes. Pap. Mich. Acad. Sci. 13:291–301.

HUBBS, C. L., and L. C. HUBBS. 1932. Experimental verification of natural hybridization between distinct genera of sunfishes. Pap. Mich. Acad. Sci. 15:427–437.

HUBBS, C. L., and L. C. HUBBS. 1933. The increased growth, predominant maleness, and apparent infertility of hybrid sunfishes. Pap. Mich. Acad. Sci. 17:613–641.

LAGLER, K. F., and C. STEINMETZ. 1957. Characteristics and fertility of experimentally produced sunfish hybrids, *Lepomis gibbosus* × *Lepomis macrochirus*. Copeia. 1957(4):290–292.

SWINGLE, H. S. 1954. Experiments on commercial fish production in ponds. Alabama Polytechnic Institute, Agricultural Experiment Station (mimeographed pamphlet). 22 pp.

11

Miscellaneous Asian Pond Fishes

Snakeheads or murrels

Sleeper gobies

Loaches

References

SNAKEHEADS OR MURRELS

In no other part of the world are so many varieties of fish cultured as in Asia. Most of the cultured Asian species have been dealt with in preceding chapters, but members of the families Channidae (snakeheads or murrels), Eleotridae (sleeper gobies), and Cobitidae (loaches) have not been covered and are treated here.

The snakeheads, only Asian species of which are discussed here (see Chapter 12 for information on snakeheads in African fish culture), are voracious and, except for the young, usually exclusively piscivorous. For this reason, they are considered undesirable in many forms of fish culture and are sometimes the objects of eradication measures. On the other hand, they are excellent food fish and are thus often selected for stocking where a predatory fish is needed in polyculture ponds.

Snakeheads possess an accessory breathing organ similar to, though not as highly developed as, that of the labyrinth fishes (Anabantidae), and they are thus able to withstand very low concentrations of dissolved oxygen. Not only are their oxygen requirements low, but snakeheads are extremely hardy with respect to all other environmental parameters.

The most widely distributed of the snakeheads is *Ophicephalus striatus,* which is native from India to China, Indonesia, and the Philippines, and reaches a maximum length of more than 90 cm. In addition to being used in polyculture over most of its range, *O. striatus* is grown in floating cages in Cambodia, as is *Ophicephalus micropeltis.* Next to catfishes of the genus *Pangasius,* snakeheads are the most important fishes for this purpose. (See Chapter 7 for a full description of cage culture.)

A larger species, *Ophicephalus marulius* (maximum length 120 cm), and a smaller one, *Ophicephalus punctatus* (maximum length 30 cm), are cultured in India and Pakistan not only in ponds and rice fields but also in irrigation wells, where few other fishes will survive. Some Indian investigators are of the opinion that *O. punctatus* is primarily insectivorous.

The northern representative of the Ophicephalidae is *Ophicephalus argus,* which ranges well into Manchuria. The extent of utilization of this species in Chinese fish culture is not known, but it is known that it has been artificially spawned by means of pituitary injection at the Tsingpu Experimental Freshwater Fish Farm of the Shanghai City Fisheries Bureau. This is a rather unusual practice, as most fish farmers in southeast Asia stock snakeheads in very small quantities, and find natural reproduction quite adequate to supply their needs.

SLEEPER GOBIES

The sleeper gobies are fully as voracious as the snakeheads, but most of them, including the cultured species, feed at a lower trophic level, consuming mainly invertebrates. The only sleeper goby currently stocked by practical fish culturists is the sand goby (*Oxyeleotris marmoratus*), which attains weights of about 900 g in 1 year when reared in ponds in Malaysia, Singapore, Cambodia, and Vietnam. It is also used in cage culture in Cambodia. *O. marmoratus* spawns in ponds without any attention on the part of the culturist.

LOACHES

The loaches, distributed throughout Europe and Asia, as well as in Morocco and Ethiopia, are a fairly large family of mostly small, often eel-like, benthic carnivores. Only a few of the loaches are large enough to be valuable as human food, and only in Japan, where *Cobitis* spp. and the dojo (*Misgurnus anguillicaudatus*) are stocked in rice fields, are they cultured.

REFERENCES

Hora, S. L., and T. V. R. Pillay. 1962. Handbook on fish culture in the Indo-Pacific Region. FAO Fisheries Biology Technical Paper, No. 14. 204 pp.

12

Culture of African Freshwater Fishes Other Than *Tilapia* spp.

HISTORY AND STATUS OF FISH CULTURE IN AFRICA

The African freshwater fish fauna is very large and diverse, reaching an apex of complexity in the Stanley Pool of the Congo River, which reportedly contains more species of fish than any other body of freshwater in the world. However, proportionately few African food fish have been cultured, even experimentally. There is, in fact, no native tradition of aquaculture in Africa south of the Sahara. Such fish culture as currently exists was largely initiated by Europeans who, probably for reasons of familiarity, have concentrated on exotic species, plus the easy-to-breed members of the genus *Tilapia* (family Cichlidae). (See Chapter 18 for a discussion of *Tilapia* culture.) A few of the native fishes other than *Tilapia* spp. have been cultured, in some cases with good results, and these are the subjects of this chapter.

The African fish farmer is confronted with a set of socioeconomic factors which are unique and rather limiting. Unlike the Asian peoples, Africans have no tradition of fish culture, but are accustomed to obtaining wild fish with hooks or nets. Thus an African who will energetically care for a vegetable garden must acquire a new set of habits before he can be effective as a fish culturist. As a consequence, fish culture programs in Africa have a history of enthusiastic beginnings and a quick deterioration, except in countries where considerable government supervision occurs. A consequence of inadequate maintenance of fish ponds may be stunting, which is a more serious matter in Africa than in similar situations in Asia, since Africans, unlike Asians, have a strong bias for large fish. Africans, in general, are also unlike Asians in their reluctance to use manure, particularly human waste, in pond fertilization. Finally, the African fish culture scene is further complicated by the political instability of much of the continent.

POTENTIALLY USEFUL NATIVE SPECIES

The prospective fish culturist in Africa must consider all the preceding factors and more, but we shall concentrate on the biology, status, and potential of those African freshwater fish which have thus far been studied by fish culturists. Table 1 summarizes the ecology, distribution, and status of propagation of the species which have been used or are being considered for use.

There is a preponderance of piscivorous fishes in Table 1. Fish culture in Africa is largely based on the prolific *Tilapia* spp., which have a pronounced tendency to overpopulate ponds and produce stunted populations. Among the possible means of coping with this problem is to stock predatory fish of suitable size to eliminate or thin out the young tilapia, thus piscivorous fishes are being carefully studied. While the attention of African fish culturists is still chiefly focused on tilapia-predator stocking,

TABLE 1. CHARACTERISTICS OF NATIVE AFRICAN FRESHWATER FISHES (OTHER THAN *Tilapia* SPP.) USED IN PRACTICAL OR EXPERIMENTAL FISH CULTURE

FAMILY	SPECIES	RANGE	FOOD HABITS	OTHER OUTSTANDING CHARACTERISTICS	PROPAGATION
Lepidosirenidae (lungfishes)	Protopterus dolloi	Congo River Basin; cultured on a subsistence basis in the Congo	Young: worms, insect larvae, small crustaceans; adults: fish, snails, etc.	Due to its air breathing capabilities, can tolerate not only water devoid of dissolved oxygen, but complete dryness	Breeding in captivity not attempted; spawns during dry season in a hole dug among marginal plants; eggs and young guarded by male
Heterotidae	Heterotis niloticus	Tropical Africa; cultured in Cameroon, the Congo, Gabon, the Ivory Coast, Madagascar, and Nigeria	Omnivorous: filters plankton during dry season, but can also consume larger food items; accepts prepared food in captivity	Aerial respiration possible through use of swim bladder	Bred in captivity and distributed to farmers in a number of countries; spawns in large nests in swamps—in July in Gambia; eggs and young guarded by parents
Gymnarchidae	Gymnarchus niloticus	Upper Nile River, Lake Chad Basin, West Africa from Senegal to Niger; cultured experimentally in Nigeria	Small fish	Navigates by means of weak electrical impulses	Not bred in captivity; spawns in large floating nests at start of rainy season; eggs and young guarded by parents

TABLE 1. (Continued)

FAMILY	SPECIES	RANGE	FOOD HABITS	OTHER OUTSTANDING CHARACTERISTICS	PROPAGATION
Citharinidae	*Citharinus citharus*	Tropical central Africa; cultured experimentally in Nigeria			Males, at least, mature in ponds, and milt may be readily obtained, but none of the family have been artificially propagated; *C. citharus* spawns in the rainy season (August–September) in Gambian swamps; the other two species spawn in winter in the Congo
	Citharinus congicus	Upper Congo River Basin, Stanley Pool; cultured experimentally in the Congo	Little known; accept dry foods	Grow very rapidly; *C. citharus* reaches 0.5 kg in ponds in 1 year	
	Citharinus gibbosus	Congo and Rwanda; cultured experimentally in the Congo			
Cyprinidae (carps and minnows)	*Barbus occidentalis*	Has been experimentally cultured in Nigeria	Omnivorous?	Very delicate with respect to handling, and highly vulnerable to predation	Breeds freely in ponds
	Barbus sp.	Experimentally cultured in Rwanda	Omnivorous?		Not bred in captivity
	Labeo forskali	Nile River Basin and tributaries of Blue Nile; suggested by M. Huet for culture in the Sudan	Omnivorous; fond of algae	Resistant to low temperatures	Breeding in captivity not attempted

247

TABLE 1. (Continued)

FAMILY	SPECIES	RANGE	FOOD HABITS	OTHER OUTSTANDING CHARACTERISTICS	PROPAGATION
Bagridae	*Auchenoglanis occidentalis*	Widely distributed, from Lower Nile River to Congo and Senegal; experimentally cultured in Togo	All kinds of animal matter, including fish	Nocturnal, requires cover during day	Not bred in captivity
	Bagrus docmac	Nile River Basin; experimentally cultured in Uganda	All kinds of animal matter, including fish	Nocturnal, requires cover during day	According to Uganda Department Fisheries "breeds freely in ponds"; spawns in May in shallow water in the White Nile River; not bred in captivity
	Chrysichthys spp., including *C. nigrodigitatus*	Congo River Basin and Atlantic drainages to the north; experimentally cultured in Nigeria and Togo; suggested for culture in Cameroon and the Congo	All kinds of animal matter, including fish	Nocturnal, requires cover during day	Not bred in captivity; *C. furcatus* and *C. nigrodigitatus* spawn in Gambian rivers during the dry season
Clariidae	*Clarias* spp., including *C. lazera*	All of Africa; cultured on a subsistence basis in the Congo and Rwanda; suggested for culture in Cameroon	Any sort of animal matter; some species, at least, accept plant matter; very voracious	Very tolerant of low dissolved oxygen concentrations, due to accessory air-breathing structure; may voluntarily travel **across** land	*C. mossambicus* eggs have been artificially fertilized; otherwise not bred in captivity; spawning usually takes place in temporary waters after rains

TABLE 1. (Continued)

FAMILY	SPECIES	RANGE	FOOD HABITS	OTHER OUTSTANDING CHARACTERISTICS	PROPAGATION
Channidae (snakeheads or murrels)	Ophicephalus spp.	Africa south of the Sahara to the Congo and Kenya; cultured on a subsistence basis in Congo	Young: earthworms, tadpoles and young fish; adults: almost exclusively fish, up to and including fish as large as themselves; appetites insatiable	Extremely hardy	Several spp. bred in aquaria; lay floating eggs; males often care for eggs but do not build a bubble nest; habits of Asiatic species are better known than those of African forms
Serranidae (sea perches)	Lates niloticus (Nile perch)	Nile, Senegal and Congo River Basins; cultured in Nigeria and Uganda	Fish	Reaches huge sizes—up to 180 cm long; very sensitive to oxygen depletion and turbidity	Easily sexed and bred in small ponds; in Nigeria, 90% of pond-bred young are females
Cichlidae (cichlids)	Astatoreochromis sp.	West Africa; being considered for use in snail control in Cameroon, the Congo, Gabon and the Central African Republic	Snails and perhaps other foods		
	Haplochromis spp., including H. carlottae and H. mellandi	Most of Africa	Probably varied; H. mellandi is very fond of snails and is stocked to control them, but also consumes insect larvae		Mouth-breeders; tendency to overpopulate ponds

249

TABLE 1. (*Continued*)

FAMILY	SPECIES	RANGE	FOOD HABITS	OTHER OUTSTANDING CHARACTERISTICS	PROPAGATION
Cichlidae (cichlids)	*Hemichromis fasciatus* (five-spot cichlid)	Central West Africa; cultured in Cameroon, Rwanda, and, experimentally in the Congo and Togo	Live animals, including small fish	Aggressive toward its own and other species	Spawn on cleaned stones; parents guard eggs and young; tendency to overpopulate ponds
	Pelmatochromis robustus	East Africa	Snails, may be stocked to control snails	Often enter brackish water	May breed like *Hemichromis fasciatus*, or may be mouth-breeders
	Serranochromis angusticeps, S. macrocephalus, S. robustus, and *S. thumbergi*	Congo River Basin and East Africa from Zambia to Okavango Swamp; cultured in Zambia and, experimentally in the Congo	Young: insects; adults: live animals, including small fish	*S. robustus* grows faster and attains largest size	Mouth-breeders; easily bred and raised; adults may be cannibalistic on fry, but nevertheless may overpopulate ponds; breed August–January in Zambia; each female produces more than one brood annually
	Tylochromis lateralis	Cultured experimentally in the Congo			May be mouth-breeders

other types of polyculture are beginning to develop. Of particular interest in African pond polyculture is the extremely adaptable *Heterotis niloticus,* which can apparently occupy any of a number of ecological niches. In regions where tilapia grow poorly, as in the western Congo, culture systems based on native fishes with habits similar to those of tilapia are beginning to be developed.

ECONOMIC BASIS OF AFRICAN FISH CULTURE

Fish culture in Africa is generally carried out on a subsistence basis, with a family or village operating a pond or ponds. Although most of the early ponds were too small, with the result that most ponds were either overharvested or subject to stunting of the stock, this approach has made real contributions to the nutrition of small groups of people and should be encouraged as part of a program to provide more protein for the people of Africa. Family or village ponds have little impact on the urban African, however. Therefore a number of African governments, often with the aid of FAO and/or the former colonial governments, are experimenting with methods suitable for commercial fish culture, often incorporating more sophisticated techniques than are feasible for most subsistence farmers. At present, commercial fish culture is poorly developed in most African countries, but this may not be the case for long. A country-by-country survey of existing fish culture practices in Africa follows.

EXPERIMENTAL AND PRACTICAL FISH CULTURE IN SEVERAL AFRICAN COUNTRIES

CAMEROON

The basic species for fish culture in Cameroon are, in order of importance, the plankton-feeding *Tilapia nilotica,* the omnivorous *Heterotis niloticus,* and the predatory *Hemichromis fasciatus,* all of which can be bred in ponds with little difficulty. The government is promoting a subsistence culture system involving all three species in the eastern part of the country.

Elsewhere, commercial culture, using only *Tilapia nilotica* and *Heterotis niloticus,* is being encouraged with some effect. A number of subsistence farmers are now selling part of their crop while continuing to retain an adequate amount for home use. Fish-rice rotation, on a 9-

month–3-month cycle, is sometimes practiced. At least one commercial culturist in Cameroon has cut down costs by allowing local women to soak cassava tubers in his ponds. This process, necessary to prepare cassava for human consumption, also provides food for the fish.

Yields attained by commercial and subsistence culturists in Cameroon are not known, but in 1969 experimental culture of *Tilapia nilotica* and *Heterotis niloticus* in five ponds at the Regional Fish Culture Center at Bangui produced 980 to 3225 kg/ha, depending on the amounts of brewery waste added as food.

The Regional Fish Culture Center is also the scene of a great deal of research which may result in the expansion, diversification, and improvement of fish culture not only in Cameroon but in the Central African Republic, the Congo, and Gabon. Among the topics under study are:

1. Combined culture of tilapia and *Hemichromis fasciatus*.
2. Biology and culture of *Heterotis niloticus*.
3. Possibility of culturing the local catfishes *Clarias* spp. and *Chrysichthys* spp.
4. Pond fertilization with locally available manures.
5. Preparation of feed from locally available ingredients.
6. Use of *Astatoreochromis* sp. for bilharzia control.
7. Pond management techniques suitable for the region.
8. Economic aspects of fish culture.

DEMOCRATIC REPUBLIC OF THE CONGO

Tilapia nilotica and *Heterotis niloticus* are nearly universal in culture, but some subsistence culturists add predatory fishes such as *Clarias* spp., *Ophicephalus* spp., and *Protopterus dolloi*. Fry of the predatory species are captured in rivers. *Heterotis niloticus* fry are principally purchased from the federal hatchery at Djoumouna, while tilapia are bred by the culturists themselves. The cichlids *Haplochromis mellandi, Hemichromis fasciatus, Serranochromis robustus, Serranochromis thumbergi,* and *Tylochromis lateralis* have been experimentally cultured in the past but are not presently used to any great extent.

In the acid waters of the western Congo, where tilapia grow poorly, the native fishes *Citharinus congicus* and *Citharinus gibbosus* are being cultured experimentally.

The most commonly used method of feeding pond fish in the Congo is by soaking cassava tubers in the pond, but manioc leaves and brewery waste are also fed.

Personnel at the Djoumouna hatchery are pursuing the same research goals outlined for Cameroon.

GABON

The government maintains six fry production centers, where *Tilapia* spp. and *Heterotis niloticus* fry are raised for distribution to farmers. Commercial and subsistence farming are expanding, and average annual production is estimated at about 500 kg/ha.

IVORY COAST

The Centre Technique Forestier tropical fish culture station in Bouaké functions as a training and research center for all the former French colonies in west Africa, including Dahomey, Togo, and Upper Volta. Emphasis is on pond fertilization and culture of *Tilapia* spp. and *Heterotis niloticus*. Fertilization with 60 kg/ha of calcium superphosphate and 50 kg/ha of ammonium sulfate doubled production in experimental ponds.

The main problem in culture of *H. niloticus* in the Ivory Coast has been high mortality due to disease. Proper feeding is of course important in promoting survival as well as growth, and experiments have been carried out with three feeds, crushed cottonseed, peanut oil cake, and rice bran. Any of these feeds enabled young *H. niloticus* to reach sexual maturity up to a year before unfed fish. Of the three foods, peanut oil cake and crushed cottonseed produced more rapid growth than rice bran, and peanut oil cake was superior to the other two in terms of survival, but only rice bran was economically feasible. The maximum yield of fry attained with peanut oil cake was 2500 kg/ha.

MALAGASY REPUBLIC

Heterotis niloticus fry are produced by the Forestry and Fish Culture Station at Ivoloina for use in stocking, and at the arboretum of Menagisy for experimental polyculture with common carp (*Cyprinus carpio*), and goldfish (*Carassius auratus*).

NIGERIA

Freshwater fish culture facilities in Nigeria are in the process of reconstruction after that country's Civil War. Prior to the war, experimental fish farming was carried on in a number of places, notably the government-operated Panyam Fish Farm on the Jos Plateau in the northern part of the country and an experimental fish farm on the Island of Buguma in the Niger Delta. Encouraging preliminary results were obtained at the

Buguma farm with brackish water fishes and invertebrates, *Tilapia melanopleura,* common carp, and the indigenous catfish *Chrysichthys nigrodigitatus.* Recent experiments in brackish water ponds involve stocking of *C. nigrodigitatus* in ponds which fry of tilapia and mullet (*Mugil* spp.) enter with the tide.

The original goal of the Panyam farm was to raise *Tilapia* spp., but results were poor and, starting in 1954, emphasis was placed on common carp, imported from Europe. Tilapia proved impossible to eradicate, and since they preyed on carp eggs and larvae and competed with fingerlings for food, they came to be regarded as pests. Experimental culture of native predatory fishes was therefore initiated.

The indigenous fish fauna of the small streams in the immediate vicinity of the Panyam Fish Farm is depauperate, consisting of four cyprinid species too small for culture as food fish. The piscivorous species *Gymnarchus niloticus* and *Lates niloticus* (Nile perch), both endemic to other parts of Nigeria, were therefore introduced. The Nile perch presently appears the more suitable for use in pond culture. *Gymnarchus niloticus* functions well as a predator and reaches marketable size rapidly, but it has yet to be bred in captivity, and collection of fry from rivers has proven uneconomic. Nile perch seem ideal on almost all counts. The only apparent limitation on their use is their low tolerance for deoxygenated or turbid water. An adequate stock can be maintained by simply placing one male and two females in a small breeding pond and periodically harvesting the 15- to 30-g fingerlings from the series of spawnings which ensues. A population of two to four Nile perch per hectare is adequate to control tilapia in carp ponds. Care must be taken that the Nile perch are smaller than the carp fingerlings, as they appear to prefer carp to tilapia. In ponds with large numbers of tilapia, Nile perch reach weights of 0.5 to 0.6 kg in 6 months.

Experiments were also conducted with nonpiscivorous native fishes. Best results were obtained with *Heterotis niloticus. Citharinus citharus* grew well, reaching 0.5 kg in a year, but as yet cannot be bred in captivity. *Barbus occidentalis* bred readily in ponds but proved to be so delicate to handle that intentional culture was discontinued, though small populations persisted in the carp ponds.

RWANDA

Common carp, *Tilapia* spp., and *Clarias* spp. have been successfully grown together at the government station at Butare. There is at present little if any commercial fish culture in Rwanda, but indications are that there is a ready market for cultured fish.

TOGO

The United States Peace Corps has been instrumental in the development and improvement of fish culture in Togo. At the Ná Fish Culture Station at Sokode, good results have already been experienced with supplemental feeding of *Tilapia* spp., stocked together with the piscivorous *Hemichromis fasciatus,* on millet, brewery waste, mill sweepings, cottonseed meal, spoiled corn meal, and manioc and ignam peels. The combination of polyculture and supplementary feeding has resulted in annual yields of 3000 to 10,000 kg/ha with food conversion ratios of 4 to 8:1. Attempts are being made to culture the native catfishes *Auchenoglanis occidentalis* and *Chrysichthys* sp.

UGANDA

Subsistence fish culture, mostly of *Tilapia* spp., is widespread and the Fisheries Department is trying to promote commercial culture as well. A demonstration farm has been set up for that purpose. Nile perch are spawned at the demonstration farm and have been found effective in controlling tilapia populations. More recently, *Bagrus docmac* has been found as effective a predator as Nile perch. Unlike most African catfishes, it breeds freely in ponds. *B. docmac* has exhibited quite satisfactory growth when stocked at different ratios with *Tilapia nilotica* or *Tilapia zillii.*

ZAMBIA

Pure commercial culture of *Tilapia* spp. is practiced to a small extent, but subsistence polyculture is more important. Generally two or three species from among *Tilapia andersoni, Tilapia macrochir, Tilapia melanopleura,* and *Tilapia mossambica* are stocked together with *Haplochromis mellandi. H. mellandi,* which feeds largely on mollusks, is stocked not only for its contribution to fish production, but for reasons of health. The human disease bilharzia, caused by a parasitic worm whose intermediate host may be any one of a variety of snails, is a serious problem in Zambia, and *H. mellandi* is an effective biological control against snails. *Haplochromis carlottae* and *Pelmatochromis robustus* are less commonly stocked for this purpose.

Procedures recommended by FAO biologists for polyculture of *Tilapia* spp. and *Haplochromis mellandi* in Zambia are as follows:

1. Drain the pond and treat the bottom with 1500 kg/ha of agricultural lime, then fill with water.

2. Stock the pond with a variety of sizes of fish at 200 kg/ha, using about the same amount of each species stocked. A piscivorous fish may also be added, in small numbers, to crop excess tilapia fry. The American largemouth bass (*Micropterus salmoides*) was the first species stocked in this capacity, but the native *Serranochromis* spp., particularly *S. robustus,* are now preferred.

3. Thereafter spread 150 kg/ha of agricultural lime over the surface every month. One week after each liming add 50 kg/ha of double superphosphate or 100 kg/ha of single superphosphate or basic slag.

4. Feed the fish once a week. Most Zambian fish culturists use plant foods, including grass, napier fodder, and chopped leaves of banana, cassava, papaw, sweet potato, cabbage, lettuce, spinach, carrots, or kale. Conversion ratios of these feeds are as high as 48:1 and, when possible, grain foods, including maize, rice, and brewery wastes, should be used instead. Household scrapings may also be fed. If feeding is not feasible, further fertilization with 1000 kg/ha of pig manure or 150 kg/ha of poultry manure per week is advisable.

The ideal size pond for subsistence culture using these methods is considered to be 0.04 ha. With good management, yields of 3000 to 6000 kg/ha may be obtained from such ponds, but in practice 1000 to 1500 kg/ha are more often produced.

Monoculture of fry for distribution to fish farmers is also practiced in Zambia. The fishes bred are *Tilapia* spp., *Haplochromis* spp., and *Serranochromis* spp. *Tilapia* spp. and *Haplochromis* spp. are spawned in 0.08- to 0.20-ha ponds well fertilized with phosphate. Adult fish weighing 0.17 to 0.45 kg, preferably less than 2 years old, are stocked at 100 to 125 pairs/ha and allowed to spawn naturally. The young may be removed from the smaller ponds with nets and stocked in rearing ponds. Otherwise, they are left in the spawning pond until after the breeding season. *Serranochromis* spp. are spawned in the same way but, unlike their relatives, are prone to cannibalize their fry once the period of parental care is over. Protection may be afforded by dividing the pond with wire mesh fine enough that the young but not the parent fish can pass, or by encouraging dense growths of vegetation.

OTHER COUNTRIES

It will be noticed that no mention has been made of countries south of the Congo in west Africa and Zambia in east Africa. To date, fish culture in these countries has involved almost exclusively tilapia and exotic species. The same applies to the Sudan, the Mediterranean countries, and some of the other nations not mentioned in this account. Other countries have been omitted because no information was available or because fish culture, of any sort, has barely been started.

PROSPECTUS

Most of the fishes discussed are lowland, warm water species. Fish culture could make an important contribution in the cool highlands of Africa, as well as the tropical lowlands, but outside of culture of trout for sport fishing in South Africa and Togo, virtually nothing has been done in these regions. Perhaps some of the species discussed would be suitable for upland culture, but there are probably other species, native to the highlands, that would reward the inquisitive fish culturist.

As mentioned, with the exception of *Tilapia* spp., most of the African fishes currently used in practical and experimental culture are high on the food chain. It is axiomatic that, to obtain the highest per hectare yields, pond fish culture should be based on plankton feeders, algae feeders and/or omnivores. Recent successes in a number of countries with *Heterotis niloticus* will hopefully remind fish culturists that Africa is no different from the other continents in harboring an abundance of such fishes. Certainly fishes of the family Citharinidae, whose culture has been sporadically attempted in the Congo and Nigeria, should be more thoroughly investigated. The same applies to the many African species belonging to the Cyprinidae and Characidae, which have been virtually ignored by fish culturists.

It is impossible to foresee the directions fish culture will take in Africa, but it is virtually certain than an increasing number and diversity of indigenous species will be cultured.

REFERENCES

Breder, C. M., and D. E. Rosen. 1966. Modes of reproduction in fishes. Natural History Press, Garden City, N.Y. 941 pp.

Maar, A., M. A. E. Mortimer, and I. V. Der Lingen. 1966. Fish culture in Central East Africa. FAO. 158 pp.

Nigerian Federal Fishery Service. 1963. Carp farming on the Jos Plateau. Bull. l'Ifan. XXV, série A.

Reizer, C. 1966. Influence de la distribution de nourriture artificielle sur la mortalité des jeunes alevins, la croissance pre-adulte et la maturité sexuelle d'*Heterotis niloticus*. FAO World Symposium on Warm Water Pond Fish Culture. FR: III/E-1.

Sterba, G. 1966. Fresh water fishes of the world. Viking Press, New York.

13

Culture of Native Freshwater Fishes of Latin America

Despite the severe protein problems of the peoples of Latin America, freshwater fish culture is almost unknown in that part of the world. Most of the attempts that have been made have involved exotic fishes, principally the common carp (*Cyprinus carpio*) and *Tilapia* spp. However, South America, and to a lesser extent Central America, support a diverse fish fauna, among which are certainly some species suitable for culture. The purpose of this treatise is to discuss the small progress made to date

258

in the culture of native Latin American freshwater fishes, and to suggest possible future avenues of investigation. Culture of carp and tilapia in Latin America will not be treated here, but are discussed in Chapters 2 and 18, respectively.

PRESENT PRACTICES

COMMERCIAL CULTURE OF PEJERREY IN ARGENTINA

The only successful commercial fish culture enterprise based on a fish native to Latin America exists in the lowlands of Argentina. There the pejerrey (*Odonthestes basilichthys*), a member of the family Atherinidae, is cultured by methods reportedly similar to those used in trout culture. Pejerrey are marketed as a luxury food and have virtually no effect on the protein supply in the area.

In 1967, pejerrey were introduced to Israel, as part of an attempt to diversify that country's production of cultured fish. The species has adapted well to local conditions and has been spawned in small experimental ponds at the Fish Culture Research Station at Dor, but commercial culture has barely begun. Pejerrey have also been introduced to Chile and Japan, where they are artificially propagated and stocked in lakes.

EXPERIMENTAL CULTURE OF *Chirostoma* SPP. IN MEXICO

The Mexican Department of Fisheries has recently begun experimental culture of two atherinids of the genus *Chirostoma*, locally known as "whitefish." One of these species, *C. estor*, reaches a length of 35 cm, can be bred in ponds, and is highly favored as a food fish. Thus the future of culture of *C. estor* and similar species in Mexico and elsewhere in Latin America looks promising.

HATCHERY PROPAGATION OF NATIVE FISHES IN SOUTH AMERICA

Of all the Latin American countries, Brazil has been by far the most active in fish culture. The first successful attempts to artificially propagate fish with the use of pituitary hormones were carried out in Brazil in the 1930s, but application of the technique in Latin America has been scant and unimaginative. The Brazilian federal fish culture stations confine themselves largely to induced spawning of various fishes for stocking, as alevins, in reservoirs, in the northeast part of the country. Whether or not this extremely low-intensity approach to fish culture materially augments the supply of fish available to the populace is open to debate. The fishes so spawned and stocked are mostly members of the family Characidae, including species of *Curimatus, Leporinus, Prochilodus,* and *Triportheus.* At least one intergeneric hybrid of *Leporinus* and *Prochilodus* has been produced with the aid of pituitary injections, but for what purpose is not known.

The characins propagated in Brazil are herbivores or low-order carnivores, but some more or less piscivorous fishes are also cultured, particularly doradid catfish of the genus *Trachycorystes.* In recent years, three species of carnivorous cichlids—*Astronotus ocellatus* (the "Oscar" of aquarists), *Cichla ocellaris,* and *Cichla temensis,* plus *Plagioscion surinamensis* and *Plagioscion squamosissimus,* two fresh water representatives of the predominantly salt water family Sciaenidae—have been bred, without hormone treatment, at the government stations. These fishes are stocked not only in reservoirs but also in tilapia ponds to control excess reproduction. *Cichla ocellaris* has been employed in this manner for some time in Colombia, but the practice may be deemphasized as use of sterile strains of *Tilapia mossambica* becomes more prevalent.

The Peruvian government maintains three fish culture stations devoted to spawning of the Pirarucu (*Arapaima gigas*), one of the world's largest freshwater fishes. Alevins of *A. gigas* are released into rivers of the upper Amazon River basin. The stocking program is similar to that of the Brazilian government, both in intent and in the doubtfulness of its efficacy.

POTENTIAL FOR AQUACULTURE OF LATIN AMERICAN FISHES

The two dominant families of fishes in the freshwaters of Latin America are the Characidae and the Cichlidae. Representatives of both families are found in all but the extreme northern and southern portions of the region, and their culture has been the subject of much speculation and some effort among local fishery biologists:

CHARACIDAE

The 1350 or so known species of Characidae include perhaps the greatest diversity of species of any family of fishes, and the great majority of these species are native to Latin America. One can easily envision an analog of southeast Asian pond polyculture, with a community of fishes comprising plankton feeders, benthos feeders, herbivores, and predators—all native South American characins. However, South American biologists complain that they are unable to spawn these fishes. This seems strange, since many of the smaller characins are popular among North American and European aquarists and, while a few are challenging to breed in captivity, many are regarded as being easy to spawn, yielding readily to such simple manipulations as a partial change of water or an increase in temperature. Even if it turns out that all the edible species of characins present insurmountable difficulties to culturists seeking to spawn them "naturally," the technology is at hand to induce spawning. It is difficult to see why Brazilian fish culturists dismiss artificially induced spawning as being inapplicable to practical fish culture when this technique, pioneered in Brazil, is being successfully and economically applied by fellow culturists throughout the world.

CICHLIDAE

The Cichlidae are only slightly less diverse than the Characidae and present no problems to the culturist seeking to breed them. Indeed, many cichlids are too prolific, and tend to overpopulate ponds until they become stunted. The only careful studies of the aquacultural potential of native American cichlids were carried out in Guatemala in the early 1960s. It was concluded that the eight species of *Cichlasoma* studied grew more slowly than imported African cichlids of the genus *Tilapia,* and were generally inferior for use in pond culture. The Guatemalan experiments by no means exhausted the possibilities for culture of American cichlids. It may even be that some of the eight species dismissed are suitable for culture. After all, fairly efficient methods of culturing *Tilapia* spp. had been worked out prior to the start of the experiments, whereas the researchers were forced to start from scratch with *Cichlasoma* spp. Certainly the American Cichlidae deserve further study by fish culturists. At present, Costa Rican biologists are planning experimental culture of *Cichlasoma* spp. native to that country but elsewhere the native Cichlids are completely neglected.

OTHER FAMILIES

Another large group of fishes which contribute to freshwater fisheries in Latin America comprises the several families of catfishes (suborder Siluroidei). As noted, catfishes of the genus *Trachycorystes* (family Doradidae) are propagated and stocked in Brazil, but Latin American fish culturists have inexplicably ignored not only other doradids but members of the widely distributed Pimelodidae and the Bunocephalidae of the upper Amazon River basin.

Other Latin American freshwater fishes with aquacultural potential include members of the widely distributed families Eleotridae (sleeper gobies), Synbranchidae (synbranchoid eels), Rhamphichthyidae (knife fishes), and Gymnotidae (gymnotid eels), plus *Lepidosiren paradoxa* of central South America, the only lungfish native to the western hemisphere.

PROSPECTUS

It is clear that the freshwater fishes of Latin America could make a much greater contribution to the nutrition of the inhabitants of the region than they presently do. It is equally clear that with the exception of the work mentioned here and incipient polyculture projects by the United States Peace Corps and the New Alchemy Institute in Costa Rica, nothing is being done to bring this about. Perhaps eventually culture of indigenous fishes will assume something approaching its potential importance, but no major advances are to be expected in the immediate future.

REFERENCES

DE MENEZES, R. S. 1966. Cría y Selección de los Peces Cultivados en Aguas Templadas en América del Sur y Central. FAO World Symposium on Warm Water Pond Fish Culture. FR: IV/R-5.

LIN, S. Y. 1963. El Fomento de la Pesca Continental Informe al Gobierno de Guatemala. FAO *Informe* (1719):16.

14

Experimental Fish Culture
in Australia

INTRODUCTION AND HISTORY

There is no commercial fish culture in Australia apart from some salt-water trout farms in Tasmania, and little has been done by way of experimental fish culture. One might suppose that brackish water pond culture similar to that practiced in Indonesia would be attempted along the northeast coast, but it has not been. To date Australian experimental fish culture has concerned itself almost entirely with freshwater fish.

A number of North American and European freshwater fish have been introduced to Australia. Among the Cyprinidae, the crucian carp (*Carassius carassius*), the goldfish (*Carassius auratus*), and the tench (*Tinca tinca*) have become widely established, while somewhat less success was achieved with acclimatization of the common carp (*Cyprinus carpio*). None of these species is highly valued as food by Australians, although the many European immigrants to the country purchase substantial amounts in the markets. Despite the prevalence of these fishes, particularly the common carp, in European and Asian fish culture, no attempts have been made to culture them in Australia.

The European perch (*Perca fluviatilis*) is better liked by Australians, but since it is not only high on the food chain but is provided at low cost by conventional fisheries there is little incentive for its culture.

Rainbow trout (*Salmo gairdneri*) and brown trout (*Salmo trutta*) were introduced to provide sport for Australian anglers, and they have been more or less successfully acclimatized in much of the southern part of the country. In most regions, hatcheries are employed to maintain the stocks, but to date there has been no freshwater commercial application of hatchery techniques.

In the second half of the nineteenth century, when most of the introductions of fish were made, there was great enthusiasm in Australia for such projects, and "Fish Acclimatization Societies" were formed in some districts. In this century, Australians have become more appreciative of the uniqueness of their native fish fauna, and efforts are being made to preserve it. All of the introduced species have been implicated, probably with some degree of justification, in the decline of populations of native fishes. Thus, while sport fishermen will surely see to it that the trout hatcheries are maintained, current emphasis in fishery management and fish culture is on native species.

EXPERIMENTAL PROPAGATION OF NATIVE FISHES

Experimental propagation of the Australian freshwater catfish (*Tandanus tandanus*) has been described in Chapter 7. Other freshwater fishes which have been considered for culture include the Murray cod (*Maccullochella macquariensis*), the callop (*Plectroplites ambiguus*), and the silver perch (*Bidyanus bidyanus*), all of which, along with *Tandanus tandanus*, have been bred in ponds at the Inland Fisheries Research Institute, Narrandera, New South Wales.

CALLOP

The most highly prized of these species are the Murray cod and the callop. The appellation "cod" is a misnomer, for both are freshwater members of the largely marine family of sea basses (*Serranidae*). Of the two, the callop appears more promising for pond culture due to its somewhat smaller size and less piscivorous feeding habits.

Artificial propagation of callop dates back to 1916. Early experimental culturists did not attempt pond breeding, but stripped and artificially fertilized eggs, using the "wet" method formerly popular among trout culturists (see p. 400). Hatching requires a little over 48 hours at temperatures of 15 to 23°C.

Essentially natural spawning has been carried out at Narrandera, using female fish which failed to respond to stripping. (Sex is indicated by distention of the ripe female's abdomen and inflammation of the cloacal area.) While the callop is mainly a river fish, it was found to ripen normally in ponds with no current. Two factors are necessary to induce spawning in ponds; an increase in temperature to at least 23.6°C, and a concurrent rise of the water level by 15 cm or more, with flooding of dry ground. The latter condition produces abundant plankton blooms to nourish the young. It is not necessary that the fish be present when the water level is raised; ripe fish stocked in a recently filled pond will respond by spawning. Spawning may be retarded for several months and then induced by meeting these two conditions, but if ripe fish are held at temperatures above 23.6°C and the water level remains constant, the eggs will eventually be resorbed.

It was found possible to induce spawning in 1-m-deep ponds as small as 36 m² in surface area, but most of the experiments were carried out in 0.1-ha ponds, 2.4 m deep throughout most of their extent, and 1.2 m deep along one side. Ponds were stocked with 1 to 20 pairs of callop. Greater numbers of young were produced in the more heavily stocked ponds, but after 5 months of rearing, without supplemental feeding, the juvenile fish were more numerous, larger, and in better condition in ponds which received only 1 or 2 pairs.

Egg retention was a common problem; the only females which shed all their ova were some of those over 3 kg in weight. It is believed that egg retention may be associated with the extent of flooding; slight rises in water level resulting in partial ovulation, whereas larger floods produce complete ovulation.

Spawning of callop in nature is usually associated with high turbidity, but this was not found to be an important condition. There was no difference in spawning success between ponds with Secchi disk readings ranging from 12 to 240 cm. Spawning invariably occurs at night.

Male callop in the Murrumbidgee River, the source of the stock used at Narrandera, mature sexually at about 33 cm long and 0.5 kg in weight, females at about 43 cm and 1.3 kg. The maximum weight obtained may be over 25 kg, but most of those taken by fishermen weigh 1.3 to 1.8 kg.

Callop would apparently prove adaptable to pond culture, as specimens held in deep and shallow ponds which did not fulfill the conditions described above thrived but did not reproduce. In nature, callop feed mainly on invertebrates, but there are indications that they would respond favorably to supplementary feeding.

MURRAY COD

Culture of the Murray cod (*Maccullochella macquariensis*) dates back somewhat further than callop culture, artificial propagation of the former species having been carried out at least as early as 1906. The details of sexing and artificial spawning of Murray cod do not differ from those given for callop. Hatching, which has been experimentally accomplished on gauze trays placed in floating cages anchored in a river, requires 8 to 9 days at 20°C. The larvae absorb the yolk sac in 4 to 7 days, after which time they will accept finely ground fish, fish eggs, or shellfish.

Although the Murray cod is a large fish, reportedly reaching a maximum weight of over 65 kg, mature specimens weighing as little as 2 kg may be obtained. It was found possible to spawn 2.3–6.8-kg specimens in a 0.6-ha pond at Narrandera, and natural spawning has on occasion occurred in smaller ponds. As with the callop, a rise in the water level, with flooding of dry ground, seems necessary to trigger reproductive behavior. It appears that a very slight change in water level may be sufficient, but in other respects the spawning requirements of Murray cod are more rigorous than those of callop. Of particular importance is temperature; flooding should coincide with the attainment of a water temperature of 20°C, but if 21° is exceeded, the eggs will be resorbed. A long exposure to warm water is not necessary to damage the ova; it is thought that even handling ripe fish, for example, in sexing, may constitute a thermal stimulus sufficient to effect resorption. Murray cod have been described as building nests, but it is likely that the "Murray cod nests" which have been observed were in fact old nests of *Tandanus tandanus*. At Narrandera, each female attached her eggs to the inside of a single fibro-cement pipe, 20 cm in diameter.

SILVER PERCH

The silver perch (*Bidyanus bidyanus*) is a representative of another predominantly salt water family, the Theraponidae. Its habits and distribution closely resemble those of the callop, but it is a smaller fish, reaching a maximum weight of less than 3 kg, and is less important in fisheries. Artificial fertilization of silver perch eggs has not been attempted, but it was unintentionally spawned at Narrandera when water was added to a pond to compensate for evaporation. Subsequent experiments suggest that the spawning requirements are similar to those of callop, but that the minimum temperature for spawning silver perch is 23.3°C, and that this temperature need be exceeded only to a depth of 90 cm. Unlike callop, silver perch may spawn in the late afternoon as well as at night.

POTENTIAL NEW SPECIES FOR CULTURE

A number of other native Australian freshwater fishes might be considered for culture. Foremost among these is the Australian bass (*Percalates colonorum*), a Serranid which occupies a niche similar to that of the black basses (*Micropterus* spp.) in North America. In nature, *Percalates colonorum* usually migrates downstream to estuarine waters to spawn, but, when denied access to salt water, it will spawn in freshwater.

Other possible pond fishes include yet another Serranid, the Macquarie perch (*Macquaria australasica*); a smaller congener of the murray cod called the trout cod (*Maccullochella mitchelli*); and the river blackfish (*Gadopsis marmoratus*), a cold water species which has fared poorly in competition with trout.

PROSPECTUS AND SPECIAL PROBLEMS

All of the species discussed are normally spring spawners. The crucial stimulus, however, is in most cases not an annual rhythm, but a rise in water level and temperature. In fact, the silver perch, and possibly some of the other species, will on occasion spawn in the fall if these conditions occur. Even Australian stocks of the European perch, which, in its native habitat, spawns in the early spring, regardless of water level, have been shown to respond positively to the addition of water to ponds. It appears that, under Australian conditions, flooding of dry ground has a more or less universal favorable effect on reproduction of freshwater fishes. Certainly manipulation of the water level should be attempted as a possible simple means of inducing or retarding spawning in any Australian fish species considered for culture.

The same factor which has made it necessary for Australian fishes to adapt to sudden changes in water level, namely irregular rainfall, creates problems for would-be fish culturists. Some areas of the country are subject to more or less annual droughts, while others are regularly flooded to the extent that entire river valleys become lakes. It may be argued by some that such hydrological irregularities render fish culture unfeasible. One need only consider the importance of fish culture in, on the one hand, Israel, and on the other, Southeast Asia, to see that this defeatist attitude is not necessarily justified. If freshwater fish culture techniques are eventually adapted to Australian conditions and fishes, the results could be most beneficial to the Australian diet, particularly in inland areas.

REFERENCES

LAKE, J. S. 1967. Rearing experiments with 5 species of Australian fresh water fishes. Aust. J. Mar. Fresh Water Res. 18:137–152.

ROUGHLEY, T. C. 1961. Fish and fisheries of Australia. Angus and Robertson, London. 343 pp.

15
Frog Culture

"FROG FARMS" IN THE UNITED STATES

The status of commercial frog culture is nebulous. On the one hand, United States government publications repeatedly advise prospective frog farmers that intensive commercial culture of frogs as food animals has yet to be achieved. On the other hand, one continually encounters individuals who claim to be operating profitable "frog farms." Such establishments generally turn out to be slightly modified shallow ponds or swamps, where frogs are harvested in much the same manner as wild frogs. In some cases, husbandry is limited to erecting a fence to retain the frogs and exclude predators, and the chief market is other would-be frog farmers. Other such farms, however, are more sophisticated and sell to restaurants and other food outlets. Intensive indoor culture methods have also been developed for several species, but at present they are applied only to the production of laboratory frogs, and opinions differ as to whether modifications of these methods could economically be applied to the culture of frogs for human consumption.

SOURCE OF STOCK

Most attempts at frog culture have been made in the United States, where frogs are among the most expensive luxury foods. Numerous species are harvested from the wild and are generally not discriminated among by buyers or consumers except on the basis of size. The largest and the most widely used in attempts at culture is the bullfrog (*Rana catesbiana*). Bullfrogs lay their eggs in shallow standing water during April in the South and May or June in the North. Hatching requires 4 days to 3 weeks, depending on temperature. The aquatic larvae, generally known as tadpoles, feed chiefly on benthic algae. In 5 months to 2 years they metamorphose into the semiaquatic and exclusively carnivorous adults, which may reach lengths of up to 20 cm.

Prospective frog farmers may obtain bullfrog stock from commercial sources, or eggs or tadpoles may be taken from the wild. Bullfrog eggs (and those of a few other large frogs) may be distinguished from those of small, undesirable species by the size of the floating egg mass, which covers about 0.5m^2. Size is also the distinguishing characteristic of bullfrog tadpoles, which are much larger than most other tadpoles of the same age.

STOCKING

Whichever life form is taken, they should be distributed around the perimeter of the body of water to be stocked. Although this method may be suitable for new operations, the culturist should endeavor to breed his own stock as soon as possible. In addition to the taxonomic uncertainty in collecting wild stock, wild tadpoles often harbor pathogenic organisms and suffer high mortalities in the late stages of metamorphosis.

All known attempts at frog farming in the United States have failed to incorporate any form of control over breeding but have allowed the frogs to spawn as they would in the wild. The resulting tadpoles have sometimes been reared separately from adult frogs, and adults have been segregated by size in the belief that extensive cannibalism would otherwise result. However, some researchers have found cannibalism rare or nonexistent among well-fed bullfrogs of all ages.

FEEDING

Feeding is the key not only to averting cannibalism, but to health and satisfactory growth of frogs. Frogs and tadpoles maintained outdoors will obtain some food naturally, but at commercially feasible population densities, the natural food supply must be supplemented.

Tadpoles, while primarily herbivorous in nature, will accept any soft animal or vegetable matter. Among the feeds which have been employed are boiled potatoes, meat scraps, and chicken viscera. An especially attractive idea is the use of the viscera and other scraps from butchered frogs as tadpole feed.

Once metamorphosis to the frog stage is complete, feeding becomes much more difficult. Adult frogs feed exclusively on moving animals. Japanese researchers have reportedly been able to induce frogs to ingest stationary silkworm pupae by mixing them with live nightcrawlers, then removing the nightcrawlers when the frogs have become accustomed to the pupae. Another Japanese method of eliminating dependence on live food involves the use of wooden trays containing about 12 mm of water. Dead silkworms or other food items are placed in the trays, which are anchored in shallow water near shore. A small motor keeps the trays oscillating slowly, so that the silkworms roll back and forth.

So far as is known, these Japanese techniques have not been tried with bullfrogs in the United States. Most American frog culturists have relied on stocking or attracting live food animals. One farmer, located near the ocean in Florida, stocked his ponds with marine fiddler crabs (*Uca* spp.), which are abundant on Florida beaches. Smaller species of frogs and their tadpoles may be eaten by bullfrogs and are sometimes stocked as food. Aquatic plants may be encouraged as shelter and food for tadpoles, crayfish, and other potential food animals. Terrestrial flowering plants serve a similar function by attracting flying insects, which are hunted by frogs on shore. More insects may be attracted by illuminating the shore of the pond at night with 100- to 200-W clear lamps. No combination of these methods has yet proven entirely satisfactory, and the difficulty of supplying adequate amounts of food remains the principal obstacle to successful bullfrog farming.

DESIGN OF PONDS

Another problem faced, and largely overcome, by early experimental frog culturists is territoriality. A large bullfrog may require about 7.5 m of shoreline as a feeding territory, a trait which severely limits the number of bullfrogs in most natural environments. Natural bodies of water, however, usually include large expanses of deep, open water which are of little use to bullfrogs during the growing season. Culturists are therefore able to maintain frogs at population densities much higher than those usually observed in nature by reducing the amount of open water in their ponds, while increasing the length and irregularity of the shoreline through construction of islands and peninsulas extending into the centers of the ponds. If natural ponds suitable for such modification are not available, the shoreline of artificial ponds may be maximized by constructing them as a series of narrow trenches. Such trenches should run north and south insofar as possible, so that vegetation on the banks will serve as shade for the frogs.

Whatever form of pond is used, a small portion of it should be deep enough to protect frogs and tadpoles from extreme heat or cold. In the South, 30 to 45 cm is adequate, but in the North, deeper water may be necessary to insure the survival of hibernating frogs. In any location, a large portion of the pond should be only 5 to 15 cm deep to facilitate the feeding behavior of frogs and tadpoles.

Predators of frogs and tadpoles are numerous, and some tadpole predators, such as large aquatic insect larvae, are virtually impossible to exclude from frog ponds. Some terrestrial predators may be kept out by enclosing ponds with a small-mesh wire fence about 1 m high, sloping outward at an angle of 35°. Birds are more difficult to exclude, but a wire net stretched above the shallows may be partially effective. No predator control method developed to date is 100% effective, and the culturist should allow for some loss.

HARVESTING

Harvesting of bullfrogs from ponds as just described is extremely inefficient. The methods employed are the same used in hunting wild frogs —fishing with hook and line, spearing, and hand capture. Live bait is occasionally used in fishing for frogs, but a more common practice is to dangle a crude lure, made of red cloth or yarn, in front of the frog to simulate a hovering insect. Spearing and hand capture are done at night with the aid of a bright spotlight, which dazes and immobilizes the frogs. Clearly, these methods must be supplanted by mass harvesting techniques if successful commercial pond culture of frogs is to become a reality.

Data on the commercial status of frog culture in the United States are few and sometimes contradictory. Most of the reports of success have come from Florida and Louisiana, where the growing season is very long, if not year-round, but commercial suppliers of frogs are located as far north as Vermont and Wisconsin.

INDOOR CULTURE OF FROGS

SPECIES CULTURED

A newer approach to frog farming, and one which largely eliminates climatic considerations, is indoor culture. Practical methods of indoor propagation and rearing of frogs for use as laboratory animals were first worked out by T. Kawamura of the University of Hiroshima, Japan. Kawamura's methods were subsequently adapted for use with American species by George W. Nace of the Department of Zoology, University of Michigan. Nace's methods have been the model for several other institutions which have undertaken to produce their own experimental animals, but to date no one has attempted indoor commercial culture of frogs as food animals. However, the species routinely cultured at Michigan include the green frog (*Rana clamitans*), the pickerel frog (*Rana palustris*), and the leopard frog (*Rana pipiens*), all commonly marketed as food, and there is no reason to believe that any of the larger American frogs could not be similarly cultured. Thus the possibility of indoor commercial production of frogs for human consumption cannot be ignored.

WATER SUPPLY

As might be expected in an indoor culture system, pains had to be taken to provide a suitable water supply for the University of Michigan Amphibian Facility. Based on the experience of Nace and his associates, there are four requirements for maintenance of a self-perpetuating frog colony:

1. The water supply must be abundant at all times; the Michigan facility uses up to 90 liters/min.

2. Line pressure should be adequate to permit individual fine control of flow through each container.

3. The pH should be slightly acid.
4. The water temperature must be constant at 20 to 22°C.

The source of water at Michigan is the city of Ann Arbor municipal system, which is unsatisfactory on the last three counts. Booster pumps, pressure regulating valves, and carefully designed plumbing have been installed to compensate for irregular main pressure, pH is maintained at 6 to 7 by the monitored introduction of acetic acid, and the temperature is regulated with the aid of heaters and industrial capacity mixing valves. A commercial culturist would of course seek to locate so as to avoid these expenditures if possible.

Ann Arbor city water is chlorinated, which would seem to present yet another problem, as it does for tadpoles. It has proved necessary to install an industrial activated charcoal dechlorinator to provide safe water for tadpoles. However, while chlorine is toxic to tadpoles at concentrations well below the 0.6 ppm found in Ann Arbor city water, adult frogs can stand chlorination up to 4 ppm. In fact, mild chlorination serves as a prophylactic measure against bacterial diseases. Adults are thus kept in water provided by lines which bypass the dechlorinator.

TADPOLE BOTTLES

Differences in water quality requirements, along with other aspects of the life history of frogs, dictate that separate housing facilities be provided for each life stage. Fertilized eggs and very young tadpoles are held at low density in shallow enamel pans. Dead embryos are removed regularly, and the water is changed at least every third day. When vigorous swimming commences, the tadpoles are transferred to special tadpole bottles (Plate 1).

A tadpole bottle is constructed by removing the bottom of a conventional 1-gal glass or plastic bottle, stoppering and inverting it. Water is supplied from below through a 10-mm glass tube and removed through a 15-mm plastic siphon tube extending to the stopper. In this manner a constant flow is maintained, and the oldest, "stale" water is removed. Flow through each bottle is adjusted so that the water is exchanged about three times daily, yet dangerous currents are not created. A circular stainless steel screen inserted over the inflow and siphon tubes prevents tadpoles from becoming trapped in the neck of the bottle. Tadpole bottles, each containing 25 to 75 tadpoles, depending on size and species, are held in racks provided with waste troughs to dispose of the water which is siphoned out.

In nature, certain individuals of each batch of tadpoles grow most

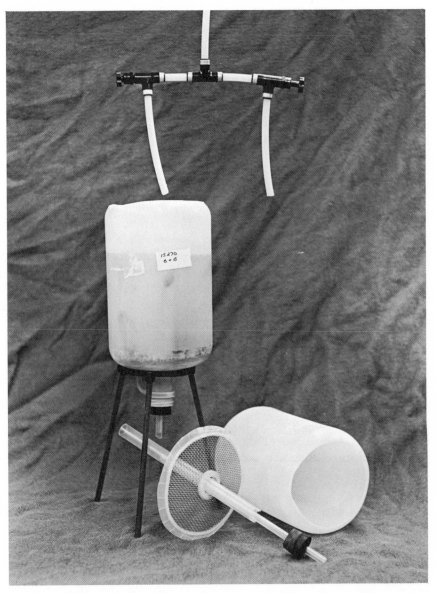

PLATE 1. Tadpole bottle used for frog culture at University of Michigan. (Courtesy University of Michigan News Service.)

rapidly at first. These large individuals release a growth-inhibiting substance which acts upon the smaller tadpoles. The result is that tadpoles metamorphose and emerge as frogs in waves, rather than all at once. While in nature this helps avert mass mortality due to predation or unfavorable conditions in the terrestrial environment, commercial culturists would not be likely to find this arrangement advantageous. Fortunately, the use of continuously flowing water, as described here, prevents the accumulation of the growth-inhibiting substance and results in relatively uniform growth and more or less simultaneous metamorphosis.

CAGES FOR METAMORPHOSED AND ADULT FROGS

Metamorphosis is considered to have occurred when the forelimbs erupt, at which time the young frogs are transferred to rectangular plastic containers lined with rubber mats and containing a few pieces of broken clay flower pots to serve as cover. These cages are placed on racks at an incline, so that one end contains water while the other is dry. The containers are cleaned and the water is changed every third day.

Transformation into the frog form need not be complete for the young to be treated as adults. Rather they are transferred to adult containers as soon as their vigor is assured. Housing for adult frogs should have the following characteristics:

1. Both aquatic and terrestrial areas.
2. Flowing water and facilities to permit flushing.
3. Construction design which permits ready access and allows cleaning with minimal handling of the frogs.
4. Closures which are strong enough to prevent the escape of frogs, fine enough to retain insects presented as food, and open enough to provide adequate ventilation.

These conditions are met by the rather elaborate two-story plastic cages used at Michigan. The opaque bottom section of each cage contains not only water but ledges so that the frogs may remain dry but secluded. Nested into the bottom sector is a dry, transparent cage with a rubber mat and clay shelters like those used in containers for newly metamorphosed frogs. A hole in the bottom of the upper section permits the frogs to move from level to level. Water is supplied through a tube in the rear of the bottom compartment and leaves via a 25-mm-high overflow tube. Tops are made of stainless steel wire cloth and incorporate an access port.

Adult cages (Plate 2), which are either 48.3 cm × 26.7 cm × 16.5 cm deep or 50.8 cm × 40.6 cm × 21.6 cm deep, are mounted on racks so

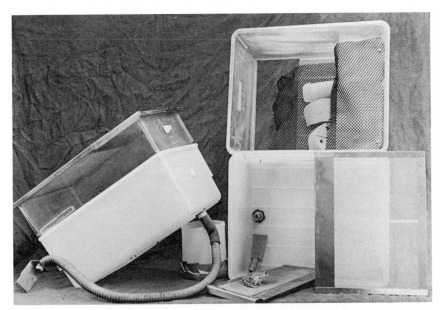

PLATE 2.   Cages for adult frogs at University of Michigan experimental frog colony. (Courtesy University of Michigan News Service.)

that each cage can be individually withdrawn for inspection, cleaning, and so on. The overflow tubes are connected to a telescoping drain so that circulation need not be interrupted during such operations.

A simpler device for holding adults, which may be more readily adaptable to commercial culture, is a 5-m-long trough, equipped similarly to the cages just described, but not incorporating the two-story structure. Such troughs may be divided into compartments, without losing the desirable feature of easy flushing, by simply leaving a 6-mm open space under each divider.

Separate housing facilities are maintained as a quarantine for newly received frogs. Such animals are first held 10 min in water containing enough calcium hypochlorite to produce a chlorine concentration of 6 ppm. Following this treatment, they are placed in isolated containers and provided with an excess of food until their health is assured.

FEEDING

Feeding is critical to the success of all forms of frog culture, and Nace's system is no exception. Tadpoles may be fed a variety of greens, but boiled romaine or escarole lettuce has been found best. Spinach is avoided, as it may cause formation of kidney stones. A good deal of judgment is required in feeding, for, although tadpoles consume great amounts of food and must be fed twice daily, they may be killed by overfeeding. The basic lettuce diet is supplemented two or three times weekly with cubes of raw or boiled liver. It should be noted that a number of other forms of food may be equally suitable from the tadpole's point of view, but that the nature of the tadpole bottles dictates a number of the properties of the food. Specifically, it must not float to the surface, produce a scum which inhibits gas exchange, or be fine enough to settle through the screen into the neck of the bottle or be flushed out.

As in outdoor culture methods, feeding presents more of a challenge once metamorphosis is reached. So far, optimal growth and rapid attainment of sexual maturity have been attained only when the frogs are fed on live insects. Nace has settled on three species, the meat fly (*Sarcophaga bullata*), the greenbottle fly (*Phoenicia sericata*), and the field cricket (*Acheta domestica*). Crickets are obtained commercially and maintained on a diet of chick mash and water, but the flies are raised in the laboratory.

Adult flies are maintained on water, sugar, and a sugar solution and allowed to deposit their eggs on a moistened mixture of sawdust and dog food, topped with several thin slices of raw liver, placed in a plastic tray. After 24 hours in a breeding cage, each such tray is placed in a 31.0 cm × 28.5 cm × 8.1 cm deep stainless steel pan, lined with paper toweling. Escape of maggots is prevented by means of a nonlethal "electric fence" created by running a 10-V current through a copper strip mounted on insulation affixed around the lip of the pan. Upon reaching full growth, the maggots migrate from the food tray into the pan, where they pupate. Pupae may be stored at 4°C for as long as 3 to 4 months, then warmed to 30°C for hatching. After hatching, the flies may again be chilled or anesthetized with CO_2 and fed to the frogs while in a torpid state.

The entire fly culture operation, which produces 25,000 flies daily, is confined to a 2.4-m × 3.0-m room. The species raised both require elevated temperatures for reproduction, thus escapees cannot become a nuisance in the Michigan climate. Potential frog culturists in tropical and subtropical climates, where meat flies and greenbottle flies might become established, might consider the possibility of relying entirely on crickets or some other species of insect.

The diet of flies and crickets has proved satisfactory for green frogs and leopard frogs, but pickerel frogs and some smaller species apparently develop a vitamin deficiency and do not survive well. Ultraviolet lighting has been found to help, but it is more efficient and equally effective to dust the food animals with powdered Pervinal, a commercial preparation containing ten vitamins, as well as calcium, magnesium, and various trace elements.

BREEDING AND SELECTION

Laboratory breeding procedures for frogs have been standardized and are well described in a number of embryology texts, thus need not be repeated here. (See references at end of this chapter.) Commercial adaptations have yet to be worked out.

Frogs have been selectively bred for some time, a process which has been greatly enhanced by the discovery that the skin markings of each frog are unique and may be used like human fingerprints as a means of individual identification. All frogs in the Michigan colony are routinely identified at metamorphosis and a complete breeding record kept for each individual, a procedure which could easily be applied by commercial culturists to selected breeder frogs (Plate 3).

POSSIBILITY OF INDOOR COMMERCIAL CULTURE

Obviously, the methods just described will have to be simplified somewhat if frogs are to be commercially cultivated indoors. The first to attempt to adapt Nace's and Kawamura's techniques to commercial culture, and also the first to culture bullfrogs indoors, is Dudley D. Culley, Jr., of the Department of Forestry and Wildlife Management, Louisiana State University. The principal difference between Culley's procedure and that used at Michigan is the source of food. Tadpoles are fed on commercially available rabbit pellets or trout chow. Adult bullfrogs have been found to prefer fish to other food organisms, thus experimental animals at Louisiana State are maintained on a diet of mosquito fish (*Gambusia affinis*) and sailfin mollies (*Mollienisia latipinna*), both of which are easily bred and maintained in captivity. Bullfrogs are also fond of tadpoles, and experiments are being conducted with excess tadpoles as food. The principal obstacle to total biological success in indoor bullfrog culture is the difficulty of breeding them, but Culley is optimistic that this problem will eventually be solved.

DISEASES

Diseases present a relatively small problem in frog culture. The most commonly reported "disease" is "red-leg," usually attributed to overcrowding. Actually there are two conditions which may give rise to a

PLATE 3. Large bullfrog showing distinctive ventral markings used to identify individuals for genetic selection. (Courtesy Dudley C. Culley, Louisiana State University.)

reddish discoloration of the legs. Red-leg in recently transported frogs most often indicates simple irritation of the skin caused by prolonged contact with a dry surface. Such irritation may, if not treated, afford access to infectious microbes, but it is not a disease symptom in itself. Infection by certain bacteria, most frequently *Aeromonas*, produces a similar response. The best preventive measure is adequate nutrition. Cures may be effected by isolating the infected individuals and treating them with such antibiotics as chloramphenicol and sulfadiazine. In severe cases, it is also advisable to keep the frogs in a salt solution approximating 25 to 30% frog Ringer's solution.

GROWTH AND DEVELOPMENT

Growth and development of well-fed frogs in both outdoor and indoor culture systems compares favorably with that observed in nature. The chief determining factors are food supply and length of the growing season. On the average, two years are required from metamorphosis to maturity in the South and four years in the North. Similar variations exist in the growth rates of tadpoles.

Of perhaps more interest to the culturist is the time required from metamorphosis to marketable size. The only reliable data of this sort come from Culley's experiments. Taking a length of 20 cm (including the outstretched legs) and a weight of 130 g as minimal for marketing in Louisiana, almost all of Culley's bullfrogs reached commercial size within 12 months of metamorphosis. The fastest growing individuals reached this size in 8 months. Through selective breeding, Culley hopes to reduce the average time required to less than 8 months.

UTILIZATION

A serious obstacle to the development of commercial frog culture in the United States is the American attitude that only the hind legs are useful as food. There is in fact an ample portion of meat on the back and front legs of any large frog, but its utilization would increase the cost of processing. Even if the back and front legs are eaten, there is still a large amount of waste. In certain Oriental cultures frogs are prepared so that the bones become soft and digestible, or even cooked with the entrails and skins intact, but it is doubtful that such practices would find favor in the United States at this time.

Culinary practices aside, if markets or uses could be found for presently wasted parts of the frog, frog culture in the United States would be closer to the threshold of economic feasibility. Mention has already been made of the practice of feeding frog entrails to tadpoles. It is also possible that frog wastes could be processed into a product similar to the marine protein concentrate (MPC) made from fish, and used as an animal feed or nutritional supplement. Efforts have also been made, with some success, to tan frog skin and use it in the manufacture of leather items.

In addition to the economic and gastronomic incentives, frog culture should be encouraged as a conservation measure. Wherever there are large populations of frogs in the United States, they are sought by hunters not only as a source of food but for sale to schools and research laboratories. Intensive hunting, along with drainage of wetlands, continues to reduce already depleted populations and some authorities foresee the disappearance of the wild frog industry within 10 years. Already American educators and scientists must import large quantities of frogs.

Disappearance or drastic depletion of frogs would mean the loss not only of an industry but of an important component of the ecosystem. Adult frogs are among the most effective insect predators, and both adults and tadpoles are important in the diets of many fish, birds, reptiles, and mammals. Tadpoles occupy a unique position in the food chain by virtue of their benthic feeding habits, which result in their recycling nutrients that might otherwise be trapped in the substrates of ponds. It is thus to be hoped that aquaculturists in the United States will make increased efforts to propagate and rear frogs for both laboratory and table use.

EXPERIMENTAL FROG CULTURE IN INDIA

Frogs support industries of some value in many of the Latin American and Asian countries, but only India maintains an extensive frog culture program, although small-scale experimental culture is reportedly being carried out in China and Cuba. The main center of experimental frog culture in India is the Pond Culture Substation of the Central Inland Fisheries Research Institute, Barrackpore, Cuttack, but preliminary work is also being done at the Freshwater Biological Station, Bhavanisagar, Madras. In both cases the species cultured are *Rana hexadactyla* and *Rana tigrina*, both of which are in high demand for export as frog legs. Officials at Cuttack are not generous with information, but it is known that some of the studies conducted concern induced spawning, food and habitat requirements, and polyculture with fish.

SPAWNING

Both species of frog, which spawn naturally during the northeast monsoon (September to November), can be induced to spawn throughout the year by the administration of frog pituitary hormones with a priming dose of progesterone. The natural rate of fertilization of *R. tigrina* is poor, so stripping of eggs and artificial fertilization have been employed. Nearly 100% fertilization has been obtained using the dry method.

There appears to be no basis for fear that either species will be a serious predator in fish ponds. Preliminary observations indicate that they feed primarily on worms, gastropods, and aquatic insects, with small fish constituting only an incidental item in the diet. If submerged weeds are encouraged by fertilization, a frog pond should produce an ample food supply.

POLYCULTURE WITH FISH

Indications are that the combination of frogs and fish would be more profitable than frog monoculture. Experimental yields obtained when small frogs were stocked with fingerlings of the Indian major carps catla (*Catla catla*), rohu (*Labeo rohita*), and mrigal (*Cirrhina mrigala*) are shown in Table 1.

TABLE 1.   STOCKING RATES AND YIELDS OF FISH-FROG CULTURE IN INDIA

FROG SPECIES STOCKED	STOCKING RATE/HA	STOCKING RATE OF MAJOR CARPS PER HECTARE	YIELD OF FROGS (KG/HA)	YIELD OF FISH (KG/HA)	TOTAL YIELD (KG/HA)
R. hexadactyla	2,000	none	259.0	0.0	259.0
R. tigrina	2,000	none	235.6	0.0	235.6
None	—	3,705	0.0	886.1	886.1
R. hexadactyla	2,000	3,705	234.8	1,611.0	1,845.8
R. tigrina	2,000	3,705	218.3	1,093.1	1,311.4

No explanation has been advanced for the higher yields of fish in ponds stocked with frogs, and they may or may not be related to the presence of frogs.

PROSPECTUS

Experimental frog culture has thus far dealt almost exclusively with the production of frogs for laboratory use or as a luxury food. If the commercial status and prospectus of such culture is uncertain, even less can be predicted about the eventual role of frogs in supplying human nutritional needs. Whatever form of frog culture one considers, an authority can be found to support any prognosis, from glowing optimism to a "can't be done" attitude. Rather than add to the confusion, let us simply acknowledge that attempts at commercial frog culture will continue, and state that its future is in the hands of a few biologists and adventurous entrepreneurs.

REFERENCES

CULLEY, D. D., and C. T. GRAVOIS. 1970. Frog culture. Am. Fish Farmer 1(10):5–10.

DIBERARDINO, M. A. 1967. Frogs. *In* F. H. Witt and N. K. Wessels (eds.), Methods in developmental biology, T. Y. Crowell Co., New York, pp. 53–74.

NACE, G. W. 1968. The amphibian facility of the University of Michigan. BioSci. 18(8): 767–775.

RUGH, R. 1965. Experimental embryology techniques and procedures. Burgess Publishing Co., Minneapolis, Minn.

U.S. Bureau of Sport Fisheries and Wildlife. 1965. Frog Raising. Fishery Leaflet 436.

Interviews and Personal Communication

NACE, G. W. Department of Zoology, University of Michigan.

16
Culture of Mullets (Mugilidae)

Of all the species of fish that inhabit estuaries, probably none is so widely distributed as the striped mullet (*Mugil cephalus*), found in tropical and semitropical waters around the world. Not surprisingly, it and several of its congeners are among the principal products of brackish water fish culture in regions as widely separated as Taiwan and Italy. However, the keen interest of fish culturists and fishery scientists in mullet is occasioned not merely by the present status of mullet culture but by the promise of even greater significance in the future.

Until recently mullets, like the milkfish (*Chanos chanos*), the only other important food fish routinely cultured in brackish water, could not be spawned in captivity. Commercial mullet culture as practiced today is thus a low-intensity operation, dependent on unpredictable natural supplies of fry, involving other species of fish, many of them anything but

285

PLATE 1. Male striped mullet (above) and female (below). (Courtesy Dr. Ziad Shehadeh, Oceanic Institute, Hawaii and I. C. Liao, Tunkang Shrimp Culture Center, Tunkang, Taiwan.)

beneficial to mullet stocks, and offering no opportunity for domestication or selective breeding.

ARTIFICIAL PROPAGATION

Artificial spawning of mullet was first achieved with striped mullet in Italy in 1930, by the use of methods similar to those employed in "stripping" trout in hatcheries (see Chapter 20). The implications for practical culture of this accomplishment were slight, however, since in most regions it is very difficult to capture ripe mullet.

A more important breakthrough was achieved in 1964 by Yun-An Tang of the Taiwan Fisheries Research Institute. Tang succeeded in inducing ovulation and successfully spawning striped mullet by injecting ripening fish with *M. cephalus* pituitary extract and the synthetic hormone synahorin. In each subsequent year, Taiwanese biologists have endeavored to improve their techniques. Current practices produce about 70% spawning of females within 20 to 24 hours of injection.

Taiwanese biologists are still dependent for experimental fish on fishermen, who are able to capture a few ripening mullet as their spawning run passes the southwest coast of the island in December or January. Females selected for induced spawning are 4 to 6 years old and average

Plate 2.   Ripe male. (Courtesy Dr. Ziad Shehadeh, Oceanic Institute, Hawaii and I. C. Liao, Tunkang Shrimp Culture Center, Tunkang, Taiwan.)

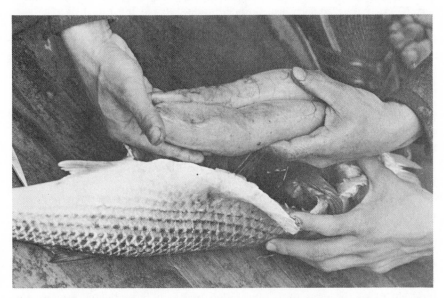

Plate 3.   Ovaries in ripe 2-kg female. (Courtesy Dr. Ziad Shehadeh, Oceanic Institute, Hawaii and I. C. Liao, Tunkang Shrimp Culture Center, Tunkang, Taiwan.)

287

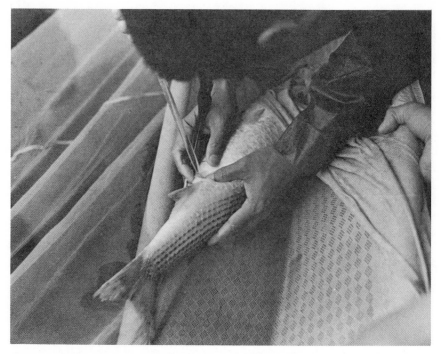

PLATE 4. Suction of oocytes from female to ascertain state of maturity prior to injection. (Courtesy Dr. Ziad Shehadeh, Oceanic Institute, Hawaii and I. C. Liao, Tunkang Shrimp Culture Center, Tunkang, Taiwan.)

55 cm in length and 2 kg in weight. Males used are 4-year-olds, averaging 50 cm long and 1.2 kg in weight. It has not been found necessary to inject males, but females receive a total of 2 to 4 pituitaries and 10 to 20 rabbit units of Synahorin, injected intramuscularly. Fractional injection is used, with 1/3 of the dosage administered at once and the rest 20 to 24 hours later. After allowing 20 to 24 hours for ovulation in holding ponds at 19 to 23°C and 32.5 to 33.0‰ salinity, fertilization is done artificially, using the "dry" method.

Eggs have been incubated in baskets suspended in large plastic tanks or in the tanks themselves with either aeration or constant circulation or both, with no difference in results. Plate 8 illustrates a hatching arrangement successfully used at the Oceanic Institute, Waimanalo, Hawaii. Raising the temperature to 22 to 24°C reduces the time required for hatching to 50 to 60 hours. In either case, hatching rates have ranged from 40 to 90%. Most of the eggs that have failed to hatch have been found to be either overripe or premature.

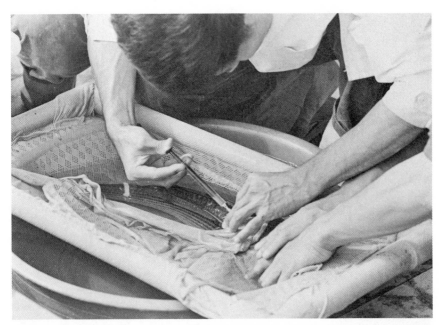

PLATE 5.   Intramuscular injection of females with mullet pituitary, synahorin (25 RU), and vitamin E, as practised in Taiwan. (Courtesy Dr. Ziad Shehadeh, Oceanic Institute, Hawaii and I. C. Liao, Tunkang Shrimp Culture Center, Tunkang, Taiwan.)

In 1968, biologists at the Fish Culture Research Station, Dor, Israel, had some success in spawning striped mullet using three fractional injections of common carp (*Cyprinus carpio*) pituitary, collected from April to July, at the peak of its potency. After the first injection with 1.6 pituitaries/kg of female mullet, the fish, which came from freshwater ponds, were placed in freshwater in 1.65-m $\times$ 1.25-m concrete tanks. The second and third injections, each of 2.0 pituitaries/kg of mullet, were given 7 and 14 hours after the fish were placed in the concrete tanks. The third injection was supplemented by 2.0 IU of luteinizing hormone. After the mullet had been in the tanks 1 to 2 days, they were transferred to similar tanks containing half strength sea water, which was gradually increased over 12 to 24 hours to full strength. Spawning occurred 17 hours later in the tank.

The eggs were placed in specially designed incubators and hatched in 36 to 44 hours at 22 to 23°C. For the first 22 hours they floated, but with the development of the embryo (visible as a darkening of the egg), they sank.

The significance of the Israeli achievement is that mullet culturists

PLATE 6. Hormone injection through polyethylene catheter. (Courtesy Dr. Ziad Shehadeh, Oceanic Institute, Hawaii and I. C. Liao, Tunkang Shrimp Culture Center,
Tunkang, Taiwan.)

need not be dependent on wild fish captured during the spawning run.
The mullet used in the experiment were captured as fry from a Mediterranean estuary and reared in freshwater ponds at Dor. To ensure ripeness, specimens used in the experiment were stocked in newly filled
ponds at 600 to 800 fish/ha in August, 2 months prior to the spawning
season. One of the obstacles to selective breeding of mullets has been
the difficulty of obtaining adequate amounts of potential spawners but,
from now on, Israeli workers should have large numbers available. The
work at Dor has also been useful in that it demonstrates that common
carp pituitaries, which are available nearly everywhere, can be used to
induce ovulation in mullet, thus eliminating the need to sacrifice mullet
for that purpose.

Another problem experienced by mullet culturists in Israel and elsewhere is the fragility of *Mugil* spp. with respect to handling. The necessity for handling during pituitary injection has been reduced by
researchers at the Oceanic Institute, Hawaii, who have developed a tech-

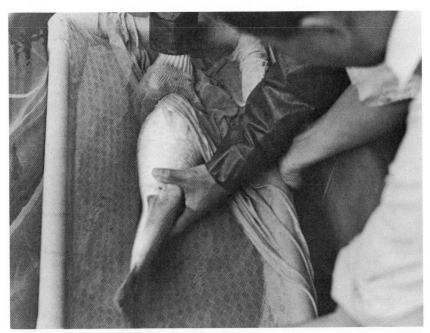

PLATE 7. Belly of female after two injections 24 hours apart. (Courtesy Dr. Ziad Shehadeh, Oceanic Institute, Hawaii and I. C. Liao, Tunkang Shrimp Culture Center, Tunkang, Taiwan.)

nique for coelomic administration of hormones via a polyethylene catheter implanted in the coelomic cavity and secured internally by means of a "feathered" sleeve. The free end, used to administer the injections, is a buoy.

By using this device three times a week to administer one mullet or salmon (*Oncorhynchus*) pituitary plus 30 rabbit units of synahorin per 800 g of female mullet, it was possible to produce egg release spontaneously or manually within 2 weeks. Fertilization was done artificially, using noninjected males, and the eggs hatched at 26°C and 32‰ salinity.

The only mullet species other than *M. cephalus* to be spawned in captivity thus far are *Mugil capito* in Israel, and *Mugil macrolepis* and *Mugil troschelli* in India. All have been spawned with the aid of pituitary injection, but in 1969, workers at the Directorate of Fisheries, Orissa State, India, captured and stripped fully mature specimens taken from the mouth of Chilka Lake (a brackish lagoon). Fertilization was successful, and hatching occurred 22 to 24 hours later at temperatures of 21 to 29°C. The hybrid ♀ *Mugil cephalus* × ♂ *Mugil capito* has been produced in Israel.

PLATE 8.   Hydrodynamic hatching baskets for mullet. (Courtesy Dr. Ziad Shehadeh, Oceanic Institute, Hawaii and I. C. Liao, Tunkang Shrimp Culture Center, Tunkang, Taiwan.)

FRY REARING

It appears that induced spawning of mullets is well on its way to perfection, but another formidable obstacle to intensive culture remains— rearing the minute, planktonic fry. Striped mullet fry commence to feed on the third day, but their dietary requirements are not well known, and it is believed that failure to provide proper food has been responsible for the universal failure, until recently, to rear them.

Most culturists count themselves lucky if they can rear newly hatched mullet fry for longer than a week. Israeli culturists have been somewhat more successful with fry of the golden gray mullet (*Mugil auratus*), captured from estuaries at weights of 0.2 to 0.5 g. Researchers at Dor have been able to raise these tiny fry to 2.0 g on a diet of fish flesh and fish flour extruded through a plastic plate with tiny perforations. Survival was greatly enhanced by fortifying this diet with algae collected from mullet ponds. Further improvement was achieved by the addition

PLATE 9. Plastic stripping from which mullet graze periphyton. (Courtesy Dr. Ziad Shehadeh, Oceanic Institute, Hawaii and I. C. Liao, Tunkang Shrimp Culture Center, Tunkang, Taiwan.)

of chironomid larvae, which the late A. Yashouv cultured at Dor. It is possible to keep 40,000 to 50,000 larvae in a flat tray 1 m in diameter and several centimeters high, filled with soil and manure, covered with water. Production is maintained during the winter by covering the trays with plastic screen to simulate a greenhouse. Using these simple techniques it is possible to harvest 0.5 kg of nourishing fish food daily from a single tray. It is theorized that these metamorphosing insects may contain some sort of growth-enhancing substance which is extraordinarily effective on cold-blooded vertebrates. Yashouv attempted to determine the minimum amount of these larvae needed to supplement the basic fish and algae diet, in the hope that it would be part of the key to successful rearing of mullet fry.

A method currently being tested at the Oceanic Institute in Hawaii utilizes buoyant plastic strips to increase the grazing surface area available to the fish. Diatoms will grow on the plastic, but improvement is hoped for as a result of imbedding nutrients in the plastic.

The greatest success in rearing mullet fry has been achieved at the Marine Fish Culture Laboratory, Tungkang, Taiwan. There I. C. Liao and his associates reared 500 *Mugil cephalus* from the egg to a length of 19 cm. As larvae they were housed in two neon-lighted concrete tanks, 20 m × 10 m × 2 m deep, with greenhouse tops. Food was provided in the form of *Brachionus,* oyster trochophore larvae, and copepods harvested from brackish water ponds, supplemented with mixed cultured diatoms. The water was static and never changed, but was aerated.

Large-scale breeding and, ultimately, selection of mullets are still not commercial realities, but it appears almost certain that major break-throughs in this area are only a few years away. When they are made, the striped mullet, and perhaps other species, will become the first marine counterparts of such truly domesticated freshwater food fish as the common carp and the rainbow trout (*Salmo gairdneri*).

MULLET IN PRACTICAL FISH CULTURE

DESIRABLE CHARACTERISTICS

Mullet now play an important role in fish culture in a number of places, notably the Mediterranean and southeast Asia. Their popularity is no accident; they possess several characteristics desirable in a fish for pond culture:

1. High quality of flesh.
2. Extreme salinity tolerance, a characteristic particularly desirable in a fish to be kept in intertidal ponds. Striped mullet have been grown at salinities of 0 to 38‰, in other words, from completely freshwater to strong sea water.
3. Wide temperature tolerance. Striped mullet survive temperatures of 3 to 35°C.
4. Low position on the food chain. Mullet are herbivores, feeding on plankton, benthic algae, and, in ponds, decaying higher plants. They thus respond well to inexpensive methods of fertilization. They also readily accept supplemental foods such as rice bran and peanut meal or cake.

SOURCE OF STOCK AND NATURAL HISTORY

Since they are not yet able to spawn mullets routinely, culturists must take advantage of the natural habits to obtain stock. The natural history of the striped mullet, which applies, with noted exceptions, to other species of *Mugil,* is as follows.

Several males and a single female spawn at sea during the cold months, laying pelagic eggs which hatch within two days. The minute, heavily pigmented fry move into estuaries and coastal tide pools in late winter or early spring, to remain there until moving offshore the following fall or winter. Mullet of all ages prefer warm, brackish water but, as mentioned, they are tolerant of a wide range of environmental conditions.

Fry feed principally on plankton and are believed to prefer diatoms and epiphytic *Cyanophyceae*. Fry of at least one predominantly freshwater species, *Mugil corsula* of India, Pakistan, and Burma, are said to prefer copepods and small insects. Adults of all species are primarily benthic feeders, consuming algae and vegetable detritus, with an incidental intake of small animals, which may be essential. Decayed higher plants are readily accepted when available. *Mugil cephalus* reaches lengths of 50 to 55 cm and weights of 1.2 to 2.0 kg in 4 to 6 years. Most of the other species are slightly smaller, although *Mugil tade* of the Indo-Pacific region reaches a maximum length of about 70 cm.

CULTURE IN ISRAEL

The most sophisticated use of mullets in fish culture is developing in Israel, where *Mugil cephalus* and, to a lesser extent, *Mugil capito* and *Mugil auratus* are used in polyculture, with common carp as the primary crop. Interest in mullet in Israel began in the late 1940s, when it was postulated that production of fish by intensive pond culture, with fertilization and supplementary feeding, as practiced in Israel at that time, could be augmented by fuller utilization of the pond environment through polyculture. A number of species were tried and rejected, but only mullet and several species of *Tilapia*, notably *Tilapia galilea* and *Tilapia nilotica*, have thus far found practical application. Today 50% of the fish farms in Israel grow mullet and/or tilapia together with carp. Nevertheless, the practice is still considered experimental, and few guidelines exist as to proper management practices.

Mullet cultured commercially in Israel are obtained as advanced fry from Mediterranean estuaries and reared alone in fertilized ponds to the second year, at which time they weigh 30 to 70 g. Stocking rates are by no means standardized, but mullets are usually stocked at 500 to 800/ha. One system successfully used in commercial culture involved 1200 carp, 1050 *Tilapia nilotica*, and 600 mullet/ha. Stocking is timed so that the mullet and tilapia reach marketable size 120 to 150 days after stocking, at the same time as the carp.

Tilapia have fitted into the Israeli scheme of fish culture very well, but success with mullet has been less than was anticipated. They seem to occupy a niche which overlaps those of both carp and tilapia, and may depress the yields of both these fishes, although the total yield of carp-tilapia ponds is usually increased by 13 to 35% if mullet are added. Moreover, mullet bring a higher price than carp or tilapia, thus many Israeli fish farmers find it advantageous to stock them, even in the rare cases where the total yield is depressed. In 1960, 116 tons of mullet, or 1.4% of the total farm fish crop, were harvested from ponds in Israel. In 1966, mullet accounted for 430 tons, or 4.6% of the crop. The general trend of mullet production is upward, although yields are poor in some years, due to low availability of fry.

In 1964, Israeli biologists began to experiment with mullet farming in the vicinity of the Dead Sea, where soils are too alkaline for agriculture and waters too saline for conventional carp culture. A series of experiments showed that these areas might be brought into food production by stocking ponds with a salinity of 36 to 145‰ with various combinations of *Mugil cephalus* and *Tilapia nilotica*. Much research remains to be done, but the best results thus far obtained were from an 0.8-ha pond stocked with 50-g mullet at 214/ha and tilapia of the same size at 139/ha. Carp were also stocked in this pond but failed to grow. The total yield of the pond at the end of a 109-day growing season was 1155 kg/ha, of which 512 kg/ha, or 44.3%, was mullet. Fish pond stocks may also profitably be skewed away from carp and toward mullet where dissolved oxygen concentrations are low, or when it is not economically feasible to feed heavily. Thus it appears that, even if breeding of mullets in captivity does not become prevalent, their importance in fish culture in Israel will continue to increase.

CULTURE IN ITALY

Methods of culture of mullets similar to those practiced in Israel are barely beginning to be developed in other Mediterranean countries, but mullet are an important food fish throughout the region. The methods used in their exploitation vary from place to place, but most take advantage of the tendency of young mullets and other fish to perform annual inshore-offshore migrations. Shoreward migrating fishes, attracted by the high temperatures, oxygen content, and fertility of shallow, brackish waters are attracted to, and frequently spend months in estuaries. Many Mediterranean estuaries can be totally or partially blocked off to facilitate capture of the fish or, in some cases, to hold them for growth. This is done with various degrees of sophistication, and in attempting to describe the techniques used one is confronted with the question "Where does fishing leave off and fish culture begin?" There is no distinct line separating the two, but for our purposes simply trapping a number of fish in a blocked-off estuary and waiting for them to grow will not be considered as "culture." If, however, certain species or individuals are selected and stocked in an estuary, or if feeding or fertilization is resorted to, it will be considered to be within the scope of this treatise.

Of the Mediterranean countries, the only one with large areas of brackish water which are so exploited as to fall under our definition of "culture" is Italy. Some of the Italian "valli," as the modified estuaries are called, produce primarily eels (*Anguilla anguilla* and *Anguilla vulgaris*); but mullet are the principal crop in the most advanced valli, located in the lagoon of Venice. Four species are commonly stocked, according to the season. The first species available in the spring is *Mugil capito,* fry of which enter the lagoon as early as February. In April, *Mugil chelo* becomes available. From July to September *Mugil saliens* is stocked, and finally, from October to December, the largest and most valued species, *Mugil cephalus,* is found in the lagoon. *Mugil auratus* may also be taken in the early spring, but is relatively uncommon, as it rarely enters waters of less than 20‰ salinity. Other fishes stocked include eels in the fall and gilthead bream (*Sparus auratus*) and "bass" (*Dicentrarchus labrax*) in the spring. The latter two species are of dubious value, as they are predators and may seriously deplete stocks of mullet fry. On the other hand, they bring the highest prices of any fish grown in the valli.

Mullet and other fish stocked in the valli are for the most part not captured by the valli operators themselves, but are purchased from specialized fry fishermen. Fry are usually not stocked directly into open water, but are placed in a "seragio," a series of parallel trenches (Fig. 1) so located as not to be exposed to the full force of the wind and accessible to supplies of both fresh and brackish water. Fry are left in the seragio until they are large enough and well enough acclimated to get along in open water. *Mugil cephalus* fry, in particular, need the shelter afforded by the seragio to protect them from the cold waters of late fall and early winter. A similar series of trenches, called a "conserva," is included on the opposite side of the valle as a wintering ground for slightly larger fish.

Still larger fish may be wintered in a "canale raccoglitore" located along that side of the valle which is best protected from wind and storms. The canale raccoglitore is wide, deep and up to several kilometers long. The banks should be heavily planted with trees as an added protection from wind. Like the seragio and the conserva, the canale raccoglitore should open into both the valle proper and a source of freshwater.

Yet another structure of similar nature is the "fossa circondaria," a peripheral canal at least 2 m deep. Sluice gates leading to all these

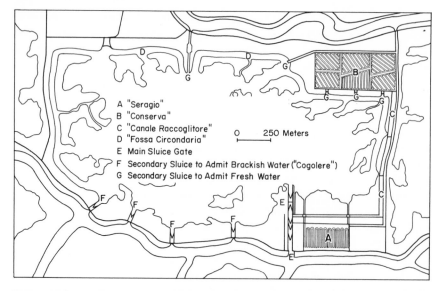

A "Seragio"
B "Conserva"
C "Canale Raccoglitore"
D "Fossa Circondaria"
E Main Sluice Gate
F Secondary Sluice to Admit Brackish Water ("Cogolere")
G Secondary Sluice to Admit Fresh Water

0   250 Meters

FIG. 1. Valle used for brackish water fish culture in the lagoon of Venice, Italy. (After de Angelis, 1960.)

shelter areas may be opened during storms or exceptionally cold weather. The deep fossa circondaria also functions as a shelter during very hot weather. In addition, it serves as a trough to divert muddy water away from the valle proper following storms.

Some valli are not as elaborately constructed as just described and may not be abundantly supplied with freshwater. However, maximum efficiency in valle culture depends on skillful manipulation of temperature and salinity, which can be accomplished only in valli abundantly supplied with freshwater and containing shelter areas for large and small fish. The best valli—those which permit maximum control of environmental parameters—are essentially ponds, entirely surrounded by earthwork. Such valli, including all the 300- to 500-ha structures in the Venice area, are often located above the high tide line and connected with the sea by long canals, from which brackish water is pumped at high tide, creating essentially an artificial tide.

Whether brackish water enters tidally or is pumped in, it does so through a series of sluice gates. The main gate only is equipped with a catching device called a "lavoriero," consisting of a series of V-shaped screens, open at the apex, which eventually funnel fish leaving the valle into a catch pond. Secondary gates are provided with similar screens called "cogolere," which are not open at the apex, to block the escape of fish without impeding the flow of water.

Valle management proceeds roughly as follows. Cogolere are removed in late February and left off through mid-April or May. During this time the sluice gates are left open constantly to take full advantage of fry which enter the valle naturally. During most of May, the sluice gates are left open only at high tide, to fill the valle as full as possible. Incoming fry are channelled off into the seragio, where purchased fry or fry captured in the lagoon are also stocked. (An exception is made for fry of gilthead bream, which are stocked in the valle proper.) The natural ascent of fry usually ceases in May, at which time the cogolere are reinstalled. During the summer, the sluice gates are opened at high tide as often as necessary to replace water lost by evaporation. Salinity is thus increased until the temperature begins to fall off in September, at which time preparations are made for fishing.

First, the water level is lowered by evaporation or, if necessary, by opening the sluice gates at low tide. Then the lavoriero is fitted to the main gate. A supply of freshwater is admitted to the conserva, which has been cleaned and weeded during the summer. The pass between the conserva and the valle proper is then opened and the mullet, attracted by the influx of freshwater, try to enter the conserva. Nets, designed to permit the passage of fish too small to be marketed, are placed across the pass. The first fish to respond to the freshwater are usually yearlings, and they are permitted to enter the conserva. Eventually marketable specimens appear, in this order: *Mugil capito, Mugil chelo, Mugil cephalus,* and *Mugil saliens.* Young *Dicentrarchus labrax* are also strongly attracted by the freshwater. Insofar as possible, they are separated and placed in a special conserva to prevent heavy predation on the mullet in the spring. Marketable sized fish may be retained or placed back in the valle proper, to be harvested later by taking advantage of their tendency to swim against the current.

Harvesting is done by opening the main sluice gate at high tide, particularly at night when there is a full moon. (Eels are harvested on moonless nights.) Such operations continue into December, by which time all the fish usually have been captured. Nets are placed in the open waters of the valli only in exceptional circumstances.

The sluice gates are usually left open and both stocking and fishing are suspended during January and February, especially if the weather is cold.

The total annual yield of valli ranges from 90 to 200 kg/ha of fish. Though mullet are the principal crop and gilthead bream and bass the most valuable, it is the number of eels harvested which makes the difference between productive valli and poor ones. Further south on the Adriatic coast of Italy, lagoon fisheries, which are managed in much less elaborate fashion than valli, yield an *average* of 200 kg/ha of fish annually. Productive valli are more lucrative than lagoon fisheries, however, since the fish produced are more uniform and because the culturist can time his harvest to coincide with the periods of greatest demand.

EXPERIMENTAL CULTURE IN OTHER MEDITERRANEAN COUNTRIES

The total brackish water area of the Mediterranean, excluding Albania and the African coast west of Egypt, is approximately 10 million ha. Most of this water is naturally productive of mullet and other fishes, but very little of it has been developed for fish culture as has the lagoon of Venice. This is in part due to the fact that most of the Mediterranean coast does not experience the extreme tidal variation characteristic of the northern Adriatic, and essential for valle fish culture. Another reason for the lack of development of fish culture in much of the Mediterranean is simply that fisheries have been successful in supplying the demand for fish in the area. Nevertheless, mullet culture should be developed in preparation for future population increase.

Topography and availability of mullet fry would appear to permit valle type culture in Cyprus, northern Yugoslavia, and much of Greece, but at present the only area outside of Italy in the Mediterranean basin where estuarine fish culture is practiced on a commercial scale is the Porto Lago Lagoon in Greece. The Porto Lago Lagoon is a 5000-ha complex comprising fresh, brackish, and salt waters. The principal fishery product is common carp, caught in the freshwater portions, but fry of *Mugil cephalus, Mugil chelo, Mugil capito,* and *Mugil saliens* are collected in brackish water and stocked throughout the area by a fishermen's cooperative.

Adoption of more sophisticated brackish water fish culture techniques, similar to those employed in the lagoon of Venice, has been suggested for the Agoulinitsa Lagoon of the western Peloponnesus, presently slated for "reclamation" for agricultural purposes, a dubious plan at best. Experimental stocking of mullet, eels, and sole (*Solea vulgaris*), occasionally augmented by shrimp, is already under way in a number of estuaries in the United Arab Republic.

Efforts are also beginning in Yugoslavia, the United Arab Republic, and France to emulate Israeli techniques of pond culture of mullet. In Yugoslavia it is hoped that mullet will be a satisfactory substitute for tench (*Tinca tinca*) in polyculture with common carp. Tench are in fairly high demand in a number of European countries, but they grow rather slowly and may compete with common carp more severely than mullet.

Experiments with pond culture of mullet in the United Arab Republic have been aimed at monoculture, though *Tilapia* spp., *Anguilla vulgaris, Synodontis schall,* and *Chrysophrys auratus* have turned up in most of the harvests. Preliminary results have not been encouraging; in no case have yields approached the levels routinely achieved in commercial culture in Israel. In tests conducted at the Mex Fish Farm in the U.A.R., the best yield achieved by stocking 22-mm *Mugil cephalus* at $2/m^3$ was 350 kg/ha. Total fish production, including the extraneous fishes mentioned previously, was 521 kg/ha. This was achieved with the aid of 30 kg/ha of phosphate fertilizer. Similar ponds left unfertilized or treated with nitrates or manures yielded considerably less. The bottoms of the ponds at Mex Fish Farms are largely composed of cockle (*Cardium*) shells, and it is possible that the resulting high lime content interfered with the action of the fertilizers. It is further possible that fertilization, by increasing the production of plankton, shaded out the benthic organisms, which are the primary source of food for mullet. H. S. Swingle of Auburn University has suggested that ponds used for culture of benthic feeders be fertilized only during the coldest times of the year, when plankton production is usually at its lowest. This technique seems worthy of trial in Mediterranean waters.

It is interesting to note in connection with mullet culture experiments in the United Arab Republic that, while *Mugil cephalus* suffered mortalities of 46.2 to 83.4%, *Mugil capito,* tested on a smaller scale, experienced no more than 33.3% mortality, suggesting that the latter species is less sensitive with regard to handling.

Another approach to low-intensity mullet culture being tested in the Mediterranean basin is the stocking of sea-caught fry in such brackish water lakes as Lake Kelbia in Tunisia, Vrana Lake in Yugoslavia, and the North Delta Lakes in the U.A.R. The type of fish community depends on the salinity. In Vrana Lake, where the salinity varies from 2 to 8‰, common carp are the principal species exploited, accounting for 87.0% of the annual fish production of 23 kg/ha. *Mugil* spp., the only fish stocked in the lake, account for 8.7%, while *Anguilla anguilla* make up the remainder. Attempts are being made to increase the productivity of this lake by fertilization with superphosphate at 40 kg/ha.

The North Delta Lakes, which reach salinities of up to 22‰, are better suited to a tilapia-mullet fishery. In these lakes, annual production of *Tilapia* spp. varies from 136 to 678 kg/ha. *Mugil* spp., which again are the only fish stocked, contribute 18 to 62 kg/ha, depending on the amount stocked, while eels account for 0.7 to 5.2 kg/ha.

At present, the only threat to mullet fisheries and culture in the Mediterranean appears to be pollution, which has caused the curtailment of some plans for brackish water fish culture in France and may be responsible for the recent increase in abundance of the inedible green alga, *Ulva,* in Italy. Domestic pollution will probably become more severe as the human population of the area increases. Bearing this in mind, some of the massive fertilization schemes which have been proposed, for instance one suggesting that 250 tons of superphosphate, a like amount of sulfuric acid, and various other substances be dumped into the Bay of Kaštela in Yugoslavia, seem as likely to lead to premature eutrophication as to any increase in fish production.

Unless pollution reaches crisis proportions, it appears likely that the present low-intensity methods of mullet culture and/or fishery management will persist, in part because, given the high availability of mullet fry in most of the area, they are adequate to supply local demand, and partly because mullet do not mature in Mediterranean estuaries, or at least not in the Italian valli. The physiological questions implicit in the latter fact should be explored in case it ever becomes necessary to intensify mullet culture in the Mediterranean.

CULTURE IN THE INDO-PACIFIC REGION

Although the importance of mullet culture, as opposed to fisheries, is greater in the Indo-Pacific region than in the Mediterranean, the variety of methods employed is equal. In some countries, particularly India, fairly primitive methods are employed with good success.

The principal culture areas in India are located in West Bengal and Kerala, but East Bengal and Madras also produce significant amounts of mullet. Mullets in India and, to a lesser extent, elsewhere in Asia appear to have more diversified feeding habits than European species. In addition to the ubiquitous *Mugil cephalus* and the similar *Mugil macrolepis,* Indian culturists make use of *Mugil dussumieri* and *Mugil troschelli,* which are more inclined than other mullets to feed on plankton as adults, as well as *Mugil tade* and the essentially freshwater *Mugil corsula,* which often feed on filamentous green algae.

Most Indian mullets spawn during the southwest monsoon. Principal fry-collecting seasons are October to April for *Mugil dussumieri,* the most commonly cultured species, and December to March for *Mugil cephalus.* In East and West Bengal and Kerala, as well as Bangladesh, fry are caught in crude seines or dip nets from shallow creeks, canals, borrow pits, or any bodies of water that can be dammed off easily.

So hardy are mullet with regard to salinity differences that fry may be transferred directly to freshwater ponds without acclimation. It is recommended, however, that acclimation be carried out over 1 to 12 days, depending on the size of the fry and the salinities involved. A reduction in salinity of 5% every 4 hours has been shown to prevent mortality of 1.5- to 4.0-cm fry.

Fry to be stocked in the state-owned mullet farms in southern India are harvested from adjoining swamps by means of manually operated purse nets. Such nets are rectangular, about 12 m × 6 m × 1 m deep. The size of the mesh diminishes from 2.5 cm at the mouth to 6 mm at the posterior end. In use the net is kept open by two men, one holding each of the wings. The top of the net is buoyed with wooden floats about 60 cm apart, while the bottom is kept flush with the substrate by the fishermen standing on it. Fry are driven into the net by means of a scare line about 30 to 60 m long, with palm leaves attached to it. When the scare line is brought up to the net, the two fishermen bring the wings together and enclose the fry.

Brackish water ponds in Bengal and Bangladesh may be stocked by simply opening the sluice gates on high tide. In some cases, additional temporary gaps may be created in the embankment so as to admit more fry.

Separate fry rearing enclosures, so common in other forms of fish culture in Asia, are rare in mullet culture in India, although in some of the brackish water farms near Calcutta young mullet are placed in isolated ponds for their first year.

Mullet are generally cultured together with other fishes, mainly pearl spot (*Etroplus suratensis*) and milkfish in southern India, and cock-up (*Lates calcarifer*) further north and in Bangladesh. Various species of prawns are often stocked in mullet ponds in both regions. Feeding is anything but intensive, consisting of no more than a small amount of rice bran every few days.

Most Indian mullet ponds are fertile enough to provide for fairly rapid growth. Table 1 illustrates average growth rates for the four most commonly cultured species. Annual yields vary from 150 to 1500 kg/ha, depending on fertility, stocking rate, and the amount of food given. Mullet may be harvested in the same year they are stocked, or allowed to grow for three years or more.

In Hawaii, the ancient techniques of mullet culture, involving chiefly *Mugil cephalus*, were similar to those practiced in brackish waters in India. A number of other species invariably entered Hawaiian ponds, the most important being milkfish, tarpon (*Megalops cyprinoides*), and ten pounder (*Elops machnata*).

In Indonesia, where the principal commercial mullet species are *Mugil dussumieri, Mugil engeli,* and *Mugil tade,* spawning occurs during the west monsoon, and fry are available in coastal waters from October to April. Females of *Mugil dussumieri* and *Mugil tade* with well-developed roe have been found inland, suggesting the possibility of intensive culture

TABLE 1.   AGE AND GROWTH OF MULLETS CULTURED IN INDIA

SPECIES	AGE (YEARS)	AVERAGE LENGTH (CM)	AVERAGE WEIGHT (KG)
Mugil cephalus	1	14	—
	2	24	—
	3	33	—
	4	39	—
	6–7	50	1.3
Mugil corsula	3	35	—
	3	45 (maximum)	—
Mugil dussumieri	1	15–19	—
	—	25 (maximum in culture; wild fish attain 40 cm)	—
Mugil tade	1	24–25	—
	2	34–36	1.4–1.8
	—	70 (maximum)	—

entirely in freshwater or, failing that, the feasibility of the Israeli method of artificially inducing spawning.

Nevertheless, most Indonesian fish culturists continue to specialize in common carp or milkfish, depending on the salinity of their ponds. Mullet, which are to some extent competitive with milkfish, are usually considered an extraneous fish in brackish water ponds in Java, the principal fish culture island of Indonesia. They are important, however, in ponds which are in the process of construction or which have weakened dikes. While construction or repair proceeds, rather than risk the loss of expensive milkfish fry, the culturist may let in a stock of "wild" fish, which usually turn out to be mostly *Mugil engeli*. Total annual yield of such operations is 100 to 150 kg/ha.

Similarly, in the Philippines, fish culturists concentrate on milkfish and, perforce, Java tilapia (*Tilapia mossambica*). Where mullet are utilized as other than an incidental component of the pond ecosystem, the intensity of operations is low, comparable to that in India or Hawaii. Annual yields average 336 kg/ha. A sophisticated polyculture system, based on Philippine conditions and using mullet, milkfish, and silver carp (*Hypophthalmichthys molitrix*) as the principal components, has been proposed but, so far as is known, not attempted on a commercial scale (Table 2).

As is true in so many forms of fish culture, it is the Chinese people who have brought mullet culture to its highest development in Asia. Mullet have always been of secondary importance to the Chinese carp complex in fish culture in mainland China, although the government of the People's Republic of China is currently looking into means of better utilizing mullets. Mullet culture has, however, long been of great importance in Hong Kong and Taiwan.

Basic fish culture techniques in Hong Kong evolved from those practiced in the adjacent regions of mainland China, but farmers in Hong Kong early added mullet to the pond ecosystem because of their ready availability, as well as to take advantage of the brackish character of local waters. Since the communist revolution in China, supplies of Chinese carp fry have become more difficult to obtain in Hong Kong, and the importance of mullet has increased. The political factor is even more important in Taiwan but, even when Taiwan was politically united with China, the difficulty of shipping Chinese carps (which seldom reproduce naturally in the small rivers of Taiwan) from the mainland made it necessary to substitute locally available fish to some extent. Recent success in artificially inducing spawning of Chinese carps has reduced the severity of this limitation on Taiwanese fish culture, but mullet are popular food fish and are firmly entrenched in local practice.

The only mullet species cultured to any extent in China, Hong Kong, and Taiwan is *Mugil cephalus*, 25- to 45-mm fry of which are available in coastal waters in late fall and winter. Fry are captured at low tide by the use of large dip nets or 2- to 4-m square umbrella nets, with a mesh size of 6 mm or less. Dip nets are usually operated by pairs of fishermen, who wade against the tide and drive fry into the nets with their feet. Umbrella nets are operated from shore with the aid of an 8-m bamboo pole. Captured fry may be stocked directly into growing ponds without acclimation or nursing or they may be overwintered in special ponds and stocked in the spring, at which time they will be about 75 mm long and weigh 2 to 4 g each.

Two types of stocking system are practiced in Hong Kong. One, in which mullet are secondary to the Chinese carp complex, has already been described under Chinese Carp Culture (Chapter 3). In brackish water ponds, 10,000 to 15,000 75-mm mullet fingerlings per hectare may be stocked in February to April, along with 1000 to 2000 Chinese carp fingerlings per hectare. Smaller numbers of larger mullet may be added in the late spring and again in the fall. The species of Chinese carps used vary according to the availability of different types of food in the pond (see pp. 105–113). The mullet stock is thinned to 3500/ha whenever they reach individual weights of 140 g. Mullet of this size may be sold or used to stock other ponds.

TABLE 2. STAGGERED STOCKING AND HARVESTING SYSTEM SUGGESTED FOR POLYCULTURE IN A ONE-HECTARE BRACKISH WATER POND IN THE PHILIPPINES

FOOD ORGANISMS	SPECIES	STOCKING			
		APPROXIMATE DATE	SIZE (NO. OF FISH/KG)	NUMBER TO BE STOCKED	WEIGHT OF STOCK (KG)
Phytoplankton	Silver carp	First week of March	200–300	2,000	8.0
	Milkfish	First week of March	300–400	1,500	4.3
	Mullet	First week of March	400–500	3,000	6.7
Macrophytes	Grass carp	First week of March	100–200	600	4.0
Zooplankton	Bighead	First week of March	100–150	300	2.0
Benthos	Common carp	First week of March	400–500	500	1.1
Nekton	Lates calcarifer (apahap)	First week of June	200–350	100	0.3
Total				8,000	26.4

TABLE 2. (*Continued*)

FOOD ORGANISMS	SPECIES	CROPPING				
		APPROXIMATE DATE	AVERAGE WEIGHT OF FISH (KG)	MORTALITY (%)	SURVIVAL (NO. OF FISH)	TOTAL WEIGHT (KG)
Phytoplankton	Silver carp	Last week of February	1.0	5	1,900	1,900
	Milkfish	First week of July	0.3	5	1,425	427
	Mullet	Last week of February	0.5	50	1,500	750
Macrophytes	Grass carp	Last week of February	2.0	5	570	1,140
Zooplankton	Bighead	Last week of February	2.5	10	270	675
Benthos	Common carp	Last week of February	0.6	40	300	180
Nekton	*Lates calcarifer* (apahap)	Last week of February	0.6	10	90	54
Total					6,055	5,126

307

TABLE 3.   FEEDING SCHEDULE IN INTENSIVELY CULTURED FISH PONDS IN HONG
KONG, WITH MULLET AS THE PRIMARY CROP

NUMBER OF DAYS AFTER STOCKING	KIND OF FEED	AVERAGE DAILY RATION (KG/HA)
1– 10	—	—
11– 30	Rice bran	1.0– 1.5
31– 60	Rice bran	1.5– 3.0
16– 90	Rice bran,	3.0– 5.0
	Peanut cake	2.0– 5.0
91–150	Rice bran,	5.0– 8.0
	Peanut cake	5.0–10.0
151–210	Rice bran,	8.0–12.0
	Peanut cake	10.0–16.0
211–300	Rice bran,	12.0–16.0
	Peanut cake	16.0–24.0

SOURCE: Ling (1966).

Intensive feeding is carried out in ponds where mullet is the primary crop, according to the schedule given in Table 3. Total amounts of feed required over the 300-day growing period are 2500 kg/ha of rice bran and 3000 kg/ha of peanut cake, sometimes supplemented with soybean cake. Additional small amounts of rice bran and peanut cake, along with human and pig manure, are added every 2 to 5 days for the purpose of fertilizing the pond, but may also be utilized as food. Emphasis on manuring, particularly with human waste, is declining due to the health problems engendered by the practice in the increasingly overpopulated area that is Hong Kong.

In Taiwan, where 1,425,217 kg of mullet were produced in 1965, stocking systems are complex. Some mullet are raised in freshwater ponds in the foothill regions, where the primary fish crop is the eel *Anguilla japonica* (see Chapter 19), while others occur as a secondary crop in milkfish ponds, along with the shrimps *Penaeus carinatus* and *Metapenaeus ensis*, but most mullet are raised in the 6000 ha of very rich brackish water ponds on the coastal plain. For reasons which are partly beyond the fish culturists' control, Java tilapia are increasingly the primary component of the harvest in some of the most fertile ponds. In the Tainan and Kaohsiung areas, tilapia account for 50% of the crop, mullet for 12%, and silver carp for 10%. Other species stocked in mullet ponds in Tainan, Kaohsiung, and elsewhere in Taiwan include eels, milkfish, big head, grass carp, mud carp, common carp, goldfish (*Ca-*

rassius auratus), and Crucian carp (*Carassius carassius*). A few mullet are still grown in rice fields, but this practice is declining, due to the increasing reliance of rice farmers on insecticides, to which mullet are extremely sensitive.

Stocking practices vary widely; a "normal" regime for a 1-ha pond from which tilapia can be excluded might include 3000 mullet, 2000 milkfish, 1000 silver carp, 1000 big head, 1000 mud carp, 500 common carp, and 250 grass carp, stocked in early spring.

A feeding schedule for such a pond might require 2000 kg of rice bran, 500 kg of soybean cake, and 36 kg of peanut meal, distributed as indicated in Table 4.

TABLE 4.   FEEDING SCHEDULE IN MULLET-MILKFISH-CHINESE CARP PONDS IN TAIWAN

February–April	Small amounts of rice bran, primarily as a fertilizer
May	Begin intensive feeding with rice bran
June	As soon as the rainy season begins and supplies of natural food begin to be diminished as a result of the decreasing salinity of the pond, add soybean cake in small pieces; continue heavy feeding with rice bran
July–September	Add small amounts of peanut meal to the ration; this is believed to stimulate growth

Fertilization is also practiced, at the culturist's discretion. Manures may be used, but superphosphate is increasingly popular. The usual dose is 1000 kg/ha, which amounts to 60 kg/ha of $P_2O_5^-$. However, production of mullet in experimental ponds has been shown to increase in linear fashion with dosages of P_2O_5 up to 180 kg/ha; what would happen beyond this point is not known. It is very likely that Taiwanese mullet growers would find it economically advantageous to increase the dosages of superphosphate in their ponds.

As one might expect, growth and production of mullet in Taiwan and Hong Kong are much better than in other Asian countries, where culture methods are not so intensive. Striped mullet attain weights of 0.3 kg after 1 year of growth, 1.2 kg at the end of 2 years, and, if left in ponds for 3 years, may be grown to 2.0 kg. The highest verified yield is 2508.8 kg/ha in a 300-day growing season from a pond in Hong Kong, but yields of up to 3500 kg/ha have been claimed, and probably achieved, in the most intensively managed ponds.

PROSPECTUS

Mullet are mainly tropical and subtropical fishes, but a few species have ranges which extend well into the temperate zones, and interest in their culture in temperate climates is increasing. Culture of striped mullet is occurring on a small scale in Japan, and the pioneer British fish culturist C. F. Hickling is seeking ways to cultivate the thick-lipped mullet (*Crenomugil labrosus*), in the United Kingdom. Little is known about the thick-lipped mullet, which ranges as far north as Plymouth in England, but it has the broad salinity tolerance and predominantly herbivorous food habits which have made *Mugil* spp. so popular with fish culturists. Ripe female thick-lipped mullet are found off England from fall through early spring, but the time of spawning is not known. Hickling plans to conduct experiments with artificially inducing spawning of thick-lipped mullet by the methods which have been so successful with *Mugil cephalus*.

The potential of mullet culture in the United Kingdom is great, for there are no cultured seafood products which meet with wide acceptance in that country today. Carp are not popular and trout are strictly a luxury item, but mullet are both well-liked by British consumers and potentially capable of being produced in large volume at low cost.

Biologists in the Soviet Union are having considerable success with artificially induced spawning of *M. cephalus* which, along with *Mugil auratus* and *Mugil saliens,* is important in both fisheries and fish culture in the Black Sea, the Caspian Sea, and the Sea of Azov.

There are still large areas of warm water where mullet culture, while not practiced on a commercial scale today, is potentially feasible, and some experimental culture is occurring. For example, in 1969, Iraq emulated its Mediterranean neighbors by inaugurating the culture of *Mugil capito, Mugil cephalus,* and *Mugil oligolepis* in the brackish water lake of Abbu-Dibis.

Encouraging results have been achieved, with little effort, in the culture of mullet in the southeastern United States, but prospects for future development of commercial mullet culture are not bright, since mullet are not normally regarded as food fish in the United States, with the exception of Hawaii. Their aquacultural potential is shown by results obtained in South Carolina and Florida. An 0.6-ha brackish water pond, 1 to 2 m deep, at Bears Bluff Laboratories, Wadmalaw Island, South Carolina, stocked by natural processes and virtually unmanaged, yielded 85 to 227 kg/ha of fish, of which 47.5 to 74.2% were *Mugil cephalus,* during five 6- to 13-month growing seasons. Similar yields were obtained in fertilized ponds used for experimental monoculture of striped mullet at Marineland Laboratories, Marineland, Florida.

Higher yields of mullet were obtained, albeit inadvertently, at the Florida Board of Conservation Marine Laboratory in St. Petersburg. The body of water involved was an oblong 5.6-ha brackish water pond, averaging 1.7 m deep, originally intended for experimental culture of pompano (*Trachinotus carolinus*) (see Chapter 27 for details), fry of which were stocked and fed on ground trash fish. The yield of pompano was disappointing, but the yield of extraneous fishes was high. Silver mullet (*Mugil curema*) and striped mullet constituted the majority of the fish population and yielded 767 kg/ha over a 2-year growing period.

Mullet culture is not developed at all in the remainder of the Western hemisphere, although its potential for alleviating the serious protein problems of Latin America is obvious. The same applies to tropical Africa, except that experiments in brackish water fish culture which were begun in 1962 on the Island of Buguma in the Niger River delta, Nigeria, included, among other fishes, *Mugil falcipinnis* and *Mugil grandisquamis*. Preliminary results were encouraging, but the Nigerian civil war caused the interruption of this project.

One of the few serious objections to mullet as a food fish is that *Mugil cephalus*, at least, carries a fluke (*Heterophyes heterophyes*) dangerous to man. Under truly intensive culture conditions it should be possible to control this parasite.

Looking at all aspects of mullet culture, one must conclude that even if it does not become possible to institute controlled breeding, the spread of the best techniques practiced today—those used in Israel, Hawaii, Taiwan, and Hong Kong—would greatly increase the importance of mullet as a source of food for man. If, as seems inevitable, researchers succeed in unlocking the secrets of spawning and rearing *Mugil* spp. on a large scale, mullet could well become the most important human food product of the estuarine environment.

REFERENCES

DE ANGELIS, R. 1960. Mediterranean brackish water lagoons and their exploitation. General Fisheries Council For the Mediterranean, studies and reviews (12).

ELLIS, J. N. 1969. Hydrodynamic hatching baskets. Prog. Fish. Cult. (April 1969): 114–117.

EL-ZARKA, S. E., and F. K. FAHMY. 1966. Experiments in the culture of the grey mullet, *Mugil cephalus* Linn. in brackish-water ponds in the UAR. FAO World Symposium on Warm Water Pond Fish Culture. FR: VIII/E-9.

FISHELSON, L., and D. POPPER. 1966. Experiments on rearing fish in salt waters near the Dead Sea, Israel. FAO World Symposium on Warm Water Pond Fish Culture. FR: VIII/E-7.

GANAPATI, S. V., and K. H. ALIKUNHI. 1952. Experiments on the acclimatization of salt-water fish seed to fresh water. Proc. Indian Acad. Sci. 35:93–109.

HORA, S. L., and T. V. R. PILLAY. 1962. Handbook on fish culture in the Indo-Pacific region. FAO Fisheries Biology Technical Paper No. 14. 204 pp.

JOHNSON, M. C. 1954. Preliminary experiments on fish culture in brackish water ponds. Prog. Fish. Cult. 16(3):131–133.

LIAO, I. C. 1969. Artificial propagation of grey mullet, *Mugil cephalus* Linnaeus. J.C.R.R. Fisheries Series, No. 8, pp. 10–20.

LING, S. W. 1966. Feeds and feeding of warm-water fishes in ponds in Asia and the Far East. FAO World Symposium on Warm Water Pond Fish Culture. FR: III-VIII/R-2.

LUNZ, G. R. 1951. A salt water fish pond. Contributions from Bears Bluff Laboratories, No. 12. Bears Bluff Laboratories, Wadmalaw Island, South Carolina.

PILLAI, G. T. 1962. Fish farming methods in the Philippines, Indonesia, and Hong Kong. FAO Fisheries Technical Paper 18, 68 pp.

TANG, Y. 1964. Induced spawning of striped mullet by hormone injection. Jap. J. Ichthyol. 12(1/2):23–28.

YASHOUV, A. 1966. Mixed fish culture—an ecological approach to increase pond productivity. FAO World Symposium on Warm Water Pond Fish Culture. FR: V/R-2.

YASHOUV, A. 1969. Mixed fish culture in ponds and the role of Tilapia in it. Bamidgeh 21(3):75–82.

YASHOUV, A. 1969. Preliminary report on induced spawning of *M. cephalus* (L.) reared in captivity in freshwater ponds. Bamidgeh 21(1):19–24.

Interviews and Personal Communications

DELMENDO, M. Philippine Fisheries Commission.

HICKLING, C. F. 95 Greenway, London N20, England.

LIN, S. Y. Professor of Fish Culture, National Taiwan University.

MOZZI, C. Institute of Hydrobiology, University of Padua, Italy.

SHEHADEH, Z. Oceanic Institute, Waimanalo, Hawaii.

YASHOUV, A. Fish Culture Research Station, Dor, Israel.

17

Milkfish Culture

The milkfish (*Chanos chanos*) is one of the fishes best suited for culture in brackish water ponds. In addition to being very euryhaline, disease resistant, of high quality as a food fish, and growing rapidly, it feeds near the bottom of the food chain, mostly on algae, so that large amounts of

313

milkfish can be supported in a restricted area. Milkfish are found in warm offshore waters of the Red Sea, the Indian Ocean from East Africa to southern Australia, and the Pacific Ocean from southern Japan to Australia on the west and San Francisco Bay to southern Mexico on the east. Despite their many desirable qualities, wide distribution, and the fact that adults are difficult to harvest by conventional fishing methods, milkfish are cultured on a large scale only in Indonesia, the Philippines, and Taiwan. Small-scale or experimental culture occurs in a few other Asian countries and in Hawaii.

NATURAL HISTORY

Milkfish spawn annually or biannually in the sea near the coast in about 25 m of water. Each female broadcasts up to 5 million pelagic eggs, which hatch in about 24 hours. The larvae seek out clear coastal and estuarine waters at least 23°C in temperature, with 10 to 32‰ salinity and an abundance of phytoplankton. On occasion they may ascend into freshwater lakes. After about a year of inshore life the young, by then about 20 cm long and 200 g in weight, move out to sea, to mature about the sixth year of life. Adults, which feed on both phytoplankton and zooplankton, reach weights of up to 20 kg.

BASIC CULTURE METHODS

Though milkfish may attain lengths of up to 1 m in ponds, they do not mature sexually in confined waters and it has not thus far been possible, even by means of pituitary injections, to spawn the species in captivity. Culture thus remains dependent on fry captured in coastal and estuarine waters. Certain areas are fortuitously located for fry collection, and localized fry industries have developed in all three of the major milkfish-producing countries. Fry are acclimatized to brackish water or freshwater and reared to market size in a series of ponds. Supplementary feeding is sometimes practiced, but the primary source of nutrition for all ages in culture is a complex of benthic algae, protozoa, and detritus encouraged by specialized pond management techniques, particularly fertilization.

MILKFISH CULTURE IN INDONESIA

Milkfish culture probably originated in Indonesia, where fish farming in saltwater ponds dates back at least to 1400. Since 1821, when the Dutch colonial government began to register fish ponds, the area devoted to aquaculture in the coastal regions has more than doubled, but the methods used in milkfish culture have remained rather primitive.

FRY COLLECTION INDUSTRY

In Java, some fry may be captured in coastal ponds or "tambaks" by merely opening the sluice gates on the incoming tide, but this is not an adequate means of obtaining stock for culture, so most milkfish farmers there, as elsewhere in Indonesia, purchase fry from dealers. The chief fry collecting areas are the north shore of Java and the coasts of Madura. Postlarvae and juveniles 15 to 25 mm long are abundant in these areas from March to May and again from September to December. October and November are the best collecting months. In addition to seasonal variations there are definite lunar and tidal periodicities in occurrence of milkfish fry; the best collections are usually made at high tide during full and new moons. Preferred locations are gently sloping sandy beaches with clear water. Collections are never made over clay bottoms, since the turbid water usual in such regions obscures the fry. For the same reason, collectors prefer calm water.

The mouths of tidal creeks are particularly favorable collecting sites, as are the leeward sides of bars, and so on, where fry often seek cover. Where there is no naturally occurring cover, collectors may lure fry by constructing low rock walls at right angles to the beach. A more common and convenient form of artificial cover is the "blabar." A blabar consists of a long palm fiber rope with strips of coconut palm, sugar cane, or banana leaves or grasses plaited into it in such a way that long garlands about 10 to 20 cm in diameter are formed. Milkfish fry may be attracted to blabars simply floated on the surface at right angles to the beach, but where fry are few a different technique is employed. One end of the blabar is tied to a wooden post driven into the beach, and 20 m or so, held by a fisherman, are paid out in a wide circle near shore. When fry are observed under the blabar, the fisherman slowly narrows the circle until the fry are concentrated in a small central space.

The actual collecting gear is primitive, consisting of no more than a net of coarse fabric, with which fry are dipped out of the sea. Good collections may also be made on ebb tides by fishermen standing in rows across creek mouths.

Captured fry are immediately transferred to earthenware jars. Acclimatization to brackish water may begin at this stage, as some collectors dilute the sea water in the jars with up to 10 parts of fresh water. Such dilution also serves to kill most unwanted invertebrates. The milkfish fry must, however, still be separated from other fishes, an operation

PLATE 1.   Catching milkfish fry on the coast of East Java. (Courtesy *Science*, photograph by R. U. D. Sterling.)

requiring considerable skill and experience. Characteristics used in identification include two black spots on the head, another in the center of the body and characteristic movements. Fry may be sold on the beach to peddlers or taken to the storage houses of fry dealers.

Flat, water-tight 15-liter baskets made of interwoven strips of bamboo coated on the inside with cement or tar are used for long-distance transport of fry. Methods used in keeping fry in good condition during transport vary widely among dealers. Generally, the baskets are filled with dilute sea water to a depth of a few centimeters and fry are stocked in densities varying with the length of the journey, but averaging 20,000 to 40,000/basket. No artificial aeration is used, but the water is changed every other day. When traveling away from the sea, salinity is maintained by adding unrefined sea salt in amounts determined by taste. On long journeys and during storage the fry are fed on rice flour, which may be slightly roasted, or finely mashed hard-boiled egg yolk.

Fry may pass through several middlemen on their way to the farmer and must be counted each time they change hands. This is usually done by individually counting a sample lot, then proceeding on the basis of volume.

More than a little art is involved in supplying fry of high quality, and some dealers are more successful than others. Their methods are understandably not publicized, but they involve storage and transport at low densities and special diets. All other things being equal, Indonesian milkfish culturists prefer fry captured nearby to those which have been transported long distances. Fry collected in the September to December season are believed to be of higher quality than those taken in March to May, but since the annual demand for fry is 20 million, all live fry are salable. Since there is demand for fry at all times, some fry dealers maintain stocks temporarily stunted by minimal feeding for year-round sale.

CONSTRUCTION AND OPERATION OF TAMBAKS

Milkfish culture in Indonesia is centered on the island of Java. Most of the culture ponds, or tambaks, are located within 1 to 3 km of the sea, except in East Java where they may be up to 20 km from salt water. Most receive water from the sea, not directly, but via tidal streams, canals, and ditches; 59% are dependent on an irrigation system. Freshwater also enters the tambaks, particularly during the rainy season, with the result that the range of reported salinities is incredible: 0 to 260‰.

Tambaks may have originated on the island of Madura in connection with an ancient salt industry, but now they are constructed specifically for aquaculture by clearing and excavating mangrove swamps. Milkfish culturists may derive supplementary income from the sale of firewood obtained in the clearing process. Tambaks are usually constructed on emerging coastlines and represent a transitional stage between mangrove swamps and agricultural land, but it is nevertheless necessary to guard against erosion, so a dense plantation of mangroves along the ponds is encouraged. Such trees may also be periodically harvested for firewood, and leaves and twigs are used as green manure in the tambaks.

Ideally, a tambak should be so situated that salinity averages 10 to 35‰. For ease in drainage and general maintenance, the bottom of the tambak should be just slightly lower than high tide. The best soils for milkfish culture are soft, jellylike, hydrophilic, and biologically active muds containing about 4% humus and large amounts of clay. Such mud, which fortunately is usual in mangrove swamps, encourages the blue-green algae preferred as food by milkfish and discourages green algae which are not only less digestible by the fish, but are associated with the malaria-carrying *Anopheles* mosquito. Maintenance of a gentle current or oscillation in water level is also conducive to development of a suitable algal flora.

The structure of tambaks varies considerably. In its simplest form, as

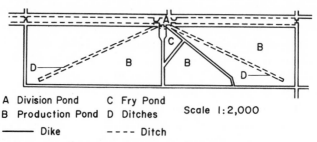

A Division Pond   C Fry Pond
B Production Pond   D Ditches   Scale 1:2,000

——— Dike   - - - - Ditch

Fig. 1.   Taman-type tambak (Indonesia). (After Schuster, 1949.)

found in west and central Java, a tambak consists of a rectangular 0.5- to 3.0-ha pond, 0.3 to 0.7 m deep, with a small wooden or bamboo sluice gate to regulate the inflow and outflow of water. The high ridges of soil frequently observed in the center of such tambaks have no function other than saving the trouble of carrying excavated mud to the embankment.

In east Java, tambaks are larger and more complex and have more elaborate sluice gates. The structure of the gates is partly a reflection of the higher degree of sophistication of milkfish culture in east Java than elsewhere in Indonesia, but heavier gates would be needed in any event to cope with the higher tides in that region. Since concrete construction is not feasible in the soft swamp soils, the gates are made of teak wood.

Tambaks constructed on elevated sites in east Java are of the Taman type (Fig. 1). The division pond (A) is connected to the main sluice gate and, through a system of secondary gates, to the fry pond (C) and production ponds (B), so that each can be individually drained and filled. During the dry season such tambaks may partially dry up and fish must be kept in the ditches (D).

The most advanced tambak design is the Porrong type, used in coastal east Java (Fig. 2). As in the Taman type, each compartment can be individually drained, but the division pond is larger, drainage is more efficient, and a 100- to 1000-m² nursery pond (D) is added. Some Porrong type tambaks may be even more complex than the one illustrated, comprising up to 10 compartments. A further refinement is the inclusion of shallow "baby boxes" 1.6 × 1.8 m, in which newly acquired fry may be placed for the first few days. Overall size of Porrong type tambaks averages 7.6 ha but may reach 30 ha.

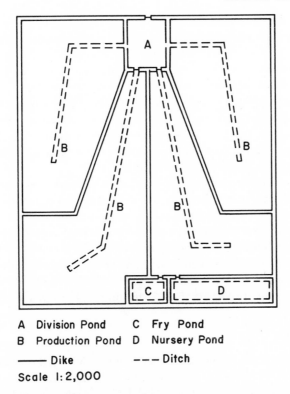

A Division Pond   C Fry Pond
B Production Pond D Nursery Pond
——— Dike   — — — Ditch
Scale 1:2,000

Fig. 2. Improved porrong-type tambak (Indonesia). (After Schuster, 1949.)

NURSING FRY AND EARLY FINGERLINGS

The depth of the compartments used for fry and early fingerlings should be less than 10 cm. Older fish are adaptable as to depth, but water less than 30 cm deep is most conducive to growth of desirable forms of algae. In ponds of low salinity the slope of the bank should be steep to discourage rooted plants. Water temperatures in tambaks vary from 24.0 to 38.5°C; pH varies from 7.1 to 7.9.

The first step in preparing for fry stocking is to drain the fry compartments and dry them for a week or more. When the bottom soil is dry it is loosened and leveled by tilling and raking. Wet, foul-smelling spots are treated with lime to prevent anaerobic decay and the production of hydrogen sulfide. Drying, tilling, and liming are usually sufficient to kill potential predators and provide an adequate growth of blue-green algae when the compartment is filled. Manuring is usually not carried out in nursery ponds, for a sudden die-off of the resulting heavy algal bloom might lead to heavy mortality of the fry. (Nursery ponds are manured in Taiwan, but perhaps algal blooms are less of a problem in that country's temperate climate.) As a final step, large palm leaves may be stuck in the banks around the perimeter to provide shade for the fry.

Stocking of nurseries is carried out in the cool hours of early morning or evening. Temperature and salinity in the fry container are equalized with conditions in the tambak by immersing the container in the tambak and periodically replacing a small portion of the water with tambak water. Fry are stocked at densities of up to $55/m^2$.

Fry in baby boxes may be fed mashed hard-boiled egg yolk, wheat starch, or fine rice bran. Older fry usually subsist on naturally occurring phytoplankton, but their diet may be supplemented with rice bran fed once or twice a day. Except for occasional feeding, management of nurseries is usually limited to changing the water twice a month on high tides.

The fry are considered ready for transfer to rearing ponds when the food supply in the nursery begins to be depleted, usually after about 30 days, at which time the fry are 5 to 7 cm long and weigh 1.4 to 3.7 g.

GROWING FOR MARKET

Production ponds are stocked at a maximum of 600 kg/ha (2000 to 10,000 fingerlings/ha). Preparation for stocking is similar to that practiced in nurseries, but manuring at about 2000 kg/ha is also carried out. Green manures only are used, as inorganic fertilizers have not been found useful. The usual fertilizer is mangrove (Avicennia), leaves and twigs of which are readily available, but occasionally rice straw is used. Although after milkfish pass 20 cm in length they become progressively more capable of consuming such comparatively tough foods as green algae and higher plants, blue-green algae are encouraged in rearing ponds as well as nurseries. The bottom of a well-managed production pond will be covered by a thick mat of such blue-green algae as Oscillatoria, Lyngbya, Phormidium, Spirulina, Microcoleus, Chroococcus, and Gomphosphaeria, as well as diatoms, including Navicula, Pleurosigma, Nastogloia, Stauroneis, Amphora, Nitzschia, and Gyrosigma. These constitute the main food of cultured milkfish, but other components of the benthic flora and fauna are also ingested, and filamentous green algae and higher plants may be eaten, particularly if they have been softened by decay. Supplementary feeding is not necessary in good ponds but may be carried out using rice bran, wheat starch, and various kinds of oil cakes.

Cultured milkfish are remarkably free of parasites and hardly any of those which have been recorded are pathogenic. The only disease known to afflict milkfish is described as a "cold." Following sudden chills, milk-fish may become lethargic, cease to feed, and develop a milky discoloration of the skin which disappears after 2 to 3 days. The "cold" itself is seldom fatal but does result in weight loss and probably increased vulnerability to predators.

Predatory fishes, crabs, and so on, may present some problems but can be controlled by adequate screens on sluice-gates and by occasional draining of the tambaks. In east Java, this is done up to four times a year, not only to control predators but to increase the fertility of the tambak and to harvest juvenile prawns for culture.

A common pest in tambaks is the snail *Cerithidea*, which has been observed at densities of up to 700/m². *Cerithidea* is reputed to compete for food with milkfish and may deplete the calcium content of tambak water in the process of shell formation. Fertilization with molasses has been recommended as a control measure, but this is uneconomic and functions only by promoting the growth of blue-green algae, which should be abundant in any event. In tambaks which are properly managed so as to maintain blue-green algae blooms *Cerithidea* populations seldom reach problematical levels. Another pest is the polychaete worm *Eunice*, which by its burrowing activities causes tambak soils to become excessively porous. No preventive or remedial measures have been devised specifically for *Eunice*, but poisoning with teaseed cake is effective in control of the related *Nereis* in Taiwan.

Despite the lack of the disease and parasite problems which beset many other fish culturists, Indonesian milkfish farmers generally experience poor survival of their stock. Normally 20 to 50% of the fry stocked are harvested, but some Chinese culturists, who are generally more skilled and meticulous than the native Javanese farmers, achieve 60 to 80% survival. So far as is known, the chief causes of mortality are increasing salinity and pollution through decay of organic detritus. Proper tambak management, as described above, will control the latter, while salinity may be held down by occasional draining and/or flushing of tambaks with freshwater.

Milkfish are occasionally grown to weights of 1 to 3 kg, but most are harvested at 300 to 800 g. The time required to reach this size depends on the skill of the culturist, but also on the location of the tambak. Porrong-type tambaks in east Java may yield three crops a year, whereas in inland tambaks up to 10 months are required for growth to marketable size.

Around 1952 a new system of milkfish culture in freshwater, which may have application in inland tambaks of low salinity, was developed in central Java. Fry are stocked directly into shallow production ponds

2500 m² or less in area. The ponds contain dilute sewage, the concentration of which is gradually increased until a bloom of blue-green algae and diatoms appears. In this enriched environment a 3-month rotation system can be used and annual yields of 5000 kg/ha are not uncommon.

The average yield of conventional milkfish culture in Indonesia has been variously estimated at from 50 to 500 kg/ha. The discrepancy in figures is in part due to variations in soil type, as some tambaks are constructed over sterile substrates of senile lateritic soil or even sand.

POLYCULTURE

Milkfish culture in Indonesia is essentially a monoculture system, and other species are not intentionally introduced, but the nature of the tambaks insures that some extraneous organisms will find their way in. Some of these may be chiefly detrimental by competing with or preying on the milkfish, but other extraneous species may be harvested and sold, notably Java tilapia (*Tilapia mossambica*), tarpon (*Megalops cyprinoides*), ten pounder (*Elops hawaiensis*), sindo (*Mugil engeli*), cock-up (*Lates calcarifer*), and erong-erong (*Therapon jarbua* and *Therapon theraps*). Total yield of extraneous fishes is usually about 16 to 35 kg/(ha)(year), but in tambaks in the early stages of construction or in those with weak dikes "wild" fish may be encouraged and 100 to 150 kg/(ha)(year) of mixed fish, chiefly *Mugil engeli,* harvested.

Of greater importance are various species of crustaceans. In particular, Penaeid shrimps, including the Indian prawn (*Penaeus indicus*), the green tiger prawn (*Penaeus semisulcatus*), the yellow prawn (*Metapenaeus brevicornis*), and *Metapenaeus ensis,* are encouraged. The methods of trapping and harvesting prawns are essentially the same as those practiced in Malaysia (see pp. 592–593). Yields vary from 25 to 400 kg/(ha)(year). Other crustaceans of commercial importance include crabs (*Scylla serrata*), primitive Mysid shrimps or "rebon," and "djembret," a mixture of small and larval decapods augmented by fish larvae. An average of 15 to 30 kg/(ha)(year) of rebon and djembret are harvested from coastal tambaks and cured by partial drying and bacterial action to produce a paste called "trassi," highly valued as a flavoring ingredient in Javanese cookery.

In practice, only 70% of the income of a typical Javanese milkfish "monoculture" operation is derived from sale of milkfish. Prawns account for 20%, extraneous fish for 5%, and by-products (rebon, djembret, crabs, mangrove wood, vegetables planted on the dikes, etc.) for the remaining 5%.

Intentional polyculture is practiced in the region of the Bengawan

Solo River, where 6000 ha of tambaks are of such low salinity (less than 8‰) as to be essentially freshwater ponds. There milkfish are sometimes cultured together with tawes (*Barbus gonionotus*). Tawes generally constitute the main crop, but some ponds contain up to 75% milkfish, depending on the demand for the two species and the availability of suitable food. Where filamentous green algae and higher aquatic plants are abundant, it is wise to stock large numbers of tawes. Not only do they thrive on such food, but partly digested plant remains in their feces provide both a source of food for the milkfish and fertilizer for blue-green algae.

HARVEST AND PROCESSING

Milkfish are most easily harvested in tambaks where strong tidal currents occur. At high tide they tend to congregate near the sluice gates and can be netted easily. Where tidal influence is not strong, tambaks may be drained for harvesting, or cast nets may be employed.

Milkfish is an important food for all economic classes in Java. Small farmers auction most of their crop as fresh fish in local markets. Larger operators usually work through middlemen and ice their fish for shipment throughout the country. Sometimes a large dealer will offer a flat rate for the entire crop of a small farmer's tambak.

In addition to being sold fresh and iced, milkfish are cured for market by boiling in brine or by smoking. Boiling, which is particularly prevalent in remote areas where ice is not available, yields an important by-product. The evaporate of the brine after boiling is a paste known as "petis," which is sold as a flavoring ingredient. Smoked milkfish is principally a luxury food enjoyed by well-to-do Indonesians.

YIELDS, PRODUCTION, AND PROSPECTUS

Although recent data are difficult to obtain from Indonesian sources, 150,000 to 200,000 ha of tambaks are believed to be under cultivation, chiefly in Java, but also in Madura, Celebes, Sumbawa, and northern Sumatra. Total annual production of milkfish is estimated at 65,000 metric tons.

Application of more sophisticated methods of milkfish culture, such as those practiced in Taiwan, would undoubtedly prove beneficial in Indonesia. This would involve financial assistance to tambak operators, establishment of research stations and, particularly, propaganda and demonstrations aimed at the milkfish farmers, who for the most part are very conservative about adopting new methods. At present, canals and sluice installations on most East Javanese tambak complexes have silted in and otherwise fallen into disrepair; thus rehabilitation must precede or accompany improvements and adoption of possible innovations.

Increasing the area under cultivation is also possible, especially since very many tambaks incorporate large areas of unproductive shallow water. It is estimated that excavating such areas could provide 14,000 ha of additional productive water in Java.

Opportunities for constructing new tambaks in Java are limited, because there are only about 10,000 ha of unexploited coastal swamps remaining on the island. In Indonesia as a whole, however, there are an estimated 6 million ha of mangrove swamps, some of which could be developed into tambaks. Construction of tambaks in much of this area may be retarded by inaccessibility and the consequent high cost of labor. Mechanization is not the answer, even if it were possible to have the necessary equipment on the spot, since heavy machinery is virtually useless in hydrophilic mud. Nevertheless, ways of expanding tambak aquaculture should be sought; if as little as 5% of the undeveloped areas were brought under cultivation, the milkfish production of Indonesia could be nearly tripled with no improvement whatsoever in culture techniques.

MILKFISH CULTURE IN THE PHILIPPINES

FRY COLLECTION

The milkfish fry industry in the Philippines differs from that in Indonesia in that most fry are raised to fingerling size (5 to 10 cm) before sale to market growers. Fry are available for collection in Luzon from March to August, with a peak in abundance during May and June. Depending on the circumstances of collection, capture may be accomplished by means of nets or traps.

Dip nets are most efficient at the peak of the season and wherever fry are highly concentrated, as in narrow tidal streams. A variety of dip net known as a scissor net is particularly effective on level shores. The net proper is triangular in shape, but only two sides have rigid frames; the third side is reinforced with rope. The poles which form the rigid sides are extended to form handles. Where they cross they can be rotated on each other like a pair of scissors, so that the angle between them, and thus the width and tautness of the net, can be altered. Where fry are dispersed, or in the mouths of large rivers, unweighted seines 1 to 5 m long and 1 to 1.3 m deep are employed.

At the mouths of tidal streams too large for efficient use of dip nets, funnel-shaped traps made of bamboo or woven palm leaves may be set facing upstream in about 1 m of water. Such traps are 2 to 5 m long, 3 to 5 m wide at the mouth, and taper to about 30 cm at the rear where a fine mesh net 0.6 m wide and 1.5 m long, spread and floated by means of two parallel bamboo poles, is attached. Traps, which may be placed close together so as to completely block off a stream mouth, are anchored to bamboo or wooden poles driven into the bottom. Sometimes traps are attached to the poles in such a manner that they float and can rise and fall with the tide.

Fry collected in the Philippines are usually only about 10 to 13 mm long, slender, and so transparent they are nearly invisible. For this reason milkfish fry are usually separated from extraneous animals not visually but by sieving through 1.5-mm mesh.

Sorted fry are transported in 15- to 30-liter unglazed earthen jars stocked at a rate of about 100 fry/liter. If they are to be taken inland, the salinity is gradually reduced by dilution to facilitate changing of water en route. It is believed beneficial to keep fry in darkness during transport, so the jars are covered. Fry are capable of surviving for up to two weeks without food, and transportation is usually arranged so as to ensure that the fry will reach the nurseries within that time. Young fry are very fragile and careless handling may result in mass mortality. Nevertheless, most consignments reach their destination with losses of only 5 to 10%.

REARING FRY IN NURSERIES

Nurseries are of various sizes, usually accounting for about 1% of the total water area of those farms which incorporate nurseries and production ponds. Individual ponds are small (500 to 5000 m²), rectangular, and laid out in a regular pattern to facilitate management. Each of a pair of growing ponds is separated by a gate from a single 20- to 50-m² catching pond. As mentioned, most fry are raised to fingerling size by specialists. In the minority of farms where both fry nursing and growing for market are carried out, ponds divided into compartments, similar to Indonesian tambaks, may be used.

The success of a milkfish nursery is in large part dependent on proper site selection. The requirements for a nursery site include:

1. An adequate year-round supply of clean brackish water.
2. Location such that ponds can easily be drained.
3. Fertile soil containing large amounts of clay.
4. Freedom from floods.

5. Accessibility to fry fishing grounds.
6. Proximity to milkfish farmers who will buy fry.

Areas free of vegetation are preferred to thickly wooded regions, which are expensive to clear. On the other hand, such areas do not yield the wealth of by-products characteristic of the ecologically integrated milkfish operations in the Indonesian mangrove swamps.

Preparation of nursery ponds begins 2 months before stocking. First the ponds are drained, then immediately tilled with wooden rakes, and leveled, so that the bottom slopes toward the sluice gate. Some culturists dig a diagonal ditch from one corner of the pond to the outlet to provide a refuge for the fry during hot weather. If predator control is not strict, however, the fry refuge may become a fry trap. After drying for 2 to 3 days to fertilize the pond and kill unwanted organisms which may be buried in the mud, enough water is admitted to cover the bottom to a depth of 3 to 5 cm. Within 3 to 7 days the benthic complex of blue-green algae, diatoms, bacteria and various animals, which is typical of well-managed milkfish ponds, will start forming. When this biological complex, known as "lab-lab" in the Philippines, is first observed, the culturist may begin to gradually add water until the pond is 10 cm deep. From then on, the water should be changed at least every 2 weeks and preferably more frequently to prevent a build-up in salinity.

When the growth of lab-lab is luxuriant, the pond is ready for stocking. Fry are stocked at 30 to $50/m^2$, using the same methods as in Indonesia.

Predator control is particularly important in nursery ponds. Despite drying and screens on sluice gates and pipes, some predatory fish, crabs, and shrimp do get in, and they should be removed as soon as they are seen. Frogs, as well as frog egg masses, are also removed. Predatory birds may sometimes be discouraged by erecting poles on the pond bank and criss-crossing the pond with strings.

Another serious cause of mortality is sudden reduction of salinity or temperature during heavy rains. This is especially critical during the first 3 to 4 days after stocking, when heavy rainfall may cause up to 80% mortality by lowering the temperature of the surface water. To guard against such catastrophes, the culturist should raise the water level in his ponds during rainy periods. It is helpful to maintain a reservoir of brackish water for this purpose and other emergencies.

In the Philippines, as elsewhere, disease is scarcely a problem in milk-fish culture. An epidemic fin rot has been recorded in fry, but it is rare. The only serious health problem in fry nursing is undernourishment. In addition to the usual symptoms, such as hollow bellies and lethargy, starved milkfish fry tend to separate from the schools in which they normally swim and develop a blackish color on their backs. The bases of the fins may appear shiny, due to protruding bones. The only "cure" for undernourished fish is an adequate growth of lab-lab, but supplementary feeding with rice bran may be undertaken as a stop-gap measure.

Within 4 to 6 weeks of stocking the fry should reach the fingerling stage (5 to 10 cm and 1.2 to 5.0 g) and must be sold to market growers or stunted for future sale. Stunting is accomplished by stocking fingerlings at a density of at least 30/m² in ponds low in lab-lab (often this can be accomplished by merely leaving them in the nursery pond) and feeding 10 kg/ha of rice bran or dried filamentous green algae daily.

The first step in harvesting fingerlings is to partially drain the pond at low tide. The fish are thus concentrated near the gate separating the nursery pond from the catching pond. On the next low tide, the gate is opened. The fry, being naturally inclined to swim against the current, swim into the catching pond, from which they may easily be seined.

Fingerlings are transported like fry, but the journeys undertaken are usually shorter. When possible, transport of fingerlings is done during the night or the cooler parts of the day.

GROWING FOR MARKET

Traditional Methods.   In the Philippines, milkfish fingerlings are not usually placed directly into production ponds; first they spend a short time in 0.5- to 5.0-ha "transition ponds," 15 to 25 cm deep. Transition ponds are prepared in the same manner as nursery ponds, but they are not stocked nearly as heavily, and green algae may be permitted to grow in them. If the growth of lab-lab becomes so luxuriant that masses of it break off and float, these should be removed, or broken up so they sink, or the fingerlings transferred to another pond, since small fish may become entangled in such floating algal masses.

After about a month in the transition pond the young milkfish are 10 to 15 cm long, have attained the "garuñgin" stage, intermediate in appearance between fingerlings and mature milkfish, and are ready for stocking in production ponds at 1000 to 2500 kg/ha.

Production ponds, transition ponds and, where used, nursery ponds may form part of a complex not unlike the Indonesian tambak (Fig. 3).

In the traditional Philippine method, preparation of the 1- to 50-ha production ponds for stocking follows the pattern described for nursery and transition ponds, but extra pains are taken to assure a good growth of lab-lab and filamentous green algae. Fertilization with green manures or copra slime at 450 to 900 kg/ha may be carried out while the pond is dry.

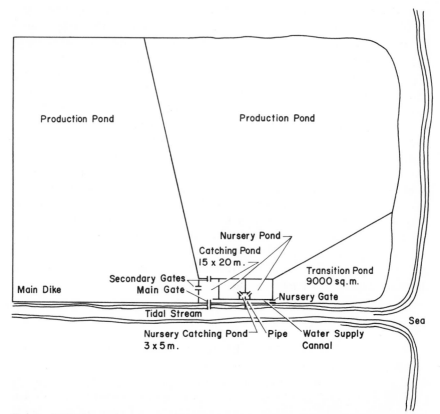

FIG. 3.   Milkfish pond system in the Philippines. (Philippines Bureau of Fisheries.)

If the algal crop of a production pond is not satisfactory, it may be supplemented by stocking with lab-lab or green algae grown in ponds set aside for this purpose. Filamentous green algae is especially well suited to this purpose since it will settle and grow on poles placed in ponds and is then easily transported from pond to pond.

Supplementary feeding is sometimes carried out, using paddy straw or hydrophilic plants such as water hyacinth (*Eichornia crassipes*), *Ruppia*, *Najas*, *Halophila*, and *Thalassia* (the latter four collectively known as "digman") in a decayed or dried form.

A more recent development is the use of fresh marine red algae (*Gracilaria*) as food for milkfish. It is in most respects a far more nourishing food than other plants commonly fed to milkfish, as it contains 1/15 to 1/10 as much water, therefore much more protein and carbohydrate per fresh weight. Table 1 shows the supposed nutritional content of fresh

PLATE 2. Philippine milkfish ponds with nursery ponds in foreground connected to estuary by filtered flume. (Photograph by J. H. Ryther.)

PLATE 3. Closer view of bamboo fiber filter used to screen water entering milkfish ponds from the estuary. (Photograph by J. H. Ryther.)

TABLE 1. NUTRITIONAL CONTENT OF PLANTS FED TO MILKFISH (PERCENT)

FOOD PLANT	WATER	ASH		FAT		PROTEIN		CARBOHYDRATES[a]	
		FRESH	DRY	FRESH	DRY	FRESH	DRY	FRESH	DRY
Gracilaria confervoides (marine red alga)	6.92	15.31	16.48	0.4	—	11.98	12.89	65.39	70.63
Chaetomorpha spp. (green algae)	85.50–91.46	2.09–2.82	19.50	0.27–0.91	0.71	2.82–3.72	27.66	7.87	52.13
Cladophora spp. (green algae)	57.20	9.90	23.16	0.84	1.96	5.16	12.07	26.90	62.81
Enteromorpha intestinalis (green alga)	81.35	6.02	32.27	0.48	2.57	3.66	19.61	8.48	45.55
Mixed blue-green algae	90.14	5.11	—	0.21	—	2.32	—	—	—
Phytoflagellates	88.02	0.72	—	1.32	—	3.91	—	—	—
Mixed diatoms	87.13	6.52	—	0.94	—	2.89	—	—	—
Eichornia crassipes (freshwater higher plant)	89.81	1.34	13.15	—	—	2.19	21.49	6.66	65.36

[a] In some cases, carbohydrate weights may include molecular water.

and dry *Gracilaria* and a number of other milkfish food plants. The water contents cited for fresh *Gracilaria* and *Cladophora,* particularly the former, seem impossibly low for aquatic plants. It is likely that the apparent inconsistency lies in the inclusion of molecular water in the carbohydrate weights.

On some Philippine farms *Gracilaria* has almost completely replaced lab-lab and other plants as a food for milkfish in rearing ponds and has resulted in such improvements in growth that an extra harvest can be scheduled. Fresh *Gracilaria* is not, however, suitable for use in inland ponds, since it will not withstand salinities below 5‰. *Gracilaria* may be added all at once, before stocking, to form a 15-cm-deep carpet on the bottom, or it may be fed in small lots throughout the period of growth. In either case, the cumulative amount fed usually amounts to 200 to 500 kg/ha. One disadvantage of *Gracilaria,* as opposed to lab-lab, filamentous green algae, and some higher plants, is that it provides no shade for the fish. This may be compensated for by raising the water level as high as possible during the summer.

Milkfish grown using these methods attain an average weight of 450 g and may be harvested for market 6 to 9 months after stocking. Mortality during the rearing period averages 50 to 70%.

Improved Methods. Around 1966, the Philippine Fisheries Commission, acting on the advice of Taiwanese FAO biologists, began to promote modernization of milkfish culture in the country, using methods which have proven effective in Taiwan. Suggested changes involve increasing scientific use of fertilizers, general adoption of supplementary feeding, more rigorous pest control, harvesting smaller fish, and population control by means of more or less continuous planting and cropping in rotation. In a large country like the Philippines, scheduling of farm operations to take advantage of the proposed new methods must take into account local climatic conditions (Table 2). Our detailed description of techniques applies to Management Program I in Table 2 but may readily be adapted to the other two programs.

Preparation of a production pond for the first growing season begins in early November with draining and drying for 2 weeks, or until the bottom soil hardens and cracks. Before filling the pond with 15 cm of water, 2000 kg/ha of chicken manure and 400 kg/ha of tobacco waste are spread on the bottom. Tobacco waste, which is plentiful in the Philippines, serves not only to fertilize the pond but, due to the nicotine content, acts as a powerful yet biodegradable natural pesticide. Saponin at 15 to 18 kg/ha or quicklime at 1000 kg/ha is equally effective. For soils lighter than silty loam, the dosage of chicken manure should be increased by 25%.

TABLE 2.   MANAGEMENT PROGRAMS FOR MILKFISH PONDS IN THE PHILIPPINES

TYPE OF MANAGEMENT PROGRAM	PERIOD FOR CONDITIONING THE POND BOTTOMS	PERIOD FOR REARING THE FISH	TYPE OF WEATHER	REGION
Type I				
First fish rearing season	November	December to April	First type: two pronounced seasons, dry from November to May, wet during the rest of the year	The western part of North and Central Luzon and of Mindoro, Panay, Negros and Palawan
Second fish rearing season	May	June to October		
Type II				
First fish rearing season	October	November to March	Second and third types: seasons not very pronounced, relatively dry from November to April, or with maximum rainfall from November to January	The eastern part of South Luzon, Samar, Leyte, Panay, Negros, Palawan and Mindanao; the western and central part of Mindanao; and the whole island of Masbate and of Cebu
Second fish rearing season	April	May to September		
Type III				
First fish rearing season	February	March to May	Fourth type: rainfall more or less evenly distributed throughout the year	The western part of South Luzon, Samar, Leyte, and Mindoro; the eastern part of North and Central Luzon; the southern part of Mindanao; and the whole island of Bohol
Second fish rearing season	June	July to September		
Third fish rearing season	October	November to January		

No more water is permitted to enter, and by December or January the pond will dry and the bottom will crack as before. At this point it is recommended to add 400 kg/ha of rice bran, along with inorganic fertilizer (100 to 150 kg/ha of 18-46-0 or 200 kg/ha of 16-20-0, 12-12-12, or 12-24-12). It seems strange that Taiwanese biologists would recommend inorganic fertilizers, which have proven inferior to manures, and in some cases actually detrimental to blue-green algae, in every experiment conducted in Taiwan or Indonesia. Perhaps it is worth noting that the FAO biologists' recommendations are repeated verbatim in pamphlets distributed in the Philippines by the petrochemical industries, which supply both inorganic fertilizers and chemical pesticides. It would appear that the thinking Philippine milkfish culturist is still very much on his own when it comes to selecting and applying fertilizers. It may be helpful as a rule of thumb to remember that 14 kg of algae are required to produce 1 kg of milkfish without supplemental feeding. If fertilization produces an excess of algae, the excess may be harvested and dried for use as a supplementary food in less rich ponds.

Following fertilization, the pond is filled to 10 to 15 cm. The tobacco waste added earlier should have eliminated polychaete worms and snails, but predatory fishes, as well as such milkfish competitors as mullets (*Mugil* spp.), scats (*Scatophagus* spp.), and chironomid larvae must still be eliminated from ponds and connecting canals. FAO and the Philippine Fisheries Commission have recommended a frightening array of chemical biocides, but saponin, which is satisfactory and readily available in the Philippines (it is a component of teaseed oil), should be chosen for the safety of the consumer and the fishpond ecosystem. A dosage of 0.5 ppm is adequate to kill unwanted animals.

Three days after poisoning, the water depth should be increased to 20 to 25 cm. A week later, after two complete changes of water, the pond may be stocked. With the last water change, the culturist should try to provide, and subsequently maintain, an environment conducive to the welfare of milkfish. This includes a temperature of 25 to 36°C, a salinity of 10 to 50‰ and a pH of 7.8 to 9.5.

The initial stocking consists of 150 kg/ha of half-grown fish (8 to 15 fish/kg), 40 kg/ha of garuñgins (30 to 60 fish/kg), and 7 kg/ha of late fry or early fingerlings (300 to 400 fish/kg). A second batch of fish, consisting of 12 kg/ha of the smallest size group, is stocked in mid-March. The goal of this sort of stocking is to keep the population of milkfish in the pond at 450 to 550 kg/ha while permitting harvest every 2 weeks throughout the growing season. Similar techniques allow 300 to 800 kg/ha. Each time a batch of fish is harvested, the total weight of fish in the pond is temporarily reduced but, as more food per fish becomes available, growth is rapid after each harvest and the loss is soon made up. More complicated pond management schemes, involving up to five stockings in a growing season, have been proposed. Figure 4 graphically illustrates the results of such a scheme.

After stocking, fertilization is continued; the FAO biologists recommend 100 kg/ha of 16-20-0 inorganic fertilizer per month. Supplementary feeding with 30 to 50 kg/ha of rice bran is recommended on cloudy days, as long as the water is clear. If feeding should induce a bloom of dinoflagellates, which are not good food for milkfish, interfere with the vision of both fish and farmer, inhibit the growth of lab-lab, and present the danger

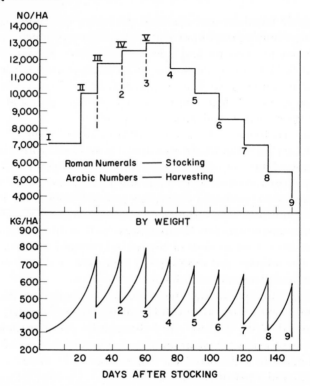

FIG. 4. Composition of milkfish population by number as proposed for management in the Philippines (above). Population dynamics of a milkfish pond stocked and harvested on a staggered basis throughout the rearing season (below). (After Tang, 1967.)

of oxygen depletion following a sudden die-off, the pond may be treated with oil cakes or a 9:1 mixture of rice bran and starch at 30 to 50 kg/(ha)(day) until the water becomes transparent. Supplementary feeding is also practiced whenever the algal pasture becomes overgrazed.

Pond preparation for the second growing season is as described above, except that before stocking the pond should be filled to 25 to 30 cm and maintained at that depth to compensate for higher air temperatures. The initial stocking in early June consists of 160 kg/ha of half-grown fish, 80 kg/ha of garuñgins, and 12 kg/ha of late fry or early fingerlings. The garuñgins stocked at this time are from the fry and fingerlings stocked in February and March. In late July, 6 kg/ha of the smallest size group are added. Most of these will be stocked again as garuñgins the following February.

Fertilization and feeding must be heavier in the second growing season.

Heavy supplementary feeding is important near the end of the growing season when lab-lab is likely to become depleted. Where the soil is heavier than silty loam, the rate of supplementary feeding should be increased by 25% throughout the second growing season.

POLYCULTURE

Such polyculture as takes place in Philippine milkfish ponds occurs more by accident than by design. Among the fish most commonly harvested together with milkfish are tarpon, ten pounders, climbing perch (*Anabas testudineus*), and hito (*Clarias batrachus*), as well as such invertebrates as the sugpo prawn (*Penaeus monodon*) and the crab *Scylla serrata*. In recent years, the Java tilapia (*Tilapia mossambica*) has become increasingly common. Although it is highly palatable, it is disliked by the milkfish farmers because it devours the lab-lab intended for their stock.

Prawns and milkfish are sometimes intentionally grown together, but the trend is for those growers who have access to good supplies of prawns to switch to monoculture of prawns, which is more profitable, though a combination of the two is more productive (see pp. 594–598).

HARVEST

Various methods are used in harvesting milkfish in the Philippines, but the rotation method of stocking and harvest already described is facilitated if gill nets are used. The size of net used depends on the size of fish it is desired to harvest (Table 3). Gill nets and seining may damage lab-

TABLE 3.   MESH SIZE OF NETS USED IN HARVESTING MILKFISH IN THE PHILIPPINES AND THE SIZE OF FISH CAUGHT

STRETCHED MESH SIZE (CM)	APPROXIMATE WEIGHT OF FISH CAUGHT (G)
7.0–7.5	285
6.5–7.0	250
6.0–6.5	222
5.5–6.0	200
5.0–5.5	182

lab, particularly if the growth is thick. To prevent such damage, the pond may be partially drained and the fish captured in a specially constructed catch pond, as is done in Indonesia. Near-total drainage is an efficient means of harvest but is believed to impart a muddy taste to the fish. Many Philippine farmers prefer to harvest milkfish at night, so they can take the freshest possible fish to market in the morning.

YIELD, PRODUCTION, AND PROSPECTUS

The average yield of traditional Philippine milkfish culture has been estimated at from 300 to 500 kg/ha. The improved methods advocated by FAO biologists have been adopted by a few farmers, and their average yield is now reported to be 1000 kg/ha. Government-operated demonstration ponds have produced yields 500% greater than the national average, and the stated goal of the Philippine Fisheries Commission is to raise average milkfish production levels to 2000 kg/ha. If this goal were realized, the national production would increase from 70,000 metric tons to 200,000 metric tons.

Expansion of the industry will also play a role in the future of Philippine milkfish culture. In 1966, there were an estimated 137,000 ha of water devoted to milkfish culture. In 1970, there were 157,000 ha, and there are still nearly 500,000 ha of undeveloped swamps potentially suitable for milkfish farming. Thus the maximum theoretical yield of Philippine milkfish culture, using improved methods is approximately 1.3 million metric tons.

The principal restraint, both to development of new areas and universal application of improved techniques, is the scarcity of fry. The fry industry had difficulty supplying the 1,370,000 fry needed in 1966, and unless the efficiency of fry capture and culture improves or new sources are found, it will simply be impossible to supply the projected demands of an expanding and increasingly efficient milkfish farming industry.

MILKFISH CULTURE IN TAIWAN

It is not known how milkfish culture got started in Taiwan. Taiwanese freshwater fish culture originated in the sixteenth century, when immigrants from the Chinese mainland brought with them the ancient technique of pond polyculture, as well as the various cyprinid fishes used in traditional Chinese pond culture. Until the development in this century of methods for induced spawning of Chinese carps, these fish could not be reproduced in Taiwan and it is possible that, to compensate for a shortage of carp fry, Chinese farmers on Taiwan adapted their techniques to suit the milkfish which, though available on the mainland, was not cultured there. Or perhaps the Dutch, who occupied Taiwan from 1624 to 1661, brought the knowledge of milkfish farming from Indonesia.

Whatever its origin, milkfish culture was well established in Taiwan by the late seventeenth century and has played an important role in the island's food economy ever since.

Despite its shorter history and the shorter growing season in Taiwan (eight months as opposed to a full year), Taiwanese milkfish culture is much more productive than that of Indonesia and the Philippines. Historically this has been due to better and more progressive farm management, and Taiwan continues to lead the world in developing improved methods of milkfish farming.

FRY COLLECTION

The best months for fry collection in Taiwan are April and May, but some fry are available as late as August. Collecting is generally best on spring tides, during full and new moons. The supply of fry varies greatly from year to year; a record 204 million were taken in 1958, but the following year only 58 million were caught. The approximate annual demand has leveled off at 160 million fry, but the catch was well below that level in five of the nine years 1958 to 1966, the most recent years for which data are available (Table 4). In years when the supply is low, fry are imported from the Philippines. It is hoped that the annual catch can be increased and stabilized by the introduction of mechanization and the development of improved methods of estimating and predicting fry crops.

As elsewhere, the traditional fry collecting gear is a triangular dip net. The Taiwanese version is 1.2 to 1.8 m wide at the front, 1.5 to 2.7 m long, and has a rigid frame on two sides only; the third side consists of a rope incorporating lead weights. A metal or bamboo receptacle is attached to the cod end of the net. All ages and sexes participate in fry fishing; the smaller nets are designed for women and children. Fry collectors may either wade or, in deep water, float on an inner tube.

A more efficient piece of gear, especially in deep water, is a sort of bag seine of varying length, with an opening in the top of the bag. Such seines may be pulled by two men wading or by one man on a raft and one on shore. Periodically the fishermen stop and dip the fry out of the bag into buckets. Where there is a strong current, the bag seine may be anchored facing into the current and operated as a trap.

Fry are stored in baskets or specially constructed cement tanks on the beach pending sale, either directly to farmers or to dealers, the latter located principally in the city of Tainan. Dealers keep fry in lots of about 20,000 in 3-m² concrete tanks fed with tap water rendered sufficiently saline by periodic addition of common salt. When fry are to be trans-

TABLE 4. AREA DEVOTED TO MILKFISH CULTURE, PRODUCTION OF MARKET-ABLE MILKFISH, AND MILKFISH FRY PRODUCTION IN TAIWAN, 1920 TO 1966

YEAR	TOTAL AREA (HA)	ANNUAL PRODUCTION (KG)	KG/HA	TOTAL FRY PRODUCTION (MILLIONS OF FRY)
1920–1929	8,000[a]	8,000,000	1,000	29.5
				Extremes
				(14–49)
1930	6,940	7,990,800	1,160	24
1931	7,420	8,209,321	1,100	24
1932	7,465	8,090,730	1,080	25
1933	8,028	6,436,742	800	25
1934	7,830	7,893,758	1,000	14
1935	7,717	9,020,544	1,170	—
1936	7,667	9,592,547	1,210	—
1937–1939	—	—	—	—
1940–1944	6,835	5,761,364	840	26
1945	6,067	3,007,306	500	25
1946	6,465	5,766,080	880	50
1947	8,698	8,190,088	940	43
1948	10,600	13,078,284	1,230	73
1949	11,154	13,348,029	1,200	77
1950	13,084	15,359,992	1,180	102
1951	13,103	14,090,760	1,080	54
1952	12,724	15,467,744	1,220	93
1953	13,457	19,324,143	1,453	96
1954	13,759	22,407,427	1,620	145
1955	13,869	26,507,347	1,900	124
1956	14,315	24,397,443	1,700	151
1957	14,337	27,033,629	1,890	148
1958	14,987	29,206,180	1,940	204
1959	15,326	25,693,506	1,480	58
1960	16,713	26,156,836	1,560	202
1961	16,600	31,740,247	1,900	135
1962	16,417	25,714,543	1,560	92
1963	15,506	25,880,540	1,800	94
1964	16,147	30,686,265	1,900	171
1965	15,616	27,562,304	1,760	92
1966	15,616	29,094,000	1,863	163

[a] Rough estimate by Yamamura (1942).

ported short distances, they are placed in waterproofed bamboo baskets; for longer trips they are stocked in lots of 3000 in galvanized cans 45.7 cm high and 33 cm in diameter. The water is periodically replaced with freshwater filtered through gauze. Correct salinity is maintained with common salt.

LAYOUT OF FARMS

The layout of milkfish farms varies from place to place in Taiwan, but the general scheme is the same. An ideal site is an extensive area of tide flat where the water depth at high tide is at least 60 cm. Other areas are suitable, but the sort of topography described facilitates excavation by eliminating the necessity of hauling dirt to and from the site and renders pumping of water unnecessary. Ideally, the soil should consist of 64 to 82% silt, 16 to 32% sand, and 2 to 4% clay. The more organic matter the better, since it has been demonstrated that there is a linear relationship between organic content of bottom soil and milkfish production.

Since the Taiwanese coast is not fringed by mangroves as are more tropical coasts, the first step in building a milkfish farm is construction of a storm dike along the low-water mark. The dike should be 3 to 5 m high, 25 to 45 m wide at the base, and 3 to 4 m wide on top and have a slope of 1:3 to 1:4. The seaward side should be reinforced with stones or bricks. Water is admitted by means of a series of gates in the dike, each connected to a canal. There should be one gate and canal for every 100 to 200 ha of ponds.

Three types of pond are necessary:

1. *Production ponds* should be at least 30 to 40 cm deep and 3 to 5 ha in area. Ponds smaller than 3 ha are costly to construct and inhibit growth of milkfish. Ponds larger than 5 ha are difficult to harvest. The bottom should slope slightly toward the adjacent passageway, which should be 30 to 40 cm deeper than the pond to facilitate drainage.

2. *Wintering ponds* must be 1 to 2 m deep. As protection against the cold (water temperatures as low as 2 to 4°C occur), a windbreak of bamboo thatched with straw or reeds, slanting toward the pond at an angle of about 30°, is constructed on the northeast side of each pond. Since wintering ponds are very heavily stocked, they should be constructed adjacent to the main canal so that well-oxygenated sea water can readily be admitted. Plastic greenhouses are also being tested. Wintering ponds are not necessary in south Taiwan.

3. *Nursery ponds* are 30 cm deep and connected by gates to the winter-

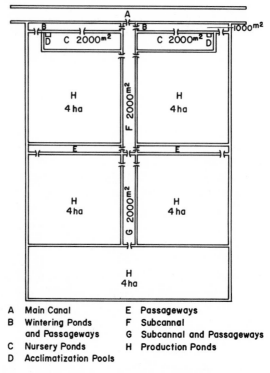

A   Main Canal   E   Passageways
B   Wintering Ponds   F   Subcannal
     and Passageways   G   Subcannal and Passageways
C   Nursery Ponds   H   Production Ponds
D   Acclimatization Pools

FIG. 5. Milkfish farm unit in Taiwan. (After Lin, 1968b.)

ing ponds. A tiny acclimatization pool may be constructed in one corner of the nursery pond.

The three types of pond are interlinked by a series of canals and passageways, the design and function of which will become clear as the techniques of culture are described.

The size of a milkfish farm is limited only by the available land and capital, but management is facilitated if the farm is broken down into 12- to 35-ha units, each containing all three types of pond and connected directly to a main water supply canal. Table 5 and Fig. 5 illustrate the layout of such a unit.

NURSING FRY

Nursery ponds to be stocked with newly caught fry in April are prepared starting in November or December. Nurseries are first drained, leveled,

TABLE 5. TYPES, NUMBERS, AND SIZES OF UNITS COMPRISING A MILKFISH FARM UNIT IN TAIWAN

TYPE OF POND AND CANAL	AREA (HA)	APPROXIMATE TOTAL AREA (%)
4 to 6 production ponds, 3 to 5 hectares each	20.00	94
2 wintering ponds	0.20	1
2 nursery ponds	0.40	2
4 passageways and refuges	0.40	2
1 subcanal	0.20	1
Total	21.20	100

and dried for 2 weeks, then filled to 5 to 20 cm and allowed to evaporate to dryness, which takes 3 to 4 weeks. Next they are manured with rice bran, which may be enriched with human waste, straw, or oil cakes, at 400 to 1000 kg/ha. Rice bran is placed in the pond in 22- to 30-kg bags, then water is added to a depth of 7 to 13 cm. When the contents of the bags are thoroughly soaked they are cut open and the fertilizer spread around. This process of evaporation, drying, and fertilization may be repeated several times before stocking. Finally, in March or early April, the ponds, along with all canals and passageways, are filled to a depth of 12 to 18 cm and poisoned with tea seed cake or tobacco waste at 150 to 200 kg/ha. The poison will dissipate within 7 to 10 days, at which time the water level is raised to 18 to 20 cm. By the time fry are stocked at 70,000 to 150,000/ha in early April there should be an abundance of blue-green algae.

By June the fry should have attained fingerling size and be ready for stocking in production ponds. At the same time, a second lot of newly caught fry are placed in the nursery ponds. The crop of algae produced by initial fertilization should last through July, but after that time the combined effects of milkfish grazing and hot weather cause deterioration of unattended algal pasture. Therefore, in July or August the ponds are once again fertilized, using rice bran, peanut cake, soybean cake, human waste, or, less frequently, pig manure, chicken manure, legume seed cake (*Leucaena glauca*), flax seed cake, sesame cake, or coconut cake. Choice of fertilizers is made principally on the basis of availability. One series of experiments was carried out to compare the effectiveness of various fertilizers in producing blue-green algae, principally *Lyngbya,* in unstocked milkfish ponds. Unfortunately, growing periods of two different lengths were used, and one of the most commonly used fertilizers, peanut cake, was not tested. As nearly as can be determined, soybean cake was the best of the fertilizers tested. Fertilizers may be applied between fry stockings or added to ponds containing fry. In the former case, the ponds may be drained. If fry are present, some of the "fertilizer" may function as food.

The Tainan Fish Culture Station has experimented with inorganic fertilizers, but they have not been able to demonstrate any advantage over organic fertilizers. The 1962 edition of the *Agriculture Handbook,* put out by the Ministry of Agriculture and Forestry of the Republic of China, includes a recommended schedule for fertilizing milkfish nurseries and production ponds, using both manures and inorganic fertilizers, but all commercial growers continue to rely on experience and natural fertilizers. The only widespread use of inorganic fertilizers is in treatment of "yellow water" caused by blooms of dinoflagellates. Ponds containing such blooms may be treated with 50 to 300 kg/ha of superphosphate or zeolite, a soil conditioner containing large amounts of silica, at 100 to 200 kg/ha.

Fry are seldom given supplemental foods. In experiments, 15- to 23-mm milkfish fed on flour, soybean meal, rice bran, and peanut meal and denied access to algae suffered much higher mortality than similar fry which were allowed to graze on benthic algae.

STOCKING GROWING PONDS AND CROPPING IN ROTATION

The distinctive feature of Taiwanese milkfish culture is the system of stocking and cropping in rotation which is just beginning to be adapted for use in other countries. Though the precise niches occupied by milkfish of various ages are ill defined, there is little doubt that the Taiwanese stocking system is essentially a polyculture scheme, and that the carrying capacity of a pond for fry, fingerlings, and adult milkfish is considerably greater than its capacity for any of these age groups alone. Production ponds, which are prepared for stocking in the same manner as nursery ponds, are first stocked in March or early April, at which time no freshly caught fry are available. Thus overwintered fish of an assortment of sizes from 5 to 100 g are stocked at 5000 fish/ha. The larger fingerlings, which constitute about 40% of the fish stocked, are from the previous summer's fry and have been overwintered by the farmer himself. The smaller fish are stunted stock obtained from nursery specialists. A portion of these fish, grown to weights of 250 to 600 g, account for the first harvests in June and July. Subsequent stockings of newly captured fry are made from April to August and harvested as 150- to 300-g adults from August to October or early November. Stocking rates vary with many

PLATE 4. Experimental overwintering milkfish pond in Taiwan sheltered by wind-break. (Courtesy Ziad Shehadeh, Oceanic Institute, Hawaii.)

factors, not the least of which is availability of fry. Table 6 illustrates a stocking system based on maximum rates in the Tainan area.

OVERWINTERING

Fry stocked in August do not reach marketable size by the end of the growing season in November and are the source of the overwintered

TABLE 6.   MAXIMUM RATE OF STOCKING FOR OVERWINTERED MILKFISH FIN-GERLINGS AND NEW FRY IN TAINAN, TAIWAN

MONTH OF STOCKING	AVERAGE WEIGHT OF A FINGERLING OR FRY (G)	STOCKING RATE (FISH/HA)	AVERAGE WEIGHT (G) ATTAINED IN THE CULTURE PERIOD OF:		
			60 DAYS	90 DAYS	120 DAYS
April	70	1,500	350	400	600
	30	2,000	—	350	400
	25	1,500	—	—	250
May	0.05	2,500	30	100	250
June	0.06	2,500	—	—	200
July	0.06	2,000	—	—	200
August–September	0.06	3,000	—	100	—

PLATE 5. Experimental overwintering milkfish pond in Taiwan sheltered by plastic greenhouse. (Courtesy Ziad Shehadeh, Oceanic Institute, Hawaii.)

fingerlings planted the following spring. Fingerlings to be overwintered are stocked at $20/m^2$ and do not grow; in fact they may lose 10 to 17% by weight. This is offset insofar as possible by feeding with rice bran on warm days.

Another problem in wintering ponds is increasing salinity. Since winter is the dry season in Taiwan, fresh or brackish water must be admitted periodically to dilute wintering ponds. The same problem occurs in production ponds, which may reach a salinity of 40‰ by March and must be freshened before stocking in the spring. If an abundant supply of water of low salinity is not available, overwintered fingerlings are first let into the passageways adjacent to the wintering ponds. (Passageways, due to the constant influx of sea water, have a salinity of about 35‰.) The gates between the wintering ponds and the passageways are then closed, while the gates between passageways and growing ponds are opened so that the salinities can be equalized and the fish can acclimate naturally to production pond conditions.

High salinity and heavy stocking are probably responsible for the occasional outbreaks of sea lice (*Argulus*) which may cause weight loss or even mortality in wintering ponds but do not normally affect milkfish in production ponds. Chlorinated hydrocarbon pesticides have been found effective in eliminating *Argulus,* but in view of the danger of residues building up in the fish, other methods of control, perhaps by salinity variation, should be investigated.

PESTS AND OTHER PROBLEMS

Predators, diseases, and parasites other than *Argulus* are not a problem in Taiwanese milkfish ponds, but a number of competitor organisms must be dealt with. Perhaps the most troublesome are larvae of the midge *Tendipes longilobus,* which compete for food with milkfish. It is estimated that during the summer a population of these Chironomids may consume 60 to 90 kg of benthic algae daily in a pond where the milkfish require 100 kg/day. Chemical pesticides are commonly employed to control this pest, but there are possibilities for biological control which should be pursued. Midge larvae are a favored food of certain fishes, notably the common carp (*Cyprinus carpio*), and some shrimps, and the possibilities of polyculture as a means of biological control are potentially great but have rarely been tested on a commercial scale.

Other food competitors include the polychaete worm *Nereis,* the snail *Cerithidea,* and Java tilapia. *Cerithidea* does not thrive in pond soils containing large amounts of organic matter, hence it is seldom a problem in ponds more than 5 years old. Where necessary, both *Nereis* and *Cerithidea* may be controlled by application of teaseed cake at 200 kg/ha or tobacco waste at 100 to 200 kg/ha.

Java tilapia were introduced to Taiwan in 1946 and have since become one of the dominant fish in the southern part of the island. Although they have greatly increased the total fish production of some badly polluted ponds which support other species in small numbers if at all, they compete for food with milkfish, have a tendency to overpopulate ponds, and are generally considered pests. Effective control measures have not been developed. Periodically, an unknown virus or bacterial disease causes mass mortality of tilapia in Taiwan but does not affect other fish. The causative factors are unknown.

The high concentration of organic matter in milkfish ponds makes pollution and deoxygenation constant preoccupations of milkfish farmers, even though milkfish can withstand dissolved oxygen concentrations as low as 1.5 ppm. The most common sign of impending deoxygenation is "brown water" caused by large populations of protozoans. When this condition occurs, the only remedy is to drive the fish from the production pond into the adjacent passageway, drain the pond, and refill it. Some culturists guard against brown water by aerating their ponds or providing a constant inflow of well-oxygenated water daily during the critical hours 2300 to 0800.

The relative efficiency of milkfish culture in Taiwan, the Philippines, and Indonesia is illustrated by the mortality rates prevailing in the three countries. While culturists in the latter two countries consider 50% survival of planted fry a good harvest, Taiwanese culturists, some of whose stock must endure the rigors of winter, expect to harvest 80% of the fry they stock. If the overwintered stock are subtracted from the total, survival averages 85 to 95%.

POLYCULTURE

Apart from Java tilapia, the fish species most commonly found in milkfish ponds in Taiwan is the striped mullet (*Mugil cephalus*), which occupies a niche roughly equivalent to that of milkfish. In ponds of low salinity, milkfish may constitute a minor component of complex polyculture systems based on tilapia and/or the Chinese carp association (see Chapters 3 and 18). Sometimes the shrimps *Penaeus carinatus* and *Metapenaeus ensis,* which may be collected along with milkfish fry, are grown with milkfish.

HARVEST, YIELD, AND PRODUCTION

Harvest of milkfish in Taiwan is selective for size, as described for the Philippines.

The average yield of milkfish ponds in Taiwan increased dramatically in the late 1940s and early 1950s, to 1800 to 1900 kg/ha, principally as a result of widespread adoption of the continuous stocking and harvesting system described. The area under cultivation has also increased, until in 1966 roughly 16,000 ha of water were devoted to milkfish culture and produced nearly 30 million kg of fish. Best yields were nearly 3000 kg/ha. (See Table 4 for a yearly listing of total and per hectare yields from 1920 to 1966.) It is not likely that increases of similar magnitude will occur in the near future, as the average yield has not increased since 1955. Further, Taiwan does not have extensive undeveloped areas suitable for milkfish culture. There are perhaps 10,000 ha of swampland available for development, but new sources of fry will have to be exploited before even these swamps can be used. To make matters worse, the cost of milkfish culture in Taiwan is increasing, since it has been found that rice bran, the best supplemental feed, is a good source of oil. Thus much of the supply is being diverted to oil production, and milkfish growers are forced to use expensive substitutes or shift to some other form of aquaculture. Nevertheless, Taiwan will continue to be one of the greatest producers of milkfish and to set the example for milkfish culturists in other countries.

MILKFISH CULTURE IN OTHER COUNTRIES

The only other area where milkfish culture has a long history is Hawaii, where the ancient method of allowing the incoming tide to stock ponds has only recently been supplanted by selective stocking.

India is fortunate in that milkfish fry are available nine months of the year, from March to August and again from October to December. Fingerlings are also available in southern India, a fact which may have implications for increasing the efficiency of culture. Milkfish culture in India was initiated in Madras in 1931 and has since become regionally important. The clearing of coastal swamps for milkfish culture also opened up possibilities for rice growing and salt drying, which activities supplement the incomes of some farmers. In India, and in Ceylon as well, milkfish have been successfully acclimated and cultured in freshwater. Ceylon is also similar to India in that fingerlings as well as fry are available. The collecting seasons are March to April and October to December.

It has been less than 15 years since it was discovered that milkfish fry and fingerlings occur in large numbers off the coast of Thailand from April to October and that fry can be taken off Vietnam from May to November. Milkfish culture is thus a rather new phenomenon in the two countries but it holds great promise.

PROSPECTUS

The expansion of milkfish culture into new areas seems almost certain. In addition to the vast areas of undeveloped swampland in Indonesia, and the smaller but significant areas in the other countries where milkfish farming presently occurs, there are many hectares of suitable shoreline in east Africa, Bangladesh, Burma, the Malay Peninsula, Cambodia, Australia, south China, Mexico, and some of the Pacific islands. Assuming the availability of fry, milkfish culture could make a significant contribution to human nutrition in all these areas.

Optimism over the expansion of milkfish culture must, however, be tempered with a healthy skepticism as to the market for its product. A case in point occurred in Kenya, where the Department of Fisheries successfully raised milkfish—only to find that the local populace would not purchase them.

It has also been suggested that milkfish could be introduced and cultured in tropical Atlantic waters, for example, on the coast of Brazil. However, it seems likely that the large brackish water fauna of the

Brazilian coast already contains fish species well suited to pond culture. Even in the unlikely event that there are no such species, the possible effects of milkfish on the local ecology should be assayed before introductions are undertaken.

The key to success in milkfish culture is of course proper pond fertilization. The methods developed to date have been largely empirically determined and can probably be improved. There is no lack of effort in this area, but it is difficult to know in which direction to move, since it is not known which species of algae and other food organisms should be encouraged to maximize milkfish production. There is an abundance of literature on milkfish feeding habits, but the end result of reviewing it is likely to be confusion rather than enlightenment, since much of it is vague and contradictory. It must also be remembered that methods of fertilization should be tailored to soil and water chemistry, thus there is no *best* system of pond fertilization. Despite these difficulties, research on nutrition and pond fertilization in milkfish culture continues, particularly in Taiwan. Much of it is perhaps handicapped by the premise that there must be methods of inorganic fertilization capable of producing optimum conditions in milkfish ponds. If such methods were developed, they might produce considerable savings in money and labor but, based on past experience, it seems likely that organic fertilization will continue to dominate practical culture. At present the true beneficiaries of research on inorganic fertilization of milkfish ponds are the chemical companies.

Immediate gains in milkfish production outside of Taiwan are most likely to be made by adapting Taiwanese methods to local conditions, a process which is already yielding good results in the Philippines. Of particular importance is the practice of continuous stocking and harvesting which, if instituted in countries with a year-round growing season, could result in yields surpassing anything achieved in Taiwan. This has already been achieved on a small scale in central Java (see pp. 321–322).

All increases in milkfish production, whether through expansion of the industry, improvement of fertilization, or intensification of stocking methods, are ultimately dependent on the availability of fry for stocking. Already shortages are sometimes experienced in Taiwan, the Philippines and Java. There are almost certainly unexploited stocks to be discovered, but the ultimate solution must be the breeding of milkfish in captivity. Even with the aid of pituitary injections, which have been the key to successful breeding of such stubborn fishes as the Chinese and Indian carps, it has thus far not been possible to mature and spawn milkfish in ponds. Milkfish breeding experiments are under way, on a small scale, in the three major producing countries, but their progress is held back by a lack of knowledge of the conditions required for natural spawning. Once this difficulty is overcome, truly intensive culture can begin, and the rate of expansion and improvement of milkfish farming may be spectacular, unless highly controlled mullet culturing methods develop so well and fast as to supply much of the market which is now being undersupplied by insufficient milkfish production.

REFERENCES

Botke, F. 1951. A short survey on fishing and transportation of Bandeng fry. Land-bouw 23:399–409.

Bureau of Fisheries, Philippines. 1951. Construction and layout of bangos fishponds. Fish Culture Series, Fishery Leaflet 3.

Hora, S. L., and T. V. R. Pillay. 1962. Handbook on Fish Culture in the Indo-Pacific Region. FAO Fisheries Biology Technical Report 14. 204 pp.

Lin, S. Y. 1968a. Milkfish farming in Taiwan—a review of practice and problems. Taiwan Fisheries Research Institute, Fish Culture Rep. No. 3.

Lin, S. Y. 1968b. Pond Fish Culture and the Economics of Inorganic Fertilizer Application. Chinese-American Joint Committee on Rural Reconstruction, Fisheries Series, No. 3.

Ling, S. W. 1966. Feeds and feeding of warm-water fishes in ponds in Asia and the Far East. FAO World Symposium on Warm Water Pond Fish Culture. FR: III-VIII/R-2.

Prowse, G. A. 1966. A review of the methods of fertilizing warm water fish ponds in Asia and the Far East. FAO World Symposium on Warm Water Pond Fish Culture. FR: II/R-2.

Schuster, W. H. 1949. Fish culture in saltwater ponds on Java. Department van Land-bouw en Visserij Publicate no. 2 van de Onderafdeling Binnenvisserij.

Schuster, W. H. 1952. Fish culture in brackish-water ponds of Java. Indo-Pacific Fisheries Council, Special Publication 2.

Tang, Y. A. 1966. Evaluation of the relative suitability of various groups of algae as food of milkfish in brackish-water ponds. FAO World Symposium on Warm Water Pond Fish Culture. FR: III/E-4.

Tang, Y. A. 1967. Improvement of milkfish culture in the Philippines. Indo-Pacific Fisheries Council Current Affairs Bulletin, No. 49.

Yamamura, M. 1942. Milkfish culture. (Unpublished paper)

Interviews and Personal Communication

Delmendo, M. Philippine Fisheries Commission.

Shehadeh, Z. Oceanic Institute, Waimanalo, Hawaii.

18

Culture of *Tilapia*

Members of the genus tilapia (family Cichlidae) have been an important source of food for man at least since recorded history began. The fish Saint Peter caught in the Sea of Galilee and those with which Christ fed the multitudes were tilapia. An Egyptian tomb frieze, dated at 2500 B.C., illustrates the harvest of tilapia and suggests that they may have been cultured. Since that time, and probably before, the various species of tilapia have been of major importance in the fisheries of their native lands, the Near East and Africa.

DISTRIBUTION OF TILAPIA SPP.

From the point of view of human nutrition, tilapia was already firmly entrenched as one of the world's most important fish by the start of the twentieth century. But with greatly increased emphasis on fish culture in this century, plus the advent of modern transportation, tilapia became even more valuable to man. Today no fish, with the probable exception of the common carp (*Cyprinus carpio*), is more widely cultured. As early as the 1920s experiments in tilapia culture were being carried out in Kenya. Tilapia had become an intercontinental traveler by 1939 when naturally propagating stocks of *Tilapia mossambica,* native to the streams of Africa's east coast, were discovered in Java. No one is quite sure how they got there, but it is likely that aquarists, who have long been intrigued by the bizarre "mouth-brooding" habits of this and most other tilapia species, were implicated. However *T. mossambica* arrived, they made themselves at home and spread rapidly throughout the island. So prevalent did they become that, despite the geographic reference in the scientific name, the generally accepted common name for *T. mossambica* is "Java tilapia." They were originally regarded as a nuisance, but when facilities for the traditional Indonesian practice of milkfish (*Chanos chanos*) culture began to deteriorate under the Japanese occupation, the less demanding nature of tilapia became apparent, and by the end of World War II tilapia were not only too well established in the waters of Java but too deeply entrenched in Indonesian fish culture for anyone to consider control measures against them.

The Japanese, always a nation cognizant of the nutritional possibilities of fish, helped spread tilapia throughout Indonesia, where today it is found in virtually every body of water, including ditches and stagnant pits where few other fish of value can be grown. The retreating Japanese also introduced tilapia to Malaya, from whence it spread throughout southeast Asia, achieving considerable importance as a food fish in virtually every country. Today, due to the enthusiasm generated by the research of such men as H. S. Swingle at Auburn University in Alabama and C. F. Hickling at the Tropical Fish Culture Research Institute in Malaysia, plus the missionary efforts of such organizations as FAO and the Peace Corps, the Java tilapia and its congeners are cultured, at least experimentally, not only in southeast Asia, but in Japan, Asiatic Russia, the Indian subcontinent, the Near East, virtually all of Africa, parts of Europe, the United States, and many of the Latin American countries.

SELECTION OF SPECIES FOR CULTURE

The Java tilapia was the first member of the genus *Tilapia* to come to the attention of large numbers of fish culturists and has remained the most widely cultured species. However, according to the late A. Yashouv of the Fish Culture Research Station at Dor, Israel, it is "the most difficult to manage" member of the genus. This is not to say that there is a "best" tilapia for culture. At least 14 species have been cultured and all share the hardiness, ease of breeding, rapid growth, and high quality of flesh which have made the Java tilapia popular. Any one of them or some as yet untried species or hybrid might be the right fish for culture in a given situation. Too often problems in tilapia culture have resulted from hasty or uninformed selection of a species. In choosing a species the culturist should take into account the factors listed in Table 1 and discussed further here.

AVAILABILITY

The Java tilapia is practically universally available, but some species are still available only within their home range or have been introduced elsewhere on a limited scale. This will probably become a less important consideration as time goes by and the more desirable species are distributed among fish culturists.

FOOD HABITS

All tilapia are more or less herbivorous, but some prefer higher plants, whereas others are adapted to feed on plankton. If the food habits of a species are not known, they may be predicted by examination of the gill rakers. Numerous long, thin, closely spaced gill rakers indicate a plankton feeder; the opposite shows that the fish consumes larger particles of food. Some tilapia are relatively omnivorous and will benefit from artificial feeding with vegetable materials and in some cases even accept animal food if nothing else is available; others are obligate herbivores. Some macrophyte feeders are sufficiently voracious to function well as biological weed controls, whereas others are useless in this capacity.

SALINITY TOLERANCE

Most tilapia are tolerant of brackish water, but some are better adapted to it than others and may thrive and even breed in sea water.

TEMPERAMENT

Tilapia are less aggressive than most carnivorous cichlids, but they may attack and nip the fins of other species, an undesirable habit if a fish is to be used in polyculture. This behavior is not necessarily species-specific; there are conflicting accounts in the literature for several species. Such factors as sex, temperature, and population density are known to affect aggression and may be involved in the reaction of tilapia to other fish.

TEMPERATURE TOLERANCE

Tilapia are essentially tropical lowland fish, but some species and some stocks withstand cool temperatures much better than others. Where no data are available, valid inferences may often be drawn from consideration of the climate of a species' native habitat.

One more word should be said in behalf of the Java tilapia and Nile tilapia. Both of these species may be and are cultured where great amounts of organic enrichment make it scarcely possible for other edible fish to survive.

SPAWNING AND GROWTH OF THE YOUNG

In many forms of fish culture, obtaining spawn is one of the most difficult tasks. Tilapia present no such problem; indeed it is difficult to prevent them from spawning. The fact that it takes no skill whatever to spawn tilapia in ponds is one of the reasons they have been widely promoted as a fish for subsistence culture, and certainly these prolific fish have enabled many an African or Asian farmer to produce his own fish without acquiring extensive skills or technological know-how.

To spawn tilapia little more is needed than a pond, preferably one with a loose, sandy bottom, and some breeding stock. The spawning pond should be stocked with 25 to 30 females/1000 m² and about half

TABLE 1. CHARACTERISTICS OF *Tilapia* SPP. USED IN PRACTICAL AND EXPERIMENTAL FISH CULTURE

SPECIES	AVAILABILITY	FOOD HABITS	SALINITY TOLERANCE	TEMPERAMENT	TEMPERATURE TOLERANCE
T. andersoni	Cultured in Katanga and Zambia; seldom mentioned in literature of fish culture	Omnivorous	Unknown	Unknown	Unknown
T. aurea	Native to West Africa from Senegal to the Chad Basin and the lower Nile, also Israel and Jordan; experimentally cultured in Alabama	Unknown	Mainly a freshwater species	Aggressive	Unknown
T. galilea	Native from Jordan to east and central Africa as far west as Liberia; cultured in Israel and the Congo	Uncertain; referred to in the literature as "omnivorous" and "strictly a plankton feeder"	May vary according to the source of stock; certainly populations in such waters as the Sea of Galilee or coastal waters would be quite tolerant, but inland populations from central Africa might not be	Unknown	Unknown; probably varies with source of stock

TABLE 1. (*continued*)

SPECIES	AVAILABILITY	FOOD HABITS	SALINITY TOLERANCE	TEMPERAMENT	TEMPERATURE TOLERANCE
T. heudeloti (black-chinned mouthbreeder) —considered by E. Trewavas, the leading authority on *Tilapia* taxonomy, to be synonymous with *T. macrocephala*	Range: coastal west Africa from Senegal to the Congo; widely cultured experimentally	Herbivorous, but not so voracious a feeder on higher plants as some tilapia, suggesting that it may utilize phytoplankton and/or algae	Mostly found in brackish water in nature	Generally considered peaceful with its own and other species	Does well in aquaria at 20–30°C
T. hornorum (Zanzibar tilapia)	Native to Zanzibar and the east African coast opposite Zanzibar; introduced to Malaysia by C. F. Hickling in the 1950s; more recently introduced to Costa Rica	Unknown; small numbers of gill rakers suggest that it is not a plankton feeder	Very wide in Zanzibar population	Unknown	Unknown
T. leucocista	Cultured in Uganda, where it is native, in the 1950s; since largely supplanted by *T. mossambica*, *T. nilotica* and *T. zillii*	Unknown	Unknown; may be relatively low in cultured stocks, which are of inland origin	Unknown	Unknown

355

TABLE 1. (*continued*)

SPECIES	AVAILABILITY	FOOD HABITS	SALINITY TOLERANCE	TEMPERAMENT	TEMPERATURE TOLERANCE
T. macrocephala —considered synonymous with *T. heudeloti*, which see	—	—	—	—	—
T. macrochir	Native to Congo, Zambesi, Kafue, and Okavango river systems; has been cultured, at least experimentally, in Cameroon, the Ivory Coast, Katanga, Rhodesia, Rwanda, Madagascar, and the Sudan	Phytoplankton feeder	Probably varies with source of stock	Unknown	Growth poor in cool climates
T. melanopleura (Congo tilapia)	Native from Upper Congo to South Africa; commercially cultured in South Africa and experimentally cultured in several Asian, European, and American countries	Strictly an herbivore, with a preference for higher plants; may be used in weed control	Unknown	Unknown	Unknown

TABLE 1. (continued)

SPECIES	AVAILABILITY	FOOD HABITS	SALINITY TOLERANCE	TEMPERAMENT	TEMPERATURE TOLERANCE
T. mossambica (Java tilapia)	Cultured in virtually all of southeast Asia, the Near East, and southern Africa; experimentally cultured in Japan, Latin America, the United States and the Soviet Union; stocks may be currently be obtained almost anywhere in the world; by far the most widely cultured Tilapia	Mainly a plankton feeder, but consumes all kinds of plants and artificial feeds of vegetable origin; in the absence of plant food will accept animal food	Very wide; found in brackish as well as fresh waters in the wild, and will even breed in sea water	Somewhat aggressive toward other species; this may be related to the deleterious effect of Java tilapia on production of catla (Catla catla) and, in some instances, milkfish	There is a great deal of experimental and observational data on this species: optimum temperatures suggested range from 22 to 30°C, but they can be cultured at temperatures as low as 15.5°C, below which they cease to feed; will not survive long below 12°C and 9°C is lethal; does not grow during the cold months in such climates as Alabama and Israel, nor above 1000 m in the tropics; its hybrids with T. nilotica are more cold resistant
T. nigra	Cultured, at least experimentally, in east Africa	Omnivorous	Unknown	Unknown	Unknown

357

TABLE 1. (continued)

SPECIES	AVAILABILITY	FOOD HABITS	SALINITY TOLERANCE	TEMPERAMENT	TEMPERATURE TOLERANCE
T. nilotica (Nile tilapia)	Range: from Syria into east Africa through the Congo to Liberia; widely introduced outside that range; probably, next to T. mossambica, the most commonly cultured tilapia	Variously reported in the literature to be a plankton feeder, an omnivore, and to feed on higher plants to the extent that it may be used in weed control, though not as effectively as T. melanopleura; requires plant food when kept in aquaria	Probably quite high, since it is found in brackish water and is generally considered one of the hardiest tilapia	Little studied; may be aggressive toward other species	Similar to T. mossambica; does well above 15.5°C; does not survive below 12°C; lethal temperatures 11°C and 42°C; hybrids with T. mossambica are more cold resistant
T. randalli	Once cultured in the Congo, now largely supplanted by other species	Unknown	Unknown	Unknown	Unknown
T. sparmanni	Native to east Africa south of the equator; cultured experimentally in Japan	Omnivorous	Unknown	Aggressive toward other species	Less tolerant of high or low temperatures than T. mossambica; does well in aquaria at 21–30°C; optimum for breeding, 23–26°C

TABLE 1. (continued)

SPECIES	AVAILABILITY	FOOD HABITS	SALINITY TOLERANCE	TEMPERAMENT	TEMPERATURE TOLERANCE
T. volcani	Native to Lake Rudolf, Kenya; a few specimens accidentally imported to Israel in 1967–1968; other supposed *T. nilotica* from Lake Rudolf may be *T. volcani*	Unknown	Unknown	Unknown	Unknown
T. zillii (Zill's tilapia)	Native to Near East and Africa north of the equator; cultured in east Africa and, experimentally, in Madagascar and Malaysia	Strictly herbivorous, mostly on higher plants; some use in weed control	Unknown	Aggressive toward other species	Optimum 22–24°C; aquarium stock does well at 23–26°C. Optimum for breeding 26°C; central African stocks do not do well below 20°C, but north African stocks withstand 14–16°C for long periods of time, although growth is greatly retarded

PLATE 1.   Nests of *Tilapia nilotica*. (Courtesy Marcel Huet, Groendendaal-Hoeilaart, Belgium.)

again as many males. If the water is warm enough (see Table 1 for suggested breeding temperatures for some species), the males will begin digging holes, perhaps 35 cm in diameter × 6 cm deep, in the pond bottom (Plate 1). A female deposits 75 to 250 eggs in such a nest then picks them up in her mouth. Next the male discharges sperm into the depression and this too is picked up by the female. Fertilization thus takes place inside the female's mouth, where hatching occurs within 3 to 5 days. The larvae are retained in the mouth until the yolk sac is absorbed, after which they may venture forth, but for 10 to 15 days they still return to the female's mouth when threatened. During this time the female eats seldom if at all. If the young are separated from the mother during brooding, they may be raised quite satisfactorily by themselves. Apparently the chief function of mouth-brooding is protection from predators.

The young tilapia mature at an age of 2 to 3 months, at which time they are 6 to 10 cm long. From then on they breed every 3 to 6 weeks as long as the water is warm. Whenever the water temperature approaches the lower limits of tolerance for a particular species, breeding activity is suspended. Thus the nonreproductive period ranges from about 2 months in subtropical climates to none whatever near the equator. As for other environmental factors impinging on spawning, it can be stated that tilapia will reproduce in almost any sort of water they can survive in. Java tilapia have been successfully spawned in water with a salinity of 35‰.

This description of spawning applies specifically to Java tilapia, but it generally fits all the cultured species except *T. heudeloti,* in which the male incubates the eggs, *T. galilea,* in which both sexes share in mouth-brooding, and *T. sparmanni* and *T. zillii,* which spawn in typical cichlid fashion, on a clean stone or other smooth object, and do not mouth-brood, although the parents do guard the eggs and young. *T. sparmanni* and *T. zillii* compensate for the lack of the protective mouth-brooding trait by producing more eggs—up to 5000 in large *T. zillii.* In nature the survivors from a few thousand eggs in the open would probably approximate the survivors of a few hundred mouth-incubated eggs and young, but in the protected environment provided by the fish culturist, more eggs are likely to mean more young. Thus *T. sparmanni* and *T. zillii* are even more prone to overpopulate a pond and produce a stunted population than are other tilapia.

THE PROBLEM OF OVERPOPULATION AND METHODS OF CONTROL

Mouth-brooders or not, overpopulation is the greatest problem encountered in raising any species of tilapia (Plate 2). Stunting is bad enough in countries such as the Philippines where tilapia as small as 75 mm may be marketed, but in places like East Africa, where large fish are preferred for the table, it can spell the difference between success and failure. The seriousness of the problem may be emphasized by describing the situation in Kivu Province of the Congo. There yields to tilapia culture of 4325 kg/ha were considered normal, which sounds quite good until it is pointed out that 70% of such yields consisted of fish less than 15 cm long. Classical methods of preventing overpopulation and stunting include separating parents and young immediately upon hatching, monosex culture, and stocking predators along with the tilapia. Selective breeding for large fish has been attempted, but with no success. Apparently environmental override genetic factors in determining size.

SEPARATION OF PARENTS AND YOUNG

Separation of adults and young could theoretically be accomplished by netting out the adults after spawning. When the brooders are sufficiently

PLATE 2. Seine haul from an unmanaged tilapia population in the Congo; note the many small fish. (Courtesy Marcel Huet, Groendendaal-Hoeilaart, Belgium.)

disturbed by nets and the like they will usually spit out the fry. The captured adults could then be placed in another pond and the spawning pond used for a rearing pond. This technique is not practicable, however, since the breeding cycles of individual adults are far from synchronous, so that in a sizable population spawning occurs more or less continuously.

It is possible to remove the young. A method practiced in Indonesia makes use of a drainable spawning pond at a higher elevation than the fry pond. When the eggs hatch, the adults are disturbed so that the larvae are released and drained into the fry pond. Different sizes of fry may then be periodically cropped from the fry pond for further use in culture.

MONOSEX CULTURE

Monosex culture is much more commonly employed. Stocks of a single sex may be obtained by sexing and sorting the stock individually. The structure of the genital papillae is indicative of sex. In males there is a single urogenital opening on the tip of the papilla; in females the genital

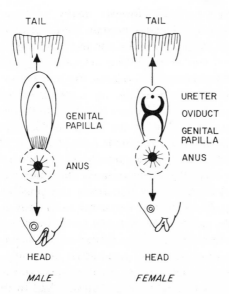

FIG. 1. Distinguishing characteristics of male and female *Tilapia mossambica*. (After Maar, Mortimor, and Van der Lingen, 1966.)

opening is separate and located on the frontal wall of the papilla, close to the apex (Fig. 1). This method has the disadvantage of being laborious. Moreover, even experienced workers can usually achieve only 80 to 90% accuracy. Since all it takes is one female inadvertently introduced into a pond of males being fattened for consumption to undo all the labor involved in sexing, easier and more effective methods for obtaining mono-sex stocks have been sought. The best results have been achieved by hybridization. Of the many interspecific and intraspecific tilapia crosses which have been attempted, at least three have produced 100% male offspring. These are:

    ♂ *T. macrochir* ♀ × *T. nilotica* (a difficult cross to produce)
    ♂ *T. mossambica* (Zanzibar stock) × ♀ *T. nilotica* (Lake Albert stock)
    ♂ *T. hornorum* × ♀ *T. mossambica*.

Better than 98% males were also obtained from the intraspecific mating

    ♂ *T. mossambica* (African stock) × ♀ *T. mossambica* (Malaysian stock).

Attempts to produce monosex stocks of tilapia have generally con-centrated on obtaining males since, unlike females, males continue to grow during breeding periods. However, males have one disadvantage: Whether or not there are females present, they optimistically construct spawning nests. The favored location for nest building is at the base of the pond bank, and the nest-building activities of a great number of males may eventually undermine the bank. For this reason there has been some speculation as to the feasibility of stocking growing ponds with females. No crosses which consistently produce high percentages of females have been reported, but at the Tropical Fish Culture Research Institute in Malacca, Malaysia, one male Java tilapia of Malaysian origin turned up which, when bred to two females of the same stock, produced 88 and 97% female offspring. Unfortunately this male was lost.

As one might gather from this discussion, the possibilities of hybridization in monosex culture have by no means been fully explored. Only a few of the possible interspecific hybrids have been produced, to say nothing of crosses between different strains of the same species, which vary in sex composition of the F_1 generation as widely as interspecific crosses. Further, the F_1 generation of many tilapia crosses are fertile and might be used in further breeding experiments. Add to this the fact that the taxonomy of tilapia is by no means cut and dried, and it can be seen that the possibilities are almost unlimited, but that duplication of a desirable hybrid may be quite difficult.

Monosex culture requires a certain amount of technical supervision to be successful. The small farmer often lacks the time, expertise, or inclination to sex fish or keep spawning records. Thus monosex culture is best suited to large commercial operations, or wherever there can be considerable governmental supervision. (A minority of authorities, notably S. Tal, Israel's Director of the Inland Fisheries, are skeptical of its practicability in any situation.) Thus while monosex culture of tilapia has been somewhat successful in Uganda, where the government has largely taken over fish culture, in the neighboring Congo, where fish culture is the responsibility of the individual farmer, it has failed.

CONTROL BY PREDATORS

Where tilapia are to be raised on small, technologically unspecialized farms, a more suitable method of population control is the stocking of predaceous fish to crop the young tilapia. The use of predatory fish is most common in Africa, where stunted fish are often simply unacceptable as food. The most commonly used predators there and in southeast Asia are catfishes of the genus *Clarias,* but eels (*Anguilla japonica*), largemouth bass (*Micropterus salmoides*), and carnivorous cichlids such as *Serranochromis robustus* and *Hemichromis* spp. have also been used.

CAGE CULTURE

Experiments at Auburn University suggested a completely different approach to population control: culture in floating cages. Nile tilapia stocked at 7000 to 15,000/ha in such cages exhibited growth and survival comparable to that obtained through pond culture but were apparently unable to reproduce. It was supposed that the eggs and sperm passed through the bottom of the net. More recently, however, Java tilapia have been found to have successfully reproduced in floating cages in Lake Atitlan, Guatemala.

STOCKING SYSTEMS

MONOCULTURE

Inclusion of a predator is one form of polyculture, a practice which is becoming nearly universal in raising tilapia. Indeed, except for very primitive subsistence culture, or in waters that will not support other edible fish, monoculture of tilapia for human consumption seems scarcely defensible. Monoculture is practiced, on a small scale, in rice fields in southeast Asia. Stocking rates for this practice are 120 to 180 fingerlings/ha. Care must be taken to use plankton- and algae-feeding species, since macrophages might destroy the rice crop.

POLYCULTURE

Since the ecological niches of the various *Tilapia* spp. are imperfectly known, and since the history of tilapia culture is so short compared to that of some other polycultural systems, species combinations and stocking practices are by no means codified. Table 2 outlines some of the stocking systems which have been commercially or experimentally applied, along with their results, when known.

It bodes well for the future of tilapia in fish culture that in almost every case where tilapia have been added to an existing pond culture community, total production has risen with no reduction in the nontilapia components of the harvest. One exception was found in India where tilapia of an unknown species depressed total production of a pond and apparently virtually eradicated milkfish and catla. Where tilapia are accused of exterminating other fishes, one must wonder whether some hyperaggressive species with a more typical cichlid temperament may not have been stocked. Tilapia have not been accused of eradicating other

TABLE 2. STOCKING SYSTEMS USED IN TILAPIA CULTURE, AND THEIR RESULTS

COUNTRY	SPECIES STOCKED	TYPE OF CULTURE	RESULTS
Cameroon	T. nilotica, Heterotis niloticus, Hemichromis fasciatus	Both subsistence and commercial pond culture	Good production and effective population control of tilapia
China	T. mossambica, striped mullet (Mugil cephalus), milkfish (Chanos chanos), Chinese carp association	Traditional Chinese pond culture	Addition of tilapia, mullet, and milkfish to Chinese carp culture increases total yield in practice, even in cases where the yield to traditional stocking was as high as 7,500 kg/ha
Costa Rica	T. melanopleura and/ or T. mossambica	Experimental culture in private and government-operated ponds	Average annual production, 2,784 kg/ha; maximum, 3,449 kg/ha
England	Tilapia sp., common carp	Experimental culture in power station cooling ponds	—
India	Tilapia sp., milkfish, catla, (Catla catla)	Growing in very fertile ponds	Yields of tilapia were quite good but the total yield of the pond was depressed; milkfish and catla were virtually wiped out, presumably by the tilapia
	Tilapia sp., Barbus sp., Nagendram fish (Osteochilus thomassi)	Commercial freshwater pond culture	—

TABLE 2. (continued)

COUNTRY	SPECIES STOCKED	TYPE OF CULTURE	RESULTS
Indonesia	T. mossambica (35%), common carp (30%), nilem (Osteochilus hasselti) (20%), gourami (Osphronemus goramy) (15%)	Short-term growing in small tropical ponds	Good production of small size fish
Israel	T. nilotica (fingerlings at 2,550–3,000/ha), common carp (2,500/ha)	Experimental pond culture	Average total yield to monoculture of carp, 4,159 kg/(ha)(yr); average total yield with tilapia added, 5,292 kg/(ha)(yr)
	T. nilotica (3,000/ha), common carp (5,500/ha)	Experimental pond culture	Yield of carp same for monoculture or polyculture, but addition of tilapia raised total yield by 13–35%
	T. nilotica, common carp	Experimental pond culture with fertilization and supplementary feeding	Monoculture of carp yielded 900–1,500 kg/ha; total production with tilapia added was over 2,500 kg/ha
	T. nilotica (1,000–1,500 fingerlings/ha), common carp (2 size groups)	Commercial pond culture	Total yield higher than can be achieved by monoculture of either species
	T. nilotica, common carp, Mugil cephalus, Mugil capito	Experimental pond culture	Addition of tilapia to carp-mullet ponds generally increased production of carp and mullet and invariably increased total production

367

TABLE 2. *(continued)*

COUNTRY	SPECIES STOCKED	TYPE OF CULTURE	RESULTS
Kenya, Uganda, and Tanzania	*Tilapia* sp., mirror carp (var. of *Cyprinus carpio*)	Freshwater pond culture	Carp recommended by Y. Pruginin to increase yields of ponds presently devoted to monoculture of tilapia
Malaysia	*T. mossambica*, *Barbus gonionotus*	Experimental freshwater pond culture with 40–120 kg/ha superphosphate	Efficient utilization of very heavy phytoplankton blooms induced by fertilization
Philippines	*T. mossambica*, milkfish	Brackish water pond culture, primarily for milkfish	Tilapia enter milkfish ponds accidentally and are often unwanted by milkfish culturists; nevertheless they may increase total production
Southeast Asia	*Tilapia* sp., common carp, *Barbus* sp.	Brackish water pond culture	—
	Tilapia sp., common carp, kissing gourami (*Helostoma temmincki*)	Sewage ponds	—
Sudan	*T. macrochir*, *T. melanopleura* (mixed ages)	Freshwater pond culture	Polyculture resulted in increased yields of both species, as compared to monoculture
Taiwan	*T. mossambica* (50% or more of harvest), *Mugil cephalus* (12%), silver carp (*Hypophthalmichthys molitrix*) (10%), remaining 28% made up of: common carp, goldfish (*Carassius auratus*), big head (*Aristichthys nobilis*), eel (*Anguilla japonica*), milkfish	Fresh or brackish water pond culture in very rich, often organically polluted ponds	Good production of fish

368

TABLE 2. (*continued*)

COUNTRY	SPECIES STOCKED	TYPE OF CULTURE	RESULTS
Uganda	*T. mossambica*, common carp	Freshwater pond culture	Yields to polyculture higher than to monoculture of either species
U.S.A. (Alabama)	*T. mossambica* (4,400/ha), channel catfish (*Ictalurus punctatus*) (1,250/ha)	Experimental culture with feeding of catfish	Production with monoculture of channel catfish, 1,400 kg/ha; production with polyculture, 1,568 kg/ha of catfish, plus 266 kg/ha of tilapia; feed conversion quotient of catfish same in either case
	T. nilotica (2,500/ha), channel catfish (7,500/ha)	Experimental culture with feeding of catfish	Production of catfish equal to or better than achieved by monoculture; total production better with addition of tilapia; food conversion quotient of catfish same in either case
	T. mossambica (2,500/ha), *T. nilotica* (2,500/ha), largemouth bass (*Micropterus salmoides*) (500/ha)	Experimental freshwater pond culture	Production, 21.2 kg/ha of bass and 2,118.0 kg/ha of tilapia
	T. mossambica (2,500/ha), *T. nilotica* (2,500/ha), *T. melanopleura* (2,500/ha), largemouth bass (500/ha)	Experimental freshwater pond culture, with addition of an unknown number of fathead minnows (*Pimephales promelas*) as food for bass	Production, 108.6 kg/ha of bass and 2,182.7 kg/ha of tilapia; fathead minnows all consumed

fish in the Philippines, but milkfish farmers there are ambivalent if not hostile toward the presence of Java tilapia. Part of their dislike of tilapia stems not from any direct effect of tilapia on fish production but because tilapia, in addition to consuming phytoplankton and naturally occurring algae, eagerly devour the elaborately cultured "lab-lab" meant for the milkfish. In other countries, though, there are numerous examples of tilapia being deliberately stocked with milkfish, usually with good results.

Surprisingly little work has been done on the combined use of two or more *Tilapia* species. Although an all-tilapia counterpart of Chinese carp culture may never be developed, certainly the variety of food habits among tilapia—plankton feeders, filamentous algae feeders, macrophages, and omnivores—points in this direction. The association in Africa of the plankton-feeding *T. macrochir* and the macrophage *T. melanopleura* also usually involves an assortment of ages and sizes, not to gain any particular polycultural advantage but because, given the prolific nature of tilapia, it is simpler. Initial stocking of growing ponds is carried out with a mixture of fry, large breeders, and various in-between ages. At harvest time all but a few of the large breeders are taken for consumption. Excess fry are also harvested for consumption or, more often, for use in stocking other ponds. Restocking may not be necessary, but if artificial feeding is employed, 10 to 20% of the estimated production by weight may be stocked after harvest.

Granted the relative simplicity of mixed-age culture, a harvest consisting of many sizes of fish may be economically disadvantageous. Stocking fish of approximately the same size will not produce a harvest of even-size fish, even if monosex culture is employed, but may more closely approach it. M. Huet of the University of Louvain, Belgium, has devised a formula for restocking single-size artificially fed tilapia ponds:

$$\text{number of tilapia to be stocked} = \frac{\text{total production}}{\text{individual growth}}$$
$$+ (\text{waste}) \left(\frac{\text{total production}}{\text{individual growth}} \right)$$

Individual growth of fish may be fixed within certain limits by referring to the normal growth of the species cultured, or it may be determined more precisely by weighing a sample of the stock at stocking and again at harvest.

Waste percentage is largely a matter of estimation or perhaps intuition. Values normally used in the application of Huet's formula are 10, 15, and 20%.

Total production equals natural production plus production due to feeding. Natural production must be known from previous culture, without feeding, in the pond to be stocked or a similar pond. Production from feeding may be obtained by dividing the weight of food used by its nutritional quotient (see p. 47). Huet does not give a method for determining production due to fertilization, but many of the commonly used organic fertilizers double as feeds and may perhaps be treated on that basis.

As an example of Huet's formula in application, let us suppose a 0.5-ha growing pond which is cropped twice a year. Its natural production is estimated at 400 kg/(ha)(yr). Now 3000 kg of cottonseed oil cakes, which have a nutritive quotient of 5, are available as feed, and waste is reckoned at 15%. Total production for 6 months thus equals

$$\frac{400 \text{ kg/(ha)(yr)} \times 0.5 \text{ ha}}{2} + \frac{3000 \text{ kg food}}{5} = 700 \text{ kg}$$

Individual growth of fish is estimated at 0.1 kg and waste at 15%. So, plugging in the data, we get

$$\text{no. of fry to be stocked} = \frac{700}{0.1} + (0.15)\left(\frac{700}{0.1}\right) = 8050 \text{ fry}$$

The formula is of course applicable only if the pond is to be restocked with the same size fish at the same species ratio as originally stocked.

The only other noteworthy example of multiple-species tilapia stocking is H. W. Swingle's work at Auburn University in Alabama. In addition to two or three species of *Tilapia,* one of which, *T. nilotica,* has unknown feeding habits, Swingle used a presumptive predator, the largemouth bass (*Micropterus salmoides*). The bass may have gotten some nutrition from young tilapia, but when fathead minnows (*Pimephales promelas*) were added to the pond, they were eradicated by the bass over the course of 7 months, resulting in a fivefold increase in bass production but only a slight increase in tilapia production (see Table 2). It has long been part of fishermen's lore, though not well substantiated, that many predators, including the largemouth bass, are reluctant to take spiny finned prey. If so, then perhaps the role of some predators in tilapia culture should be reevaluated.

Swingle has also stocked Java tilapia and Nile tilapia separately with channel catfish (*Ictalurus punctatus*), which, though not as piscivorous as the largemouth bass, might conceivably act as a predator on tilapia fry. Significant consumption of tilapia fry by the catfish was not noted, but the tilapia apparently utilized not only plankton, but wastes and excess feeds intended for the catfish, and production of catfish was substantially increased (see Table 2).

STOCKING RATES

Most of the reported stocking densities for tilapia are lower than necessary, at least where monosex culture is employed. The problem of excess small fish in tilapia ponds is more closely allied to excess reproduction than to initial overstocking. Experiments at Auburn University in which 1-year-old, 100-g *T. nilotica* were stocked in 2.02-ha ponds at various rates demonstrated increased production at each density increment up to 5039/ha, the highest tested.

Research in Uganda on the hybrid ♂ *T. mossambica* (Zanzibar stock) × ♀ *T. nilotica* (Lake Albert stock) showed that, up to a weight of 50 g, fry did not suffer retarded growth at densities as high as 8000/ha. Above that size it was found necessary to transfer them to rearing ponds at 1000 to 1500/ha if normal growth was to continue.

Much higher stocking densities were successfully employed in further experiments at Auburn University involving tilapia fingerlings. Nile tilapia fingerlings stocked at 20,000/ha and fed gave a production of 2822 kg/ha in 196 days, but those stocked at 40,000/ha produced 3699 kg/ha. Of these fish, 98.5% and 91.9%, respectively, were considered to be of usable size. Similar results were obtained with Java tilapia at densities up to 50,000/ha, but the percentage of usable fish, 99.7 at 10,000 fish/ha, declined to 67.7 at 50,000/ha, due presumably to the very prolific nature of Java tilapia. Of course it should not be assumed that more is better than less in tilapia stocking. Many factors other than density enter the picture, not the least being the question of available food supplies and its corollary, the economics of feeding.

MANAGEMENT OF TILAPIA PONDS

SEGREGATION OF AGE GROUPS

The practice of segregating different ages of fish in nursing, rearing, and growing ponds, so prevalent in culture of species where spawning is strictly controlled, is rare in tilapia culture. However, in parts of Africa, fry may be nursed from 6 weeks to 2 months, or until they begin to breed, in 0.01- to 0.1-ha ponds. This technique may become more widespread, particularly for use in monosex culture.

POND FERTILIZATION

Pond fertilization is of vital importance in culture of tilapia, particularly the plankton-feeding species, but has yet to be treated systematically. This is largely due to the short history of tilapia culture, as well as to the frequent use of tilapia in "crash" programs designed not to provide data for more sophisticated efforts but to alleviate existing critical protein shortages. In southeast Asia, tilapia may be cultured in ponds which are already heavily enriched by agricultural runoff and/or domestic pollution. Some of these waters carry such heavy loads of organic nutrients as to be uninhabitable by other desirable fishes. Fertilization would thus be superfluous, and fertilizers are set aside for use in culture of carps, milkfish, mullet, and so on. But even in southeast Asia, such ponds are the exception rather than the rule.

It may be expected that as tilapia culture matures, especially if tilapia becomes a commercial product sought by all economic classes, fertilization will be studied in some detail.

A certain amount of research on fertilization of tilapia ponds has been carried out, most of it involving Java tilapia and all of it involving the use of phosphates, which are the most effective group of fertilizers for enhancing phytoplankton production. The importance of phosphorus in production of Java tilapia was demonstrated by a series of experiments at Auburn University in which ponds received 8-8-2 (N-P-K) fertilization, 0-8-2 (N-P-K) fertilization, or no fertilization at all. Both fertilizers significantly increased tilapia production at population densities of 4942, 9884, 14,826, and 19,768/ha. Except at the highest density, the 0-8-2 (N-P-K) mixture was actually more effective than the mixture containing nitrogen compounds. Although it can be predicted that use of phosphates will enhance phytoplankton production, the digestibility of various phytoplankton differs widely. Java tilapia digest *Anabaenopsis* and *Oedogonium* well, *Botryococcus* partially, *Microcystis* and *Spirogyra* poorly, and perhaps cannot digest *Oscillatioria* or *Anabaena* at all. Unfortunately it is not yet practical to produce cultures of a particular phytoplankton species on a large scale.

Experiments in South Africa yielded the none too surprising result that 185 kg/ha of 19% superphosphate in conjunction with lime increased yields of Java tilapia by a factor of greater than 4. Basic slag, which is often used in practical fish culture in Africa, achieved similar results when applied at 225 kg/ha. More surprising is the similar effect of experimental application of 19% superphosphate at 330 kg/ha to ponds containing the macrophagous *T. melanopleura*. Over a 5-month period an increase in production of 240 kg/ha over natural production was recorded.

Phosphatic fertilization in the form of P_2O_5 applied at 40 to 120 kg/(ha)(yr) to Malaysian ponds containing Java tilapia and *Barbus gonionotus* raised the total yield of fish by 261 to 1260 kg/(ha)(yr). It should be pointed out, though, that both of these species are plankton feeders and highly tolerant of turbidity.

The effects of organic fertilizers on tilapia are not well known, but such fertilizers are fairly widely used, for example, in Indonesia, where marigold plant is used as a green manure. It is also likely that where such artificial feeds as oil seed cakes are used in tilapia culture, a large percentage of the intended feed is not consumed by the fish but functions as a fertilizer. Sewage is used to fertilize tilapia ponds in southeast Asia and has been suggested for use in semiarid regions of Africa where green manures are not abundant and animal manures are better employed in agriculture. This suggestion has met with little enthusiasm since unlike most Asians, Africans in general are as squeamish with regard to human wastes as are Europeans and Americans.

Pond fertilization in fish culture is of course not merely a matter of knowing the characteristics of the fish to be cultured, but must also take into account the chemistry of the water and soil. Thus fertilization experiments carried out with Java tilapia in, say, Alabama, are of only limited value to the prospective tilapia culturist in Uganda. As time goes on and more on-site research is carried out, scientists and fish culturists will be in a better position to evaluate the pros and cons of fertilization in culture of tilapia and all fishes.

SUPPLEMENTARY FEEDING

Supplementary feeding in tilapia culture is as important and as little understood as fertilization. Most authorities are of the opinion that supplementary feeding is essential for success in large-scale culture of tilapia, but nowhere is it carried out in a systematic manner. Among the feeds employed in Asia and Africa are rice bran, broken rice, oil cakes, flour, corn meal, kitchen refuse, rotten fruit, coffee pulp, and a variety of aquatic and terrestrial plants. In South Africa, *T. melanopleura* are provided with fodder by growing rice to a height of 20 to 30 cm, then flooding it and allowing access to the fish.

In countries where little money is available to purchase feeds, use of waste products should be encouraged. For example, in the Congo the simple expedient of throwing mill sweepings into tilapia ponds has resulted in production well above normal for the area.

There is great need for research in tilapia nutrition, not only to determine what feeds are best but also what rates of feeding are most effective.

The only worthwhile study of tilapia feeding rates was carried out on Java tilapia and Nile tilapia at Auburn University, using a mixture of 35% peanut meal, 35% soybean meal, 20% ground beef liver, 15% fish meal, and 15% distiller's dry solubles. Experimental feeding rates were 1, 2, 3, and 4% of body weight per day. Java tilapia grew best at 3% but nearly as well at 2%. Growth of Nile tilapia improved with each increment of feed. Feed conversion rates were best at 2% and 1%, respectively. The feed used seems unusual for tilapia in its high animal content; certainly such a feed would not be economically feasible where tilapia are raised to offset protein deficiencies in human diets.

USE OF TILAPIA IN WEED CONTROL

Correlated with the feeding habits of the various *Tilapia* species is their use in weed control. Five species have to date been used or tested in this capacity:

T. heudeloti fed on higher plants in experiments at Auburn University, but they did not exert effective control over any species.

T. melanopleura vies with the grass carp as one of the best agents of biological control of aquatic weeds.

T. mossambica eats filamentous algae, a principal habitat for many species of mosquito larvae, thus is often used, in combination with insectivorous species, as an agent of malaria control. For this purpose, 2500 to 5000 fish/ha are recommended. Its role in relation to higher plants is not clear. Some authorities claim that it does not consume them in appreciable quantities. However, in ponds in Texas it appeared to effectively control some aquatic and emergent plants, though it would not touch *Elodea* or *Riccia*. Consumption of higher plants probably depends in part on the availability of other foods.

T. nilotica experimentally stocked at 2500 to 5000/ha controlled filamentous algae and reduced some higher plants. It may be even more effective in malaria control than *T. mossambica* since it not only eats algae but is reputedly fond of mosquito larvae.

T. zillii in experiments in Malaysia controlled higher plants, including *Fimbristylis acuminata,* which grass carp will not touch.

Weed control by tilapia is generally a matter of stocking them in ponds where weeds interfere with fish culture. Thus stocked, they not only perform a service by eating weeds, but add to the total fish production of the pond. Tilapia have occasionally been suggested for use in control of nuisance plants in natural waterways where they are not native, as in the St. John's River of Florida where water hyacinth interferes with sport fishing and boating. Such suggestions are usually resisted, since the total effect of a voracious herbivore in a new environment is, to say the least, difficult to predict.

GROWTH AND PRODUCTION

Growth of tilapia varies greatly with stocking density, frequency of spawning, and food supply. Under very favorable conditions, individual Java tilapia may reach a weight of 850 g in 1 year; in brackish water they may reach 450 g in 8 months. But in most ponds 85 to 140 g is a more realistic weight to expect after a year if the sexes are raised together. Males grow two to three times faster than females, thus monosex culture of males produces correspondingly better growth. Monosex culture of females would also improve growth by eliminating periods of no growth associated with spawning, but no data are available on this rarely practiced technique.

Heterosis is not unknown in tilapia, but the subject has been insufficiently investigated. It has been shown that both of the reciprocal crosses of Java tilapia and Nile tilapia exhibit better growth and food conversion than either parent. The same is true of both the intraspecific crosses of Java tilapia of African and Malaysian stocks.

Since the future of tilapia culture appears to be largely bound up with polyculture, it seems superfluous to discuss production at length. The question to be asked is not "How many kilograms of tilapia can be produced in this pond?" but rather "Will tilapia add significantly to this pond's fish production?" As we have seen, in most cases the answer to the second question is yes. The amount of added production will of course vary, but even in instances such as the experimental culture of Java tilapia with channel catfish, where only 266 kg/ha of tilapia were produced, the effect of adding tilapia must be regarded as significant. Although if the total production of a fish pond were 266 kg/ha it would be a poor pond indeed, the production of tilapia in this case must be considered as supplementary to the production of the primary crop, channel catfish. The 266 kg/ha of tilapia produced, *plus* an increase in catfish production of 168 kg/ha over the 1400 kg/ha achieved by monoculture, amounts to an increase of 434 kg/ha of fish, or a total production gain of 30.3%. Addition of tilapia to carp ponds in Israel has, on occasion, resulted in production gains of more than 165%.

Where monoculture of tilapia is practiced, the "normal" figure usually cited for natural production is 500 kg/ha in the tropics and somewhat less in moderate climates. With fertilization and/or supplementary feeding yields of 1000 to 2500 kg/ha can be achieved. Huet suggests that 2500 kg/ha is a minimum desirable production in Africa but, at least in subsistence culture, smaller yields surely accrue some benefit to the culturist. Maximum yields of tilapia may be as high as 18,000 kg/ha, but reports of such yields must be taken with a grain of salt since they are likely to contain a preponderance of fish too small to be usable.

HARVESTING AND MARKETING TILAPIA

Harvesting of tilapia crops is usually done by seining or, where feasible, pond draining. However, experiments indicate that if only a portion of the stock is to be retained electrofishing may be more efficient and result in less injury to the fish.

In most places where tilapia is cultured it is marketed as fresh or iced fish, but it may also be sold frozen. The market price for tilapia varies greatly. In Israel, tilapia now brings a better price than the traditionally cultured common carp. In Africa the commercial value of tilapia is largely dependent on size (Plate 3). In southeast Asia, size is not an important factor, but the market price varies regionally. In some areas the acceptability of tilapia is lessened by its black skin, and consumption of tilapia is largely limited to people who cannot afford, say, milkfish. In other areas it is marketed as a gourmet food, which is quite appropriate, for the quality of tilapia flesh is usually very high. Tilapia too small to be marketed for human consumption need not be wasted, as they may be used as fodder for more expensive fish such as trout, as an ingredient in livestock feeds, or as bait in commercial fishing.

Experiments in Alabama indicate that tilapia could be produced in the United States at extremely low cost. Data are not available for other parts of the world, but it would appear that tilapia could compete favorably with other aquacultural and fishery products in most tropical and moderate climates.

Some authorities, however, notably S. Tal and C. F. Hickling, do not view tilapia as a good commercial proposition, at least not on a large scale. Even Tal and Hickling do not dispute tilapia's importance where there is an immediate need to feed large numbers of people or where fish can go directly from pond to pot. Whatever the commercial feasibility of tilapia culture may prove to be, it must be conceded that, thanks to subsistence culture and small-scale commercial culture, tilapia is currently one of the most important fish crops in much of the world, including most of Africa, Jamaica, southern Taiwan, and parts of Indonesia.

PLATE 3. Good sized tilapia being marketed in the Congo. (Courtesy Marcel Huet, Groendendaal-Hoeilaart, Belgium.)

DISEASES AND PARASITES

Diseases and parasites are somewhat less of a problem with tilapia than with many cultured fishes. Among the parasites found on tilapia are *Trichodina*, *Chilodon*, and *Saprolegnia*. The only disease mentioned in the literature is bacterial fin rot, but one might expect that, at least in marginal climates, tilapia would be subject to ichthyophthiriasis, the precondition for which is usually chilling. Tilapia also act as carriers of catarrhal enteritis.

STATUS AND PROSPECTUS OF TILAPIA CULTURE

The prospectus for tilapia culture can best be outlined on a regional basis.

THE NEAR EAST

The future of tilapia in Israel is debatable, but as long as the present demand is sustained it seems likely that tilapia will hold their own in fish culture as well as in fisheries. There are perhaps more institutions carrying on fish culture research in Israel than in any other country, so breakthroughs in tilapia culture techniques may well be made and applied in Israel.

The growth of fish culture in other Near Eastern countries is hampered by the lack of an aquacultural tradition and, in some cases, a shortage of water. Present emphasis in the region is on culture of the common carp, but one may expect tilapia to play some role in the development of fish culture.

AFRICA

There is perhaps more excitement over tilapia in tropical Africa than anywhere else, but results to date vary greatly, as does the prospectus. The problem of stunting, which must be reckoned with wherever tilapia are raised, is crucial in Africa due to the reluctance of most Africans to accept small fish. The future of African tilapia culture is thus dependent on the improvement and spread of population control techniques.

Culture of tilapia will undoubtedly continue to be practiced in virtually all African countries but, as in the past, success may be expected to be greatest in those countries where government supervision and assistance are greatest. Most noteworthy in this respect are Uganda and the Malagasy Republic. Government supervision is necessary not only for economic reasons but because Africa lacks a tradition of fish culture. Thus a farmer who is perfectly willing to work long and hard on a terrestrial crop is culturally conditioned to think of fish as wild game rather than as livestock to be tended, and he may let a carefully constructed and stocked fish pond deteriorate for lack of attention.

SOUTHEAST ASIA

Tilapia may be expected to continue to play an important role in fish culture, particularly in Taiwan and Indonesia. Whether significant expansion will occur depends on the results of research, on the extent to which eutrophication renders inland and brackish water unsuitable for culture of other fish, and, in some countries, on whether or not regional attitudes toward tilapia change.

UNITED STATES

The consensus of workers at Auburn University is that tilapia are not presently feasible for culture in the United States, due to the premium placed on large fish, the difficulty in overwintering (even in Alabama, tilapia used in research must be brought indoors during the winter), the doubtfulness of consumer acceptance, the availability of a wide variety of native fishes for culture, and the justifiable concern of conservationists and sport fishermen over the possible ecological effects of the intentional or accidental stocking of tilapia in natural waterways.

Large-scale culture of tilapia is not expected in the United States but, despite the warnings of ecologists, occasional introductions may be made. Java tilapia and/or Nile tilapia have already been stocked locally as food or sport fish or for mosquito control in at least six of the southern states and Hawaii. There is as yet no indication of their impact on fisheries or the ecology in any of these areas.

TEMPERATE REGIONS OF EUROPE AND ASIA

Little future is seen for tilapia in these regions, except perhaps in industrial cooling waters. Experiments along these lines are being carried out in England, the Soviet Union, and West Germany. Some success has also been reported from tilapia culture in the south of France, but details are not available.

LATIN AMERICA

In no other major area of the world is fish culture so poorly developed. Widespread protein deficiency in Latin America has in recent years prompted the various national governments and international aid agencies to investigate the potential of fish culture in the region. The most frequently suggested group of fishes for culture there are the *Tilapia* spp. The only serious objection which can be raised is that if tilapia find their way into natural waters, their effect on the local ecology could be disastrous. With this in mind studies have been undertaken in a number of countries, notably Brazil, Guatemala, and Peru, to determine the aquacultural potential of native fish species. Results to date are not encouraging, so experimental culture of tilapia is proceeding in many Latin American countries, especially Brazil and Costa Rica.

Commercial production of tilapia is already under way in Jamaica and Trinidad. In Jamaica tilapia culture is quite successful in terms of yield but shows little promise of being able to add significantly to the abundant fish supply contributed by marine fisheries. In Trinidad the situation is much the same as regards pond culture, but Trinidad and some other Caribbean islands have further resources for fish culture in the form of extensive fresh and brackish water swamps. It has been suggested that tilapia could be cultured and stocked in these swamps to supplement the take of fishermen who regularly exploit them.

CURRENT AND FUTURE RESEARCH IN TILAPIA CULTURE

There is no form of fish culture that cannot benefit from research, but tilapia culture, as a young branch of the art involving 12 or so species grown under widely differing conditions all over the world, is in particular need of information obtained through research.

GROWING TILAPIA IN THERMAL EFFLUENT

One aspect of tilapia research—the experimental culture of tilapia in industrial cooling waters—may result in still further geographical expansion of tilapia culture, but most tilapia research is aimed at improving production in the tropical and moderate climates where they are already successfully raised.

POPULATION CONTROL

Easily the most important advance which would be made would be a foolproof method of preventing overcrowding and stunting. This has usually been approached through hybridization. As mentioned earlier, there are at least three tilapia crosses that yield 100% male offspring, but one of these is difficult to produce, one involves the use of stocks from rather limited geographic areas, and one involves *T. hornorum*, which is not available to many culturists outside east Africa and Malaysia. What is needed is not only an easily obtained and duplicated all-male hybrid, but a series of them, including plankton feeders, macrophages, salinity and temperature tolerant strains. Such hybrids might also exhibit heterosis, as does the existing all-male hybrid ♂ *T. mossambica* (Zanzibar stock) × ♀ *T. nilotica* (Lake Albert stock).

TAXONOMY AND GENETICS

To achieve the ends of hybridization it will be necessary to better understand tilapia genetics, both at the specific and subspecific levels; to understand, for example, why mating male *T. mossambica* from Zanzibar with female *T. nilotica* from Lake Albert produces 100% male offspring, whereas mating the same sexes of the same species from other parts of the world may not. The attainment of this understanding will entail careful examination of the taxonomy of the genus *Tilapia*, which is currently quite confused.

ECOLOGICAL AND ETHOLOGICAL STUDIES

Concurrent with hybridization experiments should be studies to more precisely determine the ecological niche occupied by each of the cultured species. Many attempts at tilapia culture have yielded unsatisfactory results merely because the culturist did not have the necessary information to choose the right species or hybrid for the job. Once the roles of the various species are known, hybridization can proceed more intelligently and the culturist will have an even wider selection of tilapia types to choose from. One of the best means of approaching the problem of ecological niche would be to study tilapia behavior and food habits under natural or seminatural conditions.

When the roles played by the various species and hybrids are known there will be a real basis for the foundation of tilapia polyculture, with or without nontilapia species, at a level of sophistication comparable to that of Chinese carp culture. Even now, studies aimed at finding better species combinations and stocking rates for pond communities would not be amiss.

POND FERTILIZATION AND SUPPLEMENTARY FEEDING

Among the less glamorous tasks facing researchers on tilapia culture is the assessment of bodies of water for culture. It is all very fine to do research on pond fertilization in Alabama or Israel, but to attempt to extrapolate from there to a pond of unknown physical and chemical characteristics in East Africa is scarcely realistic. For this and other purposes, regional research stations should be established wherever large-scale tilapia culture is contemplated, and surveys should be made of all waters to be cultivated so that the culturist may proceed intelligently with specific management measures appropriate for his area.

Much more work needs to be done on pond fertilization and feeding in tilapia culture. Both laboratory research on nutritional needs and field work on the effects of different feeds and fertilizers in practical culture are needed. Not only the yields which may be achieved, but the protein content and marketability of the fish produced should be assessed.

With a sound basis in research, further development of tilapia culture can proceed apace. Development will surely include at least some practical culture in Latin America, gradual replacement of individually operated subsistence cultures with more efficient commercial or communal operations, education of would-be tilapia growers, particularly in Africa, as to the need for population control, proper feeding, and other management measures, and further mechanization wherever tilapia are cultured. More precise means of economic analysis of culture methods should also be applied.

At the present time there seem to be more questions than facts with regard to tilapia culture, but it is a safe bet that, for the foreseeable future at least, the tilapia complex will continue to be an important contributor to the world's protein supply.

REFERENCES

AVAULT, J. W., and E. W. SHELL. 1966. Preliminary studies with the hybrid tilapia *Tilapia nilotica* × *Tilapia mossambica*. FAO World Symposium on Warm Water Pond Fish Culture. FR: IV/E-14.

BARDACH, J. E. 1957. Marine fisheries and fish culture in the Caribbean. Proceedings of the Gulf and Caribbean Fisheries Institute, 10th Annual Session, pp. 132–137.

FUJITA, K., K. FUKUSHE, J. HATORI, K. KURONUMA, and Y. NAKAMURA. 1966. Comparison of morphology and ecology in 3 species of tilapia. Proc. Jap. Soc. Syst. Zool. 2 (1966):31–35.

HICKLING, C. F. 1966. Fish hybridization. FAO World Symposium on Warm Water Pond Fish Culture. FR: IV/R-1.

HOFSTEDE, A. E. 1949. Pond culture of warmwater fishes in Indonesia. Proc. U.S. Sci. Conf. Cons. Util. Res. 7:136–138.

HORA, S. L., and T. V. B. PILLAY. 1962. Handbook on fish culture in the Indo-Pacific Region. FAO Fisheries Biology Technical Report 14. 204 pp.

HUET, M. 1957. Dix années de pisciculture en Congo Belge et Ruanda Urandi. Trav. Sta. Rech. Groenendaal (D) 22:1–159.

LAHSER, C. W. 1967. *Tilapia mossambica* as a fish for aquatic weed control. Prog. Fish. Cult. 29(1):48–50.

LESSENT, P. 1966. Essais d'Hybridation dans le Genre *Tilapia* à la station de Recherches Pisciocles de Bouake, Côte d'Ivoire. FAO World Symposium on Warm Water Pond Fish Culture. FR: IV/E-5.

LIN, S. Y. 1968. Pond fish culture and the economy of inorganic fertilizer application. Chinese-American Joint Commission on Rural Reconstruction, Fisheries Series, No. 6.

MAAR, A., M. A. E. MORTIMER, and I. VAN DER LINGEN. 1966. Fish culture in Central East Africa. FAO, Rome. 158 pp.

PRUGININ, Y. 1966. The culture of carp and tilapia hybrids in Uganda. FAO World Symposium on Warm Water Pond Fish Culture. FR: IV/E-12.

RABANAL, H. R. 1966. Stock manipulation and other biological methods of increasing production of fish through pond fish culture in Asia and the Far East. FAO World Symposium on Warm Water Pond Fish Culture. FR: V/R-3.

SHELL, E. W. 1966. Monosex culture of male *Tilapia nilotica* (Linn.) in ponds stocked at 3 rates. FAO World Symposium on Warm Water Fish Culture. FR: V/E-5.

SREENIVASAN, A. 1966. Fish production in some rural demonstration ponds in Madras (India) with an account of the chemistry of water and soil. FAO World Symposium on Warm Water Pond Fish Culture. FR: II/E-9.

STERBA, G. 1966. Fresh water fishes of the world. Viking Press, New York.

SWINGLE, H. S. 1966. Biological means of increasing productivity in ponds. FAO World Symposium on Warm Water Pond Fish Culture. FR: V/R-1.

TREWAVAS, E. 1966. The name and natural distribution of the tilapia from Zanzibar (Pisces, Cichlidae). FAO World Symposium on Warm Water Pond Fish Culture. FR: VIII-I V/E-8.

YASHOUV, A. 1959. Studies on the productivity of fish ponds. I. Carrying capacity. Proc. Tech. Pap. Gen. Fish. Coun. Mediterr. 5:409–418.

YASHOUV, A. 1966. Mixed fish culture—an ecological approach to increase pond productivity. FAO World Symposium on Warm Water Pond Fish Culture. FR: V/R-2.

YASHOUV, A. and A. HALEVY. 1967. Studies on growth and productivity of *Tilapia aurea* and its hybrid "Gan-Shmuel" in experimental ponds at Dor. Bamidgeh 19(1):16–22.

Many Additional References Listed in

ST. AMANT, J. A., and M. C. STEVENS. 1967. Bibliography of publications concerning *Tilapia mossambica* (Peters). Resources Agency of California, Dept. of Fish and Game, Inland Fisheries Administrative Report, No. 67-3.

Interviews and Personal Communication

BROWN, R. J. Proyecto de Diversification Agricola, Turrialba, Costa Rica.

FISHELSON, L. University of Tel Aviv, Israel.

HICKLING, C. F. 95 Greenway, London N20, England.

HUET, M. University of Louvain, Belgium.

TAL, S. Director of the Inland Fisheries, Ministry of Agriculture, Tel Aviv, Israel.

YASHOUV, A. Fish Culture Research Station, Dor, Israel.

19

Culture of True Eels (*Anguilla* spp.)

Among food fishes, true eels (*Anguilla* spp.) occupy a position in some countries roughly equivalent to that held among meat animals by turkey in the United States; they may be eaten anytime but are traditionally served on certain days. In Italy, eel is the traditional dish on Christmas Eve. Since eels are primarily a fishery product in Italy, the supply cannot be controlled, and prices go up tremendously on that day. In Japan, one day in July is set aside as "eel day," and great quantities are sold, but since culture is the dominant mode of production, prices are more nearly stable.

Whatever the season, and even in countries like the United States, where eel is not liked by the majority of the populace, it is definitely a luxury food. This status is due partly to the declining success of eel fisheries, and partly to the great expense of eel culture. As will be seen, one of the chief factors contributing to the expense is changing.

NATURAL HISTORY AND COLLECTION OF ELVERS

The only country with a long history of eel culture is Japan, where *Anguilla japonica* has been farmed commercially for about 150 years and on a subsistence basis for centuries longer. It was not until after World War II that Taiwan became the second country where this species is cultured. Like all species of *Anguilla*, *A. japonica* has the rather unusual trait of being catadromous, that is, spawning occurs in the sea and the young migrate inshore to mature in freshwater. The inshore migration begins when the sea temperature first reaches 8 to 10°C in the spring, and the transparent, 5- to 7-cm-long young, called "elvers," reach Japanese shores in December to April. Since Japanese biologists have had very limited success in spawning eels, elvers collected at this time from streams and estuaries are the source of stock for culture.

The best runs of elvers in Japan occur in the east central part of the country, in Chiba, Ibaragi, and Shizuoka prefectures. Elvers are perhaps more abundant in Taiwan and when, in 1969, Taiwanese authorities relaxed the ban on export of elvers, 1 to 2 million were collected and sold to Japanese culturists, in addition to those purchased by Taiwanese eel growers. In both countries, elvers are captured in dip nets, kept in floating baskets or boxes for 4 or 5 days, then transferred to nursery ponds, where they are reared for 6 months to 2 years, or until they reach 12 to 15 cm in length, before sale to culturists. Some Japanese growers circumvent this expense by capturing eels of the same size from rivers, by the use of traps or bamboo pipes.

CULTURE IN JAPAN AND TAIWAN

REARING ELVERS

Nursing of elvers in Japan is carried out in two stages. First they are stocked at a rate of 300,000/ha in ponds 200 m² or less in area. After 3 months, the population is thinned and redistributed in 600-m² ponds. In both types of pond, they are fed daily with dried sardines or sardine meal, sometimes augmented by silkworm pupae. Young eels are transported in very little water in special baskets, 40 cm in diameter × 20 cm deep. From 35 to 70 kg of eels may be safely carried in such baskets, which are stacked one on top of another, packed with straw mats, and tied together. Ice may be placed under the top mat during hot weather.

Taiwanese nursery ponds are much more heavily stocked; figures of up to 3 million elvers/ha are cited. Foods given include worms and minced, cooked trash fish, both presented in submerged bamboo mesh baskets.

Stocking regimes for eel production ponds in Japan and Taiwan are quite different from each other, in keeping with the different aquacultural traditions of the two countries. Eel culture in Japan is essentially a monoculture system, although common carp (*Cyprinus carpio*) or striped mullet (*Mugil cephalus*) may be stocked to clean up the excess food which is a necessary evil in eel ponds. In Taiwan, however, eels are integral parts of complex polyculture systems derived from the classical method of Chinese carp culture.

STOCKING PRODUCTION PONDS

In coastal Taiwanese ponds where Java tilapia (*Tilapia mossambica*) are the major crop (see Chapter 18 for details), eels may constitute a minor component of the stock, but in another class of ponds they are the principal species stocked. Such ponds, which have smooth mud bottoms sloping toward the lower end, and walls lined with cement, brick, or stone, vary in size from 0.08 to 4 ha or more, but seldom average over 1.0 to 1.5 m deep. One such 4-ha pond was stocked in 1967 with—in addition to elvers at about 25,000/ha—silver carp (*Hypophthalmichthys molitrix*), big head (*Aristichthys nobilis*), common carp, mud carp (*Cirrhinus molitorella*), and striped mullet. Table 1 illustrates the roles of each of these species in the pond ecosystem and in the harvest.

The system outlined in Table 1 differs from the ordinary pond polyculture systems originally developed in China and currently practiced in Taiwan and elsewhere in that it is neither the natural character of the pond nor fertilization *per se* that determines the structure of the fish community. Rather, the voracious but sloppy feeding habits of the eels dictate that large amounts of food be given, and the species and numbers of other fish stocked are in turn determined by the effect of the "waste" eel food. Striped mullet, which would not ordinarily be stocked in a freshwater pond, are of particular importance, not so much for their contribution to fish production as to control the dense growths of blue-green algae which result from the superabundance of nitrogenous matter. Other species that may be fitted into this type of ecosystem are goldfish (*Carassius auratus*), which may replace silver carp, and crucian carp (*Carassius carassius*), which, as omnivores, may take advantage of any niche that is not completely filled. Grass carp (*Ctenopharyngodon idellus*)

TABLE 1.   STOCKING RATE AND YIELDS OF POLYCULTURE PONDS IN TAIWAN,
WITH EELS AS THE PRIMARY CROP

SPECIES	NICHE	NO. STOCKED	SIZE STOCKED (G)	SURVIVAL (%)	HARVEST (KG)
Eel	Principal crop, heavily fed	100,000	25	70	14,000
Silver carp	Phytoplankton feeder	8,000	10–20	50	6,000
Big head	Zooplankton feeder	1,000	10–20	100	5,600
Common carp	Utilizes excess eel food	32,000	10–20	25	4,000
Mud carp	Benthos feeder	4,000	—	100	1,600
Striped mullet	Feeds on blue-green algae	6,000	0.5	50	800
Total		151,000			32,000 (8,000 kg/ha)

may also be stocked, but the culturist will then have two species to feed, since few ponds support enough higher plants to satisfy this voracious species.

There is a trend in Taiwan toward higher stocking rates of eels. A large pond such as the one just described would probably now be stocked with double the number of elvers (50,000/ha). Ponds less than 0.5 ha in area are stocked at higher rates—up to 160,000 elvers/ha. The stocking rates thus far cited assume that running water can be provided. In small ponds, artificial aeration by means of pumps or mechanical paddles may also be resorted to. However, some eel ponds in Taiwan are stagnant, and oxygen deficiencies may develop. Stocking rates of such ponds must be less than half those cited.

Shizuoka Prefecture produces about 65% of the eels grown for market in Japan. Young eels purchased from nurseries are stocked in two types of water, ponds and running water enclosures. In addition, a closed recirculating system of the type developed by A. Saeki of Tokyo University, described in detail in Chapter 2, has been tested under experimental conditions but has yet to be adopted in practical eel culture.

Nearly all enclosures of both types have walls of concrete, stone, or occasionally wood. They vary greatly in size, from less than 0.3 ha to

PLATE 1.   Eel ponds in Taiwan showing mechanical aerator. (Courtesy Ziad Shehadeh, Oceanic Institute, Hawaii.)

more than 3 ha, but ponds average much larger. In 1964, 84% of running water enclosures were less than 0.8 ha in area, while almost half of the eel ponds were 1.5 ha or larger. Running water enclosures are much more productive than ponds, and operating costs only about $\frac{2}{3}$ as much, but pond operators are able to make up the difference by maintaining larger culture units.

The rate of stocking is determined by the rate of circulation of water and varies from 20,000 to 44,000 eels/ha. When common carp are added, they are stocked as 7-cm young at 10,000/ha.

FEEDING

In any eel culture system, it is primarily feeding which determines the success of the operation. The more animal protein that can be supplied (at least 50% is essential), the greater the weight of eels that can be produced. Food thus accounts for an average of 51 to 55% of the expenses of growing eels for market in Japan, and may run as high as 80%.

The traditional foods for eels in Japan are fresh and cooked trash fish, silkworm pupae, and, to a lesser extent, earthworms, aquatic worms, and

PLATE 2. Workers removing bottom sediment from eel ponds at end of growing season. Note stones on dykes to prevent the eels from burrowing. (Courtesy Ziad Shehadeh, Oceanic Institute, Hawaii.)

crushed mollusks. In Taiwan, trash fish are the most commonly used food, but offal from slaughterhouses and fish packing plants, ox blood, earthworms, aquatic worms, and small crabs, all chopped into small pieces, are also used. Crabs are believed to be especially effective in promoting growth. Taiwanese eel farmers feed silkworm pupae only in the last stages of culture, when the fish are being fattened for market. For this purpose, the pupae are soaked in water for an hour or so before feeding.

Conversion ratios obtained with traditional foods are extremely poor (Table 2), but recently developed artificial feeds have produced dramatic improvement. In Japan, artificial eel feed, which is produced in the form

TABLE 2.   CONVERSION RATIOS OF EEL FEEDS USED IN JAPAN AND TAIWAN

COUNTRY	FOOD	CONVERSION RATIO
Japan	Chopped fish, silkworm pupae, etc.	5.5:1
	Sardines	5.43–7.16:1
	Artificial feed (pellets)	3.03–4.36:1
Taiwan	Trash fish, etc.	10–15:1
	Artificial feed (granular)	2.1–2.6:1
	Artificial feed mixed with fish liver oil	1.9:1

of pellets, contains about 60% fish meal and 20% starch, along with vitamins, minerals, and so forth. It is expensive and while it is used exclusively by the larger culturists, small operators vary their feeding plan according to the relative cost of artificial feed and fresh fish.

The artificial eel feed used in Taiwan (Table 3) is granular in form and is of slightly more recent origin, but has already been adopted by considerable numbers of culturists, who report it to be cheaper, easier to handle, and more sanitary than traditional foods.

Whatever is fed, it is usually presented in baskets or perforated troughs, rather than being broadcast in the pond. The rate of feeding should be

TABLE 3.   COMPOSITION AND CHEMICAL ANALYSIS OF THE ARTIFICIAL EEL FEED USED IN TAIWAN

COMPOSITION	
Fishmeal	65.00%
Defatted soybean meal	10.16
Yeast powder	10.12
Fish concentrate	5.00
Starch	8.12
Multivitamins	1.00
Lysine	0.20
Antioxidant	0.20
Binding substance	0.20

CHEMICAL ANALYSIS	
Crude protein	51.91%
Crude fat	5.36
Crude carbohydrate	16.03
Ash	17.97
Moisture	8.73

PLATE 3. Cages with feed (dough consistency) lowered into eel cages. (Courtesy Ziad Shehadeh, Oceanic Institute, Hawaii.)

PLATE 4. Eels swim into cages and burrow through food. (Courtesy Ziad Shehadeh, Oceanic Institute, Hawaii.)

5 to 10% of the body weight of eels daily, depending on temperature. Below 8 to 10°C, feeding ceases.

Table 2 shows conversion ratios obtained with various foods. If the improved conversion achieved with the new artificial feeds can be sustained at low cost to the farmer, the status of eel culture may be radically changed.

GROWTH, YIELD, PRODUCTION, AND MARKETING

Even under optimum conditions of heavy feeding, high dissolved oxygen concentration, and temperatures of 20 to 28°C, eels characteristically exhibit great differences in growth between individuals (in an experiment with the Atlantic species, *Anguilla anguilla,* some individuals doubled their weight, while others, stocked at the same size, grew by a factor of 120) and special cropping systems have evolved accordingly. In Japan, about 30% of the eels stocked reach the marketable size of 100 to 200 g within 1 year after stocking. These are then harvested and replaced with an equal number of young eels. At the end of the second year, all the eels are harvested. Total survival over the 2-year period is 60 to 90%. In Taiwan, larger eels, weighing 200 g or more, are preferred by consumers. Eel ponds are selectively fished for eels on a daily to monthly basis, depending on demand, and all such specimens marketed. The other species stocked in eel ponds are harvested annually.

Yields and production of eel culture in Japan have increased considerably in the last decade. In the late 1950s, 15,000 kg/ha was considered an exceptional yield. In 1965, however, the most productive running water enclosures yielded about 45,000 kg/ha, and the average yield in running water was higher than the maximum formerly achieved. Although the lower yields realized from ponds lower the overall average yield considerably, it is still impressive (Table 4).

About 98% of the eels produced in Japan are sold domestically, mostly as fresh fish, but also canned or as a frozen, precooked item. The remainder of the crop is exported, mostly to western Europe. There is a

TABLE 4.   PRODUCTION AND YIELD OF EELS CULTURED IN RUNNING WATER ENCLOSURES AND PONDS IN JAPAN, 1964

TYPE OF ENCLOSURE	TOTAL AREA (HA)	PRODUCTION (KG)	YIELD (KG/HA)
Running water	49	1,346,117	26,360
Pond	123	664,843	6,120
Total	172	1,810,960	10,528

great demand for eels in Europe, which increases as the fisheries there decline, but Japan has not made a major effort to enter the market. To do so effectively, Japanese culturists would have to grow eels to twice the length of the 45- to 60-cm specimens currently marketed.

If the growth of eel culture in Japan has been fast, its growth in Taiwan has been spectacular. In 1965 122,753 kg were produced; by 1969 about 2 million kg were being produced annually from about 200 ha of water, for an average yield of 10,000 kg/ha. Taiwanese culturists have, since the beginning of eel culture in the country, grown somewhat larger eels than their Japanese counterparts, thus they export a significant portion of their crop.

CULTURE IN EUROPE

Imports are not adequate to satisfy the high demand for eels which exists in virtually every European country and, as natural populations decline due to increased pollution of rivers, interest in culture of the native *Anguilla anguilla* grows not only in Europe but in Israel and the United Arab Republic as well. Large numbers of eels are still taken in the Italian valli and other Mediterranean lagoons, in practices some of which constitute low-intensity aquaculture (see pp. 296–302), but intensive eel culture is a new phenomenon outside of Asia.

Commercial culture is already a reality in the Soviet Union and West Germany, where 20-cm elvers are collected along the coast for stocking. Culture proceeds much as in Japan or Taiwan, but *A. anguilla* is thus far reluctant to accept dry food, so trash fish are fed almost exclusively. Conversion ratios of 8 to 10:1 are realized, but ratios might be improved through monosex culture, since females of *A. anguilla* grow much faster than males. West German researchers have found that the shape of the culture enclosure also has an effect on growth. Eels, which are olfactory feeders, can locate food more efficiently in long, narrow ponds.

Diseases represent more of a problem with *A. anguilla* than they do with *A. japonica*. West German biologists refer to a "cauliflower" disease and a "red" disease, both of which severely affect growth and, to a lesser degree, survival, but no cures are suggested. An apparently more serious problem occurred in 1957 in the Po River delta of northern Italy, where eels constitute 70% of the production of valle fish culture. Eels there developed heavy infestations of an apparently new species of *Argulus*, and production declined by 50%. Attempts to eliminate the parasite by flushing the valli with large volumes of freshwater were only partly successful.

PROSPECTUS

It is scarcely accidental that, in Asia, eel culture has developed only in Japan and Taiwan, the two wealthiest nations on the continent. Although eels are found along the entire west coast of the Pacific and throughout the Indian Ocean (as well as on both coasts of the Atlantic), culturists in southeast Asia have been advised against eel culture. The principal objection to eels in that region is that they are extremely wasteful protein converters, thus not appropriate for culture where supplies of protein for human consumption are low. However, the greatly improved conversion rates recently achieved by use of dry foods may soon render this objection obsolete.

Another factor contributing to the high cost of eel is the inability of culturists to propagate them in captivity, although *A. anguilla* has reportedly been artificially spawned in France. The logical way to approach artificial propagation of eels would be by considering their spawning habits in nature, but these are little known. Even if they were well known, it might prove very difficult to simulate the natural conditions encountered by a catadromous fish. If efficient artificial propagation can be achieved, cultured eels may eventually contribute significantly to protein supplies in southeast Asia, Africa, and Latin America. Otherwise, eel will likely remain a luxury food.

REFERENCES

Brown, E. E. 1969. The fresh water cultured fish industry of Japan. University of Georgia College of Agriculture Experimental Station, Research Report 41. 57 pp.

Hora, S. L., and T. V. R. Pillay. 1962. Handbook on fish culture in the Indo-Pacific Region. FAO Fisheries Biology Technical Paper 14. 204 pp.

Koops, H. 1966. Feeding of eels (*Anguilla anguilla* L.) in ponds. FAO World Symposium on Warm Water Pond Fish Culture. FR: III-VIII/E-3.

Lin, S. Y. 1968. Pond fish culture and the economy of inorganic fertilizer application. Chinese-American International Commission on Rural Reconstruction, Fisheries Series. No. 6.

Ling, S. W. 1966. Feeds and feeding of warm-water fishes in ponds in Asia and the Far East. FAO World Symposium on Warm Water Pond Fish Culture. FR: III-VIII/R-2.

Interviews and Personal Communication

Mozzi, C. Institute of Hydrobiology, University of Padua, Italy.

20

Commercial Culture of Freshwater Salmonids

Genera *Salmo, Salvelinus,*
Thymallus, and *Hucho*

In 1589, when Henry IV, King of France, promised "a chicken in every pot," it seemed almost an extravagant offer. Today, in the United States and other affluent countries, a politician making the same offer would be accorded a blasé reception, for, through streamlined production methods, chicken has become one of the cheapest meats. Trout may soon achieve an analogous position among quality fish products; already commercial trout culturists are gaining production levels undreamed of by the early hatcherymen, with consequent reduction in cost per unit weight.

In the United States and some European countries trout culture has the longest history of any form of fish culture, due largely to the popularity of trout as a sport fish. At least since the appearance of Dame Juliana Berners' "Treatise on Fysshynge with an Angle" in the *Boke of St. Alban's,* published in 1486, salmonids have been *the* prestige fish among freshwater anglers. Unfortunately, since salmonids prefer environments which are sterile compared with those favored by such other popular sport fishes as the black basses and the pikes, they are also among the most easily depleted by fishing and other activities of man. As early as 1741, when Stephen Ludwig Jacobi established the first trout hatchery in Germany, anglers began to depend on culturists to augment and, too often, substitute for natural production in the maintenance of sport fisheries.

SPECIES CULTURED

The most widely cultured salmonid for sport fishery purposes is the rainbow trout (*Salmo gairdneri*). Native to the Pacific Coast drainages of North America from Alaska to Baja California, the rainbow trout has been introduced as a sport fish to virtually all suitable waters in the affluent countries. Nearly as much effort has been devoted to the propagation of the European brown trout (*Salmo trutta*), which has been distributed throughout the world and has assumed special importance in the United States, where it has been the savior of trout fishing in many streams which have become too warm and/or polluted to support native salmonids. The only other salmonid species which has received comparable attention from culturists is the brook trout (*Salvelinus fontinalis*). Originally restricted to northeastern North America, the brook trout has been introduced wherever water temperatures are cold enough to meet its demands, which are somewhat more restrictive than those of rainbow or brown trout.

Other species cultured for sport fishery purposes on a more limited basis include the Atlantic salmon (*Salmo salar*), of the European coast, Iceland, and the Atlantic coast of North America from Maine north; the lake trout (*Salvelinus namaycush*) of northern North America; the cutthroat trout (*Salmo clarki*) of western North America; the Sunapee trout (*Salvelinus aureolus*) of northern New England; and the most beautiful of all salmonids, the golden trout (*Salmo aguabonita*) of California's High Sierras.

Among the qualities that have endeared trout to sport fishermen is the high quality of the flesh. Thus it is not surprising to find that attempts at commercial trout culture were made as early as 1853 in the United States and perhaps earlier in Europe. Until recently, most of the techniques used in commercial trout culture were identical to those practiced in sport fish hatcheries. Though understanding of the rather different goals of the two types of trout culture has resulted in considerable differences in culture practices today, the commercial trout culturist should keep abreast of developments in the sport fish hatcheries, where many relevant new developments occur. There is an extensive literature on trout culture for sport fishing, but this book is concerned only with commercial trout culture. (The reader seeking a detailed treatment of trout culture for sport fishing is referred to *Trout and Salmon Culture* by Earl Leitritz and *Culture and Diseases of Game Fishes* by H. S. Davis.)

The "big three" species in commercial trout culture are the same three favored by sport fishermen, but the preeminence of the rainbow trout is more solidly established. Of all the salmonids, none is so amenable to captivity or so tolerant of different temperatures, salinities, and population densities. The brown trout is nearly as adaptable as the rainbow trout with respect to environmental parameters but is more territorial, thus does not do as well when cultured at high densities. Some also consider it slightly less desirable as food. The brook trout grows less rapidly and is the most delicate of the three species, particularly with respect to temperature. However, it is preferred by some culturists where trout are sold by weight rather than length since it is a deeper bodied fish than rainbow or brown trout. Many fishermen also consider it a superior table fish.

Virtually all large-scale commercial trout culture is based on these three species, but in regions where salmonids enter commercial fisheries young fish of other species may be cultured and released to supplement natural reproduction for commercial as well as sport purposes.

HATCHERY TECHNIQUES

GENERAL CONSIDERATIONS

Wherever rainbow, brown, or brook trout are raised the basic techniques of spawning and hatching are the same. In nature, most salmonids dig nests, called redds, in streambeds in riffle areas. The eggs are deposited in these depressions, fertilized, and buried with gravel, to hatch some months later. In no instance are cultured trout allowed to spawn in this manner. Rather, eggs are artificially obtained, fertilized, and hatched, resulting in fertilization rates of nearly 100% and hatching rates well in excess of those achieved in nature.

In general, wild rainbow trout spawn in the spring whereas brown trout and brook trout breed mainly in the fall. However, strains of these species, particularly the rainbow trout, may be found spawning at practically any time of year. Some commercial culturists take advantage of this variability by importing eggs from various parts of the world to maintain a constant supply of trout of all sizes. Others have selectively bred early or late spawning strains. Spawning time may also be regulated by artificial control of the photoperiod to simulate seasonal change or by injection of pituitary hormones. The latter technique has not found as much application among trout culturists as among growers of catfish, carp, and some other species.

The rate of hatching of eggs appears to be partially genetically controlled and brood stock is carefully selected for hatching rate as well as for production of large eggs. Large eggs, which are associated with the size of the female as well as the genetic strain, produce large larvae, which generally exhibit better survival than smaller ones.

Even trout of selected strains do not all mature at the same time and during spawning season brood females must be sorted frequently so that the very ripest individuals, as determined by how readily eggs ooze from the fish when handled, are spawned. Overripeness as well as under-ripeness is a problem, for trout will seldom release eggs naturally under hatchery conditions. Use of overripe or underripe spawners may result in lower rates of fertilization and injury to the female.

Not only must brood females be handled very carefully during sorting, precautions must also be taken with their diet. Commercial feeds containing cottonseed meal adversely affect egg production and are to be strictly avoided.

ARTIFICIAL FERTILIZATION

The process of obtaining eggs and milt is called stripping and requires a fair amount of skill to avoid injuring either fish or eggs. In some hatcheries injury to the spawners is reduced by anesthetizing them with MS-222. Where anesthesia is not employed, stripping may be carried out as a one-man operation, but less fumbling and consequent injury to the fish occurs if it is done by a two-man team. One man grasps a ripe female by the caudal peduncle and pectoral fins and holds it over a pan, while the other gently extrudes the eggs by applying pressure to the abdomen with thumb and forefinger, beginning just forward of the vent and proceeding forward to the pelvic fins. If the fish is held tail down the eggs will flow naturally into the pan. Pressure should not be applied forward of the pelvic fins, since this may damage the internal organs. Another advisable precaution is to wear wool gloves so as to be able to grip the fish firmly. Care should also be taken not to break any eggs, as the albumen from broken eggs will coat other eggs and inhibit fertilization. For this reason it is not wise to try to extract every egg from a female. Forcing out the last hundred or so eggs increases the chance of eggs being broken.

As soon as a female has been stripped, a ripe male is stripped into the pan using the same technique. The eggs and sperm are then gently mixed to effect fertilization by what is referred to as the "dry" method. At one time, a "wet" method was also popular, in which a small amount of water was added just prior to mixing the eggs and sperm, but the dry method is now almost universally preferred. If no eggs are broken it is customary to strip two or more pairs before emptying the pan, to eliminate the chance of fertilization not taking place due to a sterile male.

INCUBATION OF EGGS

Hatching is best carried out in gently flowing water with a temperature of 8 to 13°C, containing at least 7 ppm of dissolved oxygen. Several devices for artificially incubating salmonid eggs have been developed, but

PLATE 1. Vertical drip salmonid incubation system. (Courtesy Michigan Department of Natural Resources.)

most commercial hatcheries use a tray incubator. This consists of a series of metal or fiberglass trays stacked rather like the drawers in a dresser, so that any individual tray may be totally or partially removed. Eggs are placed in trays, a single layer to a tray, as soon as they are water hard-ened—within a few minutes after fertilization. Egg trays are made of a special type of wire screen with oblong openings about 15 mm × 3.5 mm. The oblong mesh retains the eggs but permits the elongate alevins to pass through.

Formerly it was necessary to inspect eggs daily and individually remove each dead one, or else fungus would gain a foothold on dead eggs and spread to smother adjacent live ones. This tedious task has been largely

PLATE 2. Egg incubation by the jar battery system. (Courtesy Michigan Department of Natural Resources.)

eliminated by using malachite green to control fungus. The first malachite green treatment takes place before the loaded egg trays are placed in the incubator. The loaded trays are stacked four high in a trough with the bottom tray supported 25 mm. off the floor of the trough. Water flow in the trough is maintained at 24 liters/min. Each stack of trays is separated from adjoining trays by a divider. The dividers act as baffles to produce an upwelling of water through each tray. The trays are left in the trough for about 9 hours, at the beginning and end of which they are flushed with 7.6 g of a solution of 43 g of malachite green in 4 liters of water, introduced at the head of the trough. At the conclusion of the second such treatment the trays are placed in the incubator. During this and all subsequent operations until hatching the eggs must be covered, since exposure to light may result in premature hatching or death.

Two types of incubator are used. In the drip incubator egg trays are alternated with perforated metal trays. Water is introduced at the top

PLATE 3.   Newly hatched trout fry. (Courtesy Michigan Department of Natural Resources.)

and allowed to drip down over the eggs. This provides enough moisture to maintain the embryos but, since the trays do not retain water, the eggs must be removed to troughs just before hatching.

In the vertical flow incubator water also enters at the top but is piped from tray to tray in such a manner as to enter each tray at the bottom. Upwelling through the eggs provides adequate aeration. Since the trays containing the egg baskets are not perforated, the eggs may be allowed to hatch and alevins left in the incubator until they begin to feed.

Both types of incubator operate on a rather small volume of water, thus it is economical to control temperature to accelerate or retard development. To provide for 20 35-cm square trays, or about 600,000 to 1,350,000 eggs, 14 liters/min, is considered adequate. (The precise number of eggs depends on size and may be outside this range for exceptionally large or small eggs or for species other than those usually cultured commercially.) During incubation treatment with malachite green is repeated twice weekly, using 3.75 g dissolved in 3 liters of water. Water

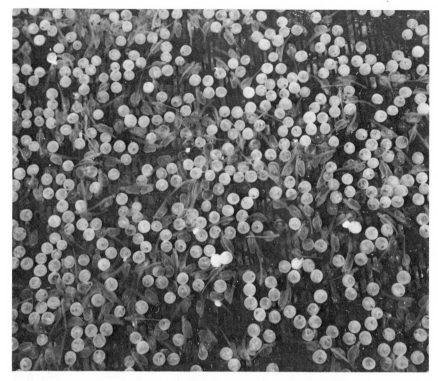

Plate 4.   Newly hatched trout fry. (Courtesy Michigan Department of Natural Resources.)

flow is reduced to 8 liters/min during treatment and the solution is allowed to drip in at about 30 ml/min for 1 hour.

The rate of hatching is determined by temperature. Table 1 shows hatching rates for the three commercially important species.

SALE AND SHIPPING OF EGGS

Some culturists grow their own stock through all phases of the life cycle, but not a few hatcheries specialize in egg production since, perhaps due to the presence of trace elements or other solutes, certain localities seem better suited to producing high percentages of viable eggs. Markets for eggs include not only commercial and experimental culturists around the world but also state fish and game departments in the United States, some of which are not able to produce eggs as cheaply as commercial culturists.

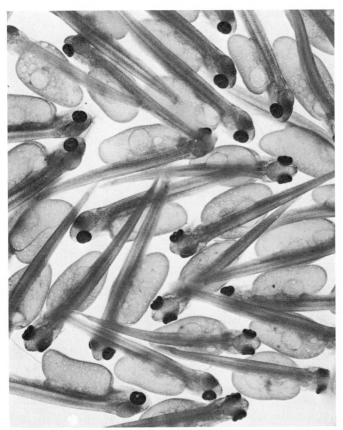

PLATE 5. Newly hatched trout fry. (Courtesy Michigan Department of Natural Resources.)

Salmonid eggs are extremely tender during some periods in their development, but they are quite tolerant of handling during the first 48 hours after water hardening and again after they are eyed. Eggs are referred to as "eyed" as soon as the eye of the embryo is clearly visible— about 2 to 3 weeks after fertilization at normal temperatures. Except for extremely short journeys, eyed eggs are chosen for shipping. Eggs which have only recently become eyed are preferred since they are less likely to hatch prematurely under the relatively high temperatures experienced in even the best shipping containers.

Salmonid eggs are shipped in trays similar to those used in hatching. Since malachite green treatment can scarcely be applied en route, all dead eggs must be removed to prevent the spread of fungus. The first step

TABLE 1.   RATE OF HATCHING, AT DIFFERENT TEMPERATURES, OF EGGS OF
RAINBOW, BROWN, AND BROOK TROUT

SPECIES	TEMPERATURE (°C)	DAYS TO HATCHING
Rainbow trout	4.5	80
	7.3	48
	10.0	31
	12.0	24
	15.7	19
Brown trout	1.7	156
	4.5	100
	7.3	64
	10.0	41
Brook trout	1.7	144
	4.5	103
	7.3	68
	10.0	44
	12.0	35

in this process is "shocking," which amounts to nothing more than agitating the eggs vigorously enough to rupture the yolk membrane of any infertile eggs but not severely enough to injure viable eggs. After shocking, dead eggs turn white and can be individually removed by siphoning. The trays are then stacked one on top of another in a plastic bag with one or two trays of some absorbent material on the bottom and a tray of ice on top. Eggs must be kept moist during shipment, but water is not added because, without aeration, it would result in smothering the eggs. When so packed and placed in some sort of rigid container with good insulating properties, eggs may be shipped anywhere in the world.

Alevins are held in egg trays or in concrete or aluminum troughs until the yolk sac is absorbed. Water flow is maintained at about the level used for hatching. Once the fry become free swimming the water supply becomes crucial to the success of the operation. This is the principal reason that large-scale trout farming is confined to a few areas; most regions of the world cannot offer water in the quantity and quality needed for commercial trout culture. The minimum acceptable flow is about 3600 liters/min and considerably more is preferable. Ideally, the water should be isothermal between 10 and 18°C, have a pH of 7.0 to 8.5, contain at least 50 ppm of dissolved solids and 5 ppm of dissolved oxygen, and be free from pollutants. Surface water may meet most of these requirements but is almost never isothermal. Growth in even the best of trout streams is retarded by warm water in summer and/or cold water in winter.

COMMERCIAL TROUT CULTURE IN THE UNITED STATES

WATER SUPPLY

Illustrative of the potential of trout culture under optimum conditions is the world's largest trout farm, the Snake River Trout Company, located near Buhl, in southern Idaho. The owner of the farm, Bob Erkins, has tapped natural springs which supply him with 240,000 liters/min of isothermal 15°C water. Erkins believes a large part of his success is due to this water supply. Certainly it has played a role in making his property "the richest food-producing acre in the world." There is more to it, however, for the same water source has been and is still used by other culturists who do not approach Erkins' production of 600,000 kg/year with 10 acres of land. This amounts to 12% of the total United States production of trout. The following comments on commercial trout culture in the United States are drawn largely from Erkins' experience.

STOCKING DENSITY

One of the first departures from tradition made by Erkins was in stocking density. It had formerly been generally believed that in order to attain good growth trout had to have considerable room. Erkins, however, recognized that since trout culture is carried out in flowing water, stocking rates should be dependent not on the volume of water in an enclosure at a given moment but on the volume flowing through that enclosure in a given time, which is a function of current velocity.

At the present time, determination of optimum current velocity is as much a matter of art as science. If the current is too swift, energy which might go into growth will be used up in swimming. On the other hand, slack water results in the accumulation of wastes. As a rule of thumb, current should be sufficient to provide at least one complete exchange of water per hour.

The amount of trout which may be held in a given enclosure depends on many factors in addition to volume of water flow, so stocking rates must be empirically determined by the individual culturist. A well-managed farm should be able to sustain 100,000 kg/ha of fingerlings. Erkins gauges his operations to produce about 2.2 kg of marketable size

408

PLATE 7.   Size grading trout. (Courtesy Michigan Department of Natural Resources.)

trout per liter of water per minute. Beginning culturists should err on the low side and aim for 1.5 to 1.8 kg/(liter)(min).

Since the holding capacity of an enclosure is determined by the weight rather than the number of fish in it, it is necessary to periodically thin out stock if growth is to be maintained. Concurrent with thinning the stock is size grading to equalize competition among the fish and produce more uniform fish at harvest time. Grading is done by netting the fish and placing them in a box with a slatted bottom. The smaller fish pass between the slats while the larger ones are retained. With small fish the process may be accelerated by lifting the grader out of the water and shaking it.

Some sport fish hatcheries use more sophisticated graders. The Murray-Hume automatic grader has the advantage of not necessitating removal of the fish from the water. It functions by forcing the fish to swim upstream until they meet a series of bars set at a predetermined distance apart. The smaller fish pass through, while the larger ones remain behind. Other grading devices which sort fish into more than two size groups consist essentially of nothing more than a series of conventional graders.

Plate 8. Weighing young trout. (Courtesy Michigan Department of Natural Resources.)

POND AND RACEWAY CONSTRUCTION

Raceways 25 to 35 m long, 3 to 10 m wide, and 0.7 to 1.0 m deep are preferred to wider ponds of other shapes for rearing fingerlings. Not only is it easier to maintain a satisfactory flow of water in raceways, but eddying is reduced. Lack of eddies allows continuous flushing of metabolic wastes, which may not only cause oxygen depletion and increase the danger of disease but are claimed to be responsible for the "hatchery taste" of some cultured trout. Where it is possible to provide rapid enough exchange of water, larger ponds (up to 3 acre feet) may be used as "finishing ponds" to grow 15-cm fingerlings to marketable size.

Raceways and ponds should be so constructed that no pond flows into another pond and any pond can be drained individually. Not only is this more convenient for the culturist, it prevents the passage of metabolic wastes and disease organisms from pond to pond. Another precaution

PLATE 9. Outdoor raceway culture of trout fingerlings. (Courtesy Michigan Department of Natural Resources.)

against disease which is necessary in some localities is to construct raceways and ponds of concrete. In areas where disease is not a major problem, earthen ponds have their proponents, who claim that trout derive substantial benefit from natural food organisms produced on an earthen bottom.

Before constructing any sort of trout culture facility, it must be determined whether the site is suitable. The soil should retain water *or* be suitable for concrete construction. Slope of the land should be 1 to 3% to permit an adequate flow of water and to allow for a vertical drop of at least a few inches at the inlet of each enclosure used to hold trout. The latter precaution is needed to insure adequate oxygenation at all times.

PLATE 10.   Feeding trout fry on ground liver. (Courtesy Michigan Department of Natural Resources.)

FEEDING

Early trout culturists usually fed beef liver or ground low-grade beef, and these foods may still be in use in some places. Fish offal has also been used but has often been associated with disease problems. Nevertheless, one of the most successful American trout farms, Trout Lodge Springs Hatchery, near Ephrata, Washington, feeds mainly ground carp and salmon cannery waste and has not experienced excessive disease problems. Diseases or no, unless the culturist is able to take advantage of a particularly cheap source of meat or fish, prepared dry foods pay for themselves in convenience.

Erkins was one of the first commercial trout culturists to convert to an exclusive diet of dry feed. The feed he adopted in 1953 contained wheat, fish meal, whey, cottonseed meal, and yeast. Since then a number of feed companies have come out with trout pellets containing as many as 30 to 40 ingredients, including vitamin B_{12}, antibiotics, and even paprika, to bring out the red spots of brown trout and brook trout. Most American trout farmers have followed Erkins' lead, thus the feed industry has vigorously attempted to improve trout feed formulas. This work continues even though existing feeds have produced conversion ratios as low as 1.5:1 in commercial culture and 1.13:1 in experimental culture of fingerlings. Table 2 lists guidelines for trout feed as determined at the U.S. Fish Farming Experimental Station in Stuttgart, Arkansas.

In practice, great variation is to be found in the protein content of trout feeds. In areas where there is a cheap and abundant supply of fish

TABLE 2.   GUIDELINES FOR COMPOSITION OF TROUT FEED, AS DETERMINED AT THE U.S. FISH FARMING EXPERIMENTAL STATION, STUTTGART, ARKANSAS

COMPONENT	AMOUNT
Total protein[a]	35–40%
Carbohydrate (allowable)	30%
Fat	8–10%
Fiber	4%
Vitamins per ton of feed	
Vitamin A	3,000,000 USP
Vitamin D_3	640,000 IC
Vitamin E	216,000 IU
Riboflavin	100,000 mg
D-calcium pantothenate	48,000 mg
Folic acid	80,000 mg
Niacin	500,000 mg
Choline chloride	1,000,000 mg
Vitamin B_{12}	20 mg
D-biotin	400 mg
Ascorbic acid	240 mg
Thiamine hydrochloride	60,000 mg
Pyridoxine hydrochloride	20,000 mg

[a] To include at least 30% fish meal.
SOURCE: Meyer (1969).

or meat, trout diets may contain as much as 60% protein. On the other hand, in the United States the tendency is to try to maximize profits by getting along with as little protein as possible. Some authorities have suggested protein levels as low as 20%.

Just what is the minimum amount of protein required to maintain trout growth and energy is not known. Research at the U.S. Bureau of Sport Fisheries and Wildlife's Eastern Fish Nutrition Laboratory at Cortland, New York, indicates that substitution of carbohydrates for protein as an energy source, which has been effective with some species of fish, does not work with trout, but that, under some circumstances, small amounts of fat may demonstrate a protein-sparing action.

Another Snake River Valley culturist, Thorleif Rangen, of Rangen, Inc., is approaching the problem of economical trout nutrition by attempting to develop a high-protein diet which contains little or no animal matter yet produces conversion ratios near 1:1. Thus far, 100% vegetable foods have been found lacking in essential amino acids, but chemical experiments with amino acid sysnthesis and the success of horticulturists in selectively breeding corn with ten times the usual amount of lysine and tryptophane suggest that Rangen's goal may be attainable.

Whereas the same feed formula may be used for trout of all sizes, particle size and rate of feeding vary according to the size of fish. Table 3 shows the size of food particles (commercial size designations) recommended for various sizes of trout.

TABLE 3. PARTICLE SIZE OF DRY FOOD RECOMMENDED FOR FEEDING TROUT OF DIFFERENT SIZES

SIZE OF PELLET (STANDARD COMMERCIAL DESIGNATIONS)	SIZE OF FISH
1	4,224–2,816/kg
1 and 2 mixed	2,992–2,464/kg
2	2,840–2,112/kg
2 and 3 mixed	1,936–1,056/kg
3	1,232– 704/kg
3 and 4 mixed	880– 352/kg
4	528– 352/kg
4 and "crumbles" mixed	528– 176/kg
Crumbles	352– 106/kg
Crumbles and 3/32-in. pellets mixed	176– 70/kg
.24 cm pellets	70– 20/kg
.40 cm pellets	20/kg and larger

In feeding very small trout it may be necessary to grind prepared food to produce tiny particles. Grinding entails both a crushing and a cutting action. Crushing breaks down cell walls and makes the food more susceptible to leaching by the water, thus should be minimized in favor of cutting.

Suggested daily feeding portions are shown in Table 4. These amounts should be divided into three or four daily feedings or, for young fry, as many as ten feedings. Young fry should be fed as much as they will consume at one time or slightly more. The opposite holds true for larger fish. It is general practice in commercial trout culture to gauge feeding by the appetites of the smaller fish in a group so as to equalize the growth rate and create a more uniform product. Feeding in excess may result in slightly improved growth, but only with the loss of uniformity, and is unlikely to compensate for the increased expenditure.

A more sophisticated feeding guide, suitable for young of any salmonid species, and taking into account differences in the nutritional value of diets, was developed by David C. Haskell of the New York State Conservation Department. Table 5 is a slightly modified form of Haskell's guide. Before using it past records of trout growth and water temperature must be consulted to compute a "hatchery constant" as follows:

Instruction for Use of Feeding Guide

I. For hatcheries with constant water temperatures:

A.
$$\frac{L_2 - L_1}{\text{days in period}} = L$$

where: L_1 = length of fish at beginning of period,
L_2 = length of fish at end of period,
L = daily increase in length.

B.   Hatchery constant = 300 × conversion × L

Note: Choice of a conversion should be based on recommendation of the feed manufacturer or past experience with the diet.

C. Enter the feeding guide under size of fish opposite the appropriate constant and read the percentage of body weight to feed daily. *Note:* It is necessary to calculate the hatchery constant only once for each species or strain, as long as there is no change in diet or water temperature.

II. For hatcheries with variable water temperatures:

A. Determine average monthly water temperature in °F.

B. Determine temperature units for the month. (TU = average monthly water temperature minus 38.6.)

C. Determine the length increase per month.

TABLE 4. RECOMMENDED AMOUNTS OF DRY FEED TO FEED RAINBOW TROUT OF DIFFERENT SIZE GROUPS AT DIFFERENT TEMPERATURES (PERCENTAGE OF BODY WEIGHT)

WATER TEMPERATURE (°C)	NUMBER PER KG: 5,592+ APPROX. SIZE (CM): 1-2.3	5,592-669 2.5-5.0	669-194 5.0-7.5	194-83 7.5-10.0	83-43 10.0-12.5	43-26 12.5-15.0	26-16 15.0-17.5	16-11 17.5-20.0	11-8 20.0-22.5	8-6 22.5-25.0	6 or - 25.0+
2- 3	2.7	2.2	1.7	1.3	1.0	0.8	0.7	0.6	0.5	0.5	0.4
	2.7	2.3	1.8	1.4	1.1	0.9	0.7	0.6	0.5	0.5	0.4
3- 4	2.9	2.4	2.0	1.5	1.2	0.9	0.8	0.7	0.6	0.5	0.5
	3.0	2.5	2.2	1.7	1.3	0.9	0.8	0.7	0.6	0.6	0.5
4- 5	3.2	2.6	2.2	1.7	1.3	1.0	0.9	0.8	0.7	0.6	0.5
5- 6	3.3	2.8	2.2	1.8	1.4	1.1	0.9	0.8	0.7	0.6	0.5
	3.5	2.8	2.4	1.8	1.4	1.2	0.9	0.8	0.7	0.6	0.5
6- 7	3.6	3.0	2.5	1.9	1.4	1.2	1.0	0.9	0.8	0.7	0.6
	3.8	3.1	2.5	2.0	1.5	1.3	1.0	0.9	0.8	0.8	0.6
7- 8	4.0	3.3	2.7	2.1	1.6	1.3	1.1	1.0	0.9	0.8	0.7
	4.1	3.4	2.8	2.2	1.7	1.4	1.2	1.0	0.9	0.8	0.7
8- 9	4.3	3.6	3.0	2.3	1.7	1.4	1.2	1.0	0.9	0.8	0.7
	4.5	3.8	3.0	2.4	1.8	1.5	1.3	1.1	1.0	0.9	0.8
9-10	4.7	3.9	3.2	2.5	1.9	1.5	1.3	1.1	1.0	0.9	0.8
10-11	5.2	4.3	3.4	2.7	2.0	1.7	1.4	1.2	1.1	1.0	0.9
	5.4	4.5	3.5	2.8	2.1	1.7	1.5	1.3	1.1	1.0	0.9

TABLE 4. *(continued)*

WATER TEMPERATURE (°C)	APPROX. SIZE (CM):	5,592+ 1–2.3	5,592–669 2.5–5.0	669–194 5.0–7.5	194–83 7.5–10.0	83–43 10.0–12.5	43–26 12.5–15.0	26–16 15.0–17.5	16–11 17.5–20.0	11–8 20.0–22.5	8–6 22.5–25.0	6– 25.0+
11–12		5.4	4.5	3.6	2.8	2.1	1.7	1.5	1.3	1.1	1.0	0.9
		5.6	4.7	3.8	2.9	2.2	1.8	1.5	1.3	1.1	1.1	1.0
12–13		5.8	4.9	3.9	3.0	2.3	1.9	1.6	1.4	1.3	1.1	1.0
		6.1	5.1	4.2	3.2	2.4	2.0	1.6	1.4	1.3	1.1	1.0
13–14		6.3	5.3	4.3	3.3	2.5	2.0	1.7	1.5	1.3	1.2	1.0
		6.7	5.5	4.5	3.5	2.6	2.1	1.8	1.5	1.4	1.2	1.1
14–15		7.0	5.8	4.8	3.6	2.7	2.2	1.9	1.6	1.4	1.3	1.2
15–16		7.3	6.0	5.0	3.7	2.7	2.3	1.9	1.7	1.5	1.3	1.2
		7.5	6.3	5.1	3.9	3.0	2.4	2.0	1.7	1.5	1.4	1.3
16–17		7.8	6.5	5.3	4.1	3.1	2.5	2.0	1.8	1.6	1.4	1.3
		8.1	6.7	5.5	4.3	3.2	2.6	2.1	1.8	1.6	1.5	1.4
17–18		8.4	7.0	5.7	4.5	3.4	2.7	2.1	1.9	1.7	1.5	1.4
		8.7	7.2	5.9	4.7	3.5	2.8	2.2	1.9	1.7	1.6	1.5
18–19		9.0	7.5	6.1	4.9	3.6	2.9	2.2	2.0	1.8	1.6	1.5
		9.3	7.8	6.3	5.1	3.8	3.0	2.3	2.0	1.8	1.6	1.6
19–20		9.6	9.1	6.6	5.3	3.9	3.1	2.4	2.1	1.9	1.7	1.6
20–21		9.9	9.4	6.9	5.5	4.0	3.2	2.5	2.1	2.0	1.8	1.7

TABLE 5. FEEDING GUIDE FOR YOUNG SALMONIDS IN HATCHERIES

LENGTH (CM):	1.9	2.0	2.2	2.3	2.4	2.5	2.7	2.8	2.9	3.1	3.2	3.4
NUMBER PER KG:	13,200	10,100	8,800	7,700	6,600	5,500	4,400	3,960	3,520	3,030	2,640	2,200
HATCHERY CONSTANT	PERCENTAGE OF BODY WEIGHT TO FEED DAILY											
2.10	2.8	2.7	2.5	2.4	2.2	2.1	2.0	1.9	1.8	1.7	1.7	1.6
2.40	3.2	3.0	2.8	2.7	2.6	2.4	2.2	2.2	2.1	2.0	1.9	1.8
2.70	3.6	3.4	3.2	3.0	2.9	2.7	2.5	2.4	2.3	2.2	2.1	2.0
3.00	4.0	3.8	3.5	3.4	3.2	3.0	2.8	2.7	2.6	2.5	2.4	2.2
3.30	4.4	4.2	3.9	3.7	3.5	3.3	3.1	3.0	2.8	2.7	2.6	2.4
3.60	4.8	4.6	4.2	4.0	3.8	3.6	3.4	3.2	3.1	3.0	2.8	2.7
3.90	5.2	4.9	4.6	4.4	4.2	3.9	3.6	3.5	3.4	3.2	3.1	2.9
4.20	5.6	5.3	4.9	4.7	4.5	4.2	3.9	3.8	3.6	3.5	3.3	3.1
4.50	6.1	5.7	5.3	5.1	4.8	4.5	4.2	4.1	3.9	3.7	3.5	3.3
4.80	6.5	6.1	5.6	5.4	5.1	4.8	4.5	4.3	4.1	4.0	3.8	3.6
5.10	6.9	6.5	6.0	5.7	5.4	5.1	4.8	4.6	4.4	4.2	4.0	3.8
5.40	7.3	6.8	6.3	6.1	5.8	5.4	5.0	4.9	4.7	4.5	4.3	4.0
5.70	7.7	7.2	6.7	6.4	6.1	5.7	5.3	5.1	4.9	4.7	4.5	4.2
6.00	8.1	7.6	7.1	6.7	6.4	6.0	5.6	5.4	5.2	5.0	4.7	4.4
6.30	8.5	8.0	7.4	7.1	6.7	6.3	5.9	5.7	5.4	5.2	5.0	4.7
6.60	8.9	8.4	7.8	7.4	7.0	6.6	6.2	5.9	5.7	5.5	5.2	4.9
6.90	9.3	8.7	8.1	7.8	7.4	6.9	6.4	6.2	5.9	5.7	5.4	5.1

TABLE 5. (continued)

LENGTH (CM):	1.9	2.0	2.2	2.3	2.4	2.5	2.7	2.8	2.9	3.1	3.2	3.4
NUMBER PER KG:	13,200	10,100	8,800	7,700	6,600	5,500	4,400	3,960	3,520	3,030	2,640	2,200
HATCHERY CONSTANT	PERCENTAGE OF BODY WEIGHT TO FEED DAILY											
7.20	9.7	9.1	8.5	8.1	7.7	7.2	6.7	6.5	6.2	6.0	5.7	5.3
7.50	10.1	9.5	8.8	8.4	8.0	7.5	7.0	6.8	6.5	6.2	5.9	5.6
7.80	10.5	9.9	9.2	8.7	8.3	7.8	7.3	7.0	6.7	6.4	6.1	5.8
8.10	10.9	10.3	9.5	9.1	8.6	8.1	7.6	7.3	7.0	6.7	6.4	6.0
8.40	11.3	10.6	9.9	9.4	9.0	8.4	7.9	7.6	7.2	6.9	6.6	6.2
8.70	11.7	11.0	10.2	9.8	9.3	8.7	8.1	7.8	7.5	7.2	6.9	6.4
9.00	12.1	11.4	10.6	10.1	9.6	9.0	8.4	8.1	7.8	7.4	7.1	6.7
9.30	12.5	11.8	10.9	10.4	9.9	9.3	8.7	8.4	8.0	7.7	7.3	6.9
9.60	12.9	12.2	11.3	10.8	10.2	9.6	9.0	8.6	8.3	7.9	7.5	7.1
9.90	13.3	12.5	11.6	11.1	10.6	9.9	9.3	8.9	8.5	8.2	7.8	7.3
10.2	13.7	12.9	12.0	11.5	10.9	10.3	9.5	9.1	8.8	8.4	8.0	7.6
10.5	14.1	13.3	12.3	11.8	11.2	10.6	9.8	9.5	9.1	8.7	8.3	7.8
10.8	14.5	13.7	12.7	12.1	11.5	10.9	10.1	9.7	9.3	8.9	8.5	8.0
11.1	14.9	14.1	13.0	12.5	11.8	11.2	10.4	10.0	9.6	9.2	8.7	8.2
11.4	15.3	14.4	13.4	12.8	12.2	11.5	10.7	10.3	9.8	9.4	9.0	8.4
11.7	15.7	14.8	13.7	13.1	12.5	11.8	10.9	10.5	10.1	9.7	9.2	8.7
12.0	16.1	15.2	14.1	13.5	12.8	12.1	11.2	10.8	10.3	9.9	9.4	8.9

TABLE 5. (continued)

LENGTH (CM):	3.6	3.7	3.9	4.1	4.3	4.5	4.6	4.9	5.1	5.4	5.9	6.0	6.1
NUMBER PER KG:	1,980	1,760	1,540	1,320	1,100	990	880	770	660	550	440	418	396
HATCHERY CONSTANT	PERCENTAGE OF BODY WEIGHT TO FEED DAILY												
2.10	1.5	1.4	1.4	1.3	1.2	1.2	1.1	1.1	1.0	0.98	0.91	0.89	0.88
2.40	1.7	1.6	1.6	1.5	1.4	1.4	1.3	1.3	1.2	1.1	1.0	1.0	1.0
2.70	1.9	1.8	1.8	1.7	1.6	1.5	1.5	1.4	1.3	1.3	1.2	1.1	1.1
3.00	2.1	2.1	2.0	1.9	1.8	1.7	1.6	1.6	1.5	1.4	1.3	1.3	1.2
3.30	2.4	2.3	2.2	2.1	1.9	1.9	1.8	1.7	1.6	1.5	1.4	1.4	1.4
3.60	2.6	2.5	2.4	2.3	2.1	2.0	2.0	1.9	1.8	1.7	1.6	1.5	1.5
3.90	2.8	2.7	2.6	2.4	2.3	2.2	2.1	2.0	1.9	1.8	1.7	1.7	1.6
4.20	3.0	2.9	2.8	2.6	2.5	2.4	2.3	2.2	2.1	2.0	1.8	1.8	1.8
4.50	3.2	3.1	3.0	2.8	2.6	2.6	2.5	2.3	2.2	2.1	1.9	1.9	1.9
4.80	3.4	3.3	3.2	3.0	2.8	2.7	2.6	2.5	2.4	2.2	2.1	2.0	2.0
5.10	3.6	3.5	3.4	3.2	3.0	2.9	2.8	2.7	2.5	2.4	2.2	2.2	2.1
5.40	3.9	3.7	3.6	3.4	3.2	3.1	3.0	2.8	2.7	2.5	2.3	2.3	2.3
5.70	4.0	3.9	3.8	3.6	3.4	3.2	3.1	3.0	2.8	2.7	2.5	2.4	2.4
6.00	4.3	4.1	3.9	3.8	3.5	3.4	3.3	3.1	3.0	2.8	2.6	2.5	2.5
6.30	4.5	4.3	4.1	3.9	3.7	3.6	3.4	3.3	3.1	2.9	2.7	2.7	2.6
6.60	4.7	4.5	4.3	4.1	3.9	3.8	3.6	3.4	3.3	3.1	2.9	2.8	2.8
6.90	4.9	4.7	4.5	4.3	4.1	3.9	3.8	3.6	3.4	3.2	3.0	2.9	2.9

TABLE 5. (continued)

LENGTH (CM):	3.6	3.7	3.9	4.1	4.3	4.5	4.6	4.9	5.1	5.4	5.9	6.0	6.1
NUMBER PER KG:	1,980	1,760	1,540	1,320	1,100	990	880	770	660	550	440	418	396
HATCHERY CONSTANT	PERCENTAGE OF BODY WEIGHT TO FEED DAILY												
7.20	5.1	4.9	4.7	4.5	4.2	4.1	3.9	3.8	3.6	3.4	3.1	3.1	3.0
7.50	5.4	5.1	4.9	4.7	4.4	4.3	4.1	3.9	3.7	3.5	3.2	3.2	3.1
7.80	5.6	5.3	5.1	4.9	4.6	4.4	4.3	4.1	3.9	3.6	3.4	3.3	3.3
8.10	5.8	5.5	5.3	5.1	4.8	4.6	4.4	4.2	4.0	3.8	3.5	3.4	3.4
8.40	6.0	5.8	5.5	5.3	4.9	4.8	4.6	4.4	4.2	3.9	3.6	3.6	3.5
8.70	6.2	6.0	5.7	5.4	5.1	4.9	4.8	4.5	4.3	4.1	3.8	3.7	3.6
9.00	6.4	6.2	5.9	5.6	5.3	5.1	4.9	4.7	4.5	4.2	3.9	3.8	3.8
9.30	6.6	6.4	6.1	5.8	5.5	5.3	5.1	4.8	4.6	4.5	4.0	4.0	3.9
9.60	6.9	6.6	6.3	6.0	5.6	5.5	5.2	5.0	4.8	4.5	4.2	4.1	4.0
9.90	7.1	6.8	6.5	6.2	5.8	5.6	5.4	5.2	4.9	4.6	4.3	4.2	4.1
10.2	7.3	7.0	6.7	6.4	6.0	5.7	5.5	5.3	5.0	4.7	4.4	4.3	4.2
10.5	7.5	7.2	6.9	6.6	6.2	6.0	5.7	5.5	5.2	4.9	4.5	4.5	4.4
10.8	7.7	7.4	7.1	6.8	6.4	6.1	5.9	5.6	5.3	5.0	4.7	4.6	4.5
11.1	7.9	7.6	7.1	6.9	6.5	6.3	6.1	5.8	5.5	5.2	4.8	4.7	4.6
11.4	8.1	7.8	7.5	7.1	6.7	6.5	6.2	5.9	5.6	5.3	4.9	4.9	4.8
11.7	8.4	8.0	7.7	7.3	6.9	6.6	6.4	6.1	5.8	5.5	5.1	5.0	4.9
12.0	8.6	8.2	7.9	7.5	7.1	6.8	6.6	6.3	5.9	5.6	5.2	5.1	5.0

421

TABLE 5. (*continued*)

LENGTH (CM):	6.2	6.3	6.5	6.6	6.8	7.0	7.2	7.4	7.5	7.7	7.8	8.0
NUMBER PER KG:	374	352	330	308	286	264	242	220	209	198	187	176
HATCHERY CONSTANT					PERCENTAGE OF BODY WEIGHT TO FEED DAILY							
2.10	0.86	0.84	0.83	0.81	0.79	0.77	0.74	0.72	0.71	0.70	0.68	0.67
2.40	0.98	0.96	0.94	0.92	0.90	0.88	0.85	0.82	0.81	0.79	0.78	0.76
2.70	1.1	1.1	1.1	1.0	1.0	0.99	0.96	0.93	0.91	0.89	0.88	0.86
3.00	1.2	1.2	1.2	1.2	1.1	1.1	1.1	1.0	1.0	0.99	0.98	0.96
3.30	1.4	1.3	1.3	1.3	1.2	1.2	1.2	1.1	1.1	1.1	1.1	1.1
3.60	1.5	1.4	1.4	1.4	1.3	1.3	1.3	1.2	1.2	1.2	1.2	1.1
3.90	1.6	1.6	1.5	1.5	1.5	1.4	1.4	1.3	1.3	1.3	1.3	1.2
4.20	1.7	1.7	1.7	1.6	1.6	1.5	1.5	1.4	1.4	1.4	1.4	1.3
4.50	1.8	1.8	1.8	1.7	1.7	1.6	1.6	1.5	1.5	1.5	1.5	1.4
4.80	2.0	1.9	1.9	1.8	1.8	1.8	1.7	1.7	1.6	1.6	1.6	1.5
5.10	2.1	2.0	2.0	2.0	1.9	1.9	1.8	1.8	1.7	1.7	1.7	1.6
5.40	2.2	2.2	2.1	2.1	2.0	2.0	1.9	1.9	1.8	1.8	1.8	1.7
5.70	2.3	2.3	2.2	2.2	2.1	2.1	2.0	2.0	1.9	1.9	1.9	1.8
6.00	2.5	2.4	2.4	2.3	2.2	2.2	2.1	2.0	2.0	2.0	2.0	1.9
6.30	2.6	2.5	2.5	2.4	2.4	2.3	2.2	2.2	2.1	2.1	2.1	2.0
6.60	2.7	2.7	2.6	2.5	2.5	2.4	2.3	2.3	2.2	2.2	2.1	2.1
6.90	2.8	2.8	2.7	2.7	2.6	2.5	2.4	2.4	2.3	2.3	2.2	2.2

TABLE 5. (continued)

LENGTH (CM):	6.2	6.3	6.5	6.6	6.8	7.0	7.2	7.4	7.5	7.7	7.8	8.0
NUMBER PER KG:	374	352	330	308	286	264	242	220	209	198	187	176
HATCHERY CONSTANT	PERCENTAGE OF BODY WEIGHT TO FEED DAILY											
7.20	3.0	2.9	2.8	2.8	2.7	2.6	2.6	2.5	2.4	2.4	2.3	2.3
7.50	3.1	3.0	3.0	2.9	2.8	2.7	2.7	2.6	2.5	2.5	2.4	2.4
7.80	3.2	3.1	3.1	3.0	2.9	2.8	2.8	2.7	2.6	2.6	2.5	2.5
8.10	3.3	3.3	3.2	3.1	3.0	3.0	2.9	2.8	2.7	2.7	2.6	2.6
8.40	3.4	3.4	3.3	3.2	3.1	3.1	3.0	2.9	2.8	2.8	2.7	2.7
8.70	3.6	3.5	3.4	3.3	3.3	3.2	3.1	3.0	2.9	2.9	2.8	2.8
9.00	3.7	3.6	3.5	3.5	3.4	3.3	3.2	3.1	3.0	3.0	2.9	2.9
9.30	3.8	3.7	3.7	3.6	3.5	3.4	3.3	3.2	3.1	3.1	3.0	3.0
9.60	3.9	3.8	3.8	3.7	3.6	3.5	3.4	3.3	3.2	3.2	3.1	3.1
9.90	4.1	4.0	3.9	3.8	3.7	3.6	3.5	3.4	3.3	3.3	3.2	3.2
10.2	4.1	4.1	4.0	3.9	3.8	3.7	3.6	3.5	3.4	3.3	3.3	3.2
10.5	4.3	4.2	4.1	4.0	3.9	3.8	3.7	3.6	3.5	3.5	3.4	3.3
10.8	4.4	4.3	4.3	4.2	4.0	3.9	3.8	3.7	3.6	3.6	3.5	3.4
11.1	4.5	4.5	4.4	4.3	4.2	4.1	3.9	3.8	3.8	3.7	3.6	3.5
11.4	4.7	4.6	4.5	4.4	4.3	4.2	4.0	3.9	3.9	3.8	3.7	3.6
11.7	4.8	4.7	4.6	4.5	4.4	4.3	4.1	4.0	4.0	3.9	3.8	3.7
12.0	4.9	4.8	4.7	4.6	4.5	4.4	4.3	4.1	4.1	4.0	3.9	3.8

TABLE 5. (*continued*)

LENGTH (CM):	8.1	8.3	8.5	8.8	9.0	9.3	9.7	10.0	10.5	11.0	11.7	12.6
NUMBER PER KG:	165	154	143	132	121	110	99	88	77	66	55	44
HATCHERY CONSTANT	PERCENTAGE OF BODY WEIGHT TO FEED DAILY											
2.10	0.66	0.64	0.63	0.61	0.59	0.57	0.55	0.53	0.51	0.48	0.45	0.42
2.40	0.75	0.73	0.71	0.70	0.68	0.65	0.63	0.61	0.58	0.55	0.52	0.48
2.70	0.84	0.82	0.80	0.78	0.76	0.74	0.71	0.68	0.65	0.62	0.58	0.54
3.00	0.94	0.91	0.89	0.87	0.84	0.82	0.79	0.76	0.73	0.69	0.65	0.60
3.30	1.0	1.0	0.98	0.96	0.93	0.90	0.87	0.84	0.80	0.76	0.71	0.66
3.60	1.1	1.1	1.1	1.0	1.0	0.98	0.95	0.91	0.87	0.83	0.78	0.72
3.90	1.2	1.2	1.2	1.1	1.1	1.1	1.0	0.98	0.94	0.90	0.84	0.78
4.20	1.3	1.3	1.3	1.2	1.2	1.1	1.1	1.1	1.0	0.97	0.91	0.84
4.50	1.4	1.4	1.3	1.3	1.3	1.2	1.2	1.1	1.1	1.0	0.97	0.90
4.80	1.5	1.5	1.4	1.4	1.4	1.3	1.3	1.2	1.2	1.1	1.0	0.96
5.10	1.6	1.6	1.5	1.5	1.4	1.4	1.3	1.3	1.2	1.2	1.1	1.0
5.40	1.7	1.6	1.6	1.6	1.5	1.5	1.4	1.4	1.3	1.2	1.2	1.1
5.70	1.8	1.7	1.7	1.7	1.6	1.6	1.5	1.4	1.4	1.3	1.2	1.1
6.00	1.9	1.8	1.8	1.7	1.7	1.6	1.6	1.5	1.5	1.4	1.3	1.2
6.30	2.0	1.9	1.9	1.8	1.8	1.7	1.7	1.6	1.5	1.4	1.4	1.3
6.60	2.1	2.0	2.0	1.9	1.9	1.8	1.7	1.7	1.6	1.5	1.4	1.3
6.90	2.2	2.1	2.1	2.0	1.9	1.9	1.8	1.7	1.7	1.6	1.5	1.4

TABLE 5. (continued)

LENGTH (CM):	8.1	8.3	8.5	8.8	9.0	9.3	9.7	10.0	10.5	11.0	11.7	12.6
NUMBER PER KG:	165	154	143	132	121	110	99	88	77	66	55	44
HATCHERY CONSTANT	PERCENTAGE OF BODY WEIGHT TO FEED DAILY											
7.20	2.2	2.2	2.1	2.1	2.0	2.0	1.9	1.8	1.7	1.7	1.6	1.4
7.50	2.3	2.3	2.2	2.2	2.1	2.0	2.0	1.9	1.8	1.7	1.6	1.5
7.80	2.4	2.4	2.3	2.3	2.2	2.1	2.0	2.0	1.9	1.8	1.7	1.6
8.10	2.5	2.5	2.4	2.3	2.3	2.2	2.1	2.0	2.0	1.9	1.8	1.6
8.40	2.6	2.6	2.5	2.4	2.4	2.3	2.2	2.1	2.0	1.9	1.8	1.7
8.70	2.7	2.7	2.6	2.5	2.5	2.4	2.3	2.2	2.1	2.0	1.9	1.7
9.00	2.8	2.7	2.7	2.6	2.5	2.5	2.4	2.3	2.2	2.1	1.9	1.8
9.30	2.9	2.8	2.8	2.7	2.6	2.5	2.4	2.4	2.3	2.1	2.0	1.9
9.60	3.0	2.9	2.9	2.8	2.7	2.6	2.5	2.4	2.3	2.2	2.1	1.9
9.90	3.1	3.0	2.9	2.9	2.8	2.7	2.6	2.5	2.4	2.3	2.1	2.0
10.2	3.2	3.1	3.0	2.9	2.8	2.8	2.7	2.6	2.5	2.3	2.2	2.0
10.5	3.3	3.2	3.1	3.0	3.0	2.9	2.8	2.7	2.5	2.4	2.3	2.1
10.8	3.4	3.3	3.2	3.1	3.0	2.9	2.8	2.7	2.6	2.5	2.3	2.2
11.1	3.5	3.4	3.3	3.2	3.1	3.0	2.9	2.8	2.7	2.6	2.4	2.2
11.4	3.5	3.5	3.4	3.3	3.2	3.1	3.0	2.9	2.8	2.6	2.5	2.3
11.7	3.7	3.6	3.5	3.4	3.3	3.2	3.1	3.0	2.8	2.7	2.5	2.3
12.0	3.7	3.7	3.6	3.5	3.4	3.3	3.2	3.0	2.9	2.8	2.6	2.4

TABLE 5. (continued)

LENGTH (CM):	12.9	13.1	13.3	13.6	13.9	14.2	14.6	15.0	15.5	15.9	16.2	16.5
NUMBER PER KG:	42	40	37	35	33	31	29	26	24	22	21	20
HATCHERY CONSTANT	PERCENTAGE OF BODY WEIGHT TO FEED DAILY											
2.10	0.42	0.41	0.40	0.39	0.38	0.37	0.37	0.36	0.35	0.33	0.33	0.32
2.40	0.47	0.47	0.46	0.45	0.44	0.43	0.42	0.41	0.39	0.38	0.38	0.37
2.70	0.53	0.52	0.51	0.50	0.49	0.48	0.47	0.46	0.44	0.43	0.42	0.42
3.00	0.59	0.58	0.57	0.56	0.55	0.53	0.52	0.51	0.49	0.48	0.47	0.46
3.30	0.65	0.64	0.63	0.62	0.60	0.59	0.57	0.56	0.54	0.53	0.52	0.51
3.60	0.71	0.70	0.69	0.67	0.66	0.64	0.63	0.61	0.59	0.57	0.56	0.55
3.90	0.77	0.76	0.76	0.73	0.71	0.70	0.68	0.66	0.64	0.62	0.61	0.60
4.20	0.83	0.82	0.80	0.78	0.77	0.75	0.73	0.71	0.69	0.67	0.66	0.65
4.50	0.89	0.87	0.86	0.84	0.82	0.80	0.78	0.76	0.74	0.72	0.71	0.69
4.80	0.95	0.93	0.91	0.90	0.88	0.86	0.83	0.81	0.79	0.77	0.75	0.74
5.10	1.0	0.99	0.97	0.95	0.93	0.91	0.89	0.86	0.84	0.81	0.80	0.78
5.40	1.1	1.0	1.0	1.0	0.99	0.96	0.94	0.92	0.89	0.86	0.85	0.83
5.70	1.1	1.1	1.1	1.1	1.0	1.0	0.99	0.97	0.94	0.91	0.89	0.88
6.00	1.2	1.2	1.1	1.1	1.1	1.1	1.0	1.0	0.99	0.96	0.94	0.92
6.30	1.2	1.2	1.2	1.2	1.1	1.1	1.1	1.1	1.0	1.0	0.99	0.97
6.60	1.3	1.3	1.3	1.2	1.2	1.2	1.1	1.1	1.1	1.1	1.0	1.0
6.90	1.4	1.3	1.3	1.3	1.3	1.2	1.2	1.2	1.1	1.1	1.1	1.1

TABLE 5. (continued)

LENGTH (CM):	12.9	13.1	13.3	13.6	13.9	14.2	14.6	15.0	15.5	15.9	16.2	16.5
NUMBER PER KG:	42	40	37	35	33	31	29	26	24	22	21	20
HATCHERY CONSTANT	PERCENTAGE OF BODY WEIGHT TO FEED DAILY											
7.20	1.4	1.4	1.4	1.3	1.3	1.3	1.3	1.2	1.2	1.1	1.1	1.1
7.50	1.5	1.5	1.4	1.4	1.4	1.3	1.3	1.3	1.2	1.2	1.2	1.2
7.80	1.5	1.5	1.5	1.5	1.4	1.4	1.4	1.3	1.3	1.2	1.2	1.2
8.10	1.6	1.6	1.5	1.5	1.5	1.4	1.4	1.4	1.3	1.3	1.3	1.2
8.40	1.7	1.6	1.6	1.6	1.5	1.5	1.5	1.4	1.4	1.3	1.3	1.3
8.70	1.7	1.7	1.7	1.6	1.6	1.6	1.5	1.4	1.4	1.4	1.4	1.3
9.00	1.8	1.7	1.7	1.7	1.6	1.6	1.6	1.5	1.5	1.4	1.4	1.4
9.30	1.8	1.8	1.8	1.7	1.7	1.7	1.6	1.5	1.5	1.5	1.5	1.4
9.60	1.9	1.9	1.8	1.8	1.8	1.7	1.7	1.6	1.6	1.5	1.5	1.5
9.90	2.0	1.9	1.9	1.8	1.8	1.8	1.7	1.7	1.6	1.6	1.6	1.5
10.2	2.0	2.0	1.9	1.9	1.8	1.8	1.8	1.7	1.7	1.6	1.6	1.6
10.5	2.1	2.0	2.0	2.0	1.9	1.9	1.8	1.8	1.7	1.7	1.6	1.6
10.8	2.1	2.1	2.1	2.0	2.0	1.9	1.9	1.8	1.8	1.7	1.7	1.7
11.1	2.2	2.2	2.1	2.1	2.0	2.0	1.9	1.9	1.8	1.8	1.7	1.7
11.4	2.3	2.2	2.2	2.1	2.1	2.0	2.0	1.9	1.9	1.8	1.8	1.8
11.7	2.3	2.3	2.2	2.2	2.1	2.1	2.0	2.0	1.9	1.9	1.8	1.8
12.0	2.4	2.3	2.3	2.2	2.2	2.1	2.1	2.0	2.0	1.9	1.9	1.8

TABLE 5. (continued)

LENGTH (CM):	23.8	22.6	21.6	20.8	20.1	19.4	18.9	18.4	17.5	17.5	17.2	16.8
NUMBER PER KG:	7	8	9	10	11	12	13	14	15	17	18	19
HATCHERY CONSTANT	PERCENTAGE OF BODY WEIGHT TO FEED DAILY											
2.10	0.22	0.24	0.25	0.26	0.27	0.28	0.28	0.29	0.30	0.30	0.31	0.32
2.40	0.26	0.27	0.28	0.29	0.30	0.31	0.32	0.33	0.34	0.35	0.35	0.36
2.70	0.29	0.30	0.32	0.33	0.34	0.35	0.36	0.37	0.38	0.39	0.40	0.41
3.00	0.32	0.34	0.35	0.37	0.38	0.39	0.40	0.41	0.42	0.43	0.44	0.45
3.30	0.35	0.37	0.39	0.40	0.42	0.43	0.44	0.46	0.47	0.48	0.49	0.50
3.60	0.38	0.40	0.42	0.44	0.46	0.47	0.48	0.50	0.51	0.52	0.53	0.54
3.90	0.42	0.44	0.46	0.48	0.49	0.51	0.52	0.54	0.55	0.57	0.58	0.59
4.20	0.45	0.47	0.49	0.51	0.53	0.55	0.56	0.58	0.59	0.61	0.62	0.63
4.50	0.48	0.51	0.53	0.55	0.57	0.59	0.61	0.62	0.64	0.65	0.67	0.68
4.80	0.51	0.54	0.56	0.59	0.61	0.63	0.65	0.66	0.68	0.70	0.71	0.73
5.10	0.54	0.57	0.60	0.62	0.65	0.67	0.69	0.70	0.72	0.74	0.75	0.77
5.40	0.58	0.61	0.63	0.66	0.68	0.71	0.73	0.75	0.76	0.78	0.80	0.82
5.70	0.61	0.64	0.67	0.70	0.72	0.75	0.77	0.79	0.81	0.83	0.84	0.86
6.00	0.64	0.67	0.70	0.73	0.76	0.78	0.81	0.83	0.85	0.87	0.89	0.91
6.30	0.67	0.71	0.74	0.77	0.80	0.82	0.85	0.87	0.89	0.91	0.93	0.95
6.60	0.70	0.74	0.78	0.81	0.84	0.86	0.89	0.91	0.93	0.96	0.98	1.0
6.90	0.74	0.78	0.81	0.84	0.87	0.90	0.93	0.95	0.98	1.0	1.0	1.0

TABLE 5. (*continued*)

HATCHERY CONSTANT	LENGTH (CM): 16.8	17.2	17.5	17.5	18.4	18.9	19.4	20.1	20.8	21.6	22.6	23.8
NUMBER PER KG:	19	18	17	15	14	13	12	11	10	9	8	7
	PERCENTAGE OF BODY WEIGHT TO FEED DAILY											
7.20	1.1	1.1	1.0	1.0	0.99	0.97	0.94	0.91	0.88	0.85	0.81	0.77
7.50	1.1	1.1	1.1	1.1	1.0	1.0	0.98	0.95	0.92	0.88	0.84	0.80
7.80	1.2	1.2	1.1	1.1	1.1	1.0	1.0	0.99	0.95	0.92	0.88	0.83
8.10	1.2	1.2	1.2	1.1	1.1	1.1	1.1	1.1	0.99	0.95	0.91	0.86
8.40	1.3	1.2	1.2	1.2	1.2	1.1	1.1	1.1	1.0	0.99	0.94	0.90
8.70	1.3	1.3	1.3	1.2	1.2	1.2	1.1	1.1	1.1	1.0	0.98	0.93
9.00	1.4	1.3	1.3	1.3	1.2	1.2	1.2	1.1	1.1	1.1	1.0	0.96
9.30	1.4	1.4	1.3	1.3	1.3	1.3	1.2	1.2	1.1	1.0	1.0	0.99
9.60	1.5	1.4	1.4	1.4	1.3	1.3	1.3	1.2	1.2	1.1	1.1	1.0
9.90	1.5	1.5	1.4	1.4	1.4	1.3	1.3	1.3	1.2	1.1	1.1	1.1
10.2	1.5	1.5	1.5	1.4	1.4	1.4	1.3	1.3	1.2	1.2	1.1	1.1
10.5	1.6	1.6	1.5	1.5	1.5	1.4	1.4	1.3	1.3	1.2	1.2	1.1
10.8	1.6	1.6	1.6	1.5	1.5	1.5	1.4	1.4	1.3	1.2	1.2	1.2
11.1	1.7	1.6	1.6	1.6	1.5	1.5	1.5	1.4	1.4	1.3	1.2	1.2
11.4	1.7	1.7	1.7	1.6	1.6	1.5	1.5	1.4	1.4	1.4	1.3	1.2
11.7	1.8	1.7	1.7	1.7	1.6	1.6	1.5	1.5	1.4	1.4	1.3	1.2
12.0	1.8	1.8	1.7	1.7	1.7	1.6	1.6	1.5	1.5	1.4	1.3	1.3

D.
$$\frac{TU/\text{month}}{\text{length increase}/\text{month}} = TU/\text{in.}$$

Repeat this procedure for a number of months to obtain an average TU per inch of growth. This figure should remain constant for any particular species or strain of trout, within the temperature range of 38.6° to 60°F, as long as the diet is not changed or the fish do not become stressed.

E.
$$\text{Projected daily length increase} = \frac{\text{expected } TU/\text{month}}{TU/\text{in. of growth}} \div 30 \text{ days}$$

F.   Follow I-B above.

GROWTH

Depending largely on diet and temperature, marketable size trout (20 to 35 cm in the United States) can be produced in 7 to 14 months of intensive culture, starting with the egg. A few culturists specialize in rearing 2.5- to 16.5-cm fingerlings for sale to other trout farmers. Growth is fastest during the fingerling period, thus such culturists may produce salable fish within 1 to 8 months.

DISEASES

Due to the relatively long history of salmonid culture, a comparatively large amount is known about the diseases of these fishes. However, there has been too much emphasis on cure and not enough on prevention. Further, money has literally been thrown down the drain by the needless application of medication. The following hygienic measures should be applied by every culturist:

1.   Limiting shipping of live trout at all stages. Not only is there the possibility of infecting one's own stock with someone else's disease or contaminating a disease-resistant strain, but diseases may be introduced to new parts of the world. A number of diseases which create problems for American trout culturists are virtually unknown in Europe and vice versa.

2.   If trout are purchased from another culturist, demand a pathological inspection of the stock.

3.   Quarantine all diseased fish in ponds which are not connected with enclosures containing healthy fish.

4. Use disinfectants on ponds, nets, and all equipment exposed to disease.

There is still much to be learned about the diseases of salmonids and, while various government agencies are doing their share, it would behoove the large commercial enterprises to involve themselves in pathology research. No one is better equipped to understand the problems of trout culture than the culturist himself.

HARVEST AND PROCESSING

As mentioned, there is a considerable market for trout eggs and some opportunity for sale of fingerlings, but the primary product of American trout culture is edible size trout. Some of these are sold for stocking in fish-out ponds, where the public is allowed to angle for them for a fee, but most are processed and marketed iced or frozen. Canadian trout farmers are also considering offering smoked trout.

Small farmers may sell their crop to a central processor, but the large enterprises usually do their own processing. The most highly mechanized operation of this sort is Erkins'. After harvest by seining or pond draining, the trout are transported to the processing plant in specially constructed tank trucks with an oxygen circulating system. (Rangen, Inc., has pioneered a double floored tank truck. Since during transport the trout congregate near the bottom, this has doubled the carrying capacity.) Upon entering the plant, the trout are killed by electrocution to minimize lactic acid, graded—by machine—for the last time, and passed on to an Erkins innovation called an "Eviscerator," which can automatically dress up to 56 trout/min, leaving heads and tails intact. After a careful inspection, the fish are prechilled and given a final washing in circulating refrigerated water, boned, packaged in individual polyethylene bags, and frozen at $-22°C$. No trout spends more than 30 min from pond to freezing room, thus ensuring a high-quality product.

PROSPECTUS

The Snake River Trout Company and other American producers processed 5.0 million kg of trout in 1969. The majority of this came from the northern and western states, but the greatest growth of the industry is currently occurring in the southern states where, although the climate is too warm to support salmonids in streams, there are numerous large, cold, isothermal springs suitable for trout culture.

PLATE 11. Trout eviscerator developed at Snake River Trout Co. (Courtesy Robert A. Erkins.)

Two factors impede further growth of commercial trout culture in the United States. One is the lack of aggressive and imaginative marketing which has long characterized the American fishery and fish culture industries. The other is competition from other countries which are able to supply trout to the American market cheaper than domestic producers. Most notable in this respect are Denmark and Japan, the only two countries which produce more trout than the United States (10.0 million kg and 5.1 million kg in 1966 and 1967, respectively).

COMMERCIAL TROUT CULTURE IN FRESHWATER IN JAPAN

Culture of rainbow, brown, and brook trout in Japan in most respects does not differ essentially from that practiced in the United States. However, fry are kept in shallow running water ponds with a layer of pebbles on the bottom. All life stages receive a diet different from that fed elsewhere (Table 6).

Diets fed to larger trout vary, but the optimum food is considered to contain 60% protein, 25% fat, 10% carbohydrate, 5% minerals and a variety of vitamins. The diet of trout of all ages is supplemented by live insects lured into the ponds at night by lamps.

TABLE 6. STOCKING AND FEEDING SCHEDULE FOR RAINBOW, BROWN, AND BROOK TROUT IN JAPAN

LENGTH OF FRY (CM)	SIZE OF POND (M²)	DEPTH OF POND (M)	NUMBER OF TROUT	FEED
Less than 2.5 (sac fry)	1.4	0.3	—	None
2.5– 3.0	1.4	0.3	—	Fresh animal spleen and liver crushed into a juice; earthworms and *Gammarus* may also be fed
3 – 5	1.4	0.3	—	Cattle liver—50%, meat or fish powder—20%, chopped fish—10%, live chrysalids—20%; all ingredients chopped and kneaded into a soft mass
5 – 7	42	0.4	10,000	Chopped fish—40%, live chrysalids—40%, cattle liver—20%; scattered on surface
7 – 9	80	0.65	10,000	Flour, rice bran, and greens—20%, various animal foods—80%
9 –15	210	1.0	10,000	Flour, rice bran, and greens—20%, various animal foods—80%
15 –20	400	1.3	10,000	Flour, rice bran, and greens—25%, various animal foods—75%
20 –25	400	1.3	400	Flour, rice bran, and greens—25%, various animal foods—75%
25 –30	400	1.3	200	Flour, rice bran, and greens—25%, various animal foods—75%
30 –40	400	1.3	100	Flour, rice bran, and greens—25%, various animal foods—75%

COMMERCIAL TROUT CULTURE IN DENMARK

ADVANTAGES ENJOYED BY DANISH CULTURISTS

Denmark is not favored with an abundance of isothermal springs, so the efficiency of Danish trout culture can never approach that of the best American farms, but Danish culturists do enjoy a number of advantages:

1. They are so far able to operate with earthen ponds and not suffer severe disease problems. Thus construction costs are far less than in the United States. Danish culturists also assert that earthen ponds produce tastier trout.

2. The nearby Baltic Sea and North Sea fisheries provide an excellent source of cheap food in the form of trash fish.

3. The proximity of trout farms, seaports, and research stations in a small country with excellent roads minimizes transportation expenses.

4. Other industries and the government have taken an active interest in trout culture.

WATER SUPPLY

Due to the scarcity of spring water, Danish trout culturists are forced to rely upon surface water. Unpolluted streams are abundant only in one area of the country, Jutland, and virtually all of the 600 or so Danish trout farms are located there. The produce of these farms includes live and iced trout for sale in Europe, frozen fish for overseas export, and canned trout. Over 90% of the production of edible size trout is exported. Not all the farms include hatchery facilities and some of the smaller operations specialize in producing eyed eggs, fry, and fingerlings. The majority of the eyed eggs exported are brown trout, but 85 to 90% of the edible size fish produced are rainbow trout.

HATCHERY TECHNIQUES AND FRY REARING

Hatchery techniques are essentially the same as those practiced in the United States. Unlike fingerlings and adults, fry are kept in concrete, plastic, or fiberglass tanks up to the size of 5 cm. If kept in earthen ponds, fry are susceptible to "whirling disease," caused by the sporozoan *Lentospora cerebralis*. Young trout are not kept in this sort of container past the fry stage because it is believed to render them more susceptible to fin rot.

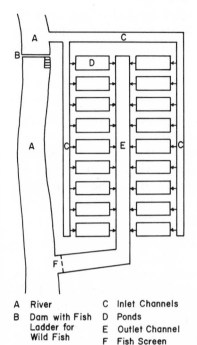

A   River   C   Inlet Channels
B   Dam with Fish   D   Ponds
    Ladder for   E   Outlet Channel   FIG. 1.   Layout of a Danish trout farm. (After
    Wild Fish   F   Fish Screen   Bregnbølle, 1966.)

FARM LAYOUT AND POND CONSTRUCTION

A typical Danish trout farm might be constructed as shown in Fig. 1.
A stream is dammed (the culturist is legally responsible for any damage
to the stream bank or to wild fish stocks) and diverted into a series of
ponds (an average farm would consist of 35 to 60 ponds rather than the 18
shown here), each 30 m long × 10 to 12 m wide, connected to a central
outlet channel. Production area is increased by stocking trout in the outlet channel, which is screened off from the stream. The danger of disease is
lessened by stocking the ponds at densities considerably lower than those
used in the United States.

FEEDING

Trout in ponds are fed mainly on minced fish, usually herring and sand
eels. If herring are used for a prolonged period of time, the culturist
must occasionally add a little vitamin B_1 (thiamine hydrochloride) to
counteract the thiaminase in herring, which breaks down vitamin B_1. Fry
in tanks are fed a dry food which, although more expensive than fresh

fish, is made in Denmark using Danish fish meal and is still relatively cheap. Increasing use is being made of pellets for feeding fingerlings and adults as well. The trout are fed as much as they will eat as long as the water temperature remains below 20°C.

GROWTH

Conversion ratios are high—5:1 to 7:1—and growth is slow by American standards. Fry hatched in March or April will reach 8 to 15 cm by November, but then growth practically ceases for the winter. The amount of time required to reach marketable size varies greatly as the 15 or more countries to which Denmark exports differ greatly in the size of trout preferred, but virtually no trout younger than 14 months are marketed and some must be retained for 2 years before being sold.

TROUT FARMERS' ASSOCIATION

Danish trout farmers maintain direct ties with trout research through a research station established by an association of trout farmers. Initially the farmers sustained all the costs, but now the government pays the wages of the scientific staff. Opinions differ as to the value of the association, to which most of the nation's trout culturists belong. One of its principal functions is to diagnose and treat diseases, and many farmers undoubtedly find it comforting to have a trained biologist at their beck and call to attend to sick fish. However, some observers have suggested that the time spent running back and forth to fish farms might better be spent in research aimed at increasing production.

MARKETING AND PROSPECTUS

One thing the association does not do is act as a sales organization. Another group has been formed for this purpose but only 35% of the farmers belong. The rest market their fish through various export firms. Better marketing organization might benefit the Danish trout culture industry, but only if production is increased. Increased production will come only as a result of improved techniques or expansion into brackish and salt water, since virtually all the suitable freshwater is being used.

TROUT CULTURE IN BRACKISH AND SALT WATER

BRACKISH AND SALTWATER TROUT FARMS IN DENMARK

Both the Japanese and the Danes have limited supplies of freshwater available for trout farming, so it is not surprising that these two countries have pioneered culture of these extremely euryhaline fishes in brackish and salt waters. Already there are eight productive Danish trout farms located on a fjord where salinities reach 10‰ or more, as well as one which uses sea water of about 30‰ salinity. The water supply for these farms is driven through the ponds by a turbine or propellor. The low sea temperatures in winter raise the danger of mortality, so that freshwater must be used.

SALTWATER TROUT CULTURE IN JAPAN

The warmer waters of Japan have permitted one farm, located on Okachi Bay, Miyagi Prefecture, to take the process one step further. One-year-old rainbow trout about 23 cm long are acclimated to sea water, stocked in huge floating net cages anchored about 150 m from shore, and fed on pellets containing fish meal, starch, and vitamins. At the end of about 9 months they have reached a length of 40 cm and are marketed. These trout of course derive a great deal of nourishment from the sea, so that conversion ratios are actually less than 1:1. Net cages are now also being employed to grow brown trout and rainbow trout off Tasmania and Norway, Atlantic salmon and rainbow trout off Scotland, and have seen successful experimental use in freshwater with brook trout in Quebec.

LOW-INTENSITY TROUT CULTURE

ESTUARINE CULTURE IN DENMARK

Most brackish and saltwater trout culture schemes are less intensive than those just described. One of the most successful is that carried out by the Danish Ministry of Fisheries which, in 1963, with the financial support of the Danish Sport Fishermen's Association, began to release pre-adapted two-year-old brown trout into fjords. It was hoped that the trout would remain in the fjords for a year or so, fattening themselves for market with no further input of money or labor and be catchable in sufficient quantities to pay for the operation.

The initial releases took place in eight fjords of varying configurations. In three which were relatively closed and sheltered the trout for the most part remained within 12 km of the release site, grew very rapidly, and were recaptured a year later by commercial fishermen in large enough numbers to yield better than 100% profit over the cost of rearing and stocking. Ecological research continues with the aim of determining what sort of fjord is most conducive to growing brown trout, but it has already been estimated that 25% of the 220,000 ha of Danish estuarine waters are suitable for this sort of culture.

Experiments are also being conducted to determine the feasibility of stocking one-year-old fish. Mortality would surely be higher, but rearing costs would be reduced. Growth would also be enhanced, since two-year-old fish reach sexual maturity during their year in the sea. Thus much energy that might have gone into growth is diverted to maturation. By stocking one-year-olds and harvesting just before the onset of maturation this problem could be circumvented.

SALTWATER CULTURE IN JAPAN

Japanese biologists have tested a more sophisticated version of the Danish technique which might be applicable in a wider variety of waters. Rainbow trout, while being acclimated to sea water, were trained to associate the sound of a buzzer with feeding at certain hours. When the trout were acclimated to full strength sea water they were placed in the sea in a net cage off Takamatsu on the island of Shikoku. Training was continued, using an automatic feeding device. When they were thoroughly trained, the net cage was removed and the trout set free to roam the sea at will. However, they continued to return daily to be fed. Reinforcement of the learned behavior was continued until the trout were harvested.

Using this technique it was possible to harvest 1.2-kg fish after a year and 2.0- to 2.5-kg fish after 2 years. The fish used in the experiments were a special fast-growing strain developed at the University of Washington. (Details on selective breeding of trout follow.) Nevertheless, it is truly remarkable to achieve such growth using inexpensive methods. The process might be intensified by training different species, strains, or age groups of fish to feed at different times of day and to respond to different wavelengths of sound.

FRESHWATER CULTURE IN CANADA

Low-intensity methods may also be effective in freshwater. For example, in Manitoba and Saskatchewan, wheat farmers, who have long unsuccessfully tried to drain the many small, shallow, unproductive lakes, or "potholes," which dot that part of Canada, now stock them with rainbow trout. Fry purchased from hatcheries and stocked in the spring exhibit survival rates of up to 69% and average about 28 cm in length when harvested in the fall. Supplementary feeding is not necessary, since the trout nourish themselves on the abundant crustacean *Gammarus lacustris*. Predation is no problem, since the lakes freeze solid in winter, and thus support no permanent fish populations. The dollar return of this extremely simple form of fish culture is about two and one-half times as much per hectare as the return to wheat farming. Total trout production in the region is expected to eventually increase to 4.5 million fish/year.

CULTURE OF ANADROMOUS SALMONIDS

TO SUPPLEMENT FISHERIES

As mentioned, all salmonids are highly euryhaline, and most, if not all species exhibit anadromous traits wherever they have access to the sea. By far the most important anadromous salmonids from the point of view of commercial fisheries or fish culture are the Pacific salmons (genus *Oncorhynchus*), which are treated in Chapter 21, but some anadromous species of *Salmo, Salvelinus,* and *Thymallus* are also important as human food. By far the most studied of these, due primarily to their great popularity with sport fishermen, are the rainbow trout, the anadromous form of which is often referred to as "steelhead," and the Atlantic salmon. Both are extensively propagated in hatcheries and released to augment sport fisheries. There is at present no commercial fishery for steelhead, but Atlantic salmon are fished commercially in Europe and are propagated for that purpose in several countries. The greatest success in culture of this most difficult of all salmonid species has been achieved in Iceland and Sweden. In Iceland 7.6% of stocked smolts (young salmon about to migrate to sea) return to the fishery, averaging about 7.6 kg in weight. Until recently the Swedish fishery was threatened by the damming of rivers to produce hydroelectric power. Now the fishery is stabilized and at least 20% of the catch is derived from hatchery-raised smolts. Salmon hatched in Swedish hatcheries also contribute to Danish and Finnish fisheries.

A large part of the success of the Swedish hatchery system is attributed to the development by Aktiebolaget Ewos, a Swedish subsidiary of Astra Pharmaceutical Products, Inc., of a special dry food (Table 7) acceptable to Atlantic salmon, which usually refuse conventional trout pellets. This diet, which incorporates fish protein concentrate (FPC), has also been found suitable by the Maine Department of Inland Fisheries and Game for rearing the landlocked strain of Atlantic salmon, as well as lake trout, which present similar feeding problems.

Recently Poland has imported eyed Atlantic salmon eggs from Canada for restocking rivers where the species has been exterminated. A trout hatchery is also being built on the Vistula River, where a commercially important run of anadromous brown trout is threatened by dams and pollution.

In Japan, the Dolly Varden (*Salvelinus malma*), a highly anadromous

TABLE 7. COMPOSITION OF A DIET FOR ATLANTIC SALMON DEVELOPED BY AKTIEBOLAGET EWOS OF SWEDEN, AND SIZE OF PARTICLES TO BE FED TO FISH OF DIFFERENT SIZES

	FEED ANALYSES		FEED GRADES			
	STARTER FEED (TYPE F-48) (%)	GROWER FEED (%)	GRADE AND SIZE	PARTICLE SIZE (CM)	NUMBER OF FISH PER KG	FISH LENGTH (CM)
Crude protein	58	47.0	Starter			
Fat	8	4.9	1	0.025–0.075	4,994–2,497	2.5– 3.3
Ash	15	10.4	2	0.075–0.15	2,497–1,250	3.3– 4.1
Water	7	6.5				
Fiber	1	1.5	Grower			
Carbohydrate	11	29.7	2	0.075–0.15	1,250– 332	4.1– 7.1
Vitamins[a]	—	—	3	0.15 –0.23	332– 66	7.1–10.9
Minerals[b]	—	—	4	0.23 –0.41	66– 23	10.9–15.0

[a] Vitamin A, vitamin D, vitamin E, vitamin K (menadion), vitamin B_1 (aneurine), vitamin B_2 (riboflavin), vitamin B_6 (pyridoxine), vitamin B_{12} (cyanocobalamin), niacin, calcium pentothenate, vitamin H (biotin), folic acid, inositol, p-Aminobenzoic acid, choline, vitamin C (ascorbic acid).
[b] The product contains all necessary minerals and trace elements.

species usually scorned by sport fishermen, is raised in hatcheries and stocked in rivers to support a commercial fishery.

The grayling (*Thymallus thymallus*) is generally restricted to wilderness waters, but in Sweden, where grayling spawning rivers have been obstructed by power dams, they are bred in hatcheries and stocked.

MORE INTENSIVE CULTURE OF STEELHEADS AND ATLANTIC SALMON

A refinement of the hatchery method of supplementing fisheries for anadromous salmonids has been suggested by Ichthyological Research Corporation of Palo Alto, California. They propose to rear steelhead in a hatchery situated at the head of a short artificial stream, preadapt 15- to 20-cm fingerlings to sea water, and release them in the stream. These fish would theoretically enter the sea, to spend the next 2 to 5 years concentrating nutrients. With sexual maturity they would return to the hatchery where they could be harvested immediately with no competition from sport fishermen and none of the deterioration usually associated with long stays in fresh water. By spawning and releasing fish year-round it might be possible to achieve year-round harvests as well. Sport fishery research in Washington has shown that steelhead runs can be artificially established in the manner proposed with 7 to 14% returns.

More intensive culture of a normally anadromous salmonid species is being explored at the University of Rhode Island, where biologists have succeeded in rearing Atlantic salmon, spawned in freshwater, from an average weight of 0.04 to 0.30 kg in 6 months, using plastic-lined pools containing sea water. Improvement is expected as techniques are perfected.

CULTURE OF FRESHWATER SALMONIDS TO SUPPLEMENT COMMERCIAL FISHERIES

Propagation of salmonids to bolster commercial fisheries is not limited to anadromous species. Lake trout have long been stocked for this purpose in the North American Great Lakes. The program was limited and of doubtful value until the parasitic sea lamprey (*Petromyzon marinus*) virtually eradicated the lake trout in the upper Great Lakes, necessitating a crash lake trout culture-lamprey control program. This joint effort by the United States and Canadian governments succeeded in reestablishing the fisheries to some extent. If lake trout populations eventually return to anything approaching their former magnitude, it is doubtful whether the present large-scale hatchery program (more than 6 million eggs are produced in Michigan alone) will continue to be justifiable.

Techniques for hatchery propagation of lake trout are essentially the same as those described earlier in this chapter for other salmonids. Until recently, the major source of spawners was fish taken in nets set near the spawning grounds, but the distinct threat that the sea lamprey would completely eliminate the lake trout necessitated the development of hatchery brood stocks.

The incubation period of lake trout eggs is about 2 months at 8.5°C, 4 months at 4°C, and over 5 months at 2°C. The hatching rate of artificially fertilized eggs averages about 65%.

Newly hatched lake trout fry develop slowly and do not absorb the yolk sac until they are about a month old. Their first food should be finely ground meat, such as beef liver or heart. Later, they may be gradually shifted to a conventional hatchery trout diet.

In the Great Lakes, lake trout usually inhabit depths of 30 to 100 m, but, fortunately, lake trout culturists have not found it necessary to simulate this aspect of the natural habitat. Hatchery ponds should, however, be covered or otherwise shaded, since exposure to strong sunlight may cause cataracts.

Lake trout have been stocked at all ages from the alevin stage to two years. Survival of course increases with age. Experiments in Lake Superior indicate that survival is markedly improved by rearing through the first winter, thus most lake trout are now planted in the spring of their second year, at a length of 10 to 13 cm. Rearing to maturity requires 4½ to 6½ years or more.

SELECTIVE BREEDING AND HYBRIDIZATION

Of the many experimental techniques being applied in efforts to increase trout production probably none has greater potential than selective breeding. Sport fish hatcheries have long practiced selective breeding, at least in a rudimentary manner, for high fecundity, large egg size, high hatching percentage, rapid growth, early maturity, high temperature tolerance, and disease resistance. To facilitate hatchery operations, selection has also been carried out for time of spawning, so that there now exist stocks of rainbow trout which, without manipulation, will spawn in any month of the year. In recent years strains have also been developed which have a pronounced tendency to leap out of the water when hooked. These "jumpers" are preferred by sport fishermen and are thus sought by fish-out pond operators, as are albino or golden rainbow trout and other "novelty" strains. Cultured trout are far enough removed from wild genotypes that trout, particularly rainbow trout, can take their place alongside the common carp (*Cyprinus carpio*) and a few species of ornamental fish as our only truly domesticated aquatic livestock.

Selection is a never-ending process, but it does not take long to achieve some worthwhile results. For example, the California Department of Fish and Game has been able to:

1. Increase the number of rainbow trout spawning at 2 years of age from 53% to 98% in three generations.

2. More than double the average weight of yearlings in five generations.

3. Increase egg production by 2-year-old females fourfold in six generations.

Less attention has been paid to selective breeding for characteristics desirable in commercial culture, although most of the characteristics sought by sport fishery culturists would be advantageous to any trout grower. In recent years, a number of workers have sought to improve the commercial breed of rainbow trout. The most famous result of such work is certainly the "supertrout" developed by one of the pioneers in selective breeding of salmonids, Lauren R. Donaldson of the College of Fisheries, University of Washington. After 38 years of selection for 10 characteristics, Donaldson's strain exhibits superior hatching qualities, grows to a length of 67 cm in 3 years, and is so tolerant of high temperatures and certain pollutants that it can be grown in water one would not normally consider fit for the species.

Danish biologists have also developed fast-growing strains of rainbow trout, some of which surpass Donaldson's fish with respect to egg size. These and other selected strains are now available to culturists throughout the world. It should be emphasized, however, that each of these strains is specifically adapted to certain conditions and that no two environments offer the same conditions. It should therefore be the responsibility of each individual culturist not only to start with the best stock obtainable, but to practice selection so as to improve his own strain. Thus far, only a few of the better farmers have undertaken this task.

It is possible that some of the good results claimed for selective breeding should be attributed to other factors. For example, the role of endocrines in determining egg size is not known. But this sort of uncertainty does not negate the importance of selection as an integral part of any efficient farming operation.

Hybridization does not appear to hold as much promise as selection for commercial trout culture. Sport fishery culturists have produced virtually

PLATE 12. Wild and mass-selected, hatchery-fed rainbow trout at the University of Washington School of Fisheries, Seattle. Fish are two years old, and the large one is the result of over 30 years of selective breeding. (Reproduced with permission of *Science*.)

every possible hybrid of rainbow, brown, brook, cutthroat, and lake trout, and Atlantic salmon, but only one or two have proved useful. The only existing hybrid of interest as a food fish is an intraspecific rainbow trout cross developed by Donaldson. By crossing anadromous and land-locked strains, he was able to produce a fish which consistently returns to spawn at two years of age, whereas only a minority of wild steelhead return that early. This hybrid has been used to establish runs in several coastal streams which did not previously support steelhead.

There are other wild strains of trout which should be investigated by culturists. For instance, the deep water Kamloops strain of rainbow trout (*Salmo gairdneri kamloops*) native to certain lakes in British Columbia consistently outgrows all other strains. Before the decimation of the lake trout in the Great Lakes by the sea lamprey there was a subpopulation of a subspecies known as the siscowet (*Salvelinus namaycush siscowet*) in Lake Superior. These fish had a much higher fat or oil content than ordinary lake trout. In fact, large specimens contained up to nearly 70% oil, perhaps the highest oil content of any fish in the world. There may also be genetic factors involved in the extremely large size attained by steelhead in a few streams such as British Coumbia's Kispiox River and Alaska's Russian River, or by the Lahontan strain of cutthroat trout (*Salmo clarki henshawi*) found only in two Nevada lakes. Hybridization of these strains with domestic stocks might produce beneficial results.

TROUT CULTURE IN THERMAL EFFLUENT, RECIRCULATING WATER SYSTEMS, AND TANKS

Although trout are generally considered cold water fish, the development of temperature-resistant strains raises the possibility that they might be grown in thermal effluent, particularly in areas where normal winter water temperatures greatly retard growth. Heated water might well be employed in conjunction with a recirculating system. Such a system, using small tanks supplied with 14°C water, is presently being successfully used to rear trout fry and fingerlings (species unknown) in Austria, but commercial applications to date have been infrequent.

One such operation exists in Canada, where Sea Pool Fisheries, located in Clam Bay, near Lake Charlotte, Nova Scotia, plan to annually market 1.8 million kg of rainbow trout, Atlantic salmon, brook trout, chinook salmon (*Oncorhynchus tschawytscha*), and possibly Arctic char (*Salvelinus alpinus*) reared in a semiclosed system in which over 90% of the water is recycled. The basic techniques of filtration, temperature regulation, and so on, are like those used in Columbia River salmon hatcheries (see pp. 487–488), but the facilities are adapted for growth of marketable size fish (up to 7 kg in the case of chinook salmon). Eggs and young fish are maintained in conventional hatchery facilities, while larger fish inhabit a series of large fiberglass pools clustered in groups of three to ten around single filter pools.

A principal feature of Sea Pool Fisheries is the gradual increase in salinity of the water with the age of the fish. By 8 months, at which age rainbow and brook trout have reached marketable size (about 0.2 kg), they are maintained in full strength sea water. In addition to the enhancement of growth and reduction of disease problems associated with high salinity, the sea water brings in with it many organisms which serve as dietary supplements for the trout.

Another interesting development occurred in 1969, when the United States Bureau of Commercial Fisheries began an aquaculture training program for American Indians of the Lummi Reservation near Bellingham, Washington. Part of the training involves raising of trout and Pacific salmon to the smolt stage in a recirculating system using 90 to 95% recycled water. Water contaminated by raw sewage passes through swimming pool filters and a series of ultraviolet lights at 4 to 10 liters/min, then through a chiller at 400 to 500 liters/min so that a constant temperature of 10°C is maintained. Another filter is used to treat water that has circulated through the hatchery once or more. The concentration of ammonia nitrogen is monitored daily to determine the amount of new water that must be fed into the system.

PLATE 13.   Silo capable of rearing 20,000 trout, Benner Spring Fish Research Station, Bellefonte, Pennsylvania. (Courtesy Pennsylvania Fish Commission.)

A potentially revolutionary breakthrough in trout culture was recently made by the Pennsylvania Fish Commission at its Benner Spring Research Station at Bellefonte. There biologists have succeeded in rearing 20,000 rainbow trout, weighing at least 2720 kg, in a cylindrical fiberglass tank 5 m high and 2.3 m in diameter with a capacity of 20,640 liters. The use of this tank has resulted in the production of more protein per acre than has any other known food production method. If such devices are eventually applied commercially it will be possible to institute trout culture at springs and other water sources located where terrain, space, soil type, or some other factor makes it impossible to construct ponds or raceways. A further advantage is that both construction and maintenance costs for such tanks are much lower than for conventional hatcheries.

TROUT IN POLYCULTURE

All of the trout culture systems thus far discussed are monoculture systems. Unlike most instances of single-species culture, trout monoculture is not ecologically inconsistent, since these essentially cold water species normally inhabit relatively sterile environments with limited species associations. Nevertheless, trout do have some potential for use in polyculture, particularly as predators to control excess reproduction of small wild fishes which compete for food with cultured species. Trout are sometimes preferred for such purposes over such traditional pond predators as the northern pike (*Esox lucius*) and the largemouth bass (*Micropterus salmoides*) since trout forage in open water, as well as along shorelines and weed beds.

Trout are most commonly used as predators in conjunction with common carp, particularly in Poland and Czechoslovakia, where 1-year-old rainbow trout are stocked in ponds at 1200 to 1500/ha, or about 10 to 15% of the carp population, and are harvested as 2-year-olds. Growth depends principally on the availability of carp fry and other natural foods. In fertile ponds, trout thus stocked may contribute 20 to 50 kg/ha over and above the normal production of carp. Production may be increased by stocking the orfe (*Leuciscus idus*) and the roach (*Rutilus rutilus*), which do not compete with carp, as food for the trout.

The application of trout in pond culture may be wider than was previously assumed. Russian biologists have found that, as long as dissolved oxygen concentrations remain above 5 ppm most of the time, rainbow trout stocked at low densities do well at 16 to 18°C, continue to feed at up to 24°C, and can withstand temperatures as high as 28°C for short periods.

In 1964 biologists of the Soviet Union's All-Union Institute of Pond Fisheries conducted an experiment with rainbow trout in a more complex polyculture system. Two fertile ponds containing 4 year classes of common carp, 2 year classes of grass carp (*Ctenopharyngodon idella*), and 1 year class of silver carp (*Hypophthalmichthys molitrix*) were stocked with 2-year-old rainbow trout as well. The food niche of the trout remained intact, as they ate aquatic and terrestrial insects, *Daphnia*, carp fry, and frogs, most of which were not utilized by the other species. At harvest, the trout contributed 30.8 and 36.5 kg/ha, respectively, or 4.7% and 6.3% of the total fish production of the two ponds. Brown trout have been recommended for similar use in polyculture in the Soviet Union.

It has been reported that the taimen (*Hucho taimen*) and the huchen

(*Hucho hucho*) are being investigated for fish cultural purposes in China and Spain, respectively. Details are not available, but these large, typically solitary and piscivorous salmonids are probably being considered for use as predators.

PROSPECTUS

Clearly, trout culture is undergoing a period of expansion. Not only are individual farms increasing their output and new farms being established in areas where trout culture is traditional, but whole new regions of the world, including the oceans, are being opened up for trout farming. It is conceivable that in the not-too-distant future rainbow trout will be as nearly universal in fish culture as common carp or *Tilapia mossambica*. Certainly the day of trout as a luxury food is nearly over and its day as a staple protein source is approaching.

REFERENCES

ANON. 1969. World's largest trout farm. Am. Fish Farmer 1(1):6–9.

ANON. 1970. Washington trout hatchery. Am. Fish Farmer 1(2):13–16.

BREGNBALLE, F. 1963. Trout culture in Denmark. Prog. Fish Cult. 25(3):115–120.

BROWN, E. E. 1969. The fresh water cultured fish industry of Japan. University of Georgia, Agricultural Experimental Station, Research Rep. 41.

BURROWS, R. E., and B. D. COMBS. 1968. Controlled environments for salmon propagation. Prog. Fish Cult. 30(3):123–136.

BUTERBAUGH, G. L., and H. WILLOUGHBY. 1967. A feeding guide for brook, brown, and rainbow trout. Prog. Fish Cult. 29(4).

DAVIS, H. S. 1953. Culture and diseases of game fishes. University of California Press, Berkeley. 332 pp.

ERICKSON, J. D. no date. Trout production opportunities and problems in the west. Snake River Trout Company, Buhl, Idaho. (Mimeographed.)

ESCHMEYER, P. H. 1968. The lake trout (*Salvelinus namaycush*). U.S. Bureau of Commercial Fisheries, Fishery Leaflet 555.

GUNSTROM, G. K. 1970. Canadian Mariculture Facility begins operation. Am. Fish Farmer, 2(1):8–11.

HASKELL, D. C. 1959. Trout growth in hatcheries. N.Y. Fish Game J. 6(2):204–237.

LEITRITZ, E. 1960. Trout and salmon culture (hatchery methods). California Department of Fish and Game, Fishery Bulletin No. 197.

LOCKE, D. O., and S. P. LINSCOTT. 1969. A new dry diet for landlocked Atlantic salmon and lake trout. Prog. Fish Cult. 31(1):3–10.

MEYER, F. P. 1969. Commercial fish production in the U.S. and its relationship to the feed industry. Feedstuffs 41(7):27.

OVCHNYNNYK, M. 1961. Soviet fish culture. *In* Atlantic Ocean Fisheries. London, pp. 267–273. Fish. News Inter. (Books), London.

PHILLIPS, A. M., and D. R. BROCKWAY. 1959. Dietary calories and the production of trout in hatcheries. Prog. Fish Cult. 21(1):3–16.

SCHEFFER, P. M., and L. D. MARRIAGE. 1969. Trout farming. U.S. Soil Conservation Service, Leaflet 552.

SMITH, T. E. 1965. Operation steelhead. Ocean Sci. and Eng. 2:1229–1239.

Interviews and Personal Communication

BERTELSEN, P. C. Danish Ministry of Fisheries.

BREGNBALLE, F. Danish Trout Research Station, Brøns, Denmark.

CLEAVER, F. Oregon Fish Com., Portland, Ore.

DONALDSON, L. College of Fisheries, University of Washington.

FUJIYA, M. Japanese Fisheries Agency.

RANGEN, T. Rangen, Inc., Buhl, Idaho.

21

Culture of the Pacific Salmons (*Oncorhynchus* spp.)

The seven Pacific salmons (*Oncorhynchus* spp.) together comprise one of the world's most important fisheries. In the United States, where they are commercially the second most important fishery product, over 140,000 metric tons were landed in 1965. In Japan, 150,000 metric tons were landed that year, and Canadian, Russian, and South Korean boats also bring in substantial catches. In the last few decades, salmon populations and catches have declined in almost all of the major fishing areas as a result of overfishing, pollution, and damming of rivers. Among the remedial measures taken are various forms of salmon culture involving improved or artificially constructed spawning areas, fry hatcheries, fingerling rearing stations, and the beginnings of true husbandry.

The need for artificial propagation of salmon is intensified by the peculiar reproductive biology of *Oncorhynchus* spp., which, with the possible exception of certain landlocked stocks of *O. masu*, inevitably die after spawning. As a consequence of this fact of salmon life, an individual which is allowed to spawn is unlikely to contribute directly to the fishery. Ripe and spent spawners are potentially harvestable, but often only at the expense of great amounts of labor. The value of ripe and spent fish is further vitiated by the deterioration of the tissues, which results from the cessation of feeding with the attainment of maturity. Nevertheless, ripe fish are eaten in Japan, and spent salmon from hatcheries are beginning to find a market in the United States. However, it appears unlikely that large numbers of natural spawners will ever enter the commercial picture. Thus the management of Pacific salmon fisheries, even more than that of other fisheries, depends on intelligent regulation of the harvest. As long as natural reproduction is a major source of fishery stocks, this necessitates that a certain number of fish be allowed annually to escape the fishermen. This number is not fixed but depends on the survival rate of the progeny to adulthood; under natural conditions this is but a fraction of 1%. If this figure could be substantially increased by artificial propagation, then less spawners would be needed, and the fishery quota could be correspondingly increased.

The pressure placed on existing salmon stocks has been intensified, first by the introduction of high seas fishing techniques to what was initially an inshore fishery and, more recently, by the entry into the fishery of South Korea. High seas fishing increases the likelihood that fish taken by fishermen of one nation were destined to spawn in another country's waters. Salmon fishery management has thus become an international issue. Although the attitudes toward the resource of the five salmon fishing nations differ considerably, fair cooperation, with probable benefit to the fishery, has been achieved.

In addition to international regulation of the fishery, large numbers

PLATE 1. Pink (humpback) salmon. (A) Adult female; (B) adult male; (C) breeding female; (D) breeding male. (Courtesy U.S. Bureau of Fisheries Bulletin, XXVI, 1906.)

of scientists in all four of the major salmon-producing countries are working on the various aspects of Pacific salmon culture. In this chapter we concentrate largely on methods developed in the United States and Canada, but only because we have not had full access to the Japanese and Russian literature. Any reader who can avail himself of the latest developments in Japan and the Soviet Union should do so, to gain a more complete picture of the state of the art of Pacific salmon culture.

THE SPECIES OF PACIFIC SALMON AND THEIR CHARACTERISTICS

CHINOOK SALMON

The largest of the Pacific salmons is the chinook salmon or king salmon (*Oncorhynchus tshawytscha*), which occasionally reaches weights of 45 kg. It spawns mainly in large rivers, from northern California to northwestern Alaska. It is less common and usually smaller along Asiatic shores, although Russia's Kamchatka River supports a sizable run. Over 90% of the commercial catch is landed at North American ports.

The life history of the chinook salmon is the most variable among the Pacific salmons. Maturity and spawning occur within 2 to 7 years of

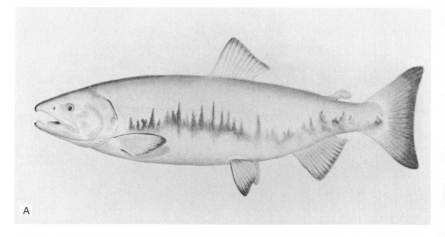

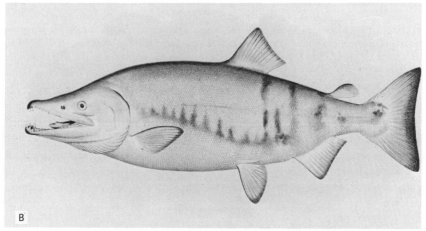

PLATE 2. Chum (dog) salmon. (A) Breeding female; (B) breeding male. (Courtesy U.S. Bureau of Fisheries Bulletin XXVI, 1906.)

hatching, usually after 4 years. Most spawning streams support two or more annual runs, in summer, fall, and/or winter, but some of the larger streams also support spring runs, which characteristically travel further upstream than summer, fall, and winter fish. The usual figure cited for the duration of the freshwater nursery period is 6 months, but it may vary from 90 days to 2 years.

COHO SALMON

The coho salmon or silver salmon (*Oncorhynchus kisutch*), which may reach 15 kg, is found from Monterey Bay, California, to Hokkaido,

Japan. Like the chinook salmon, it is more common and larger in American waters. Coho salmon usually mature at 2 to 4 years of age (extremes 1 to 5 years) and ascend streams in late fall or winter, after most other *Oncorhynchus* spp. have completed spawning. The freshwater nursery period of the coho salmon averages about 1 year at the southern limits of its range and 2 years at the northern limit. It is generally considered the hardiest of the Pacific salmons with respect to environmental conditions.

Chinook and coho salmons are taken mostly by trollers, who find the effort worthwhile since these two species, which usually reach the market smoked or as salmon steaks, bring a higher price than their congeners. Chinook and coho salmons are also the overwhelming favorites of sport fishermen, and in Alaska and Oregon anglers may purchase a commercial fishing license and market their catch. Not a few sport fishermen finance their recreation in this manner.

SOCKEYE SALMON

The sockeye salmon or red salmon (*Oncorhynchus nerka*) spawns from northern California north to the Bering Sea and south to Hokkaido, but only in streams that have one or more lakes in their courses. Spawning occurs in, above, or occasionally below such a lake and the young spend their first 2 to 3 years in the lake, followed by 1 or 2 years in the sea. Some waters, landlocked or otherwise, also contain populations of non-anadromous sockeye salmon, sometimes referred to as kokanee. Sea-run sockeye salmon occasionally reach lengths of over 100 cm; kokanee are usually much smaller.

Like the chinook and coho salmons, sockeye salmon are most abundant in American waters; before the inauguration of the high seas fishery 64% of the catch was landed in the United States. In some areas, notably the Bristol Bay region of Alaska and the Fraser River of British Columbia, it is the most important species. Sockeye salmon generally are delivered to canneries and are considered the finest salmon for that purpose.

CHUM SALMON AND PINK SALMON

The lower grades of canned salmon consist chiefly of chum salmon or dog salmon (*Oncorhynchus keta*) and pink salmon or humpback salmon (*Oncorhynchus gorbuscha*). The chum salmon is the larger of the two, attaining lengths of nearly 1 m and weights of 20 kg, whereas the pink salmon reaches no more than 4 to 5 kg. The chum salmon is also the most widely distributed of the Pacific salmons, ranging from San Francisco, California, north to Alaska, and south to Korea. Runs also occur in

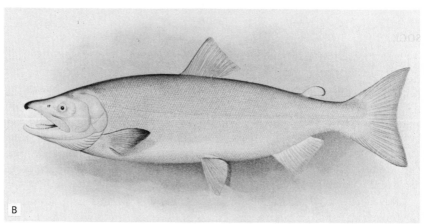

456

PLATE 3. Sockeye (red) salmon. (A) Adult female; (B) adult male; (C) breeding female; (D) breeding male. (Courtesy U.S. Bureau of Fisheries Bulletin, XXVI, 1906.)

tributaries of the Arctic Ocean from Canada's Mackenzie River on the east to Russia's Lena River on the west. Throughout most of its range it is represented by both summer and fall spawning stocks. The distribution of the fall spawning pink salmon is almost identical, but it is not found south of the San Lorenzo River, California, nor east of Alaska's Colville River on the Arctic Coast.

Both species spawn in very small streams as well as large rivers, typically, especially in the case of the pink salmon, not far from the sea, or even in the intertidal zone. Young of both species drift down to the sea within the year of hatching, if not immediately, to return as adults 2 years after hatching in the case of the pink salmon, or in 2 to 6 years for chum salmon.

Of the two species, the chum salmon is the more important in the Japanese fishery, while American, Canadian, and Russian inshore fishermen land more pink salmon. Collectively, chum and pink salmon are the most commercially important of the Pacific salmons, and both suffer from overfishing throughout their range.

CHERRY SALMON

There are two additional species of *Oncorhynchus* found only in Asiatic waters. The cherry salmon (*Oncorhynchus masu*), the most warmth-resistant of all the Pacific salmons, ranges from Korea north to the Kamchatka Peninsula and spawns chiefly in the summer, during its third

PLATE 4. Coho (silver) salmon. (A) Adult male; (B) breeding male. (Courtesy U.S. Bureau of Fisheries Bulletin, XXVI, 1906.)

or fourth year of life. *Oncorhynchus rhodurus,* sometimes also called cherry salmon, is comparatively little known and apparently restricted to southern Japan. It spawns in November at an age of 3 to 5 years. Both species are small, slow-growing fish, seldom exceeding 50 cm in length, and are of comparatively minor importance in commercial fisheries.

NATURAL REPRODUCTION OF PACIFIC SALMONS

The mechanics of reproduction are, in general, the same for all *Oncorhynchus* spp. Some time prior to spawning mature adults develop pro-

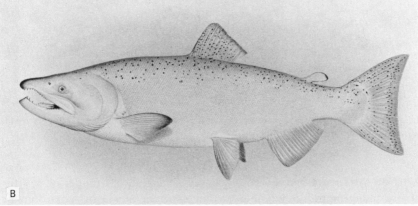

PLATE 5. Chinook (king) salmon. (A) Adult female; (B) adult male. (Courtesy U.S. Bureau of Fisheries Bulletin, XXVI, 1906.)

nounced secondary sexual characteristics, cease feeding, and move first into estuaries, then into streams. After reaching the spawning grounds, which may be located anywhere from the intertidal zone (some chum and pink salmon) to 4000 km upstream (chinook salmon in the Yukon River, Alaska), the females select suitable spawning sites and, turning on their sides, excavate circular egg pits in the gravel with undulations of body and tail. Eggs are deposited in three or more such pits, which collectively comprise the nest or "redd." Fertilization is accomplished by one or more males at the time the eggs are released. Fertilized eggs are buried immediately by the female in the same manner in which the pits were excavated, except that her digging motions become much more rapid. They

PLATE 6. Natural spawning of chum salmon showing female excavating a redd. (Courtesy W. J. McNeil.)

hatch in the stream bed where the larvae remain for a number of months before emerging to immediately seek an estuarine or marine environment (chum and pink salmons), or spend up to 2 years in freshwater (cherry, chinook, coho, and sockeye salmons).

FISHWAYS OR FISH LADDERS

The Pacific salmons have been cultured to augment the results of natural spawning runs, for transplantation into waters where salmon are not native, and for commercial rearing to marketable size, but the first is the prevalent rationale. Among the man-made devices used to assist salmon spawners, the simplest in principle are the fishways or fish ladders commonly incorporated in power dams built on salmon streams. These and devices designed to aid young salmon in their seaward migration do not fall within our definition of aquaculture, and they are not discussed here.

ARTIFICIAL AND IMPROVED SPAWNING CHANNELS

Fishways cannot compensate for the many kilometers of spawning riffles inundated by power dams, nor are they usually highly efficient in maintaining spawning populations in remaining riffle areas. The loss of spawning grounds to impoundments and the mediocre success of fishways may thus be added to pollution, overfishing, and mechanical destruction of spawning grounds as initiative factors in the construction of salmon hatcheries and artificial spawning channels. Of the two techniques the hatchery method is much the older, but artificial spawning channels involve considerably less expenditure of money, materials, and labor, and some authorities believe they offer greater potential benefits to the fishery.

RATIONALE AND FACTORS INVOLVED IN CONSTRUCTION

The environmental prerequisites for successful spawning of the species in question must be known in order to construct a satisfactory artificial spawning channel. The first clue to these requirements is the observed preference of the spawners. The types of spawning areas generally selected by the five North American species of *Oncorhynchus* are outlined in Table 1.

It should not be assumed that the data in Table 1 constitute the last word in the determination of optimal spawning conditions for Pacific salmons. There may be a number of factors involved which are not directly concerned with survival to emergence of the larvae from the stream bed. For example, safety from predators of the spawners and competition with other species may enter into the picture. Further, the best natural spawning area is likely to be suboptimal in some ways, thus potentially susceptible to improvement. This raises the possibility of augmenting salmon production by construction of artificial spawning channels in areas relatively unmodified by man, as well as where dams and the like have obstructed or obliterated natural spawning areas.

This approach would appear to be particularly appropriate to culture of pink and chum salmons, for which species mortality prior to emergence appears to be the chief factor limiting abundance. It is unusual for spawns of these fishes to experience less than 70% mortality during their time in the stream bed, and mortalities in excess of 95% are common.

Spawning channel construction and improvement may be less appropriate as conservation measures for chinook, coho, and sockeye salmons, whose preemergent mortality is believed to be less important than that

TABLE 1. SPAWNING REQUIREMENTS OF THE NORTH AMERICAN PACIFIC SALMONS (*Oncorhynchus* SPP.)

SPECIES	LOCATION OF REDD	SUBSTRATE COMPOSITION	STREAM FLOW OR DISCHARGE	WATER TEMPERATURE	DIMENSIONS OF REDD
Chinook salmon (*Oncorhynchus tshawytscha*)	Main channels	55–95% medium and fine gravel (no more than 15-cm diameter); tolerates no more than 8% silt and sand	0.5–2.0 ft³/sec	—	Vary from 1.2 to 9 m in diameter; average size in Columbia River basin, 3.25 m² for spring-run fish, 5.1 m² for fall and summer fish
Chum salmon (*Oncorhynchus keta*)	Near shore, often in very shallow water; in the U.S.S.R., summer-run fish do not usually locate over upwellings of intragravel water, whereas fall-run fish do	Gravelly; in the Columbia River basin, chum salmon strictly avoid areas where there is poor circulation of water through the streambed	0.1–1.0 m/sec	Spawning recorded at 0.5–16.0°C	Average 2.5 m long, 30–40 cm deep in U.S.S.R.; average size in Columbia River basin, 2.25 m²; 3 m² considered optimal
Coho salmon (*Oncorhynchus kisutch*)	Usually in small streams or narrow side channels of large streams; locations at head of riffles preferred; may spawn in extremely shallow water	Prefers small–medium gravel, but very adaptable; will tolerate up to 10% mud	3.4 ft³/sec preferred (since coho salmon usually spawn in late fall, after rains, this discharge is not typical of year-round conditions in most coho spawning streams)	—	Average 2.8 m² in Columbia River basin
Pink salmon (*Oncorhynchus gorbuscha*)	Main channels, often over upwellings	Medium gravel	0.03 m/sec or more	About 12°C	Average 1.1 m², 9.3 cm deep in southeastern Alaska
Sockeye salmon (*Oncorhynchus nerka*)	In shallows of lakes or in small streams directly above lakes	No more than 1% particles 15 cm or more in diameter; fine-medium gravel preferred	Varied	—	Average 1.75 m² in streams in Columbia River basin; larger and more irregular in lakes; Kokanee redds considerably smaller

suffered in freshwater nursery areas. However, preemergent mortalities of up to 96% have occasionally been recorded for chinook salmon, and several artificial channels have been constructed expressly for use by this species. Coho salmon eggs and young appear to experience lower mortality in spawning beds than the other *Oncorhynchus* spp., and no spawning channels have been built primarily for this species.

A necessary adjunct of any program designed to enhance the success of salmon spawning is a strong program of research into the factors limiting survival of the eggs and young. This, in essence, is a never-ending process for, as Dixon MacKinnon of the Canadian Department of Fisheries put it, "Virtually all biological research on salmon has some application to the problem." Research to date indicates that the chief limiting factors are spawning bed stability, water quality, and, in some cases, meteorological conditions. Following is an outline of the action of each of these and other factors, and ways in which artificial spawning channels may aid in their control.

Spawning bed stability is a function of both stream-bed particle size and current velocity. A stream bed composed largely of silt, sand, or fine gravel is subject to constant shifting and disturbance, which may result in eggs and larvae being washed out to perish in the open water. Even stream beds composed of cobbles or large particles may be unstable during severe flooding. This is a particularly severe problem in southeastern Alaska, where torrential fall rains are common. In an artificial spawning channel both variables may be controlled by selection of predominantly large substrate materials and diversion of flood waters through an overflow channel or natural stream channel.

Water quality covers a multitude of variables, but of primary importance is dissolved oxygen content, which should be at least 6 ppm in the intragravel water for good survival of salmon embryos. If concentrations of dissolved oxygen are reduced by the action of pollutants in the vicinity of the spawning area, an artificial spawning channel will not alleviate the situation, but in most salmon spawning streams pollutants are not the chief factor determining oxygen content.

One might suppose that in a cold, unpolluted, relatively sterile, and swiftly flowing salmon stream the concentration of dissolved oxygen would be at or near saturation, and this is usually the case for surface water in the riffles where salmon spawn. However, eggs and larvae are found at various depths in the stream bed where there is no source of aeration. Oxygen dissolved in intragravel water can originate only in the surface water, which typically circulates into and out of the bed at a rate partly determined by stream velocity but also by the gradient, profile, and particle size of the bed. The more rapid the circulation, the less the likelihood that the dissolved oxygen in a given volume of intragravel water will be severely depleted before the water reaches an area of upwelling. Circulation of intragravel water may be increased in an artificial channel by proper gradient selection, construction of baffles and other such devices, and selection of large-diameter bed materials which afford maximum porosity. (There is a limit to the size of particles which may be profitably employed. If the particles are too large, spawners may have difficulty construcing redds, thus the eggs may be inadequately buried and subject to being washed out.)

Droughts may drastically reduce the dissolved oxygen content of intragravel water, both by restricting circulation and by allowing decaying salmon carcasses to remain in the stream rather than being washed out. Further, droughts may expose some eggs and larvae to dehydration. Artificial spawning channels are constructed so that there are no areas of extremely shallow water and flow can be maintained by pumping, or diversion from the main stream channel may relieve the impact of droughts.

The chief meteorological limiting factor other than flooding and drought is freezing of the intragravel water. The incidence of freezing may be reduced by any measures taken to ensure good circulation and constant water depth and flow.

Other factors which may affect survival of eggs and preemergent fry, and which may be controlled to some degree in artificial spawning channels, are redd superimposition, temperature, and predation. The phenomenon of redd superimposition becomes a problem when the number of female spawners approaches the number of spawning sites. Late spawners, in the process of redd excavation, may uncover eggs of previous spawners. Under some circumstances, the results may be a reduction in the number of fry produced. If access to an artificial spawning channel can be controlled, or if the channel is artificially stocked, a predetermined optimum number of spawners may be admitted.

Temperature plays a very important role in reproduction of all fish, including the Pacific salmons. In artificial spawning channels located below impoundments it is often possible to draw on water sources with different temperatures and thus exercise some control over this variable.

Predation has seldom been found to be an important limiting factor in survival of eggs and preemergent larvae, though it may be very important in later life stages. It has been experimentally shown that the large particles deliberately chosen for use in improved spawning beds may facilitate the activities of predatory fishes, particularly sculpins (*Cottus* spp.), which feed heavily on eggs and young of salmon when given the opportunity. There have, however, been no reported instances of serious predation problems in actual operation of improved spawning channels.

If predation were thought to be important, predator access could often be more easily controlled in an artificial channel than in a natural one.

DESIGN AND OPERATION OF EXISTING CHANNELS

With the foregoing factors in mind, at least 20 artificial spawning channels for Pacific salmon have been constructed since the mid-1950s in the United States and Canada, plus one for Atlantic salmon (*Salmo salar*) in Newfoundland. Similar channels may exist in Japan or the Soviet Union, but we have no indication that this is the case, so this discussion will be limited to North American structures.

Although some salmon spawned in artificial channels have certainly entered the commercial catch, the technique is still essentially experimental and, other than the general stipulations already indicated, there are few rules for success. The existing structures vary from 60 m to 0.7 km in length, 4 to 11 m in width, and 10 to 76 cm in depth and are designed to handle 100 to 10,000 fish. Much larger channels are being planned and constructed. Existing channel gradients vary from 0.0006 to 0.002 and average discharges from 2 to 130 ft^3/sec. It is now generally believed that higher gradients, with the attendant better circulation, produce better results and the gradient of most new channels is toward the upper end of the cited range. Stream beds are constructed of various proportions of particles ranging from 6 mm to 15 cm in diameter. Rather than describe each of the existing channels, we have selected for discussion a few which are illustrative of key points.

A primitive form of the artificial spawning channel is the "improved" spawning channel, two of which were built in 1961 by the U.S. Forest Service and Bureau of Commercial Fisheries in intertidal sections of Harris River and Indian Creek, near Ketchikan, Alaska. The principal causes of mortality in spawning areas of Indian Creek are flooding and freezing, whereas in Harris River poor circulation of intragravel water may be the chief limiting factor in survival of salmon embryos. Accordingly, the design of the Indian Creek channel emphasized construction of a 12- to 15-m-wide flood plain, on both sides of the main channel, to carry excess water while maintaining a fairly constant, moderate flow in the 360- × 4.5-m spawning area, and modification of the Harris River involved chiefly the removal of sand and silt.

Since the natural substrate of Indian Creek is largely composed of particles measuring 5 to 15 cm in diameter, the stream bed was not extensively modified, but baffle boards were installed to increase intragravel circulation. The improved channel was definitely attractive to pink salmon, which account for most of the run; spawning occurred throughout the modified area at discharges of 5 to 1000 ft³/sec. Less than 10% of the spawners utilized the flood plain. Some damage occurred as a result of spawners undermining the banks of the spawning area, which was depressed 46 cm below the surface of the flood plain, but this occurred only when the banks were not protected with a layer of rubble.

Serious damage did occur, however, as a result of severe flooding shortly after the completion of spawning. The modified channel was designed to withstand discharges of at least 1000 ft³/sec, which left some room for doubt as to its stability, since peak instantaneous discharge of Indian Creek in most years is between 1500 and 2000 ft³/sec. In October, 1962, two unprecedented floods occurred, with peak discharges of 2700 and 6400 ft³/sec, respectively. The result was some erosion of the spawning area, removal of some of the baffle boards and burial of most of the rest, filling in of a settling pool at the head of the channel, and deposition of 30 to 45 cm of gravel over the entire length of the channel. The precise effects of this catastrophe are unknown, but estimated survival to emergence of the larvae was 12%, or no better than might be expected under natural conditions in a good year.

In 1962, the settling pool was cleaned out, the gravel deposited by the previous year's floods was removed, and the basic configuration of the channel was restored, but missing baffle boards were not replaced and no attempt was made to remove fine materials from the spawning bed. Once again, the modified channel was attractive, drawing 1.5 times as many spawners as the most densely populated unmodified areas of Indian Creek or Harris River, but survival of the spawn was only 10%.

In the Harris River, 1400 m² of stream bed were improved by hydraulic flushing of fine particles. Sand and silt reentered during the 1961 floods, and estimated survival of that year's spawn was only about 0.1%.

The results of the Indian Creek-Harris River experiments are inconclusive but indicate that channels constructed in natural stream beds, without adequate means of controlling flow, run the risk of being damaged or destroyed by high water. This is particularly true in southeastern Alaska, where unstable runoff patterns are the rule, but would apply nearly anywhere. It might also be added that structures such as the Indian Creek channel, in which the stream course is necessarily straightened and the banks heaped with loose gravel, are esthetic disasters, and as such should not be tolerated, particularly in such a beautiful area as the Pacific Northwest of the United States and Canada.

The first large artificial spawning channel was constructed by the Canadian Department of Fisheries in 1954 at Jones Creek, British Columbia, to compensate for spawning grounds lost to a hydroelectric project. The Jones Creek channel is 610 m long, 4.25 m wide, and handles a controlled discharge of 25 to 30 ft³/sec during the spawning period. The

PLATE 7. Artificial spawning channel at Indian Creek, Alaska. (Courtesy U.S. Forest Service.)

spawning bed consists of a 30- to 45-cm deep layer of 0.6- to 4.3-cm gravel.
A few chum and coho salmon spawn in the Jones Creek channel, but the principal species is the pink salmon. Natural runs of pink salmon occur in Jones Creek only during the odd-numbered years; results of the first four such runs to utilize the artificial channel are shown in Table 2. Survival was four to six times the previous average for the stream. (Cur-

TABLE 2.   FRY SURVIVAL AND RETURN OF ADULT PINK SALMON TO THE JONES CREEK, BRITISH COLUMBIA, ARTIFICIAL SPAWNING CHANNEL

YEAR	TOTAL NUMBER OF SPAWNERS	NUMBER SPAWNING ABOVE COUNTING FENCE	EGGS DEPOSITED ABOVE COUNTING FENCE[a]	FRY OUTPUT	PERCENTAGE OF SURVIVAL
1955	400	400	428,000	158,436	37.0
1957	1,456	1,056	947,000	363,169	38.4
1959	2,604	2,119	1,519,000	958,581	63.0
1961	5,000	4,388	3,789,300	1,100,000	30.0

[a] Based on an average of 1700 eggs per female.
SOURCE: Croker and Reed (1963).

rent results are much poorer, since the channel has been subject to heavy sedimentation.)

The preliminary success achieved at Jones Creek was the impetus for construction in 1959 to 1961 of a larger and more elaborate channel at Robertson Creek on the west slope of Vancouver Island. A plan and profile of this channel (actually six artificial spawning areas, with a combined length of 662 m, separated by a total of 518 m of relatively unmodified stream) are shown in Fig. 1. Further specifications are as follows:

Cross-sectional shape:	trapezoidal
Width, bottom:	10.7 m
top:	12.5 m
Water depth:	43–62 cm
Gradient:	0.0006
Substrate particle diameter:	19–102 mm
Depth of graded substrate:	38–61 cm
Discharge, minimum:	20 ft^3/sec
maximum:	250 ft^3/sec controlled by dam, submerged pipe, and valve
average:	130 ft^3/sec

Emphasis at Robertson Creek has been on transplantation of fertilized pink salmon eggs. Exceptionally high survival rates of 90 to 95% to emergence have been consistently recorded, but returns of adults have been as low as might be expected under natural conditions, from 0.009 to 0.03%. The adult statistic may be due to early emergence of fry in the relatively warm water of Robertson Creek, with consequent poor early growth in the ocean. It may also have to do with some inherent weakness of fry hatched from artificially fertilized eggs, as similar problems have been observed in connection with hatcheries (see pp. 472–474), or some as yet unknown factor may be at work. The survival rate of eggs and larvae, on the other hand, would surely not have been as high had natural spawning been allowed to occur. However, very respectable survival rates to emergence of 68.4 to 86.1% have been recorded for naturally spawned sockeye salmon since 1965 in an artificial channel constructed by the International Pacific Salmon Fisheries Commission on Weaver Creek, British Columbia (Fraser River system). A preliminary estimate of the benefit-cost ratio for the Weaver Creek channel is 7:1. The estimated benefit-cost ratio for the same agency's Pitt River pink salmon channel, which, like the Robertson Creek channel, receives plants of artificially fertilized eggs, is 14:1.

A few chinook and coho salmon spawners have been stocked in the Robertson Creek channel. Problems have been experienced with mortality

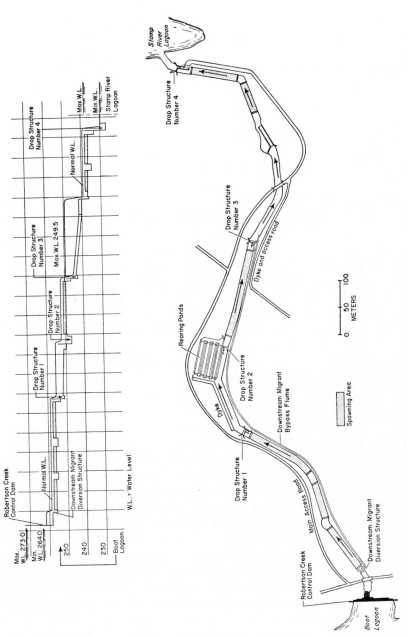

Fig. 1. Robertson Creek artificial spawning channel. Plan and profile.

469

of the adults, but over 60% of the progeny of those adults which lived to spawn are estimated to have survived to emergence. This compares more than favorably with results obtained in channels constructed specifically for chinook salmon in California and Oregon.

The Department of Fisheries envisions much larger artificial channels in British Columbia. One, on Babine Lake, is billed as the world's largest such structure. (A similar claim advanced by the British Columbia Hydro Power Authority and Department of Recreation and Conservation for the Meadow Creek channel, located on a tributary of Kootenay Lake, is of no concern to us, since this channel is utilized only by kokanee, for which there is no commercial fishery.) The Babine channel is designed for use by sockeye salmon and will be able to accommodate 240,000 spawners. The rationale for its construction is that the natural spawning streams from which Babine Lake receives sockeye salmon fry are too small and environmentally unstable to produce a number of fry approaching the carrying capacity of the lake. Experimental releases in Babine Lake of fry from the nearby Fulton River artificial spawning channel disclosed no significant difference in survival over 5 months of rearing, or up to the time of seaward migration. It is estimated that the combination of spawning in the artificial channel and natural rearing in the lake will add 1 million salmon to the Skeena River fishery at a benefit-cost ratio of 4:1.

PACIFIC SALMON HATCHERIES

HISTORY

Hatcheries have a much longer history and have been more thoroughly evaluated than artificial spawning channels. A technique for hatchery propagation of anadromous salmonids was first developed in Canada around 1857 and soon spread to the United States. Early emphasis in both countries was on chinook and coho salmon, and these species were propagated and stocked continuously thereafter, despite the lack of any evidence of success until recently.

The failure of the early hatchery programs in North America was largely due to ignorance of the life history of chinook and coho salmon. The discovery, in the 1940s, that chinook salmon spend their first 90 days to 2 years of postemergent life in freshwater and the subsequent realization that coho and sockeye salmon require a similar freshwater nursery period have radically changed the ecological basis of Pacific salmon culture in North America. Today it can be asserted that hatchery propagation of chinook and coho salmon is an effective fishery management technique.

In 1877, the American technique of salmon culture was introduced to Japan, where it was applied principally to chum salmon, the most commercially important species in that country. The history of salmon culture in Japan parallels the American experience of early failure and subsequent success, though for different reasons.

Hatchery propagation of salmon was slower to develop in Russia, but with the emphasis, under the Communist regime, on development of hydroelectric power, it became imperative that some sort of salmon culture program be initiated. Today, the Soviet Union leads the world in stocking of artificially propagated salmon. Annual releases of pink and chum salmon fry in that country number about 600 million, while about 400 million are released in Japan. The combined annual production of fry of these two species from hatcheries and artificial spawning channels in Canada and the United States totals less than 100 million.

Hatchery propagation of other *Oncorhynchus* spp. is largely confined to the United States. Chinook and coho salmon are the principal species cultured; 81 hatcheries in California, Oregon, and Washington release about 200 million fish annually. Anadromous sockeye salmon are bred in at least one hatchery in the United States, but culture of sockeye salmon is principally a Canadian concern and to date artificial spawning channels, rather than hatcheries, have been emphasized in that country. Most salmon biologists believe that hatchery culture, using present methods, is not an efficacious way of increasing sockeye salmon production.

BASIC HATCHERY TECHNIQUES

Techniques of artificial fertilization and hatching of Pacific salmon eggs are generally the same as those used in hatchery culture of freshwater trout (see Chapter 19), except that female spawners are customarily killed and the eggs surgically removed. Given the peculiar reproductive cycle of *Oncorhynchus* spp., this results in no loss of reproductive potential, and produces fewer damaged ova. Brood fish are generally ripe individuals selected from natural runs or artificially established runs which enter the hatchery. In Japan, salmon may be caught just as they ascend rivers and held in ponds until they ripen. More often, chum salmon eggs artificially hatched in Japan are supplied by Fishermen's Cooperatives, who are allowed to take salmon from rivers on the condition that they fertilize their eggs and turn them over to government hatcheries. This financially and biologically economical practice now has an analog in

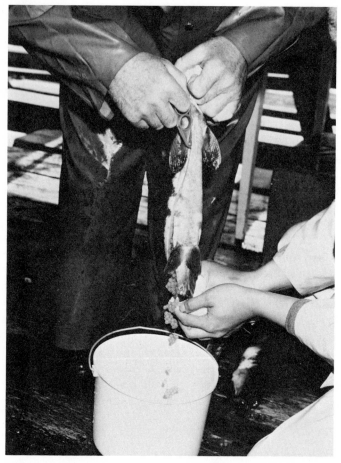

PLATE 8. Making incision in the belly of a pink salmon for spawn taking. (Courtesy J. M. Olson.)

Oregon and Washington, where spent fish from chum salmon hatcheries are sold as food.

COMPARISON OF HATCHERY-REARED AND WILD FRY

The largely temperature-dependent processes of hatching and rearing to the free-swimming stage (corresponding roughly to the time of emergence in nature) may take several months. Both processes are normally carried out in troughs or drip-type incubators (see pp. 400–404), but considerable criticism of these devices has recently been voiced, particularly for culture of chum and pink salmon, which are usually transferred directly from the incubator to estuarine waters. The environment in a hatchery tray is of course very different from that in a stream, and it is believed by critics of tray incubation that this may be an important factor in the repeatedly observed poor survival of hatchery-bred fry. Research in Japan indicates that less than 0.5% of hatchery chum salmon fry return as adults, as compared to the 1 to 5% returns normally anticipated for wild fry.

Suggested changes in incubation technique have to do with substrate texture and water velocity. Currents in standard hatchery incubators are ten times as swift as those found in high-quality natural or artificial spawning beds, while the more or less smooth, two-dimensional bottoms of the trays offer no shelter comparable to that afforded by the irregular conformation of a gravel bed. The net effect may be that hatchery fry must exert considerably more effort than wild fry in maintaining their positions, resulting in poor utilization of the yolk and reduced stamina.

These factors were apparently considered first by Russian biologists, but it is not known if any practical application in the Soviet Union has resulted. More recent work in Canada and the United States has been extended to the field in several locations in Alaska, British Columbia, and Oregon.

At the Nanika River hatchery of the Canadian Department of Fisheries, sockeye salmon eggs are kept in trays until they reach the eyed stage then removed and hatched in a gravel bottom, controlled flow channel. In 1968 to 1969, the first season of operation, survival from the eyed stage to emergence was 60 to 80%, and the young which left the channel were comparable in size to wild fry.

A more thorough comparison of channel-reared versus natural fry was made for chum salmon from the Big Qualicum River, British Columbia. No differences in length, weight, length-weight relationships, or chemical composition were observed in young salmon captured during the downstream migration nor in fish reared in saltwater ponds for 10 weeks. At Hooknose Creek, British Columbia, 70% of chum and pink salmon fry artificially incubated in gravel were adjudged very similar to wild fry in rate of development, but growth of the remainder was retarded.

Laboratory studies at Oregon State University have indicated that salmon eggs and alevins develop best when arranged in gravel in a single layer perpendicular to a water flow with a velocity of 50 to 220 cm/hour, and the university's Netarts Bay hatchery seeks to provide such conditions on a fairly large scale. Water is pumped from a natural chum salmon spawning stream to elevated circular tanks and fed by gravity to a series of covered hatchery tanks, with a total surface area of 35.5 m². Flow within the tanks is directed so as to proceed at 50 to 100 cm/hour in both upward and downward directions through a substrate of 0.6 to 1.8-cm crushed rock. At this rate it is believed that a discharge into the hatchery of 1 ft³/sec would be adequate for 5 to 10 million fry.

Fertilized eggs are placed in the tanks in screen bottom trays. On hatching, the alevins drop through the screen and onto the rocks. The layer of rocks need not be deep; it has been found that growth of alevins on the surface of a gravel bed is as good as that of buried alevins, provided they are shielded from light. The tanks are allowed to overflow so that emergent fry may escape in a natural manner and, after traversing about 45 m of shallow ditches, pass directly into Netarts Bay.

The 1969 to 1970 season was the first season of operation for the hatchery so a complete evaluation cannot be essayed for several years, but initial rates of hatching and emergence are encouraging (Table 3).

TABLE 3.   SURVIVAL OF PINK AND CHUM SALMON EGGS IN TANKS AT THE NETARTS BAY HATCHERY, OREGON

SPECIES	MONTH SPAWNED	NO. EGGS STOCKED	NO. EMERGENT FRY	PER-CENTAGE SURVIVAL	DURATION OF FRY EMERGENCE
Pink salmon	September	600,000	393,000	65.6	Early December– mid- February
Chum salmon	November	280,000	225,000	80.4	February 8– March 13

SOURCE: McNeil (1970).

The better survival of chum salmon than pink salmon was attributed to poor water circulation in pink salmon tanks, which was corrected by the time chum salmon tanks were placed in operation. The source of eggs may also have been a factor; chum salmon eggs were obtained from Whiskey Creek, which supplies water to the hatchery, while pink salmon eggs had to be imported from Alaska. On the other hand, chum salmon survival might have been even higher were it not for 5000 eggs removed for experimental use. The behavior patterns of fry emerging from the tanks were quite similar to those of wild fry.

FRY REARING

As one might expect, young of chinook, coho, and sockeye salmons must be reared for some time in freshwater if stocking is to be effective. It has been shown that, for sockeye salmon at least, feeding must begin very early in the rearing process. If food is not present in quantity immediately upon absorption of the yolk sac, impaired growth will result. For this reason, food is usually presented to alevins as well as fry and fingerlings, although actual feeding by alevins may be insignificant.

The general policy with regard to chum and pink salmons seems to be to release the fry upon emergence. It matters little whether they are released directly into an estuary or into a stream; in the latter case they will make their way to the sea with little delay. There is, however, some evidence that artificially produced chum and pink salmon fry may also benefit from short-term rearing in freshwater.

The first efforts along these lines were apparently some experiments carried out in the early 1960s by the Washington State Department of Fisheries, in which pink salmon fry reared in freshwater for a short while returned to their hatchery as adults at the excellent rate of 0.7%.

The Washington experiments might have led to practical application were it not for the difficulty of feeding the fry. On the one hand, a diet of brine shrimp (*Artemia*) resulted in a survival rate of almost 90% but poor growth; on the other hand, fry fed on meat and fish suffered losses of 40 to 50%, but attained sizes two to three times as large as those fed on brine shrimp. Returns from stocking indicate that overall survival, from egg to adult, did not differ between the two groups.

An experiment in freshwater rearing of chum salmon fry was carried out in 1965 by researchers at the Sakhalin Branch of the Soviet Pacific Fisheries Research Institute. A 620-m² controlled-flow pond with a maximum depth of 50 cm and a bottom similar to that of a natural chum salmon spawning area was stocked with 1.68 million fry and harvested after 37 days. The fry apparently obtained much of their food from natural sources, but supplemental feed was provided in the form of walleye pollock (*Theragra chalcogrammus*) eggs, placed on underwater feeding tables two or three times daily. Growth was excellent, and at the time of release it was determined that the stomachs of the pond-reared fry were three times as full as those of comparable fry from conventional hatcheries or natural stocks. It is not known if further use was made of this technique.

Applied rearing programs for Pacific salmon fry and fingerlings have usually been carried out at low levels of intensity. One of the most extensive such operations was the controlled natural rearing program of the State of Washington's Department of Fisheries. Under this program, salmon fry were stocked in protected natural and seminatural bodies of fresh and salt water, and allowed, after a suitable time interval, to escape to the sea. The waters stocked ranged from 0.4 to 240 ha in surface area; in 1964, 8 saltwater areas, totalling 174 ha, and 25 freshwater areas, totalling 785 ha, were in use.

Predatory fishes were usually eliminated by use of rotenone or toxaphene, and in some cases it was thought necessary to regulate the water level. Otherwise, management of the rearing areas was slight, though some attempts were made in small ponds to artificially turn over the water by means of aeration, and supplemental feeding and/or fertilization with one or more of a bewildering variety of substances was sometimes employed. In 1946, chinook, coho, chum, and pink salmons were all stocked in controlled natural rearing areas (Table 4), and a few sockeye salmon had been stocked in previous years, but fall chinook and coho salmon predominated, largely due to their paramount importance in the increasingly valuable sport fishery.

TABLE 4.   NUMBERS OF SALMON STOCKED BY THE WASHINGTON STATE DEPARTMENT OF FISHERIES IN "CONTROLLED NATURAL REARING AREAS" IN 1964

TYPE AREA	NO. OF AREAS	TOTAL AREA IN USE (HA)	SPECIES AND NO. STOCKED				
			COHO	FALL CHINOOK	SPRING CHINOOK	CHUM	PINK
Fresh-water	25	785	1,759,536	3,210,572	105,860	—	—
Salt water	8	174	1,006,351	2,510,472	—	1,524,216	1,539,136
Total	33	959	2,765,887	5,721,044	105,860	1,524,216	1,539,136

SOURCE: Crutchfield et al. (1965).

An economic analysis in 1965 of the controlled natural rearing program indicated a benefit-cost ratio of only 0.13:1. Although resources such as salmon populations can never be adequately evaluated in monetary terms, this definitely cast doubts on the advisability of the program. Biological and economic evaluations were also made of individual rearing areas. Standing crops of coho salmon, where they could be determined, ranged from 1.1 to 407 kg/ha; populations just prior to seaward migration ranged from 29 to 39,840 smolts/ha. Less data are available for the other species stocked, but coho salmon are generally assumed to be the most productive species in environments such as those presented by the controlled natural rearing areas.

In several cases there is reason to believe that more intensive management would have resulted in increased production, but application of the appropriate measures, particularly fertilization and water level control, would often have conflicted with recreational and other uses of the bodies of water involved. On the basis of individual economic evaluations of 29 of the rearing areas, it was recommended to the Department of Fisheries that the controlled natural rearing program be discontinued in at least 12 with a combined surface area of 288 ha. This recommendation, along with consideration of the marginal character of some of the other areas, the costs of maintaining stations to take spawn from the adults which return to the rearing areas and transporting the fertilized eggs to hatching facilities, and the lack of control by the state over shoreline development, pollution, and so on, eventually led to a reevaluation of the entire program and rejection of the concept of controlled natural rearing, except for experimental purposes or where it can be combined with conventional hatchery operations.

The Oregon Fish and Game Commission has experimented with somewhat more intensive pond rearing of chinook and coho salmon. Preliminary indications are that freshwater ponds show promise, provided the fish can be given supplementary feeds. The results with saltwater ponds are less auspicious because of the prevalence of the as yet incurable disease vibriosis.

The Washington Department of Fisheries has also reared fair numbers of salmon, mostly fall chinook salmon, in hatcheries, but, as of 1963, overall average returns to the hatcheries were only 0.1%. In several instances, however, returns of 0.8 to 1.0% were recorded at different hatcheries, suggesting that the rearing process might be improved considerably. Two major improvements have been made since the inception of the fall chinook salmon rearing program: First, long, narrow raceways, as used in trout hatcheries, have been abandoned in favor of recirculating ponds, which produce stronger fingerlings, for reasons analogous to those favoring "natural" spawning beds over raceways for growth and health of alevins (see pp. 472–474). Second, research has indicated that the optimum length of the rearing period is about 90 days. Experiments in Washington showed that young fall chinook salmon liberated after 17, 45, and 91 days of rearing returned at rates of 0.01, 0.1, and 0.7%, respectively. In a similar experiment, survival of fingerlings reared for 90 days was 21 times greater than that of fry. Survival was little increased by further extension of the rearing period, and improvement was more than offset by increased feeding costs. Coho salmon fry released at a size that would comprise 77/kg showed a return of 1 to 2%, whereas 7 to 8% of fish released at twice that size returned.

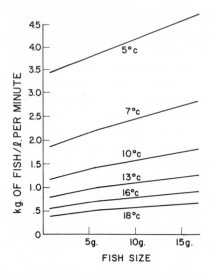

FIG. 2.   Carrying capacity of oxygen-saturated water at normal activity level of fingerling chinook salmon as affected by water temperature and fish size. (After Burrows and Combs, 1968.)

Stocking rate is a crucial factor in any form of fish culture, including hatchery culture of salmon. Although many variables are involved in determining maximum and optimum population densities in hatcheries, and each hatchery must work out its own requirements, Fig. 2, which illustrates the carrying capacity of oxygen-saturated water for chinook salmon fingerlings, may be useful.

Many other factors affecting survival of hatchery-reared chinook and coho salmons are being studied by fishery biologists in Washington and Oregon, among them nutrition, genetics, time, place, and manner of release, and acclimatization to sea water.

Studies of young salmon adapted to salt water are presently being made at the Netarts Bay hatchery. Under study, in addition to fall chinook salmon, are the presumably more salinity-tolerant hybrids of chinook salmon with chum or pink salmon. Pilot production studies are just under way, but laboratory experiments have already shown that chinook salmon fry reared in water of about 15‰ salinity for 20 days or longer will survive in full-strength sea water. Growth of such fish is inhibited, however, thus an intermediate exposure to water with a salinity of 25‰ is being tested. The effects of rearing to a larger size before introduction to salt water and nutrition in salt water are also being studied. Preliminary indications are that the hybrids ♀ *Oncorhynchus gorbuscha* × ♂ *Oncorhynchus tshawytscha* and ♀ *Oncorhynchus keta* × ♂ *Oncorhynchus tshawytscha* may adapt rapidly to sea water and render gradual acclimatization unnecessary.

Rearing of sockeye salmon fry is done relatively seldom. The only sizable sockeye salmon culture program in the United States goes on at the Leavenworth National Fish Hatchery, Leavenworth, Washington, where conventional Pacific salmon hatchery techniques are used up to the time of absorption of the yolk sac. Free-swimming fry are kept in indoor troughs until about the first of April, when the weather starts to warm up, then stocked in outdoor ponds. The young are not overwintered at Leavenworth but are released in October into nearby Lake Wenatchee. Sockeye salmon have been successfully cultured for the first full year of life at the Winthrop National Fish Hatchery and released in the spring for their seaward migration.

DISEASES AND NUTRITION

Two problems which have beset Pacific salmon culturists since fry rearing first became part of the culture process are disease and nutritional deficiencies. Among the diseases observed in hatchery populations of Pacific salmon are bacterial cold water disease, bacterial gill disease, columnaris and a similar saltwater disease caused by the bacterium *Sporocytophaga*, furunculosis, vibriosis, kidney disease, mycobacteriosis or "fish tuberculosis," protozoan diseases, white-spot or coagulated yolk disease, worm infections, and a number of specific virus diseases. Some of these, particularly the virus diseases, are presently incurable, but most can be treated with fair to excellent results.

The traditional approach of salmon culturists to disease problems is drug treatment (the reader interested in specifics is referred to Davis, 1953) but present emphasis is on preventive measures. Chemical prophylaxis, as applied in trout hatcheries, is used, and vaccination is beginning to play an important role, but genetic selection and nutrition are increasingly being recognized as major factors in disease prevention.

One link between nutrition and pathology of young salmon was made clear when the practice of feeding the carcasses of spent spawners and other fresh offal was generally abandoned. The incidence of disease, particularly mycobacteriosis and some of the viruses, was dramatically reduced. The concept of using salmon carcasses as feed is extremely attractive economically, since at a hatchery there will always be large supplies available annually at no cost. They can now be safely converted to feed by means of a process developed and used at the University of Washington. The carcasses are made into meal and combined with an ortholisate also derived from the spent salmon. The ortholisate may be prepared very rapidly by use of synthetic enzymes, but acid digestion by natural salmon enzymes takes only a matter of hours. The two ingredients

INGREDIENT	PERCENTAGE	SPECIFICATIONS
Meal mix		
Herring meal	28.0	Canadian or domestic, minimum 70% protein, *Full meal,* containing the herring solubles
Cottonseed meal	15.0	Prepressed, solvent extracted, not more than 0.04% free gossypol, minimum 50% protein
Dried whey-product	5.0	Foremost MNC or equivalent
Shrimp or crab meal (preferably shrimp)	4.0	Maximum 3% salt (NaCl), crab meal to contain minimum 30% protein
Wheat germ meal	4.0	Minimum 25% protein and 7% fat
Corn distiller's dried solubles	4.0	
Vitamin premix	1.5	See Table 7
Wet mix: Two or more of the following six fish products, provided that none shall exceed 15% of the total diet, and 1/32- and 3/64-inch pellets shall contain at least 7.5% tuna viscera	30.0	
Albacore tuna viscera (*Thunnus alalunga*)		Without heads and gills, with livers
Turbot (*Atheresthes stomias*)		Whole
Salmon viscera		Without heads and gills, with livers, pasteurized
Herring (*Clupea pallasii*)		Whole, pasteurized
Dogfish (*Squalus acanthias*)[a,b]		Whole, with livers
Hake (*Merluccius productus*)[a,c]		Whole, pasteurized
Kelp meal	2.0[d]	Algit

TABLE 5. *(continued)*

INGREDIENT	PERCENTAGE	SPECIFICATIONS
Soybean or herring oil	6.0[b,c]	Stabilized with 0.333% BHA-BHT (1:1); soybean oil to be fully refined; herring oil to contain less than 5 ppm DDT (including analogs), less than 2% free fatty acids, and not to be alkaline reprocessed
Choline chloride	0.5	Liquid, 70% product
	100.0	

[a] Not to be used in 1/32- or 3/64-inch pellets.
[b] Delete 0.3 parts oil for every 10 parts dogfish in total diet.
[c] Add 0.5 additional parts oil for every 10 parts hake in total diet.
[d] May be deleted from 1/32-inch pellets.
SOURCE: Leith (personal communication).

are combined with a commercial congealing agent and blended to make a highly nutritious and palatable feed.

Most chinook, coho, and sockeye salmon hatcheries now feed prepared, pelleted, and powdered feeds in appropriate sizes. The prevalent feed in hatcheries in the United States is the Oregon Moist Pellet, whose composition is shown in Table 5. (The Oregon Fish Commission subsequently developed a starter mash for fry, the composition of which is shown in Table 6.) Experiments conducted at various hatcheries in Oregon indicate that chinook and coho salmon fed the Oregon pellet diet return to the hatchery at significantly higher rates than fish fed on fresh meat.

Vitamins are of course important in nutrition and disease prevention in Pacific salmons, and their requirements have been fairly well worked out. Hatcheries usually meet these requirements by adding a preprepared crystalline vitamin premix to the feed. The composition of this premix and other parts of hatchery diets changes continually as more and more is learned about the nutritional requirements of salmonids. The composition of the premix used in Oregon as of November, 1970, is shown in Table 7, while Table 8 lists the known symptoms of vitamin deficiencies in salmonids.

Most of the important advances in nutrition of Pacific salmons have been the result of work done by Oregon State University's Seafood Technology Group or by Dr. John Halver and his staff at the United States Bureau of Sport Fisheries and Wildlife's Western Fish Nutrition Labo-

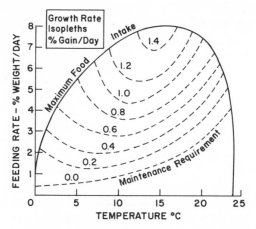

Fig. 3. Relationship of feeding rate, temperature, and growth in yearling sockeye salmon. (After Brett, Shelbourne, and Shoop, 1969.)

ratory, Cook, Washington, and further important developments may be expected from these sources.

The quantity of feed is as important as the quality; overfeeding is not only uneconomic but may result in water pollution, while the effects of underfeeding are obvious. The optimum feeding rate is partially determined by water temperature, over which the culturist often has little or no control. Figure 3 illustrates the relationship of feeding rate (with a prepared feed similar to the Oregon Pellet), temperature, and growth in weight of yearling sockeye salmon.

From Fig. 3 it would appear that the optimum temperature for growth is in the neighborhood of 15°C, and this figure does coincide with a general physiological optimum which allows greatest tolerance to oxygen debt, maximum sustained swimming speed, and maximum total metabolic activity. However, it should be borne in mind that this temperature is also more favorable for most disease organisms than the colder temperatures customarily encountered by salmon in nature.

EVALUATION OF HATCHERY PROGRAMS AND THEIR POTENTIAL

Economic and biological evaluations of hatchery propagation programs for fishery animals have generally been negative, as they have been for the efforts of the early Pacific salmon culturists in Canada, the United States, and Japan. Recent improvements in technique, however, have greatly altered the picture. Let us take, for a theoretical example, the chinook salmon, which has been more thoroughly studied than any other salmon from a hatcheryman's point of view. If we assume that the average fecundity of a chinook salmon female is about 4000 eggs, and that virtually 100% of these are fertilized, we have, in nature, 4000 embryos per spawning pair. Under normal conditions, 99% of these will perish before reaching the sea as smolts. On the average, 90% of the 40 remaining smolts will not survive to sexual maturity, leaving only 4 fish. The old salmon fishery management policy was to attempt to predict the size of the annual run and harvest 50%, leaving 50% to reproduce. Applying this to the remnant of our original 4000 embryos, and assuming a 1:1 sex ratio, we are left with 2 of the potential 4000 fish for human use, and 2 to perpetuate the species. Since actual survival varies greatly from year to year, depending on factors which are little understood, the task of the fishery manager is not enviable. In an exceptionally good year, the spawning streams may be crowded with excess fish, and the manager may be accused of wasting fish. In a poor year, a high fishery quota may result in depletion of the stock, and again it is the fishery manager who is "to blame" for not being omniscient.

Let us now assume that our hypothetical pair of chinook salmon spawners were artificially bred in a hatchery. We may safely allow for only 40% mortality to the smolt stage (in practice, much better results are often achieved), which leaves 2400 fish, instead of 40, to go to sea. Even if survival of the hatchery-reared smolts is only half as good as that of natural smolts (5% instead of 10%), the result will be 120 mature adults. The fishery manager now has 118 salmon to put on the table, in the can, on the hook, or wherever we want them, and a far greater margin of error to work with.

Have these theoretical calculations any basis in reality? Yes, according to an economic study conducted by the United States Bureau of Commercial Fisheries. During 1961 to 1964, BCF biologists fin-clipped 30 million chinook salmon smolts liberated from 12 hatcheries in Oregon and Washington which, together, account for 95% of the total hatchery production of the species. Starting in 1963, they examined an average of 23% of the total sport and commercial catch of chinook salmon from Monterey Bay, California, to Alaska for marked fish. It was concluded that the Spring Creek hatchery alone had produced 22% of the 1965 catch of the Columbia River gill net fishery, plus 15% of the sport catch. In concrete terms, the Spring Creek hatchery had contributed about 675,000 kg of 3-year-old chinook salmon to the Columbia River fisheries, while other hatcheries contributed as much as 225,000 kg each. The benefits of the program are also felt outside Oregon and Washington and, in 1966, it was calculated that 25,000, or 11.1% of the 225,000 4-year-old chinook

TABLE 6. FORMULA FOR THE "OREGON STARTER MASH" SALMON FRY DIET, NOVEMBER, 1970 (INGREDIENT SPECIFICATIONS SAME AS IN TABLE 5)

INGREDIENT	PERCENTAGE
Meal mix	
Herring meal	46.0
Wheat germ meal	10.0
Dried whey-product (MNC)	10.0
Corn distiller's dried solubles	4.0
Vitamin mix	1.5
Wet mix	
Albacore tuna viscera	8.0
Turbot, salmon viscera, or herring	8.0
Kelp meal (Algit)	2.0
Soybean or herring oil[a]	10.0
Choline chloride	0.5
	100.0

[a] To contain 0.333% BHA-BHT (1:1).

TABLE 7. OREGON VITAMIN PREMIX FOR SALMON DIETS, GUARANTEED MINIMUM ANALYSIS AND VITAMIN SOURCE LIMITATIONS, NOVEMBER, 1970

VITAMIN	GUARANTEED MINIMUM ANALYSIS PER POUND	SOURCE LIMITATION
Ascorbic acid	27.0 g	
Biotin	18.0 mg	
B_{12}	1.8 mg	
E	15,200.0 IU	Water dispersible, alpha tocopheryl acetate
Folic acid	215.0 mg	Not zinc folate
Inositol	17.0 g	Not phytate
Menadione	180.0 mg[a]	Menadione sodium bisulfite complex or menadione dimethylpyrimidinol bisulfite
Niacin	5.7 g	
d-Pantothenic acid	3.2 g	Calcium pantothenate or choline pantothenate
Pyridoxine	535.0 mg	
Riboflavin	1.6 g	
Thiamine	715.0 mg	

[a] The biological activity of 180 mg of menadione is required.

484

TABLE 8.   VITAMIN DEFICIENCY SYNDROMES IN SALMONIDS

VITAMIN	SYMPTOMS
Thiamin	Poor appetite; muscle atrophy; convulsions; instability and loss of equilibrium; edema; poor growth
Riboflavin	Corneal vascularization; cloudy lens, hemorrhagic eyes; photophobia; dim vision; incoordination; abnormal pigmentation of iris; striated constructions of abdominal wall; dark coloration; poor appetite; anemia; poor growth
Pyridoxine	Nervous disorders; epileptiform fits; hyperirritability; ataxia; anemia; loss of appetite; edema of peritoneal cavity; colorless serous fluid; rapid postmortem rigor mortis; rapid and gasping breathing; flexing of opercles
Pantothenic acid	Clubbed gills; prostration; loss of appetite; necrosis and scarring; cellular atrophy; gill exudate; sluggishness; poor growth
Inositol	Poor growth; distended stomach; increased gastric emptying time; skin lesions
Biotin	Loss of appetite; lesions in colon; coloration; muscle atrophy; spastic convulsions; fragmentation of erythrocytes; skin lesions; poor growth
Folic acid	Poor growth; lethargy; fragility of caudal fin; dark coloration; macrocytic anemia
Choline	Poor growth; poor food conversion; hemorrhagic kidney and intestine
Nicotinic acid	Loss of appetite; lesions in colon; jerky or difficult motion; weakness; edema of stomach and colon; muscle spasms while resting; poor growth
Vitamin B_{12}	Poor appetite; low hemoglobin; fragmentation of erythrocytes; macrocytic anemia
Ascorbic acid	Scoliosis; lordosis; impaired collagen formation; altered cartilage; eye lesions; hemorrhagic skin, liver, kidney, intestine, muscle
p-Aminobenzoic acid	No abnormal indication in growth, appetite, mortality

SOURCE: Halver (1970).

salmon taken in the Canadian troll fishery had originated in the hatcheries under study. The overall benefit-cost ratio of the hatchery program was estimated at 2.3:1.

Similar evaluations of coho salmon hatchery programs were initiated in Oregon in 1965. The following year, the Columbia River experienced its largest coho salmon run in history, an event which was generally attributed to improved hatchery techniques. A tentative economic evaluation of the better coho salmon hatcheries suggests benefit-cost ratios (not taking account of amortization) of 3.5 to 5.5:1.

Proponents of chum and pink salmon hatcheries are optimistic that artificial propagation can also contribute effectively to North American fisheries for these species. Speaking of chum and pink salmon, William J. McNeil, head of Oregon State University's Pacific Fisheries Laboratory, has asserted that "Hatcheries and other artificial methods can increase egg to fry survival by 4 to 80 times over that in natural streams." Even if we assume, as we did for chinook salmon smolts, that hatchery fry of chum and pink salmon are only half as viable as wild stock, the potential for supplementation of the fishery is obvious. McNeil has calculated that in nature a high quality spawning bed, receiving a flow of 10 ft^3/sec, may, with an optimum density of spawners, produce 2.2 million fry/ha. In an improved spawning channel under similar conditions, this production might be increased by a factor of 5. But, based on experience at Oregon State's Netarts Bay facility, a well-designed hatchery, using the same amount of water and far fewer adult fish, might produce 220 million fry/ha. The latter figure, based on a small prototype study, is of course highly speculative but indicates that tremendous improvement over present methods of salmon production is conceivable.

Again there are supporting data, this time from the Japanese chum salmon fishery. On the island of Hokkaido there are approximately 160 chum salmon spawning streams; 65 of the larger ones, which together contribute 98% of the total run, have hatcheries on their courses. Of the chum salmon which enter these 65 streams, an average of 57.7%, or approximately 56% of the total Hokkaido run, are spawned artificially at the hatcheries. It has been estimated that this 56% of the spawners contribute an average of 77.2% of the adult chum salmon returning to Hokkaido. No data are available for the success of natural reproduction of chum salmon in Hokkaido, but if eggs from undammed and unpolluted streams in southeastern Alaska are used as a basis of comparison, it can be shown that the rate of return from hatchery-produced eggs in Hokkaido is 80% greater. Even in Honshu, which is near the southern limits of the range of *Oncorhynchus keta,* and where pollution, damming, and other detrimental human interventions are common, hatchery reproduction is 21% more successful than natural reproduction in southeastern Alaska streams.

As elsewhere, hatchery fry in Hokkaido suffer greater mortalities in nature than wild fry, in this case by a factor of 2 to 2.5. If this mortality could be substantially reduced, then the already impressive contribution of Japanese chum salmon hatcheries could be increased. Japanese econo-

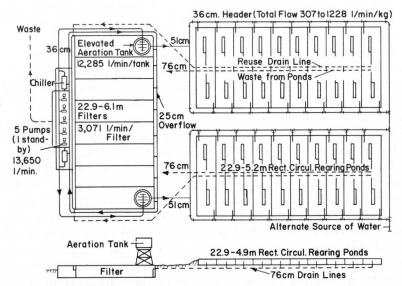

FIG. 4.   Recirculating water and environmental control system for rearing salmonid fishes. (After Burrows and Combs, 1968.)

mists currently claim a benefit-cost ratio of 14 to 20:1. At present, it is estimated that chum salmon runs in Hokkaido can be maintained by investing only 6.2% of the average value of the annual commercial catch.

ENVIRONMENTAL CONTROL IN HATCHERIES

Future developments in Pacific salmon hatchery techniques are likely to emphasize environmental control. At present, the salmon culturist is much more at the mercy of the elements than the pond fish culturist. Far greater control could be exercised over a number of environmental variables if hatchery waters were recirculated. It would then be possible to regulate the water temperature, maximize dissolved oxygen content, remove ammonia, and sterilize the water as a prophylactic measure.

Figure 4 is a schematic drawing of a recirculating hatchery system designed to provide a controlled environment for salmonids. Water which has been through the system once is collected from the rearing ponds in the reuse drain line and passed to filters containing a 1.2-m-deep layer of sharp rock covered by a 0.3-m layer of crushed oyster shell. The rock and oyster shell bed is ideal for the growth of nitrifying bacteria, which convert nitrogenous wastes to nitrates, harmless within the recirculating system. The oyster shells also prevent the accumulation of free carbon dioxide. This technique, though commonplace in sewage treatment, has rarely been applied in aquaculture.

Water from the filters is pumped under pressure through aspirators into the aeration tank and returned to the rearing ponds by gravity flow. In this way, the dissolved oxygen content can be maintained at a minimum of 6 ppm without causing supersaturation with nitrogen. Surface agitation, as used in conventional hatcheries, is equally effective as a means of aeration but, in a recirculating system, also allows possibly lethal concentrations of nitrogen and carbon dioxide to be built up.

Temperature control devices may be built into the filtration system, and these can be automated at little additional cost. Ultraviolet sterilization may also be incorporated in the filtration process. Sterilization is not feasible in conventional hatcheries, but with a recirculating system such as described, where only 2 to 10% of the total volume of water need be replaced to make up for evaporation, leakage, and so forth, ultraviolet radiation may economically be employed. The reader interested in more detailed information on the design of hatchery recirculating systems should consult Burrows and Combs (1968).

Systems such as the one just described undoubtedly have the capability to increase the number and size of young salmon produced, with the desirable side effect of reducing pollution from hatcheries. Before salmon culturists commit themselves to wholesale adoption of water recirculation and environmental control, however, they should pause to consider the possibility that young salmon reared under uniformly favorable conditions may be less well equipped to cope with the natural environment than the less numerous survivors of more primitive propagation systems.

HATCHERY-SPAWNING CHANNEL COMBINATIONS

Although artificial spawning channels and hatcheries have been discussed separately, it is not to be assumed that these two approaches to Pacific salmon conservation are mutually exclusive. On the contrary they may profitably be combined where more spawners are available than can be handled at a hatchery, or where it is desired to improve the quality of artificially produced fry. An outstanding example of the integration of various methods of salmon culture and fishery management is the Big Qualicum River Development Project of the Canadian Department of Fisheries.

The Big Qualicum River, which flows 4.3 km from Horne Lake into Qualicum Bay on the east coast of Vancouver Island, has in the past supported a fair natural run of chum salmon and a few chinook and coho salmon. Its potential for producing salmon fry has, however, historically been reduced by annual scouring floods. In 1963, this problem was eliminated by the installation of a control dam on Horne Lake, and a bypass for flood water from a tributary stream. Discharge, which formerly varied annually from 15 to 7000 ft³/sec, is now controlled at 40 to 700 ft³/sec. In addition to enhancing the survival of natural spawns, this has made it possible to construct and maintain a 0.3 km long, 12 m wide artificial spawning channel, two spawning and egg incubation stations, and several rearing ponds. In addition, pink salmon have been introduced to the watershed. In 1969, the combined production of chum salmon fry from the hatchery, the artificial channel, and natural spawning areas was a record 53.5 million. Such combinations of responsible human intervention with optimum use of the natural environment should be encouraged wherever Pacific salmons are spawned.

PROSPECTUS FOR PACIFIC SALMON HATCHERIES

In 1963, it was calculated that to sustain anything approaching the all-time-high North American salmon catch of 152 million fish (all species combined) would require 266 billion eggs, at hatchery survival rates. Obviously no system of artificial propagation could begin to fertilize and hatch that many eggs. It was assumed that hatcheries, while even then locally important, were destined to play only a supplementary role in the overall salmon production picture. The calculations leading to the figure given were, however, based on the old 50–50 escapement formula. As we have seen, if hatcheries are a major source of stock, it should be possible to harvest well over 50% of a run without depleting the source. At the rates of survival typical of the best present-day hatcheries, a fishery yield of 152 million fish could theoretically be maintained through the fertilization of 4 to 5 billion eggs—the product of perhaps 3 million pairs of spawners.

The spectacular improvements in techniques for artificial propagation of *Oncorhynchus* spp. should delight anyone concerned with the future of the Pacific salmon fisheries, and maximum use should be made of them, but they should not be construed as an excuse for environmental complacency. Even if it becomes possible to support all the commercial and sport salmon fisheries solely on the basis of artificial propagation, the continuing destruction of natural salmon runs by pollution, dams, and thoughtless urbanization should be resisted. Natural runs should be maintained not only for esthetic reasons but also to preserve wild stocks and gene pools. It is not inconceivable that some sort of unforeseen calamity, local or widespread, could negate many of the gains thus far made in hatchery culture of *Oncorhynchus* spp. In such an event, the continued existence of large natural runs might avert a fishery disaster. In maintaining the stocks of Pacific salmon, as in all ecologically based ventures, we should remember that diversity tends toward stability.

Improvements in salmon propagation techniques have served to increase or renew interest in a number of other culture and management techniques, including transplantation, genetic selection, and commercial salmon farming. Transplantation of fish to new waters does not of itself fall within our definition of aquaculture, but in the case of the Pacific salmons this practice is so intimately connected with culture as to merit treatment here.

TRANSPLANTATION OF PACIFIC SALMON STOCKS

WITHIN THE NATURAL RANGE OF *Oncorhynchus* SPP.

Since the early days of Pacific salmon culture, *Oncorhynchus* spp. have been introduced all over the world in regions far removed from their native haunts but, until recently, more notable results have been obtained from introductions within the coastal tributaries of the North Pacific. Such local transplantations may be carried out not only to introduce a species to a body of water where it is not native but to bolster production of weak runs. This sort of stocking is particularly applicable to pink salmon management, since pink salmon have a rigid two-year life cycle, and in most streams either the odd or even year run predominates; in some pink salmon streams one or the other cycle may be completely missing. Attempts have been made, with at least tentative success, to establish runs in such cases.

One of the earliest introductions of a Pacific salmon species to a new body of water occurred in 1923, when fry of anadromous *Oncorhynchus masu* were stocked in the landlocked freshwater Lake Biwa, Japan, where they became established. Since that time, attempts have been made to establish most if not all the *Oncorhynchus* spp. in watersheds where they are not native. Until recently, young fish have been stocked, but today it is not unusual to plant eyed eggs in an artificial spawning channel, as in the introduction of pink salmon to the Big Qualicum River, British Columbia, or to propagate and rear the salmon entirely within a hatchery, in which case the stream serves only as a water supply and a highway for the spawners.

Another early and notable instance of successful transplantation took place in the Cedar River, Washington. In 1932, Loyd Royal, then chief biologist for the Washington Department of Fisheries, obtained and hatched sockeye salmon eggs from the Baker River in northwestern Washington and stocked the fingerlings in the Cedar. If the planting were successful, it was reasoned that substantial numbers of adult sockeye salmon should have homed to the river starting in 1935. Few appeared, and the experiment was written off as unsuccessful. Then, unexpectedly, in 1960, 12,000 sockeye salmon entered the river. Apparently a small run had been present but undetected for 25 years. Subsequent runs have numbered as high as 75,000 and are considered to contribute importantly to commercial fisheries.

What makes the Cedar River run all the more surprising is the present status of the watershed. In the years between the initial stocking and the present, the banks of the Cedar River, which flows through the city of Seattle, have become the most heavily industrialized area in the Pacific Northwest, accommodating, among other industries, the giant Boeing Aircraft complex.

The nursery area for Cedar River sockeye salmon is Lake Washington, which is nominally a "warm water" lake. Like most such lakes in temperate regions it stratifies during the summer, remaining warm only in the surface layer, or hypolimnion. Beneath the hypolimnion is a layer of water called the thermocline, characterized by temperatures which decline rapidly with depth. The thermocline is little used by warm water fishes, but is well suited to young sockeye salmon. Possible utilization by salmon of this largely unexploited portion in other lakes should be explored.

Transplantation, in combination with hatchery techniques, may even be employed to establish salmon runs in waters which are unsuitable for natural propagation. If conditions can be artificially provided such that hatching of eggs and rearing of the young salmon can be carried out, all that is necessary is that ripe spawners be able to reach the hatchery. Such methods have considerable potential in culture of sockeye salmon, which have the most restrictive spawning requirements among the Pacific salmons. A case in point is the establishment by United States Bureau of Sport Fisheries and Wildlife personnel of a sockeye salmon run in the Methow River, Washington. Unlike all natural sockeye salmon spawning streams, the Methow River has no lakes on its course. The run, which originates at the Winthrop National Fish Hatchery, is maintained solely by artificial propagation and rearing of the offspring until time for seaward migration.

OUTSIDE THE NATURAL RANGE OF *Oncorhynchus* SPP.

No sooner had it been demonstrated that Pacific salmon eggs could be artificially fertilized and hatched, than demands for stock issued from all over the world. The result was a plethora of ill-advised stocking programs, which accomplished little more than to feed large numbers of *Oncorhynchus* eggs and fry to various predators. For example, between 1873 and 1934 more than 6 million chinook salmon, mostly fry, were stocked in the Great Lakes of the United States and Canada, but there is no indication that the survivors ever successfully reproduced. The introduction of chinook salmon to New Zealand resulted in a sport fishery which is sustained entirely on the basis of natural reproduction, but all the other early transplants failed.

It is understandable, then, that more than a few eyebrows were raised when the Michigan Department of Conservation announced its intention to stock coho salmon in Lakes Michigan and Superior, starting in 1966. By now, all fishery and fish culture workers are aware that it appears that this introduction will be successful beyond the fondest hopes of its original proponents. It may be instructive to examine the factors which differentiate the stocking program initiated in Michigan from earlier attempts to introduce the coho salmon and other Pacific salmons in the region.

Certainly the time was ripe for introduction of a predatory fish. Construction in 1932 of the Welland Canal, which circumvents the impassable Niagara Falls, opened the four upper Great Lakes to invasion by a number of previously absent fish species. One of these, the parasitic sea lamprey (*Petromyzon marinus*), eventually became so abundant as to effect the virtual extinction of the lake trout (*Salvelinus namaycush*) and the burbot (*Lota lota*), the major piscivorous species in lakes Huron, Michigan, and Superior. After years of research, an effective method of controlling lamprey populations was discovered and implemented, but by that time the valuable commercial and sport fisheries for lake trout had been destroyed. At the same time, another invader, the alewife (*Alosa pseudoharengus*), finding itself virtually free of predation, began to multiply at an enormous rate. By 1966, it was estimated that 90% by weight of the total fish population of Lake Michigan was composed of alewives. Annual production was estimated at 90 million kg, much of which was wasted in periodic mass mortalities.

Alewives have virtually no value as fishery animals, but it was reasoned that they would provide an excellent source of forage for the populations of predatory fishes which could be established now that the lamprey menace was eliminated. Of course lake trout figured prominently in the list of prospective alewive predators, but lake trout are confined to very deep water for most of the year. To exert control over the alewife at all levels, it was decided to stock a relatively shallow water predator. Already, the pink salmon, accidentally introduced to Thunder Bay, on the Canadian side of Lake Superior in 1956, was establishing runs of both sides of the lake. The pink salmon showed promise of importance as a commercial fish but is not generally considered a good sport fish. Consequently, several other species, including the native rainbow trout (*Salmo gairdneri*) and brook trout (*Salvelinus fontinalis*), as well as the chinook salmon, coho salmon, *Oncorhynchus masu,* and the striped bass (*Morone saxatilis*), were considered.

It was eventually decided to concentrate on coho salmon, continue normal hatchery operations with rainbow trout and brook trout, and postpone introduction of the other species. Selection of the coho salmon as the chief agent of alewife control was based on its desirability as a sport fish, its hardiness in egg and fry stages, the availability of eggs, and the belief that it is one of the least expensive of the salmonids to propagate and rear.

Although sea-run coho salmon stock has never been successfully adapted to freshwater, Michigan officials profited from the experience of earlier culturists by correcting several mistakes, as follows:

1. Stock was selected, insofar as possible, from environments similar to the new environment. In at least one instance eyed eggs were obtained from a run which enters the Swanson River on Alaska's Kenai Peninsula. A fair percentage of the coho salmon hatched in the Swanson River never return to the sea but pass their lives in freshwater lakes.

2. Fingerlings rather than fry were planted. The Oregon Fish Commission, which supplied the first lot of eggs, stipulated as a precondition that all young be reared at least until they averaged 55/kg.

3. The early Pacific salmon transplants were often of small numbers of fish scattered among many bodies of water. This time, initial plantings consisted of no less than 200,000 fish each, all released in one place. In this way it was felt that individuals which were poorly endowed genetically to cope with the new environment could be weeded out without seriously reducing the reproductive potential in a given spawning stream. Only three of the most promising streams were stocked, and fingerlings were released where they would have easy access to the lake.

4. It has been found that Michigan hatchery water supplies are deficient in iodine. Inadequate supplies of this trace element might well be crucial to eggs and fry of anadromous salmonids, so iodine was added at al hatcheries where coho salmon were kept.

5. The original intent of stocking Pacific salmons in the Great Lakes had been to establish natural spawning runs. Physiological difficulties in adaptation to freshwater notwithstanding, Michigan officials also hoped for natural spawning runs but did not intend to rely solely on natural reproduction for maintenance of fisheries. It was estimated that to sup-

port significant fisheries for coho salmon, rainbow trout, and brook trout in the upper Great Lakes would require production of 40 million yearlings annually, but that natural reproduction could not be relied on to provide more than 25% of that total. Accordingly, spawn-taking stations and hatchery facilities were constructed on all the streams to be stocked.

The primary aim of the Michigan Department of Conservation was to provide a new sport fishery. Only if considerable stock was left over was it intended that coho salmon contribute to the rejuvenation of the upper Great Lakes commercial fishery. The rapid growth (30 g to 1.8 kg in 7 months) and good survival of the first plants suggested that there might soon be more than enough fish to go around. Results in subsequent years have done nothing to dim the early optimism, as coho salmon have been taken in unexpected numbers and in sizes averaging better than normal on the Pacific coast. Though the specter of mercury pollution clouds the Great Lakes commercial fishery picture, it appears that there will be no problem in allotting adequate numbers of coho salmon to sport fishermen, commercial fishermen, and hatcheries. It is still too early to say whether coho salmon will be able to reproduce satisfactorily, by natural or artificial means, without spending time at sea, but indications to date are favorable, although pesticide residues in spawning streams may pose a threat.

The original stocking program continues to expand into Michigan streams, while the other Great Lakes states and the Canadian province of Ontario, encouraged by the apparent success of Michigan's efforts, have proceeded to stock coho salmon in all of the Great Lakes, even grossly polluted Lake Erie, with generally positive preliminary results. Interest has been generated in the coho salmon overseas as well, and hatchery stocking programs have been set up in Chile and Spain.

Michigan has followed up the coho salmon program by introducing chinook salmon to Lakes Michigan and Superior. The initial stocking of 850,000 fingerlings in April, 1967, was made solely to provide a source of eggs for future operations, but it is hoped that in a few years it will be possible to open fisheries. Since the chinook salmon is larger and more exclusively piscivorous than the coho salmon, it may prove to be even more valuable as an alewife predator. Further, its shorter average residency in streams offers promise of economies in hatchery operation, as well as the establishment of natural runs in streams unsuited for year-round habitation by salmon.

There continues to be some interest in introducing *Oncorhynchus masu*, or perhaps *Oncorhynchus rhodurus*, to the Great Lakes, and the Ontario Department of Lands and Forests has undertaken one experimental stocking of *Oncorhynchus masu* in a small lake. Indications from this preliminary study and observations of *O. masu* in Japanese waters are that it would occupy about the same biological and economic niches as the coho salmon but might adapt more readily to a strictly freshwater life cycle.

Chum and pink salmon have rarely been stocked in waters distant from their native habitats, but in the 1930s and again in the 1950s Russian biologists, traditionally the world's primary enthusiasts of fish transplantations, distributed these species from Pacific waters to the European waters of the Soviet Union. The early introductions were unsuccessful, but later transplants, of fish from Sakhalin Island to the Barents and White seas, were accompanied by the construction of three hatcheries on the Kola Peninsula, and appear to have been effective. Fry releases at these hatcheries have increased from 3.5 million in 1959 to 30 million in 1962, and pink salmon now appear not only in Barents Sea catches, but in the fisheries of Finland, Norway, Iceland, and the United Kingdom. Since 1962, a similar program has been carried out in the Caspian Sea, and it appears that it may also be successful.

SELECTIVE BREEDING

Among salmonid fishes, selective breeding has been most highly developed in the rainbow trout (see Chapter 20). In 1949, Lauren Donaldson of the University of Washington's College of Fisheries, who was chiefly responsible for the development of the famous "supertrout," decided to begin similar experiments with *Oncorhynchus* spp. The first step was to establish spawning runs in a small stream which flows into Union Lake, on the Washington campus. Runs of chum, coho, sockeye, and fall chinook salmon were successfully developed, but only the chinook salmon have returned in numbers sufficient to permit selection. The survivors of the 1949 plant of chinook salmon returned as 4-year-old adults in 1953 and provided twice as many eggs as were needed to maintain the run. Thus it was possible to cull up to 50% of the spawners. Eventually it became possible to stabilize the production rate at 250,000 selected fingerlings/year. Excess eggs and fingerlings are stocked in streams where it is hoped they will contribute to fisheries. Donaldson and his co-workers have selected for the following five characteristics.

1. Better growth and larger size. Size selection, both advertent and inadvertent, has long taken place at salmon hatcheries. Under hatchery conditions, one male can be used to fertilize the eggs of many females,

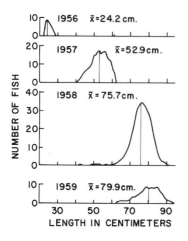

NUMBER OF FISH

1956 x̄=24.2 cm.

1957   x̄=52.9cm.

1958 x̄ = 75.7cm.

1959 x̄=79.9cm.

30   50   70   90
LENGTH IN CENTIMETERS

Fig. 5. Length distribution of 1955 brood year chinook salmon that returned to collecting ponds at the University of Washington in four different years. (After Donaldson and Menasveta, 1961.)

thus it has become customary to release or destroy all small males. At the Spring Creek hatchery in Washington this practice alone has resulted in the establishment of a run of fall chinook salmon which average larger in size than local wild fish and return in three years instead of four. The shift to a three-year cycle of course increases the chances of survival and provides a theoretical 25% increase in the number of fish that can be produced.

Donaldson has continued to select males for size, and also selects large females and those which produce large eggs. The bulk of the selected stock now return after 3 years at sea, and some have returned after only 2 years, or even 1 year, an event unprecedented in the history of chinook salmon culture. The 1- and 2-year-old fish average considerably smaller than the older fish, but, even in the early years of the project, the length of the 3-year-olds did not differ greatly from that of 4-year-olds (Fig. 5).

The average size of returning 3-year-olds has increased over the years, albeit with great fluctuations, some of which are now felt to be due to injudicious choice of marking methods in certain years, particularly the removal of the pectoral fins of 1963 brood-year fingerlings (Fig. 6).

2. Fecundity. The more eggs per female, the less females are needed to maintain a run, and the more that can safely be harvested by the fisheries. The average number of eggs per female in Donaldson's experiments increased by 34.5%, from about 3800 in 1960 to about 4900 in 1965 to 1967.

3. Time of return. It has been demonstrated that salmonid stocks can be selected to spawn early or late in the season. Since optimum conditions are unlikely to prevail throughout the spawning period, this has some implications for survival. It has also been suggested that salmon might

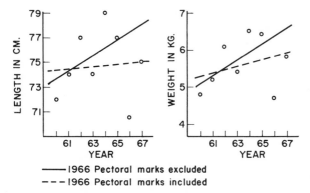

FIG. 6.   Average size of 3-year-old chinook salmon which returned to collecting ponds at the University of Washington. (After Donaldson, 1970.)

be selected, for the sake of convenience to hatchery workers, to spawn during a short period of time rather than dribbling in over a fairly long season. Before placing much emphasis on this sort of selection, culturists should pause to consider that every time the spawning season is shortened, the potential for a population disaster is increased.

4. Environmental tolerance. Better results have been achieved in this regard with rainbow trout, but there is evidence that the tolerance of chinook salmon for high temperatures, and probably other environmental variables, can be increased.

5. Disease resistance. Again, most of the progress has been made with trout, but the vigor of Donaldson's chinook salmon has certainly been a factor in the deemphasis of drug therapy as a solution to disease problems in Pacific salmon hatcheries.

Further selection is exercised at the fry and fingerling stages. Since each batch of embryos, representing the offspring of one spawning pair, is usually maintained separately to the free-swimming stage, batches of eggs and larvae exhibiting poor survival or other undesirable characteristics may be rejected. For economic reasons selected lots of fingerlings must be combined for rearing, so selection at that stage is confined to the routine hatchery process of size grading (see p. 409).

One possible result of a selection program such as that just described is better survival, and this has been achieved. Unselected wild stock have continued to return to the Washington campus on a 4-year cycle at a rate of about 0.1%. Survival from egg to adult of the selected 3-year fish, however, has averaged 1.0 to 3.25%.

The effects of genetic selection are often slow to be felt. Although the history of Donaldson's fall chinook selection program spans only 21 years, as compared to the 38 years he has devoted to the rainbow trout, it has already been shown that genetic selection stands to contribute greatly to the Pacific salmon fisheries. Donaldson and others have also studied hybridization, but at the present time it appears to offer less promise. Of the several hybrids of *Oncorhynchus* spp. which have been produced, only the crosses of chinook salmon with chum or pink salmon, as mentioned, have thus far shown promise for fish culture.

SALMON FARMING

In Donaldson's opinion, demonstration of the possibilities in selective breeding of chinook salmon points the way toward true husbandry of *Oncorhynchus* spp. Intensive culture of freshwater trout is a sizable industry in several countries, but farming of anadromous salmonids has been slow to develop, despite the great improvements in hatchery techniques.

Extensive but unsuccessful experiments in salmon farming were carried out in various locations around the perimeter of Puget Sound, Washington, during the 1950s. The plan was to plant hatchery-reared chinook and coho salmon fingerlings in saltwater lagoons, where it was hoped that they would grow to marketable size on natural food, with no supplementation. In fact, however, the food supply in most of the lagoons was inadequate, which, along with heavy mortalities due to high temperatures, oxygen depletion following die-offs of algae, and disease, eliminated any chance of the operators realizing a profit. Attempts were also made at about the same time to raise salmon in concrete raceways, but the high summertime incidence of vibriosis rendered such schemes unfeasible. A team of microbiologists at Oregon State University under the direction of John Fryer are currently attempting to develop an oral vaccine for vibriosis which would render the prognosis for future experiments of this sort more favorable.

To date, intensive commercial culture of Pacific salmon occurs only in Japan. At Okachi Bay, Miyagi Prefecture, 9-month-old salmon (species not known), measuring about 14 cm in length and weighing about 75 g, are stocked in large floating net cages anchored about 150 m from shore. The fingerlings are fed for 9 months on pellets made of fish meal and starch with added vitamins and harvested at a length of about 35 cm and a weight of 0.5 kg. The conversion ratio achieved in this operation is 1.3:1.

Recent experiments by workers at the Seattle Biological Laboratory of the National Marine Fisheries Service (formerly Bureau of Commercial Fisheries) in collaboration with Ocean Systems, Inc. suggest that net cages could be similarly used along the North American coast (for a general discussion of fish culture in floating cages, and its advantages, see p. 559). In Puget Sound, coho salmon fingerlings have been reared from an average weight of 0.02 kg to 0.28 kg in 6 months. A conversion ratio of 1.5:1 was obtained, using conventional pelleted feed, the incidence of disease was negligible, and all indications are that the system would be commercially feasible. Such an undertaking has since been initiated by Ocean Systems, Inc. who experimentally marketed their first cage-reared salmon in 1972.

Actual commercial application would require the construction of private hatcheries for propagation of salmon and rearing to the fry stage. From the hatchery facilities, the fry would be transferred to nursery ponds where they could gradually be adapted to salt water while being reared to fingerling size. The adaptation process could be eliminated if the growing cages were located in freshwater, which seems feasible at least for chinook, coho, or sockeye salmon. The first application of these techniques may come not only on the Pacific coast but also in Nova Scotia, where Sea Pool Fisheries, Ltd., plans to include hatchery-spawned and reared chinook salmon among its products.

The prospective market size of salmon reared in net cages is 0.20 to 0.35 kg, at which weight they might be more competitive with cultured trout than with fishery-caught salmon. The flesh of the coho salmon experimentally reared by Bureau of Commercial Fisheries biologists was adjudged superior to trout, a factor that lends further economic credibility to commercial net cage culture schemes.

The large-scale culture in Japan of yellowtail (*Seriola quinqueradiata*) has shown that it is possible to profitably rear large fish in net cages, and it is not inconceivable that Pacific salmon might be reared to a size comparable to those harvested by commercial fishermen. In some places this or other cage culture programs might be facilitated by mixing heated water from power plant effluents with sea water to maintain optimum growing temperatures year-round.

PROSPECTUS

The prospectus for culture of the Pacific salmons is one of growth, expansion, and intensification. It is to be hoped that those responsible for the management of this huge resource will have the wisdom not to concern themselves with determining *the* most efficient way to produce salmon. It is the combination of aggressive conservation of natural runs, maximum supplementation by means of hatcheries and artificial spawning channels, and encouragement of true salmon farming that is likeliest to perpetuate *Oncorhynchus* spp. in their role as important contributors to human well-being.

REFERENCES

BAMS, R. A. 1967. Differences in performance of naturally and artificially propagated sockeye salmon migrant fry as measured with swimming and predation tests. J. Fish. Res. Board Can. 24:1117–1153.

BAMS, R. A. 1970. Evaluation of a revised hatchery method tested on pink and chum salmon fry. J. Fish. Res. Board Can. 27(8):1429–1452.

BEVAN, D. E. 1963. An experiment to improve an Alaskan salmon spawning area. Proceedings of the 13th Alaskan Science Conference.

BORGESON, D. P., and W. H. TODY (Eds.). 1967. Status report on Great Lakes fisheries. Michigan Department of Conservation, Fisheries Management Rep. No. 2.

BRANNON, E. L. 1965. The influence of physical factors on the development and weight of sockeye salmon embryos and alevins. International Pacific Salmon Fisheries Commission, Project Rep. 12. 26 pp.

BRETT, J. R., J. E. SHELBOURNE, and C. T. SHOOP. 1969. Growth rate and body composition of fingerling sockeye salmon, *Oncorhynchus nerka*, in relation to temperature and ration size. J. Fish. Res. Board Can. 26:2363–2394.

BURROWS, R. E., and B. D. COMBS. 1968. Controlled environments for salmon propagation. Prog. Fish Cult. 39(3):123–136.

CROKER, R. S., and D. REED (Eds.). 1963. Report of Second Governor's Conference on Pacific Salmon, Seattle, Wash., Jan. 7–10, 1963. State Printing Plant, Olympia, Washington.

CRUTCHFIELD, J. A., K. B. KRAL, and L. A. PHINNEY. 1965. An economic evaluation of Washington State Department of Fisheries controlled natural rearing program for coho salmon (*O. kisutch*). State of Washington, Department of Fisheries, Research Division. 27 pp. (Mimeographed.)

DAVIS, H. S. 1953. Culture and Diseases of Game Fishes. U. of California Press, Berkeley. 332 pp.

DISLER, N. N. 1953. Development of autumn chum salmon in the Amur River. Trudy Soveshchaniya po Yoprosan Losoevoge Khozgaistva Dal rego Vostoka 9:129–143. (Trans. IPST Cat. No. 763.)

DONALDSON, L. R. 1966. Selective breeding of salmonoid fishes and application of the results to management. Paper presented at 11th Pacific Science Congress. Tokyo, 1966.

DONALDSON, L. R. 1970. Selective breeding of salmonoid fishes. *In* W. J. McNeil (Ed.), Marine Aquiculture. Oregon State University Press, Corvallis, pp. 65–74.

DONALDSON, L. R., and D. MENASVETA. 1961. Selective breeding of chinook salmon. Trans. Am. Fish. Soc. 90(2):160–164.

GUNSTROM, G. K. 1970. Canadian mariculture facility begins operation. Am. Fish Farmer 2(1):8–11.

HALVER, J. E. 1970. Nutrition in marine aquiculture. *In* W. J. McNeil (Ed.), Marine Aquiculture. Oregon State University Press, Corvallis, pp. 75–102.

HALVER, J. E., E. T. MERTZ, D. C. DELONG, and R. E. CHANCE. 1960. The vitamin and amino acid requirements of salmon. Fifth Industrial Congress on Nutrition, Abstracts, No. 191.

HUBLOU, W. F., J. WALLIS, T. B. MCKEE, D. K. LAW, R. O. SINNHUBER, and T. C. YU. 1959. Development of the Oregon pellet diet. Res. Briefs, Fish Comm. Ore. 7(1):28–56.

International Pacific Salmon Fisheries Commission. 1970. Annual Report—1969. 33 pp.

Japan Fisheries Resource Conservation Association. 1966. Propagation of the chum salmon in Japan.

KISABURO, T. 1965. Returning effect and its limit resulting from the artificial propagation of chum salmon in Hokkaido. Bull. Jap. Soc. Sci. Fish. 31:323–326.

LEITRITZ, E. 1960. Trout and salmon culture (hatchery methods). California Department of Fish and Game, Fishery Bull. 107.

LUCAS, K. C. 1960. The Robertson Creek spawning channel. Can. Fish. Cult. 26:3–23.

MCDONALD, J. G. 1969. Distribution, growth, and survival of sockeye fry (*Oncorhynchus nerka*) produced in natural and artificial stream environments. J. Fish. Res. Board Can. 26:229–267.

MCLARNEY, W. O. 1964. The coastrange sculpin *Cottus aleuticus:* structure of a population and predation on eggs of the pink salmon *Oncorhynchus gorbuscha.* M.S. Thesis, School of Natural Resources, University of Michigan. 83 pp.

MCNEIL, W. J. 1962. Variations in the dissolved oxygen content of intragravel water in 4 spawning streams of southeast Alaska. U.S. Fish and Wildlife Service, Special Scientific Report—Fisheries, No. 402. 15 pp.

MCNEIL, W. J. 1964. Redd superimposition and egg capacity of pink salmon spawning beds. J. Fish. Res. Board Can. 21(6):1385–1396.

MCNEIL, W. J. 1969. Survival of pink and chum salmon eggs and alevins. *In* T. G. Northcote (Ed.), Symposium on Salmon and Trout in Streams, H. R. McMillan Lectures in Fisheries. University of British Columbia, pp. 101–117.

MCNEIL, W. J. 1970. Culture of salmon acclimated to sea water. Paper presented to Western Division American Fisheries Society and Western Association of State Game and Fish Commissioners, July 13–16, 1970. Victoria, British Columbia.

MCNEIL, W. J., and W. H. AHNELL. 1964. Success of pink salmon spawning relative to size of spawning bed materials. U.S. Fish and Wildlife Service, Special Scientific Report—Fisheries, No. 469. 15 pp.

NECE, R. E. 1961. Hydraulic design and construction details of salmon spawning channel improvement areas on Indian Creek and Harris River. University of Washington, Fisheries Research Institute, Circular 145. 4 pp.

NIKOL'SKII, G. V. 1961. Special Ichthyology. Israel Program for Scientific Translations, Jerusalem.

PALMER, D. D., H. E. JOHNSON, L. A. ROBINSON, and R. E. BURROWS. 1951. The effect of retardation of the initial feeding on the growth and survival of salmon fingerlings. Prog. Fish. Cult. 13(2):55–62.

PHILLIPS, R. W., and E. W. CLAIRE. 1966. Intragravel movement of reticulate sculpin, *Cottus perplexus,* and its potential as a predator on salmonid embryos. Trans. Am. Fish. Soc. 95(2):210–212.

TODY, W. H., and H. A. TANNER. 1966. Coho salmon for the Great Lakes. Michigan Department of Conservation, Fisheries Management Report No. 1.

VANSTONE, W. E., J. R. MARKERT, D. B. LISTER, and M. A. GILES. 1970. Growth and chemical composition of chum (*Oncorhynchus keta*) and sockeye (*O. nerka*) salmon fry produced in spawning channel and natural environments. J. Fish. Res. Board Can. 27:371–382.

WAHLE, R. J. Salmon hatcheries, an encouraging supplement to a Pacific coast fishery. Unpublished manuscript, U.S. Bureau of Commercial Fisheries.

WORLUND, D. D., R. J. WAHLE, and P. D. ZIMMER. 1969. Contribution of Columbia River hatcheries to harvest of fall chinook salmon (*O. tshawytscha*). U.S. Bureau of Commercial Fisheries, Fishery Bull. 67(2).

Interviews and Personal Communication

BOYD, F. C. Canadian Department of Fisheries.

DONALDSON, L. R. College of Fisheries, University of Washington.

LEITH, D. A. Program Leader, Nutrition-Physiology Investigations, Oregon Fish Commission, Research Laboratory, Sandy, Oregon.

MCNEIL, W. J. Head, Pacific Fisheries Laboratory, Marine Science Center, Oregon State University, Newport, Oregon.

WEED, J. Washington Department of Fish and Game.

22

Culture of Coregonid Fishes in the Soviet Union

The whitefish and ciscoes (*Coregonus* spp.) and the sheefish or inconnu (*Stenodus leucichthys*) were formerly considered to constitute the family Coregonidae but are now placed in the Salmonidae, along with the trouts, salmons, chars, and graylings. Culture of cold water fishes for food is an offshoot of sport fish culture, thus culture of the coregonids is less advanced than that of the trouts and salmons, which include some of the world's most popular game fishes.

The whitefish group does contain a number of important food fishes but, although various whitefish were cultured in the United States and Europe during the hatchery craze of the late nineteenth and early twentieth centuries, it is only in the Soviet Union that hatchery culture of coregonids on a large scale persists. Some species of *Coregonus* are artificially propagated in Sweden for stocking in rivers where natural spawning is hindered or prevented by power dams. The Russian coregonids include species occupying a wide variety of ecological niches, thus Soviet culturists also increasingly employ them in pond culture. Table 1 lists some of the characteristics of the species cultured in the Soviet Union.

TABLE 1. SPECIES OF COREGONID FISHES CULTURED IN THE SOVIET UNION, THEIR DISTRIBUTION AND CHARACTERISTICS

SPECIES	DISTRIBUTION	SIZE	HABITAT	FOOD
Coregonus albula (European cisco)	British Isles to the Baltic Sea, also in some lakes in the Volga Basin; introduced to many parts of the U.S.S.R.	To 46 cm	The oceans and warm and cold lakes; marine populations anadromous; cultured in ponds	Zooplankton
Coregonus autumnalis (Arctic cisco)	Arctic Ocean and its tributary rivers, also Lake Baikal	To 60 cm and 2.5 kg	Mostly found in or near estuaries, also rivers and lakes; anadromous	Zooplankton and fish fry
Coregonus lavaretus (common whitefish)	Northern Europe south to Switzerland and east to Siberia; introduced to Japan and southern U.S.S.R.	To more than 50 cm	Many subspecies, each with a characteristic habitat; found in the sea, rivers, and deep and shallow waters of lakes; marine forms anadromous; lacustrine forms may be anadromous; cultured in ponds	Zooplankton as young; mostly benthos as adults

TABLE 1. *(continued)*

SPECIES	DISTRIBUTION	SIZE	HABITAT	FOOD
Coregonus muksun (muksun)	Baltic Sea to Siberia	To 70 cm	Bays and estuaries; anadromous	Mostly benthos, some zooplankton, also dead fish
Coregonus nasus (broad whitefish)	Siberia to northern Canada	To 16 kg	Rivers, lakes, and occasionally estuaries	Benthos
Coregonus peled (peled)	Siberia; introduced widely in U.S.S.R. as far south as the Ukraine	To 50 cm and 5 kg	Lakes and rivers; cultured in ponds	Planktonic crustaceans
Stenodus leucichthys (inconnu)	Arctic Ocean and its tributaries; Alaska from the Yukon River north; Caspian Sea	To 1 m and 40 kg	Oceans, rivers, and lakes; usually anadromous	Fish

Hatchery culture of coregonids is not dissimilar to trout culture (see Chapter 19) in its methodology. The major function of coregonid hatcheries in the Soviet Union is to produce fry for stocking both as a means of replenishing stocks depleted by overfishing and/or dams which interfere with spawning migrations and for introduction to new waters. No recent data are available, but Table 2 lists numbers of fry stocked for 1954. The effects of this program are not known.

TABLE 2.   NUMBER OF FRY OF FIVE SPECIES OF COREGONID FISHES STOCKED IN THE SOVIET UNION IN 1954

SPECIES	NUMBER OF FRY STOCKED (MILLIONS)
European cisco	108.6
Arctic cisco	131.0
common whitefish	127.2
peled	1.2
inconnu	2.8
Total	370.8

SOURCE: Ovchynnyk (1963).

Coregonus muksun and *Coregonus nasus* are not used for stocking natural waters, but Soviet hatcherymen have succeeded in hybridizing these species with *Stenodus leucichthys* to produce a large fish said to show promise for stocking in the lakes, rivers, and reservoirs of Siberia. Another promising coregonid hybrid is *Coregonus albula* × *Coregonus lavaretus maraenoides,* which matures in the rather short time span of 2 years and is suitable for culture in ponds 5 ha in area or larger.

Although the peled has been considered for use in monoculture, the principal use of coregonids in pond culture is to supplement production of rainbow trout (*Salmo gairdneri*) or common carp (*Cyprinus carpio*). The European cisco, which feeds on zooplankton, is the species most frequently stocked, but the peled, also a zooplankton feeder, or the benthos-feeding broad whitefish or common whitefish may be stocked in addition to, or in place of, the peled. In the vicinity of Leningrad, common whitefish mature in 2 years in ponds, as opposed to the 4 to 5 years required in nature.

Most of the Russian coregonids are native to northern waters, but the European cisco is found as far south as the Volga Basin and is stocked in ponds with common carp in the Ukraine. Soviet biologists have had considerable success in acclimating cold water fishes to the warm waters of the southern Soviet Union and the common whitefish and peled are now also stocked in Ukrainian ponds, where they do well at temperatures of up to 25°C.

Coregonid fishes seem firmly entrenched in Soviet fish culture, both for intensive cultivation as a luxury food and for low-intensity polyculture in new impoundments, where they are stocked as part of an artificial ecosystem. Both types of culture are limited by the normally slow maturation of coregonids; at present insufficient data are available to predict whether culture of coregonids, in or out of the Soviet Union, will expand.

REFERENCES

NIKOL'SKII, G. V. 1961. Special Ichthyology, 3rd ed. Translated from the Russian, Israel Program of Scientific Translation, Jerusalem.

OVCHYNNYK, M. 1963. Soviet fish culture. Fish. News Inter. 2(3):279–282.

23

Culture of Smelts and Ayu

Though the smelts (family Osmeridae), small, slender fishes found in marine and freshwaters throughout much of the northern hemisphere, are popular food fishes over most of their range, they have generally been successful in maintaining their numbers in the face of fishery exploitation. Thus smelt culture, though practiced on a small scale in Japan and the Soviet Union, has not assumed great commercial importance.

The ayu (*Plecoglossus altivelis*), sometimes classified in the Osmeridae and sometimes in its own family, Plecoglossidae, presents a different picture. This smeltlike fish is considered a gourmet delicacy in east Asia and is the basis for a profitable, though fairly primitive, form of fish culture in Japan.

Ayu are anadromous, but pass most of their one-year life cycle in freshwater. After spending their first few months of life in the sea (or in Lake Biwa, where ayu are landlocked), 20- to 30-mm fry enter rivers in the spring and gradually migrate upstream, feeding on benthic organisms, principally algae, as they go. By late summer, they have reached lengths of up to 25 cm and begin to migrate back to the lower reaches of the river, where spawning occurs in September to November. Each female deposits 10,000 to 100,000 tiny, extremely adhesive eggs on a gravel bed in 30 to 50 mm of swiftly flowing water. After spawning, the adults die. The eggs hatch in 10 to 24 days at 15 to 23°C, and the newly hatched fry are carried out to sea.

Ayu have been spawned in captivity, but results have been poor, and the practice has not yet been adopted by commercial culturists. Artificially spawned ayu have occasionally been used to stock new waters. For this purpose, ripe fish, captured by means of traps, seines, cast nets, or hook and line, are stripped and the eggs fertilized, using the dry method (see p. 400 for a description of these techniques). Fertilized eggs, in lots of about 20,000, are distributed on 36- × 24-cm mats made of hemp palm bark and placed in floating boxes until they are eyed. Eyed eggs are quite hardy and, when covered with moss, will withstand shipment to distant hatcheries, where they may be hatched in flowing water tanks and the fry stocked in streams. It has been reported, perhaps erroneously, that in Taiwan, ayu eggs have been collected on sunken mats similar to those described above.

Practical ayu culturists in Japan neither strip eggs from adult fish nor collect them from streams but rather rely on fry collected from the sea, estuaries, and rivers during January to May as a source of stock. The rivers of Tokushima Prefecture, located on the Island of Shikoku in the Inland Sea, are the source of two-thirds of the estimated 1500 tons of fry collected annually. About half of these fry are shipped elsewhere in Japan for growing.

Growing of marketable size ayu is a short process and, since stocks of fry are not available year-round, it is often practiced as a seasonal sideline by culturists whose principal product is some other fish. Fry 50 to 65 mm long are preferred for stocking and can be grown to marketable fish 17.5 to 23.0 cm long, weighing about 0.1 kg each, in 90 days at optimum temperatures of 21.0 to 23.5°C.

Ayu are grown in ponds as small as 15 m² in area, raceways, and, very rarely, rice fields; best results have been achieved in raceways, where heavy feeding is less likely to pollute the water. They have also been grown experimentally in a closed recirculating water system designed by A. Saeki of Tokyo University (see pp. 55, 58 for description), but the device has yet to be adopted by commercial ayu culturists. Whatever type of enclosure is used, cultured ayu are heavily fed on prepared foods and harvested from April to September.

The poor conversion by ayu of artificial feed (feed accounts for 56.1% of the average ayu grower's budget) is the principal contributing factor to the extremely high price of cultured ayu. Nevertheless, the demand is increasing rapidly, as are yields and production totals. In 1960, the average yield was 3846 kg/ha; by 1964 it had risen 103% to 7810 kg/ha. The Japanese government ceased to keep production figures on ayu after 1964, but it is estimated that 1810 metric tons were produced by culturists in 1965 and 2420 metric tons in 1967. The majority of this production comes from Honshu, in the vicinity of the cities of Kobe, Kyoto, and Nagoya, but Tokushima Prefecture is also an important culture region.

Ayu culturists count on heavy feeding to produce rapid growth, so that their product hits the market before the fishery product. By August and September, when the market may be flooded with both cultured and wild ayu, the price, while still high, may be less than one-fourth that in April. The better restaurants constitute the principal market, but at such times ayu may also be purchased for home use.

As long as supplies of fry hold out, the prospectus for ayu culture as a profitable business looks bright. If artificial spawning of ayu becomes commercially feasible, expansion appears even more likely, and prices may be reduced. The principal stumbling block is the difficulty of providing enough natural food during October to November, when the fry are too small to utilize artificial feeds and pond water is too cold to maintain abundant benthos.

Biologists at a number of fisheries stations in Japan have sought to circumvent this problem by controlling the photoperiod to speed up sexual maturation and spawning so that fry will be on artificial feed by the time cold weather sets in. Success was achieved by shading ponds containing brood stock to shorten the photoperiod during May, returning to a natural regime until mid-June, then suddenly or gradually shortening the daily photoperiod by 5 to 6 hours. Fish so treated spawned about 2 months ahead of schedule, but the quality of eggs was low, as can be seen in Table 1.

TABLE 1.   QUALITY OF EGGS AND LARVAE PRODUCED BY NATURALLY AND ARTI-FICIALLY RIPENED AYU (*Plecoglossus altivelis*)

	EGGS SHED BY LIGHT-TREATED AYU	EGGS STRIPPED FROM RIVER AYU
Diameter of egg (mm)	0.78–0.98	0.83–1.20
Hatching (%)	65.17	74.39
Size of fry (mm)	3.8 –6.3	6.0 –7.3
Mortality immediately after hatching (%)	47.0	0.1

SOURCE: Kuronuma (1966).

If the process of inducing early spawning is perfected, it will not only increase the likelihood of artificial spawning assuming an important role in commercial culture of ayu but will give the culturist a further advantage over the fisherman by enabling him to produce fish for market before the fishery opens. A similar advantage may be attained by means of delayed spawning, which has already been used by some culturists to supply ayu as late as New Year's Day. Spawning may be delayed very simply, by lighting the culture area so that the photoperiod is extended to 14 to 18 hours from July or early August to October. Ayu thus treated have been kept alive until February to April of the following year without experiencing reproductive activity.

Outside of Japan, ayu are found in Korea, China, and Taiwan but are not cultured in these countries. It is possible that in Taiwan there would be a demand for such a luxury product as cultured ayu, but it seems unlikely that ayu culture will excite much interest elsewhere in Asia at this time.

One species of true smelt, the wakasagi or pond smelt (*Hypomesus olidus*), is also cultured in Japan. It is an annual, anadromous fish similar to the ayu but reaches only half the ayu's size. Wakasagi, which spawn in January to March, have been artificially spawned and used to stock streams in the manner described for ayu. Wakasagi fry may also be stocked, along with other fishes, in farm ponds used as irrigation basins for rice fields. Since wakasagi are plankton feeders, they are considered most suitable for deep ponds of this type.

Another smelt, *Osmerus eperlanus,* is the object of hatchery culture for stocking of natural waters in some parts of the Soviet Union. *O. eperlanus,* like most smelts, is fished during its spawning run, and fishermen on the Neva River achieve at least a theoretical economy by supplying eggs and milt from their catch for hatchery use in providing fry to be stocked back into the river.

The immediate prospects for widespread culture of smelt and smelt-like fishes seem remote, since the individual fish are rather small, the important fisheries are not seriously threatened, and with the notable exception of the ayu, most species bring low prices.

REFERENCES

BROWN, E. E. 1969. The fresh water cultured fish industry of Japan. University of Georgia, College of Agriculture Experiment Station, Research Rep. 41.

CRAWFORD, D. R. 1926. Hatching of Japanese ayu eggs in the United States. Copeia 1926(152):113–114.

HORA, S. L., and T. V. R. PILLAY. 1962. Handbook on fish culture in the Indo-Pacific region. FAO Fisheries Biological Technical Paper 14. 204 pp.

KURONUMA, K. 1956. Fish production in fresh water units of Japan. Proc. 8th Pac. Sci. Cong. IIIA, 1089–1104.

KURONUMA, K. 1966. New systems and new fishes for culture in the Far East. FAO World Symposium on Warm Water Pond Fish Culture. FR: VIII-IV/R-1.

NIKOL'SKII, G. V. 1961. Special ichthyology, 3rd ed. Translated from the Russian, Israel Program for Scientific Translation, Jerusalem.

24

Culture of Cyprinids Native to
Europe and Asiatic Russia

SPECIES CULTURED

No family of fishes has contributed so much to aquaculture as the Cyprinidae, 55 species of which are treated in this text. With the exception of the common carp (*Cyprinus carpio*), which holds the distinction of being the world's most widely cultured fish, monoculture of Cyprinids is rare. More often they are stocked as components in polyculture systems which reach a zenith of sophistication in the traditional practices of Chinese carp culture. A large percentage of the cultured cyprinids are native to China, southeast Asia, or the Indian subcontinent; all of these species have been treated under Chinese Carp Culture (Chapter 2) and Culture of the Indian Carps (Chapter 3). The common carp is accorded a chapter to itself (Chapter 1), and the few cultured African cyprinids are discussed along with other fishes under Culture of Native Freshwater Fishes of Africa (Chapter 12). No cyprinids are native to Australia or South America. North America boasts a large number of native cyprinids, but most of them are small and none are esteemed as food fishes, nor have any of them been cultured for that purpose. There remain to be discussed a number of commercially valuable cyprinids native to Europe, the Near East, and Asiatic Russia. Biologists in the Soviet Union have been especially active in culturing these species, both for use in experimental polyculture and for stocking in natural waters, and the bulk of this chapter will deal with Russian work. Table 1 lists the species of cyprinids (other than those discussed in other chapters) which are cultured in the Soviet Union and their outstanding characteristics. Readers interested in a particular cyprinid species should consult the index if it is not to be found here.

PROPAGATION AND STOCKING OF ANADROMOUS FORMS

A number of Russian Cyprinidae are atypical for the family in being more or less anadromous. Some of these species spend most of their time in fresh or brackish water and spawn in the lower reaches of rivers; others feed well offshore in the open saline waters of the Caspian, Aral, and Black seas and make spawning migrations of 1000 km or more. The Aral Sea race of the shemaia (*Chalcalburnus chalcoides*) has become so adapted to salt water that it can reproduce successfully in water with a salinity of 11‰.

All the anadromous cyprinids are threatened, at least locally, by the proliferation of hydroelectric dams on Russian rivers. Among the species which are commercially valuable, six—the shemaia, the roach (*Rutilus rutilus*), the cut-tooth (*Rutilus frisii*), the bream (*Abramis brama*), the vimba (*Vimba vimba*), and the Aral barbel (*Barbus brachycephalus*)—are propagated in hatcheries and released in the hope of ameliorating the situation. Maintenance and improvement of natural spawning grounds are also practiced for roach and bream, and special floating spawning beds have been constructed for bream. The numbers of four of these species stocked in 1954 are listed in Table 2. No more recent data are available, but it can be stated that three changes have been made:

1. More fish of all species are stocked.
2. The cut-tooth and the Aral barbel have been added to the list of species stocked.

TABLE 1. CULTURED CYPRINID FISHES (OTHER THAN *Cyprinus carpio*) NATIVE TO EUROPE AND ASIATIC RUSSIA, THEIR DISTRIBUTION, CHARACTERISTICS, AND ECONOMIC VALUE

SPECIES	DISTRIBUTION	SIZE	HABITAT AND SPAWNING HABITS	FEEDING HABITS	ECONOMIC VALUE
Abramis brama (bream)	All of Europe east of the Pyrenees and north of the Alps, also basins of the Black, Caspian, and Aral seas; widely introduced in Asiatic Russia	To 70 cm, usually not over 45 cm	Freshwater and semianadromous populations, in deep water; spawns in lower reaches of rivers over weeds, April-July at 17–20°C	Fry: zooplankton; young and adults: benthic animals, especially chironomid larvae; large individuals occasionally take fish	Extremely important commercially
Barbus brachycephalus (Aral barbel)	Basins of Aral and Caspian seas	To 1 m and 20 kg	Open waters, anadromous, spawns June-August	Young: benthic invertebrates, especially chironomid larvae; adults: carnivorous, mostly on aquatic insects	Very important commercially in Aral basin; mostly eaten dried; superior quality for this purpose
Chalcalburnus chalcoides (shemaia)	Basins of Black, Caspian, and Aral seas	To 40 cm	Open waters of seas and some freshwater lakes; winters in streams; spawns in fresh or brackish water in May at 18–20°C	Plankton	Saltwater stocks commercially important; mostly eaten dried; superior quality for that purpose
Leuciscus idus (ide or orfe)	Europe north of the Alps and east of the Rhine to Kolmya River, Siberia	To 75 cm and 4 kg; usually smaller	Upper layers of sluggish rivers and lakes; spawns April-May at 6°C or more	Young: zooplankton; adults: benthic invertebrates, mostly insects; occasionally fish	Commercially important in U.S.S.R. only
Rutilus frisii (cut-tooth)	Basins of Black and Caspian seas	To 60 cm and 6 kg	Mostly freshwater rivers and lakes; spawns April-May over stones	Mollusks, insect larvae, and crustaceans	Commercially important; stocks declining

514

TABLE 1. (continued)

SPECIES	DISTRIBUTION	SIZE	HABITAT AND SPAWNING HABITS	FEEDING HABITS	ECONOMIC VALUE
Rutilus rutilus (roach)	Europe east of Pyrenees and north of Alps; Siberia and Aral Sea Basin	Freshwater stocks to 30 cm or more, migratory stocks to 50 cm and 1 kg	Freshwater stocks in sluggish rivers and lakes, in deep waters, often hide in vegetation; spawns in May at 10°C; migratory stocks in fresh and salt water, may winter in rivers; spawn March-May in sluggish water in lower reaches of rivers	1st year: zooplankton; 2nd year on: aquatic vegetation and benthic animals; large roach may eat fish	Extremely important commercially, particularly in Caspian Sea Basin; cultured in France
Tinca tinca (tench)	Lowland rivers and lakes of Europe, Asian basin of Arctic Ocean; widely introduced in Asia, Australia, North Africa, and North America	To 70 cm and 7.5 kg, usually smaller	Stagnant waters over soft bottoms, prefers heavy growths of vegetation; very tolerant of low O_2 concentrations; spawns on vegetation in May-July at 19–20°C	Fry: zooplankton; young and adults: benthic invertebrates, algae, and vegetable detritus	Important in pond culture, not in fisheries
Vimba vimba (vimba)	Basins of North, Baltic, Black, and Caspian seas	To 35 cm or more	Open waters; winters in rivers in basins of Black Sea and Sea of Azov; spawns on stones in streams and rivers during June-July at 18–25°C	Young: zooplankton; adults: benthic invertebrates	Important commercially, high-quality food fish

515

TABLE 2. NUMBERS OF LARVAE AND YOUNG OF NATIVE CYPRINID FISHES
(OTHER THAN *Cyprinus carpio*) STOCKED IN THE SOVIET UNION IN 1954

	LARVAE	YOUNG FISH
SPECIES	(MILLIONS)	
Bream (*Abramis brama*)	222.9	762.3
Roach (*Rutilus rutilus*)	209.1	3,358.3
Shemaia (*Chalcalburnus chalcoides*)	—	0.8
Vimba (*Vimba vimba*)	7.2	2.3

SOURCE: Ovchynnyk (1963).

3. More young fish and less larvae are stocked, as the latter practice
has been shown to result in poor survival.

POND CULTURE

Sufficient data are not available to estimate the value of rearing and stocking anadromous cyprinids, but there can be little doubt of the success
of the Soviet pond stocking program. Among the fishes discussed in this
chapter, five, the tench (*Tinca tinca*), the roach, the ide or orfe (*Leuciscus
idus*), the vimba, and the shemaia, are of some importance in pond
culture.

TENCH

Some artificially propagated tench (*Tinca tinca*) are stocked in natural
waters in the Soviet Union, and artificial ponds are built for them to
spawn in Spain, but their primary use in Russia and other European
countries is as a supplementary fish in common carp ponds. Tench have
been widely introduced outside Europe but have not achieved much
importance, though they are or have been occasionally cultured in Australia, India, Indonesia, Israel, Japan, and Tunisia.

In Europe, the general practice is to stock 90% common carp and 10%
tench, but more complicated polyculture schemes may also include tench
(see p. 229). Carp-tench culture has been criticized by a number of biologists as being an inadequate means of increasing pond productivity,
but tench possess several desirable attributes, and operators persist in
growing them. Among the advantages of tench for culture, alone or with
carp, are:

1. In feeding, tench probe more deeply into the mud than carp, and
so increase pond productivity by recycling nutrients.

2. Although the food habits of tench overlap with carp to some extent, they utilize natural food items left untouched by carp. (It is not clear just what are the food habits of tench. In Russia they are supposed to eat mainly benthic invertebrates, but Asian stocks commonly consume algae and vegetable detritus, and are said to derive only one-third of their nourishment from animal matter.)

3. Tench may consume partially digested pieces of artificial feed found in carp feces.

4. The oxygen consumption of tench is very low.

5. Tench brings 20% more in the market than does carp.

Disadvantages of tench are:

1. Tench are usually able to reproduce naturally in fish farm ponds, and may thus overpopulate them, in which case they begin to compete with carp.

2. Tench are very easily injured by handling.

3. Tench grow slowly and never obtain the size reached by carp; market demand is chiefly for 120 to 300-g fish.

4. Tench prefer to spend much of their time concealed in aquatic vegetation or buried in mud. Many may remain so hidden when a pond is drained, and thus be lost to the culturist.

It is likely that there will continue to be a small but steady demand for tench in Europe. Since the habits of tench ensure that a major tench fishery will not develop, it is likely that certain culturists, particularly in France and Yugoslavia, where tench culture is traditional, will continue to supply this demand.

In nature, tench spawn on vegetation in still water, during May to June when the temperature reaches 19 to 20°C. The culturist may spawn them under essentially the same circumstances in special 0.5- to 1.0-ha spawning ponds or in ponds also used for the rearing of fingerling or one-summer-old carp. In the latter case, 1 to 2 pairs/ha of 3- to 4-year-old breeders are stocked and the resulting young reared along with the carp until the end of the second summer, at which time the species are separated. If a special spawning pond is used, the bottom should be soft but not too muddy, and it should contain a fair amount of submerged vegetation. Under these circumstances, 20 to 40 pairs/ha may be stocked. The sexes may be distinguished by the size and structure of the pelvic fins, which in the male are larger and have a thickened second ray (Fig. 1).

Whichever method is used, the culturist will experience difficulty in collecting the small, secretive one-summer-old tench. The only way to achieve success is by keeping the drain channels clean at all times and

Fig. 1. Sex distinctions in the tench (*Tinca tinca*). Note the larger pelvic fins with thickened second ray in the male. (After Nikolskii, 1961.)

draining the pond at night, keeping the water level constant by day. Drainage should be so slow as to require 3 to 4 nights to completely drain a 2-ha pond. A process for artificially spawning tench and rearing them in incubators has been developed in Hungary which would eliminate this necessity, but it has yet to be put into practical application.

Tench are not fed directly but nourish themselves on excess carp feed and naturally occurring food items. Growth is slow at best (Table 3); maximum yields of tench in carp-tench ponds reach 80 kg/ha, but 20 kg/ha is more typical.

TABLE 3.   AGE AND GROWTH OF TENCH

AGE (NUMBER OF SUMMERS)	WEIGHT (G)
1	2–20
2	25–150
3	150–450
4	500 or more

Harvesting adult tench is slightly easier than collecting the young; the same techniques and precautions must be used.

OTHER SPECIES

Perhaps the most important cyprinid in Soviet fisheries is the roach (*Rutilus rutilus*); catches exceed even those of the wild common carp. While roach are widely distributed in Russian inland waters, growth in freshwater is very slow, and the fishery is based almost entirely on anadromous populations. Thus it is not surprising to learn that roach culture in the Soviet Union is largely confined to hatchery propagation and stocking of the anadromous form, though freshwater roach are occasionally stocked in ponds. Inclusion of roach in pond polyculture is more common in France, where the roach occupies a position similar to that of the tench. Roach may also be stocked as a forage fish for trout, pike, and others, but more often they are considered to be pests, and nowhere can the species be said to be important in intensive aquaculture. The ide (*Leuciscus idus*) has a long history in fish culture, probably due more to the esthetic qualities of a golden variety, the orfe, native to Bavaria, than to its value as a food fish. Commercial fishing, stocking, and pond culture of the ide are almost entirely confined to the Soviet Union, and there it is of minor importance. It is likely to remain relatively unimportant since it does not seem to present any particular advantages for use in aquaculture.

The vimba and the shemaia enter the polyculture picture, as benthos feeder and plankton feeder, respectively, in the southern Soviet Union. Sufficient information is not available to assess their value in pond culture.

PROSPECTUS

The ranks of the Cyprinidae, in Europe and elsewhere, have by no means been exhausted by fish culturists. In the Soviet Union alone, there are at least 25 species of known commercial value which have not been cultured. Culture of many other species, including most of those mentioned in this chapter, is far from perfected. Fish farmers may never come up with another cyprinid as amenable to culture as the common carp, but they are sure to experiment with new and unusual species, and to succeed with some of them.

REFERENCES

BREDER, C. M., JR., and D. E. ROSEN. 1966. Modes of reproduction in fishes. Natural History Press, Garden City, N.Y. 941 pp.

LOZANO Y REY, L. 1935. Los peces fluviales de España. Mem. Acad. Cien. Exac. Fis. Nat. Madrid, Ser. Cien. Nat., Vol. 5. 390 pp.

NIKOL'SKII, G. V. 1961. Special ichthyology, 3rd ed. Translated from the Russian, Israel Program of Scientific Translation, Jerusalem.

OVCHYNNYK, M. M. 1961. Development of some marine and inland Russian fisheries, and fish utilization. *In* Atlantic Ocean fisheries, pp. 267–273. Fish. News Inter. (Books), London.

OVCHYNNYK, M. M. 1963. Soviet fish culture. Fish. News Inter. 2(3):279–282.

STERBA, G. 1966. Fresh water fishes of the world. Viking Press, New York.

WOYNAROVICH, E. 1966. New systems and new fishes for culture in Europe. FAO World Symposium on Warm Water Pond Fish Culture. FR: VIII-IV/R-3.

YASHOUV, A. 1966. Mixed fish culture—an ecological approach to increase pond productivity. FAO World Symposium on Warm Water Pond Fish Culture. FR: V/R-2.

25

Culture of Sturgeon

Fish culturists and fishery managers must repeatedly acknowledge that one man's gourmet dish is another man's garbage. The failure of the common carp (*Cyprinus carpio*) in the United States and the milkfish (*Chanos chanos*) in Kenya are only two of the many reminders that regional and cultural biases are fully as important as biological and economic efficiency in determining the fate of a fishery or fish culture enterprise.

HISTORY OF THE STURGEON IN THE AMERICAN GREAT LAKES

EARLY DESTRUCTIVE PRACTICES

Few fishes illustrate the vagaries of human taste better than the sturgeons (family Acipenseridae). In Russia and other European countries, sturgeon

520

have been prized for centuries not only for their flesh but especially for their roe, which constitutes the true caviar, the gourmet food par excellence. One might suppose then that the many Europeans who emigrated to the United States and Canada would have regarded the North American sturgeons, and particularly the extraordinarily abundant lake sturgeon (*Acipenser fulvescens*) of the Great Lakes, as a valuable resource. Nothing could be further from fact. Prior to 1885, Americans (who, perhaps significantly, included few persons of Russian ancestry) regarded sturgeon as scarcely edible. In addition, lake sturgeon were accused (justly) of damaging nets intended for the capture of smaller fishes and (unjustly) of being highly predatory on the spawn of other fishes. Even the roe was considered worthless.

If there was such a thing as "management" of the lake sturgeon during that time, its goal was extermination. If sturgeon were used at all it was as fuel or fertilizer. Often their carcasses were not even accorded such minimal dignity, and great piles were burned on the beaches. Such treatment of any animal would stand as a shameful example of needless destruction, but sturgeon are for a number of reasons particularly susceptible to depletion by man:

1. Owing to their great size (to nearly 3 m in length and over 120 kg in weight for the lake sturgeon and to 4 to 5 m and $1\frac{1}{2}$ tons for the European *Huso huso*), populations of sturgeon are never very great in number, compared to those of other fishes.

2. Sturgeon grow very slowly; few of any species enter the breeding population before they are at least 6 years old; females of *H. huso* in Russia's Ural River mature at 18.

3. Sturgeon are unusual among very large fishes in preferring shallow water where they are, of course, very vulnerable to man.

4. Exposure to human predation is particularly severe during spawning, which takes place in shallow riffles.

Thus it is not surprising that, as early as 1850, the abundance of the lake sturgeon was noted to have drastically decreased in some areas. The senseless waste of sturgeon now has ceased, but pollution of the Great Lakes and damming of their tributaries still militate against the lake sturgeon's success so that, although it is not yet extinct, its population today is certainly less than 1% of its original numbers.

DEVELOPMENT OF A FISHERY

Attitudes toward the lake sturgeon started to change around 1855, when the first American caviar began to be produced at Sandusky, Ohio. In

1860, smoked lake sturgeon entered the market, and the flesh of sturgeon, which had once been unsalable, began to increase in price until today it brings the highest price of any freshwater fish in the United States and Canada. The oil and isinglass (a high-quality gelatin obtained from the swim bladder) also became commercially valuable, and by 1880 the lake sturgeon supported an important fishery. Between 1885 and 1890, the sturgeon fisheries of Lake Erie and Lake Huron began a rapid decline, soon to be followed by all the other major producing areas, until by 1915 the total catch was scarcely 10% of that taken during the 1880s.

ATTEMPTS AT ARTIFICIAL PROPAGATION

Once the lake sturgeon had been proven "valuable," after all, the cry went up to save the species. At the time, American fish culturists, their egos inflated by recent breakthroughs in hatchery spawning of trout and other fishes, were advocating artificial propagation and release of young fish as a panacea for declining fisheries. The lake sturgeon was not spared their ill-founded optimism and, for over 40 years, culturists, first in the United States then in Canada, persisted in experimenting with its artificial propagation.

The first sturgeon to attract fish culturists in the United States was not the lake sturgeon but the Atlantic sturgeon (*Acipenser oxyrhynchus*), a marine species which enters rivers from Labrador to the Gulf of Mexico. In 1875, Seth Green, the pioneer American fish culturist, hatched a number of eggs of this species in the Hudson River, using the floating screen-bottomed boxes he had earlier developed for use with shad (*Alosa sapidissima*). Green's eggs, and the sperm to fertilize them, were obtained by surgical removal of the gonads. Until the development of hormonally induced spawning, both American and Russian biologists usually found it necessary to resort to this tactic, it being virtually impossible to obtain eggs from female sturgeon by "stripping," as is done with salmonids, pikes, and other fishes (see p. 400 for details of this technique).

In 1888, the United States Fish Commission began efforts, which continued sporadically for the better part of a decade, to propagate the Atlantic sturgeon of the Delaware River. All attempts failed, due largely to the difficulty of simultaneously securing ripe fish of both sexes.

Experiments with the lake sturgeon were begun at the United States Fish Commission station at Alpena, Michigan, around 1883, and were continued for about 30 years by federal and state agencies and private individuals in various locations throughout the Great Lakes basin. Considerably greater success was experienced in obtaining and fertilizing the eggs of this species and, although the incidence of *Saprolegnia* was unusually high, some hatches were obtained and the fry stocked in various locations. There is no indication that any of these stockings contributed to the fishery. Two attempts were made to ripen sturgeon in captivity but both failed. At the height of enthusiasm over sturgeon culture, in 1898, the United States government seriously considered establishing a hatchery on Lake Erie or Lake Ontario, but by 1912 the Fish Commission had ceased to experiment with sturgeon.

Only two attempts at sturgeon culture were made in Canada. In 1924, C. P. Paulson, superintendent of a federal hatchery at Gull Harbour on Lake Winnipeg, Manitoba, succeeded in fertilizing and hatching lake sturgeon eggs without killing the fish and subsequently released about 8000 fry. Paulson's experience was similar to that of other workers in that he was not able to strip sturgeon, but he was able to secure a few of the eggs expelled by a ripe female which was being lifted from the water, a method which would scarcely be suitable for large-scale operations.

More extensive efforts were made by W. J. K. Harkness, of the University of Toronto, who devoted a lifetime to the study of the lake sturgeon. In 1924, after two previous unsuccessful attempts, he was able to obtain a plentiful supply of ripe sturgeon by journeying to the spawning grounds on the remote Gull River, a tributary of Lake Nipigon, Ontario. Eggs obtained and fertilized by sacrificing the adults were hatched in 5 to 8 days at 24 to 25°C. The fry were reared to the free-swimming stage, but efforts to feed them were unsuccessful, and by 30 days after hatching almost all had died. By 1926 Harkness, too, was forced to give up the idea of artificial propagation of lake sturgeon.

STURGEON CULTURE IN RUSSIA

HISTORY

In Russia, sturgeon have historically been accorded treatment more in keeping with Henry Wadsworth Longfellow's appellation "Mishe-Nahma . . . king of fishes." Nevertheless, Russian biologists have for some time had cause for concern over sturgeon populations. The decline in sturgeon catches seems to be directly related to emphasis by the Soviet regime on hydroelectric power, and the consequent construction of dams, sometimes with inadequate fishways, on the major Russian rivers. Overfishing is probably also a factor.

Since the Soviet Union supplies 94% of the world's sturgeon, it is not surprising that concern has been translated into action. Actually, experimental sturgeon culture in Russia dates back to before the revolution, the first incubation station having been established in 1913, but it is only since the development of hormonally induced spawning that large numbers of sturgeon have been hatched and stocked.

SPECIES PROPAGATED

Of the 25 or so species of sturgeon in the world, 13 are native to the Soviet Union, but only five predominantly southern species are presently cultured. These are the beluga (*Huso huso*) (not to be confused with the whale known as beluga in Canada), the sterlet (*Acipenser ruthenus*), the Russian sturgeon (*A. guldenstadti*), the thorn sturgeon (*A. nudiventris*), and the starred sturgeon (*A. stellatus*). The principal fisheries are located in the Black Sea, the Sea of Azov, the Caspian Sea, and the Aral Sea. *A. nudiventris* is rare in the Black Sea and the Sea of Azov but, in general, all five species are found in all these waters, since Soviet biologists have been most zealous in introducing edible fishes to new waters.

ARTIFICIAL SPAWNING AND HATCHING

Techniques used in spawning sturgeons at the ten or so hatcheries, located principally in the Astrakhan region in the Volga Delta, do not differ appreciably from species to species. Wild fish with the gonads in the final stage of development are selected as breeding stock. Although, contrary to experience in the United States, Russian culturists have found it possible to ripen about one-third of adult sturgeon in captivity by holding them in running water under conditions which duplicate nature as closely as possible, particularly with respect to temperature and chemical content of the water, the vast majority of sturgeon in Russian hatcheries are induced to ripen with pituitary injections. Sturgeon pituitary only is successful and both sexes must be injected, either in the dorsal musculature or the peritoneal cavity. Injected fish are held in tanks and checked periodically until they are judged to be ripe.

Ripe spawners are killed, slit open, and the gonads removed. Prior to fertilization, eggs are mixed with silt to eliminate their adhesive qualities which, while functional in nature, reduce the efficiency of hatchery incubating devices. In *A. guldenstadti* and *A. stellatus*, eggs from the middle and rear portions of the ovaries have been found to give the best results. After mixing with sperm for 4 to 5 min, the eggs are placed in Yuschenko incubators, or similar devices with automatic stirrers, for hatching. One Yuschenko incubator can handle about 300,000 beluga eggs, 350,000 eggs of *A. guldenstadti,* or 550,000 of *A. stellatus.* Hatching time of course varies with temperature. Table 1 shows sample hatching rates for the cultured species.

TABLE 1.   HATCHING RATES, AT DIFFERENT TEMPERATURES OF EGGS OF THE
STURGEONS CULTURED IN THE SOVIET UNION

SPECIES	TEMPERATURE (°C)	TIME TO HATCHING (HOURS)
A. guldenstadti	11.9	236
	23.0	96
A. nudiventris	19.5	120
A. ruthenus	10.0	264
	14.0	168
A. stellatus	19.8	100
	18.0	89
	23.0	50–61
H. huso	12.6–13.8	192

FRY REARING

The greatest mistake of the early American fish culturists was the naive assumption that fishery production could be materially increased by releasing fry in great numbers, an assumption that does not even hold true with respect to natural spawning. Russian sturgeon culturists at first repeated this mistake; in 1954, 176.4 million sturgeon larvae were stocked in rivers. Current emphasis, however, is on methods of rearing fry to sizes at which they are presumably more capable of coping with the natural environment.

Newly hatched larvae are stocked at densities of 5000 to 20,000/m² in shallow troughs or basins filled with running water, or in screen-bottomed boxes floated in ponds. Once the yolk sac has been absorbed, the fry are fed on small Daphnia and another cladoceran, Moina rectirostris. After 10 days of intensive feeding, the fry average about 300 mg in weight and may be stocked in ponds. Of course it is possible to stock fry directly into ponds without nursing them in basins, but survival of such fry to fingerlings suitable for stocking is only 5 to 10%, compared to 80 to 90% for nursed fry. Some culturists compromise and nurse fry for 5 to 6 days, at the end of which time they average 30 to 40 mg in weight. The survival rate of such fry in ponds is about 60%. However long they are kept in nurseries, fry are protected as much as possible from light, which has a negative effect on the development of most species (A. stellatus is an exception).

Fry are next placed in ponds for growing to the 75- to 100-mm long fingerling stage, which generally takes 1 to 3 months on a mixture of Daphnia and oligochaete worms, both of which are cultured for this purpose; neither food alone is adequate. At normal survival rates, the contents of each basin suffice to stock 1 ha of growing pond. Overall survival is often very good; in 1964 one hatchery produced 3 million fingerlings from 8 to 9 million eggs.

STOCKING

Rather than simulate nature by liberating fingerlings at the hatchery and allowing them to find their way to the estuaries, Russian culturists transport the young sturgeon to brackish water in special live boats. It has been found that this greatly reduces losses to predators. The first stocking of this sort was done in 1955; in 1960, 18 million young were released; in 1965, 52 million; the goal of the program is to stock 170 million sturgeon annually. Since the species stocked require 8 to 15 years to reach maturity, the returns cannot yet be assessed, but such information as has been accumulated suggests that 3% of the stocked fish survive to adulthood.

POND CULTURE

One species, the sterlet, because of its relatively small size (usually no more than 80 cm) and tolerance of freshwater, is also grown to marketable size in ponds, either alone or in combination with carp. It has been suggested that sterlet also be stocked in combination with pike-perch (*Lucioperca lucioperca*). The sterlet represents a rare opportunity for the pond culturist; no other fish which can survive and grow in a freshwater pond possesses its combination of large size, high value, and low position on the food chain. (It feeds mainly on insects.) Still more valuable to pond culturists may be a hybrid sturgeon created by crossing ♂ *Acipenser ruthenus* × ♀ *Huso huso,* then backcrossing the male hybrids with pure line *Huso huso* females. The resulting animal is said to combine the high growth rate of the beluga and the euryhalinity of the sterlet. Apart from these animals, the large size and anadromous habits of sturgeon would seem to insure that the role of sturgeon culture will continue to be to bolster fisheries.

PROSPECTUS

Although the results of the experiments just described are only beginning to come in, Russian sturgeon culturists are pushing ahead with expansion of their operations. Current plans call for construction by 1975 of a sturgeon hatchery on the Amur River in Siberia. The hatchery, the first in that part of the Soviet Union, is designed to produce and release 1 million fry of the kaluga (*Huso dauricus*) annually, using techniques recently worked out at the University of Vladivistok. If efforts with the kaluga are successful, similar techniques may be applied to such other commercially important but depleted sturgeons of Asian Russia as the Siberian sturgeon (*Acipenser baeri*), the Amur sturgeon (*Acipenser schrencki*), and the shovelnose sturgeon (*Pseudoscaphirhynchus kaufmanni*). It is reported that officials of the People's Republic of China are also considering the kaluga as a potential species for culture.

At least three species of sturgeon (two of them rare) are native to Japan, but experimental sturgeon culture, which started in that country in 1964, involves sturgeon (species not known) imported from the Soviet Union. Biologists at the Enoshima Aquarium have achieved the rather startling result of maturing sturgeon in four years by means of a special diet and increased temperatures. It is hoped that this is the first step in the eventual entry of Japan into the caviar market.

In addition to the nations thus far mentioned, sturgeons are found in a number of European countries, but, as far as we know, only Rumania and Iran have culture programs. Presumably in the immediate future, fishery biologists outside the Soviet Union will be watching the progress of the Russian and Japanese experiments. If sustained and profitable fisheries of any magnitude are achieved others may be moved to follow the Russian lead. Otherwise, the future of sturgeons in fisheries and fish culture does not look bright.

REFERENCES

HARKNESS, W. J. K., and J. R. DYMOND. 1961. The lake sturgeon. Ontario Department of Lands and Forests. 121 pp.

NIKOL'SKII, G. V. 1961. Special ichthyology, 3rd ed. Translated from the Russian, Israel Program of Scientific Translation, Jerusalem.

OVCHYNNYK, M. 1963. Soviet fish culture. Fish. News Inter., 2(3):279–282.

RODD, J. A. 1925. Propagation of sturgeon. *In* Annual Report of Fish Culture, Fisheries Branch, Ontario Department of Marine and Fisheries, 1924.

26

Culture of Miscellaneous Anadromous Fishes (Shad and Striped Bass)

Some of the world's important cultured food fishes, most notably the Pacific salmons, are anadromous. Logistic difficulties in their culture have been overcome by shortcircuiting their migratory habits or, more frequently, by confining husbandry to the early stages of life and using the ocean as a rearing ground. Culture of the Pacific salmons, other anadromous salmonids, coregonids, smelts, anadromous cyprinids, and sturgeons are covered in Chapters 21, 20, 22, 23, 24, and 25, respectively. Culture of many other anadromous fishes has been attempted, particularly in North America. During the late nineteenth and early twentieth centuries, fish culturists in the United States and Canada experimented with propagation of virtually every edible native anadromous fish and some exotic species in the belief that artificial propagation was the key to the solution of all fishery problems. Their expectations proved to be far too optimistic, and most of their programs have long since been discontinued. Today only the Atlantic salmon (*Salmo salar*), the rainbow trout (*Salmo gairdneri*), the five Pacific salmons (*Oncorhynchus* spp.), and the striped

528

bass (*Morone saxatilis*) are cultured as anadromous fishes in North America.

Striped bass, which are native from the Gulf of St. Lawrence to the St. John's River, Florida, and have been successfully introduced on the Gulf and Pacific coasts of the United States, as well as in some inland waters, are usually thought of as sport fish, but their commercial importance is increasing, particularly along the Atlantic coast of the United States. In a number of localities, successful reproduction is threatened by pollution or obstruction of spawning streams, and hatchery propagation has been suggested as a possible means of compensating for these problems. Already hatcheries are in operation on the Roanoke River in North Carolina and the Santee-Cooper River in South Carolina, while experimental culture is under way in Maryland and Florida. The efficacy of striped bass culture in maintaining fishery stocks is by no means established, but American experiments have begun to attract attention abroad, and in 1968 experimental culture was instituted in the Soviet Union.

Artificial propagation of striped bass necessitates capture and hormonal injection (see p. 90) of partially ripe females during the spring spawning season. Success to date has been extremely variable, and optimum dosages and techniques have not been worked out.

At 14.5 to 21.0°C, the eggs hatch in about 2 days, during which time they must remain suspended off the bottom. Suspension is also necessary for high survival of the very delicate larvae during the first 5 to 7 days of life. Jar culture (see pp. 92–93) is thus virtually universal in striped bass hatcheries; a current velocity of about 0.3 m/sec is considered adequate. Once the larvae become fry, rearing becomes somewhat easier, but supplying adequate amounts of suitable feed is problematical.

Pioneer North American fish culturists much earlier turned their attention to the American shad (*Alosa sapidissima*), valued both for its flesh and for its eggs, which are rated second only to sturgeon roe for the preparation of caviar. There is no evidence that the shad fishery on the Atlantic coast, where it is native, ever benefited from stocking. Attempts to introduce American shad to other waters were similarly unsuccessful, with the notable exception of plants in Pacific Coast streams, where self-sustaining runs now occur from California to southeastern Alaska.

Despite the poor success of American shad culture, a program of propagation of the Indian shad (*Hilsa ilisha*) has been begun at the Central Inland Fisheries Research Institute, Barrackpore, India. In the first season the success of artificial fertilization was generally adequate, ranging from 75 to 90%, but hatching rates were extremely variable (2 to 65%). Larvae are reared for the first 7 to 17 days in cloth cages fixed in rivers near the bank then stocked as 4- to 7-mm fish in nursery ponds. So far, the first spawning has yielded about 600,000 35-mm, 1-month-old fry.

REFERENCES

BARKULOO, J. M. 1970. Taxonomic status and reproduction of the striped bass (*Morone saxatilis*) in Florida. U.S. Bureau of Commercial Fisheries, Technical Paper 44.

NICHOLS, P. R. 1966. The striped bass. U. S. Bureau of Commercial Fisheries, Fishery Leaflet 592.

27

Culture of Pompano

COMMERCIAL VALUE

A number of fishes of the genus *Trachinotus* (family Carangidae) are considered to be among the finest food fish in various parts of the world. The only species experimentally cultured to date is the pompano (*Trachinotus carolinus*), which inhabits the coastal waters of the Atlantic from Massachusetts to Brazil. Adult pompano are not abundant anywhere within this range, so it is not surprising that they are among the costliest of seafood. In Florida, for example, pompano comprise about 1% by weight of the total commercial catch of marine fishes but account for about 5% of the dollar value. Successful commercial culture of pompano would seem to offer little likelihood of changing their status as a luxury

531

PLATE 1. Pompano (*Trachinotus carolinus*) fry. (Courtesy Marine Research Laboratory, Florida Department of Natural Resources.)

food, at least not at first. It is perhaps for this reason that there has been almost no interest in pompano culture outside of the United States.

In the United States, pompano are fished commercially from Virginia to Texas, but 90 to 95% of the catch is landed in Florida. With the exception of recently initiated projects in California and Louisiana, all efforts at pompano culture have been made in Florida.

There are three species of *Trachinotus* (Plates 1 and 2) found in Florida waters, but pompano play by far the largest role in fisheries and have enjoyed a similar dominance in aquacultural practice and planning. One of the other two species, the palometa (*Trachinotus goodei*), is of marginal size for culture, reaching a maximum length of about 33 cm. On the other hand, the permit (*Trachinotus falcatus*) is a very large fish, attaining lengths of about 1 m and weights of more than 22 kg. There has been some speculation that permit might therefore grow more rapidly than pompano, but the possibility has not been investigated. The maximum size of the pompano is not known, since large specimens are often confused with permit, but it certainly does not attain the size of large permit.

PLATE 2. Permit (*Trachinotus falcatus*) fry. (Courtesy Marine Research Laboratory, Florida Department of Natural Resources.)

Apart from size, the three species are similar in most respects and possess the same advantages for culture. From an economic viewpoint, they are attractive to fish culturists by virtue of the high, stable demand and the erratic, and usually inadequate supply of the fishery product. If commercial pompano culture were to become very efficient, culturists might be able to sell pompano cheaper than fishermen, as the expense of harvesting wild pompano is very great. Biologically speaking, pompano are favored for culture by virtue of their hardiness and by the ease with which they adapt to confinement and artificial feeds.

HISTORY OF EXPERIMENTAL CULTURE

Pompano culture, despite the efforts begun in 1957 by government agencies, universities, and private corporations, is still in the experimental stages. In only one instance has a crop of cultured pompano been marketed, and that particular crop represented only one of six attempts by the same organization over a six-year period. Large numbers of researchers, representing the U.S. National Marine Fisheries Service, the Florida Board of Conservation, the National Science Foundation, the University of Miami's Institute of Marine Sciences, the Battelle Memorial Institute, Armour and Company, and the United Fruit Company, are currently or have been involved in experimental pompano culture. Nevertheless, the problems which confronted the first pompano culturists in 1957 remain unsolved.

LIFE HISTORY, HABITS, AND HABITAT

Some of the difficulties encountered in experimental culture of pompano may be due to the meager knowledge of the environmental requirements of *T. carolinus*. Tolerances, insofar as they are known, have been largely determined experimentally, with captive animals. The following conclusions have been reached with regard to various environmental parameters:

Salinity. Pompano of all ages are normally found in waters of high salinity; juveniles in Tampa Bay, Florida, have been found to prefer water with a salinity of 32‰. Nevertheless, they are highly tolerant of gradual salinity changes and can be adapted to water that is fresh or nearly so.

Temperature. Pompano suffer thermal shock, from which they may recover, at about 12°C; the lower critical temperature is about 10°C. The upper tolerance limit for adults is thought to be about 38.5°C, but very small juveniles have been found in tide pools with temperature exceeding 45°C. Most culture experiments have been conducted at 21 to 25°C, but biologists of the University of Miami's Institute of Marine Sciences reported rapid growth and low mortality of young at 34°C.

Dissolved oxygen. Stress occurs at concentrations below 3 ppm, and 2.5 ppm is lethal. Ponds used in pompano culture should always contain at least 4 ppm of dissolved oxygen.

pH. Pompano are able to withstand any extreme of pH likely to be encountered in a saltwater pond.

Turbidity. Like pH, turbidity is not usually critical. Pompano will survive in water so turbid that visibility is virtually zero.

It should be unnecessary to point out the difference between critical limits and optimum values for environmental parameters, but it may be that part of the reason for the poor success in pompano culture to date is overemphasis on the hardiness of the species and neglect of possible synergistic effects.

The life history of pompano is scarcely known. Ripe, flowing adults have never been observed, but from available evidence it appears that the species has an extended spawning period—February to September in the southeast United States and perhaps year-round in more southerly waters. At least part of the population apparently migrates northward in the spring and returns south in the fall, but it is not known whether this is directly connected with reproduction. It is speculated that spawning occurs well offshore, as this is where the smallest juveniles have been captured. Juveniles are found on Florida beaches from April to mid-November. Maturity is probably reached in the second year of life. The longevity of pompano is unknown; the best guess hazarded by biologists is 3 to 4 years.

The food habits of pompano are known only in a general way. Young have been observed to eat benthic and pelagic invertebrates and larval fishes. Adults have similar food habits but take fish more frequently and are thought to rely less on benthic organisms.

ATTEMPTS AT ARTIFICIAL PROPAGATION

Certainly lack of knowledge of the habits of pompano is a contributing factor in the inability of culturists to spawn pompano, or, for that matter, any of those fishes with small, pelagic eggs and larvae. Attempts at hormonal induction of spawning had all been unsuccessful until 1969, when biologists at the United States Bureau of Commercial Fisheries Laboratory at St. Petersburg, Florida, succeeded in obtaining and fertilizing eggs by the use of human chorionic gonadotropin. For unknown reasons, none of these eggs hatched. Most workers, however, are confident that the problems of spawning pompano will eventually be solved, but they predict that more serious difficulties will be encountered in rearing the larvae to the fry stage.

CAPTURE OF WILD STOCK

For the present, would-be pompano culturists must rely on wild stock. As mentioned, pompano fry are found along Florida beaches from mid-April to November, or as long as surface temperatures remain above 21°C. Even though optimal conditions for pompano culture are found only in extreme southern Florida, particularly the Florida Keys, the largest and purest stocks of juveniles are found on the east coast, from Daytona north into Georgia. This led to an initial concentration of pompano culture in northeast Florida, but since good methods of transporting live young pompano are available, this need no longer be the case.

Beaches with a gradual slope produce the best catches of fry. Particularly good locations are long, shallow miniature sounds bounded by small sand bars parallel to the surf line. Haul seining in such areas at low tide with nets 5 to 8 m long and 1 to 2 m deep has produced up to 10,000 juveniles in a single haul, and total yields of up to 80,000.

The size of juveniles taken varies. Those taken in April are 15 to 30 mm long; by July to August fish of the same age are 75 to 130 mm long. However, later influxes of younger juveniles virtually ensure that a large range of sizes will be included in most catches. These should be sorted before stocking for, although pompano are not cannibalistic, the smaller fish are likely to fare poorly in competition for food.

Another sorting problem involves separating the species of *Trachinotus*. This is particularly important in southern Florida, where permit and palometa outnumber pompano. Table 1 lists the characteristics used in distinguishing among the three species.

EXPERIMENTAL CULTURE TECHNIQUES

RESULTS

Stocking and feeding techniques and other pompano culture practices have barely begun to be worked out. Researchers have done little to hasten standardization of methods; most experiments have thus far been carried out in anything but a systematic manner. Despite the limited knowledge of suitable culture techniques, cultured pompano have repeatedly been found to grow faster than wild fish in similar environments.

Several types of enclosure have been employed in experimental pompano culture, including concrete or wooden tanks, raceways, floating net cages, and estuarine impoundments, but most of the investigations to date have been carried out in 1- to 2-m-deep earthen ponds supplied with tidal water. In one of the most successful experiments, at Marineland, Florida, juveniles weighing 0.4 to 9.4 g each were stocked in fertilized ponds at 1324 to 4942/ha and fed as much ground whole trash fish as they would consume daily. Average mortality was 18.7%, food was converted at 6.1:1, and 270 to 438 kg/ha of pompano were produced in 65 to 133 days of growth. The greatest production occurred in the most heavily stocked ponds, indicating that the stocking densities used did not approach the maximum. On the other hand, none of the fish, in either the lightly or heavily stocked ponds, reached the preferred commercial

TABLE 1. DISTINGUISHING CHARACTERISTICS OF FRY OF THE THREE SPECIES OF *Trachinotus* TAKEN IN FLORIDA AND GEORGIA

SPECIES	DORSAL SOFT RAYS	ANAL SOFT RAYS	BODY DEPTH (ORIGIN OF SEVENTH DORSAL TO FIRST ANAL SPINE) IN STANDARD LENGTH	COLOR OF ANAL FIN LOBE
Pompano (*Trachinotus carolinus*)	24–25	21–22	2⅓ or more times	Yellow or lemon yellow; tip of lobe cinnamon in some individuals
Permit (*Trachinotus falcatus*)	17–21	16–19	2¼ or less times	Bright orange or red; almost black in some dark-bodied individuals
Palometa (*Trachinotus goodei*)	17–21	16–19	2⅓ or more times	Clear with black on the anterior edge of the lobe

size of about 340 g; the largest individual harvested weighed 268 g. All the experiments were, however, prematurely terminated due to mass mortalities.

It is probable that, once the causes of such mortalities are better understood and controlled, and feeding techniques are improved, pompano can be grown to marketable size within one year at densities greater than those used in most experiments to date. It may be found advantageous to raise them in a series of ponds such as those sometimes employed in culture of milkfish (*Chanos chanos*) in southeast Asia (see Chapter 17). Juveniles could be stocked in nursery ponds, in which suitable food organisms had previously been cultured, for the first few weeks of life then successively introduced to a series of growing ponds, in each of which the population density could be controlled.

More effort has been devoted to the study of feeding captive pompano than to stocking densities (Plate 3), but the results are no more conclusive. Early culturists used mostly ground trash fish, but during the extended

PLATE 3.   Juvenile pompano feeding on a fish cake in a rearing pen. (Courtesy Biology Laboratory, Natural Marine Fishery Service, St. Petersburg, Fla.)

periods of very windy weather which commonly occur in Florida, supplies of fish become scarce, so attention was soon focused on prepared foods. Commercial fish meal was usually found to produce poor results, but floating trout feeds are readily accepted by pompano and have produced the best growth of any feeds used so far. Other foods which have been tested include shrimp, shrimp meal, crab wastes, and tankage. Frozen adult brine shrimp (*Artemia*) are considered valuable in the diet of very young fish.

PROBLEMS

Whatever food is used, cultured pompano are still subject to such apparently diet related problems as fatty or discolored livers, enlarged kidneys and gall bladders, swollen abdomens, fluid in the body cavity, exophthalmus, and hemorrhages of the skin and muscles. It is to be hoped that such problems will become less frequent when special diets, based on the specific nutritional requirements of pompano, are developed. Unfortunately, the nutritional needs of pompano are known only in a very general way. It is thought that an ideal diet would include about 30% animal protein, 18% carbohydrates, 10% bulk, and less than 10% fat. Young pompano will consume up to 30% of their body weight in ground fish daily, but satisfactory growth has been achieved with commercial trout feed by feeding 10% of the weight of pompano daily. Several small feedings are preferred to one large one.

Thus far, pompano culturists have not suffered greatly from many of the problems which afflict growers of other fishes. Disease, in particular, has had small effect on cultured pompano. High mortalities due to sporozoans and monogenetic trematodes have occasionally been reported, but most of the parasites commonly found on pompano have caused little or no difficulty. This does not necessarily mean that pompano are highly resistant to diseases and parasites. It may be that, after a few more years of culture at high population densities, epizootics will appear.

Mass mortalities due to cold spells or oxygen depletion have occasionally occurred, the latter usually as a result of pollution by uneaten feed, particularly fish meal. Wiser use of food and the adoption of pellets rather than powdered feeds should reduce the incidence of oxygen depletion.

Perhaps the most serious problem in pompano culture, apart from lack of knowledge of optimum stocking densities and feeding regimes, is predation. Screening is currently the standard preventive method, but it does not stop the entry of predatory fishes in larval form, and in ponds where water enters through pipes, screens may significantly restrict the flow of water. If pompano continue to be grown in tidewater enclosures, some combination of screening, poisoning, and periodic draining will probably be adopted.

PROSPECTUS

If culturists succeed in producing large quantities of pompano on a regular basis, marketing should be no problem, since, as mentioned, pompano command a high, stable price and fishery catches seldom satisfy the demand. In addition to the established market for fresh pompano, new markets could be developed in the restaurant and frozen food trades.

The market situation notwithstanding, the prospectus for pompano culture is uncertain. A number of improvements in basic techniques must be made before commercial culture can be sustained. In addition to further development of stocking and feeding methods, it is imperative that practical culturists be able to spawn pompano in captivity. A few small operators might be able to profitably rear ocean-caught juveniles for market, yet the undependable supply of such stock precludes large-scale culture of pompano on such a basis. Further, to depend on wild juveniles might result in the depletion of natural pompano populations.

Controlled spawning of pompano would aid culturists in avoiding temperature problems, which have plagued pompano growers in northeast Florida, by eliminating their dependence on natural populations. Culture operations could then be shifted with no loss in efficiency to extreme southern Florida or to locations where it would be possible to take advantage of power plant discharges and other sources of thermal effluent.

There are several other species of *Trachinotus* found on the Atlantic and Pacific coasts of North and South America that might eventually be cultured, but *T. carolinus* is the only one currently popular among culturists and has, in fact, been transplanted to the Pacific Coast of the United States for purposes of experimental culture. Polyculture of pompano with other fishes and invertebrates has been suggested. In an experiment conducted by the United States Bureau of Commercial Fisheries in Tampa Bay, Florida, striped mullet (*Mugil cephalus*), spotted seatrout (*Cynoscion nebulosus*), blue crab (*Callinectes sapidus*), and American oyster (*Crassostrea virginica*) grew to commercial size in ponds stocked with pompano.

Still other possibilities for the future of pompano culture could be described, but, given the state of the art, they are no more than speculation. A great deal of work remains to be done; in some vital areas, for example, larval culture, the surface has yet to be scratched. Thus it seems that if there is ever to be large-scale commercial culture of pompano, it is still well in the future.

REFERENCES

BERRY, F., AND E. S. IVERSEN. 1967. Pompano: biology, fisheries, and farming potential. Proceedings of the Gulf and Caribbean Fisheries Institute, 19th Annual Session, pp. 116–128.

FIELDING, J. R. 1966. New systems and new fishes for culture in the United States. FAO World Symposium on Warm Water Pond Fish Culture. FR: VIII/R-2.

FINUCANE, J. H. 1970. Pompano mariculture in Florida. Am. Fish Farmer 1(4):5–10.

MOE, M. A., R. H. LEWIS, and R. M. INGLE. 1968. Pompano mariculture: preliminary data and basic considerations. Florida Board of Conservation, Technical Series 55.

28

Culture of Miscellaneous Brackish Water and Inshore Marine Fishes

Whereas a great diversity of freshwater fishes are grown in confinement, practical culture of marine and brackish water fishes is largely restricted to southeast Asia, where the milkfish (*Chanos chanos*) and various mullets (*Mugil* spp.) have been raised in brackish water ponds for centuries. (See Chapters 17 and 16 for descriptions of milkfish and mullet culture, respectively.) In recent years, the Japanese yellowtail (*Seriola quinqueradiata*) has been added to the list of marine fishes which are cultured on a large scale. (See Chapter 29 for a description of yellowtail culture.)

Culture of yellowtail is a startling achievement, since in nature they are extremely active, wide-ranging, pelagic fish, which one would not expect to respond favorably to confinement. Their successful culture has stimulated speculation and experimentation with culture of other pelagic fishes (see Chapter 30) but the majority of efforts in marine fish farming continue to involve inshore and estuarine species which can be adapted to brackish water. Large-scale studies have been carried out with pompano (*Trachinotus carolinus*) in the United States, and with various flatfishes in the United Kingdom; this work is described separately in Chapters 27 and 31, respectively. Here we wish to deal with the other species of brackish water and inshore marine fishes which have been cultured.

CULTURE OF INSHORE MARINE FISHES IN JAPAN

In addition to yellowtail, the Japanese have succeeded in commercially culturing puffers (*Fugu rubripes* and *Fugu vermicularis*), red porgy (*Chrysophrys major*), black porgy (*Mylio macrocephalus*), and a few other fishes. Of these fishes, which annually account for up to 25% of Japan's production of cultured marine fish, the most important are the puffers.

PUFFERS

Unlike yellowtail, puffer is anything but a staple food and brings a very high price from gourmets in Japan, Korea, and China. Puffers are seldom eaten in other countries, largely because of the presence in some of the tissues, particularly ovary, liver, intestines, and skin, of a neurotoxin about 13 times stronger than potassium cyanide. This substance, known as tetrodotoxin, cannot be broken down by cooking. However, its presence in muscle tissue is rare and, when properly prepared, puffers are reportedly completely safe to eat.

Originally, puffer farmers utilized 3- to 4-year-old wild fish as a source of stock, and this practice is still carried on to some extent, but, since the first success in artificial propagation of puffers in 1960, most culturists have started to rely on artificially produced stock. Brood fish, usually *Fugu rubripes,* are selected from the commercial fishery catch during the 2- to 3-week spawning period in May or June. As a rule, only about 2% of the fish taken are ripe enough for use as breeders, but each female produces 300,000 to 500,000 eggs, and the rate of hatching is quite high. Eggs and milt are obtained by hand stripping, and artificial fertilization, usually by the wet method (see p. 400 for details) is done on board the fishing boat.

Once the egg membranes have hardened, the zygotes are packed in sea water in polyethylene bags, at about 50,000 eggs/liter. Gaseous oxygen is routinely provided, and if the time of transportation is to be longer than 48 hours and/or the water temperature is over 20°C, an antibiotic such as streptomycin may be added. Most of the eggs are taken to Yamaguchi Perfecture to be hatched and reared to "seedling" size at a propagation center maintained by the prefectural government.

Puffer eggs are hatched in shallow tanks, framed nets, or hatchery jars. In the first two, running water or aeration is supplied, the eggs are spread out in a single layer at 1 to 3/cm², and 40 to 60% hatching can be expected within 10 days at 15 to 19°C. Jar hatching (see pp. 92–93 for details of this technique) is much more efficient, resulting in hatches of 90% or greater when eggs are stocked at 5000 to 10,000/liter and the water changed daily.

Newly hatched puffer larvae, which are strongly phototaxic, are kept in standing filtered sea water for the first week of life. It is necessary to change the water daily at first, and up to eight times a day by the end of the week. Almost any size tank may be used, from 50 liters to 5000 liters, as long as there are adequate facilities for changing the water and maintaining the temperature at 17 to 20°C. With the absorption of the yolk sac and the beginning of active feeding, running water is supplied. At this point, it is necessary to initiate a complicated regime of feeding and population density control if as many as 10% of the postlarvae are to survive to seedling size (25 mm and 3 g or more). Table 1 outlines feeding and stocking practices for puffers from hatching to the seedling stage.

Among the foods used are barnacles. To be edible by larval fishes, barnacles must be no older than the first nauplius stage. To assure a steady supply of this age group, barnacles must be cultured. Sets of barnacles are easily obtained by suspending bamboo sticks in the sea near the surface during summer and early fall. The sticks, with barnacles attached, are then kept in saltwater tanks until nauplii are needed. Exposure to air, followed by placing the sticks in tanks containing larvae, usually results in discharge of nauplii within 10 to 20 min. If this is done 3 to 4 times daily during the daylight hours, a density of about 100 first nauplii/liter can be maintained in the larva tank.

After about the twentieth day, cannibalism may become a problem, but it is greatly reduced once the young fish start taking minced fish flesh. Often, however, it is difficult to wean the young from live food. The effect of cannibalism is also lessened by reducing the population density after the twentieth day and by providing three or four small feedings rather than one large one daily.

Puffer growers purchase seedlings in July and stock them in ponds or floating net cages anchored in at least 5 m of water. In either case, the

TABLE 1.   FEEDING SCHEDULE FOR YOUNG CULTURED PUFFERS IN JAPAN

NO. OF DAYS AFTER HATCHING	LENGTH OF FISH (MM)	POPULATION DENSITY (FISH/LITER)	FEED	AMOUNT FED DAILY (AMOUNT OF BRINE SHRIMP EXPRESSED AS GRAMS OF DRIED EGGS)
0–6	2.6–2.8	10–100	None	—
7–10	3.5–4.0	10–100	Nauplii of barnacles	Maintained at 100/liter
11–14	—	10–100	Nauplii of barnacles and brine shrimp (*Artemia*)	Barnacles as above plus 2–5 g of brine shrimp/ 10,000 fish
15–20	—	10–100	Nauplii of brine shrimp	8–12/10,000 fish
21–25	—	5–10	Nauplii of brine shrimp, cultured copepods (*Tigriopus japonicus*) and minced fish flesh	8–12/10,000 fish plus an empirically determined amount of other foods
26–30	—	0.5	Nauplii of brine shrimp, cultured copepods (*Tigriopus japonicus*) and minced fish flesh	5–8/10,000 fish, plus an empirically determined amount of other foods
31–44	—	0.5	Minced fish flesh	Amount determined empirically
45–50	25 or more (seedling size)	0.5	Minced fish flesh	Amount determined empirically

dissolved oxygen content of the water must be above 4 ppm, the temperature must not exceed 20°C or drop below 10°C, the salinity must be at least 22‰ and turbidity must be slight. The stocking density in ponds should be about 0.05 kg/m³ but may be considerably higher in net cages, say 0.08 to 0.5 kg/m³.

Puffers are fed on trash fish (horse mackerel, anchovies, sand eels, etc.). Fish which are not fresh, or which are excessively fatty, are to be avoided or nutritional diseases may result. At first the food is minced, but after the puffers reach lengths of 100 mm, the food fish are chopped. The quantity of feeding is determined empirically, but the frequency is regulated seasonally, as shown in Table 2, and is suspended whenever the water temperature falls below 14°C.

TABLE 2. FREQUENCY OF FEEDING, AT DIFFERENT TIMES OF YEAR, FOR CULTURED PUFFERS IN JAPAN

MONTHS	NO. OF FEEDINGS DAILY
July-August	4
September-November	3
December	2
January-March	1
April-June	1–2

The minimum marketable weight of about 0.8 kg is reached in 1½ years after stocking, though many fish are fed for up to 2½ years, at which time they may weigh 2 kg. The survival from stocking to marketing averages 50 to 70%, and the overall rate of food conversion is about 4:1. Though puffer may cease to grow or even lose weight during the winter, cultured puffers are marketed during the winter, which is the off season for puffer fishermen, thus prices are high.

During the spring, some culturists also obtain 1-year-old fishery-caught puffers, weighing about 0.2 kg each, and feed them, in the manner just described, until they attain marketable size. Very large puffers may be marketed by growers who stock 1.5- to 2.5-kg fish and feed them for a few months.

Apart from nutritional diseases, cultured puffers may suffer from vibriosis, ichthyophthiriasis, and various fungus diseases. Parasites reported include the flatworm *Diclidophora tetrodonis,* which attaches to the gills, and may cause serious problems, and the arthropods *Argulus scutiforma* and *Pseudocaligus fugu.* The reader interested in treatments for some of these diseases and parasites is referred to *Culture and Diseases of Game Fishes* by H. S. Davis.

As of 1965, 40 operators were engaged in puffer farming in Japan. From 35.4 ha of water they produced 91 metric tons of fish, for an average yield of about 230 kg/ha.

RED PORGY

Like the puffer, the red porgy brings high prices in Japan, not so much due to its gastronomic or nutritional value but because of the ancient association of the red porgy with good fortune. It is thus traditionally served at birthday celebrations, weddings, and other such occasions.

Red porgy farming was first attempted in 1887, and attempts at artificial propagation have been made sporadically since 1902. However, the first success in artificial propagation was not achieved until 1962, and it is only within the last few years that small-scale farming of the red porgy has become a commercially feasible proposition.

As is the case with puffer farming, culture of the red porgy is dependent on wild brood stock. Males are easily distinguished from females, particularly during the spawning season (late April to early June), by the more angular shape of the head and their darker coloration. It is possible to obtain viable genital products from red porgies kept in captivity, but eggs of such fish exhibit substantially lower hatching rates, so artificial fertilization, usually by the dry method (see p. 400), is done on board the fishing boat.

As soon as possible after fertilization, the zygotes are washed with clean sea water, to remove all extraneous matter. Great care must be exercised in transporting the fertilized eggs, which are very sensitive to temperature changes, bright light, and physical shock.

As noted, culture of the red porgy is a recent development, so facilities and procedures are not standardized. The following is a general description of one set of techniques which has been successfully employed. Certainly they will be improved as culturists become more experienced and as biologists gather more information on the ecology and physiology of the red porgy.

Hatching is carried out in a tank 2 m × 1 m × 1 m deep, supplied with running or recirculated sea water and housed in a small building roofed with translucent plastic plates. The tanks are illuminated, but the intensity of light is not allowed to exceed 3000 lux.

Red porgy eggs are pelagic and must float for proper development, thus the most critical factor in hatching is the specific gravity of the water, which must not be allowed to fall below that of the eggs (1.0245 at 15°C). This corresponds to the rather high salinity of about 33.5‰. Where such water is not readily available, it is recommended that suitable sea water be collected and stored, since chemical control of specific gravity is not always satisfactory.

Larvae are reared to "seedling" size in 5-m × 1-m × 1-m tanks, similar to those used for hatching, but supplied with a sand filter at the outlet to prevent loss of fish. A recirculating system is preferred to running water, since such predators as large copepods may be more easily excluded from a closed system.

The larvae absorb their yolk sacs wthin three days of hatching. Feeding occurs at this stage, but it is an incidental result of the larvae swimming about with their mouths open. There is no evidence that they are able to perceive or deliberately attack food organisms. Therefore, starting on the third day after hatching, the culturist must start supplying copious amounts of tiny food animals. Nauplii of copepods are perhaps best suited for this purpose, but it is not often possible to provide adequate quantities. Acceptable substitutes include the blastula and gastrula stages of sea urchins, oyster larvae, and rotifers, all of which may be cultured.

Active feeding begins on the fifth or sixth day after hatching. Larvae 5 to 10 days old do best on a diet of copepods, which must usually be captured from the sea. After the tenth day, red porgy larvae are large enough to take brine shrimp nauplii. Growth rates and survival of larvae fed on brine shrimp nauplii vary greatly, apparently with the quality of brine shrimp, and it is recommended that a small sample of larvae be fed on brine shrimp for a few days before an entire population is switched off copepods.

On about the twentieth day after hatching, at which time the larvae are about 10 mm long, they assume a benthic life habit. For the first few days of benthic life small polychaetes (3 mm long or less) and powdered, freeze-dried shrimp meat kneaded with freshwater are suitable foods. For the rest of the culture period, until they attain seedling size (2 to 3 cm), the young porgies are fed minced meat of dry, white fish. Oily, fatty fish are to be avoided at all times.

The most important environmental factor in larval culture is organic pollution, which may be caused by dead larvae and/or overfeeding. In addition to such routine fish hatchery procedures as periodic siphoning, small crabs or other shellfish may be stocked in the tanks as biological pollution controls.

Light is as important to larvae as it is to eggs, but the optimum level varies with the size and age of the fish and, apparently, with individual factors which are not understood. Illumination control is simplified by lighting different sections of the tank at intensities varying from darkness to 3000 lux and letting the larvae express their own preference.

Some seedlings are released along the coast of the Inland Sea, in the hope of augmenting the fishery, but most are sold to fish farmers for further culture to marketable size. Farming of red porgy is usually done in floating net cages. Seedlings fed on fresh fish flesh and/or commercial fish food in pelleted form reach saleable size in 12 to 18 months. No serious diseases or other causes of mortality have thus far been reported.

BLACK PORGY

The black porgy is a less prestigious fish than either the puffer or the red porgy, but is nonetheless a fine food fish. Since it is somewhat hardier than the other saltwater fishes farmed in Japan and can be spawned in captivity, it may eventually assume greater importance in aquaculture.

Breeders are captured by angling and acclimatized in recirculating tanks. The spawning season is March to May in southern Japan and April to June in the central part of the country, and naturally occurring water temperatures at those times are acceptable for artificial propagation. Care must be taken, however, that the salinity remains between 25 and 33‰. Males mature naturally under these conditions, but females require treatment with the hormone preparation synahorin.

Ordinarily, eggs may be stripped and fertilized, using either the wet or dry method, 40 to 50 hours after injection. Black porgy eggs show a marked reduction in fertility if, after stripping, they are left for long without coming into contact with sperms. Thus fertilization should be done as soon as possible, and certainly not later than 20 min after stripping.

Hatching and rearing of the larvae are carried out in the same tank. Tanks used for this purpose are divided into four chambers separated from each other by gates and equipped so that running water can be supplied. The pelagic eggs are stocked in still water in the lowermost chamber of the tank, in quantities such that, at expected hatching rates, a population density of 3000 to 10,000 larvae/m² will result when all chambers are occupied. The tanks are housed in sheds with transparent roofs which may be shaded or screened when necessary. The water temperature should not be allowed to exceed 20°C during the hatching period, and the salinity should be maintained at at least 25‰.

Under such conditions, hatching usually occurs in about 40 hours. Occasionally, a lot of eggs exhibits a low hatching rate, and pollution resulting from the decay of dead eggs may be detrimental to larvae or live eggs. Therefore, samples of each lot of eggs are periodically inspected for viability. If it appears that the hatching rate will be less than 50%, the live eggs are removed as soon as they attain the "eyed" stage and placed in clean sea water in another tank.

Running water is advantageous to nearly all marine fish larvae from the standpoint of health, but on the other hand it may wash away such tiny creatures as black porgy larvae. The four-chambered structure of the rearing tank is an effective compromise. When the water in the hatching chamber becomes too dirty for optimum growth of the larvae, the gate into the next chamber is opened, thus permitting access to clean water and providing more room for the larvae. This process is repeated twice more, or until the full tank is being utilized. By that time (10 to 18 days after hatching) the larvae are strong enough that running water can be introduced without danger of loss. At first the running water is turned off at night, when the larvae are less active, but after about the twenty-fifth day it is left on around the clock.

Desirable temperature and salinity are the same as for hatching, but illumination requires special attention, particularly 5 to 7 days after hatching, when the larvae need a bright light for feeding. The intensity of light at this time should be 5000 to 6000 lux at the surface. This is gradually reduced to 1000 lux by the time the larvae are 1 month old. Lights should be focused from overhead; light entering from the side or bottom of the tank hinders feeding. After the first month, a minimum water depth of 0.7 m is maintained.

Though the yolk sac is seldom completely absorbed before the third day of life, active feeding may begin by the second day, thus black porgy culturists start providing food, in the form of trochophore larvae of oysters, 2 days after hatching. A number of transitions to successively larger items of food must be effected before the larvae reach seedling size at the age of 35 to 40 days. Table 3 illustrates a representative feeding regime.

Black porgy farming may be done in floating net cages or in ponds. Cages are stocked at 50 to 70 fish/m³, while ponds receive 6 to 7 fish/m³. There is an increasing tendency to stock black porgies in combination with other fishes.

Black porgies tolerate a wider range of environmental conditions than puffers or red porgies. They are particularly hardy with respect to temperature, but feeding, and therefore growth, ceases below 10°C. Temperatures below 5°C are considered potentially dangerous. No serious diseases have yet been reported.

Black porgy are quite omnivorous; they have been successfully raised on trash fish, mussels, the meat of pearl oysters, silkworm pupae, and commercial fish food pellets. The general practice is to feed chopped

TABLE 3.   FEEDING SCHEDULE FOR YOUNG CULTURED BLACK PORGY IN JAPAN

AGE OF LARVAE (DAYS)	FEED	NO. OF DAILY FEEDINGS	AMOUNT FED PER FEEDING
0–1	None	—	—
2–7	Trochophore larvae of oysters	1	10 larvae/ml, gradually increasing to 100/ml
8–14	Trochophore larvae of oysters, gradually changing to barnacle nauplii	1 of oyster larvae, 3–4 of barnacles	100 oyster larvae/ ml, and 0.1 barnacle nauplii/ml
15–17	Barnacle nauplii	3–4	0.1/ml
18–32	Barnacle nauplii and marine copepods, (Freshwater copepods may be substituted.) gradually changing to 100% copepods	3–4 of barnacles and 1 of copepods (freshwater copepods must be added more often, as they do not survive well in sea water)	0.1 barnacle nauplii/ ml and an empirically determined number of copepods
33–34	Copepods	1	Empirically determined
35–40 (seedling stage)	Minced fish meat and/or pelleted commercial fish feed	Empirically determined	Empirically determined

trash fish or dry feed, whichever is cheaper, though there may be a slight problem in conditioning the fish to accept the dry feed. The conversion rate of fresh fish by black porgy is 3 to 4:1, and marketable size (about 150 g) is reached in 16 to 20 months.

Some farming is or was done with 1- or 2-year-old wild seedlings. The 1-year-old seedlings are seined from shallow coastal areas during late May to late July, at which time they are 1 to 4 cm long, and reach commercial size in 15 to 18 months. The 2-year olds, 10 to 15 cm in length and 30 to 50 g in weight, are caught along the coast with nets or hook and line from May to July and are ready for market after 6 months of feeding. The supply of 1- and 2-year-old seedlings is rather small, and as artificial propagation of black porgy becomes more prevalent and efficient, their use by fish farmers is expected to cease entirely, if it has not already.

Certain other marine fishes, including filefish (family Balistidae) and "sea eels" (family Congridae?), are reported as being cultured in Japan, but details of these practices are not known. In 1965, 126 operators were engaged in culture of marine fishes other than yellowtail and puffer. From 157.3 ha of water, they produced 101 metric tons of fish, for a rather low average yield of about 64 kg/ha.

CULTURE OF COD IN NORWAY

The first recorded large-scale success in artificial propagation of a marine fish was achieved with the Atlantic cod (*Gadus morhua*) in southern Norway over 85 years ago. The Flödevigen Biological Station, near Arendal, where cod were first bred, is still in operation. Since 1950 its operations have been intensified, until now 100 to 150 million cod larvae are produced every 2 years. The survival rate from egg to 5-day-old larvae is approximately 90%.

The larvae are released nearby in the heavily fished Oslofjord. Contrary to general experience with this sort of program, Norwegian fishermen and biologists agree that the program of artificial propagation and stocking contributes greatly to the success of local fisheries.

Cod have not been widely cultured outside of Norway, but in the Soviet Union a new experimental fish culture station on the Barents Sea is slated to begin both artificial propagation and farming of Atlantic Cod and the closely related haddock (*Melanogrammus aeglefinus*).

BRACKISH WATER POND CULTURE

All told, the rearing of fish in brackish water impoundments is of greater significance than fish farming in truly marine environments. Brackish water fish culturists are almost always at least partially dependent on tidal flow for a supply of water and sometimes for fish stocks as well. The continual influx of water from the sea renders it nearly impossible to completely control the species composition of brackish water impoundment communities, thus culturists usually find themselves harvesting a number of fishes other than those they intended to grow. Some of these fishes may eventually prove to be more valuable than currently favored crop species. They are worthy of study by the fish culturist simply by virtue of their presence, growth, and survival in the artificial environment he has provided.

IN THE MEDITERRANEAN REGION

Fishes are most extensively harvested from brackish waters in the Mediterranean Sea and in Southeast Asia. Management practices are almost nil in most parts of the Mediterranean area, but in certain locations in Italy, particularly the lagoon of Venice, fishermen have developed elaborate schemes for trapping, growing, and harvesting fish. The principal fish crop in the Italian lagoons is mullet, but two species of predatory fish, *Dicentrarchus labrax* and the gilthead bream (*Sparus auratus*), are also encouraged and subjected to a certain amount of manipulation. The methods used in managing these species are described in Chapter 16. Among the incidental species, which enter lagoons and are harvested as a matter of course, the most commercially valuable is the eel (*Anguilla anguilla*). Others include gudgeon (*Gobius* sp.); a number of flatfish including *Pleuronectes flesus, Solea vulgaris,* and *Bothus podas;* various species of silversides (*Atherina*), particularly *A. mochon;* croaker (*Umbrina cirrhosa*); and porgy (*Dentex dentex*).

IN SOUTHEAST ASIA

Brackish water fish culture is more intensive in southeast Asia than in the Mediterranean area, involving fertilization and sometimes supplementary feeding. The traditional crops are milkfish and mullet. In many areas, the introduced Java tilapia (*Tilapia mossambica*) has become a third major crop by virtue of the inability of culturists to materially reduce its numbers. (See Chapters 17, 16, and 18 for detailed accounts of the culture of milkfish, mullet, and tilapia, respectively, in southeast Asia.) Many more species enter brackish water ponds; some piscivorous ones are considered mainly destructive, to be eradicated if possible, whereas others are too small to be of direct significance to the culturist, but a few are encouraged and harvested.

Of particular interest is the pearl spot (*Etroplus suratensis*) of India, Pakistan, and Ceylon, the only Asian brackish water fish which can be easily made to spawn in confinement. In nature, pearl spot attach their eggs to the underside of submerged objects in 1 m or less of fresh or brackish water. Culturists in southern India (the only area where pearl spot are cultured) take advantage of this trait by erecting platforms made of slabs of stone or slate in their ponds. Like all cichlids, pearl spot care for their young, thus special spawning ponds are not usually provided, though the efficiency of propagation could perhaps be increased in this manner.

Pearl spot are mainly herbivores, and feed mostly on green and blue-green algae and decaying plant remains. Of lesser importance in the diet are soft macrophytes, zooplankton, insects, worms, and fish eggs. Pearl spot attain lengths of 10 to 12 cm in ponds in 1 year and have been reared to 20 cm.

Some of the generally unwanted piscivorous fishes are also of actual or potential importance, particularly in ponds where young or stunted tilapia or other small fish compete severely with desired species. Perhaps the most widely distributed piscivore is the cock-up (*Lates calcarifer*). Cock-up are generally considered unsuited for culture with milkfish or mullet, but have been recommended, along with groupers (*Epinephelus* spp.) for certain polyculture schemes in the Philippines. Where crustaceans are extremely abundant, cock-up may derive up to 75% of their food from them, and perhaps thereby spare other fish to some extent.

Monoculture of cock-up in both fresh and brackish water occurs in India, Pakistan, and Thailand. Growth and production are dependent on the amount of food (usually trash fish or offal) which can be supplied; lengths of up to 30 cm and weights of up to 0.5 kg may be attained in 1 year.

Another potentially valuable predatory fish is the ayuñgin (*Therapon plumbeus*), which has been suggested for culture in the Philippines, but so far remains an incidental product of brackish water pond culture. Its Indonesian congeners *Therapon jarbua* and *Therapon theraps* suggest similar potential.

Other widespread piscivores are the tarpon (*Megalops cyprinoides*) and the related ten-pounders (*Elops* spp.), both of which have reputations for extreme voracity. Food studies have shown that the tarpon deserves this reputation, but that the predatory capacities of ten-pounders are popularly exaggerated. *Elops machnata* has for centuries been important in pond culture in Hawaii where fry are trapped on the incoming tide, but it and its congeners are of only incidental importance in southeast Asia.

Tarpon are cultivated only in India and Ceylon, where they are grown in freshwater ponds. Larvae and fry are collected from estuaries and transferred directly to freshwater without acclimation. The young tarpon attain lengths of 35 to 40 cm in the first year. Under very favorable conditions in brackish water, this size has been reached in 18 weeks. The principal use of tarpon is as a converter of coarse fish in heavily populated ponds. Even this role seems of doubtful value as, in the Western world at least, tarpon itself is regarded as a coarse fish and scarcely considered edible.

The most complete survey of brackish water pond fish fauna has been carried out in the tambaks used to rear milkfish in Java. Among the fishes present and already discussed are mullet, *Therapon* spp., cock-up, grouper, tarpon, and ten-pounders. Others of possible value as food fish include bonefish (*Albula vulpes*), synbranchoid eels (*Monopterus alba*), barracuda (*Sphyraena jello*), spotted scats (*Scatophagus argus*), and various catfishes, including *Plotosus canius, Plotosus anguillaris, Arius leptaspis,* and *Arius maculatus.*

Under normal circumstances, with milkfish as the principal crop, these fishes may contribute an additional 30 kg/(ha)(year) to the yield of a tambak. In order to avoid possible losses of expensive milkfish, tambaks in the early stages of construction or those with weakened dikes are often stocked exclusively with "extraneous" fishes, in which case yields of 100 to 150 kg/ha may be expected.

IN THE UNITED STATES

Culture of marine and brackish water fishes has only recently begun to attract the attention of large numbers of workers in the United States. A pioneer in American marine fish culture was G. Robert Lunz, of Bear's Bluff Laboratories, Wadmalaw Island, South Carolina. Lunz's experimental culture methods, while anything but intensive, illustrate the potential of saltwater pond fish culture in the southeastern United States.

Lunz maintained a 0.6-ha saltwater pond, averaging about 0.75 m deep, equipped with an automatic sluice gate, and allowed it to be stocked with fish on the incoming tide. On five different occasions from 1947 to 1951 he was able to produce 55 to 130 kg of marketable fish, not counting oysters, crabs, and shrimp, in growing periods of 6 to 13 months. Average production, prorated for a 12-month period, was 206 kg/ha. This was achieved without species selection, predator or disease control, fertilization, or feeding.

The only fish that comprised a significant proportion of every harvest at Bear's Bluff Laboratories were mullet. Spotted sea trout (*Cynoscion nebulosus*), spot (*Leiostomus xanthurus*), and black drum (*Pogonias cromis*) were usually present in significant amounts, while ladyfish (*Elops saurus*), and red drum (*Sciaenops ocellata*) were each important in one harvest. Also present in small quantities on some occasions were weakfish (*Cynoscion regalis*), Atlantic croaker (*Micropogon undulatus*), silver perch (*Bairdiella chrysura*), pigfish (*Orthopristis chrysopterus*), sea bass (*Centropristes* spp.), and sheepshead (*Archosargus probatocephalus*).

More recent attempts at brackish water aquaculture in the United States have concentrated on high-priced luxury products such as pompano and various species of shrimp (Chapters 27 and 32, respectively). In nearly every case where these animals have been experimentally cultured, large amounts of extraneous fishes, sometimes exceeding the cultivated species in numbers and weight, have been harvested. In one case where spotted sea trout invaded a shrimp pond and wiped out the shrimp, 448 kg/ha of sea trout were produced. This species would appear to have excellent potential for culture, were it not for its extreme fragility with respect to handling. Other species of potential value which have been produced include gag (*Mycteroperca microlepis*), gray snapper (*Lutjanus griseus*), crevalle jack (*Caranx hippos*), spotfin mojarra (*Eucinostomus argenteus*), kingfish (*Menticirrhus* spp.), pinfish (*Lagodon rhomboides*), great barracuda (*Sphyraena barracuda*), and flounder (*Paralichthys* spp.).

The extraneous species are mentioned in the hope that some of them will be found suitable for intensive culture. North American mariculturists, seemingly obsessed with the notion of producing a gourmet product, may be overlooking a number of valuable sources of food. A first step toward remedying this situation was recently taken by the Louisiana Wildlife and Fisheries Commission in cooperation with Louisiana State University. Their preliminary experiments with Atlantic croaker indicate that, although croakers of the size usually taken by commercial fishermen are of little value as human food, the species can be reared to a marketable size. Fingerlings 25 mm long, when stocked in ponds, produced 300 kg/ha of marketable size fish in one growing season, without feeding. Current studies are concerned with the feasibility of supplemental feeding and overwintering in ponds. Similar experiments have recently been initiated with red drum.

PROSPECTUS

The ancient history of brackish water pond culture in southeast Asia, the Mediterranean, and Hawaii notwithstanding, culture of estuarine and inshore marine fishes is in its infancy. Both intensive monoculture, as recently practiced in Japan, and low-intensity polyculture have great potential. Efforts should be made to introduce the latter technique to all coastal regions of the world, as it has the capability of producing large amounts of protein for human consumption with a minimum of expense and technological know-how. Both types of culture stand to benefit from a thorough investigation of the species which might be used. Certainly we have thus far utilized only a tiny minority of the cultivable fishes of the inshore environment.

REFERENCES

DAVIS, H. S. 1953. Culture and diseases in game fishes. U. of California Press, Berkeley, 332 pp.

DE ANGELIS, R. 1960. Mediterranean brackish water lagoons and their exploitation. General Fisheries Council for the Mediterranean, Studies and Reviews, No. 12.

HORA, S. L., and T. V. R. PILLAY. 1962. Handbook on fish culture in the Indo-Pacific region. FAO Fisheries Biology Technical Paper 14. 204 pp.

LUNZ, B. R. 1951. A salt water fish pond. Contribution from Bears Bluff Laboratories, No. 12. Bears Bluff Laboratories, Wadmalaw Island, South Carolina.

SCHUSTER, W. H. 1949. Fish culture in salt water ponds on Java. Dept. van Landbouw en Visserij, Publicatie no. 2 van der Orderafdeling Binnenvisserij.

Interviews and Personal Communication

FUJIYA, M. Japanese Fisheries Agency.

29

Japanese Yellowtail Culture

We remarked in the introduction (and we are by no means the first to make this observation) that aquaculture has lagged behind agriculture in its development; that man is essentially still a hunter of aquatic organisms rather than a farmer. This is especially true with respect to marine fishes. While there are ancient historic examples of culture of freshwater and brackish water fishes, as well as marine invertebrates, marine fishes were strictly a fishery product until this century.

It is not surprising that the first people to practice culture of marine fishes were the Japanese, who derive a greater proportion of their food from the sea than any other major nation. Japanese researchers have at one time or another studied virtually every edible native marine fish with respect to its potential for culture, and a number of species are experimentally or commercially cultured in Japan today, but the first species with which they had success on a commercial scale, and still by far the most important, is the yellowtail (*Seriola quinqueradiata*). A small, low-intensity yellowtail culture enterprise has existed on the island of Shikoku in the Inland Sea since 1928, but the species did not assume major importance as a cultured fish until the 1960s.

SUITABILITY FOR CULTURE OF YELLOWTAIL

The yellowtail has a number of seeming disadvantages as a species for culture. It is a wide-ranging, fast swimming, pelagic fish, which one would suppose not to be amenable to the close confinement necessary for fish culture, but, surprisingly, this has not presented a problem. It is also a highly piscivorous species, and one which biologists have thus far been unable to propagate artificially, even experimentally. Both of these factors are economically disadvantageous for yellowtail farmers; nevertheless, they are able to compete favorably with yellowtail fishermen.

COLLECTION OF FRY

Since yellowtail have not been bred in captivity, wild stock are the source of fry for culture. Adult yellowtail, which are found off Okinawa in March, migrate north in the spring, to spawn off southern Kyushu in late April or early May. Soon after spawning, larvae less than 15 mm long are brought near the coast by the Kuroshio current, captured by surrounding with fine mesh nets, and sold to fry specialists, who rear them to a size suitable for farming.

A special license is required to capture and sell yellowtail larvae, and the field is dominated by a few large operators, each of whom must abide by a catch limit set annually. The total limit for 1967 was 17 billion larvae. Thus overfishing of yellowtail larvae is not a problem, yet the demand for yellowtail has increased to the extent that the strictly regulated fishery is not able to supply it. The obvious solution to the problem is to develop means of artificially propagating yellowtail, but there is no assurance that this will be achieved in the near future.

The first task facing the fry specialist is to grade the larvae into small, medium, and large categories (see p. 409 for description of the method of grading small fish). If this is not done, up to 50% mortality may result from cannibalism.

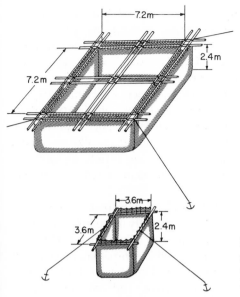

FIG. 1. Floating net cage used in growing yellowtail in Japan. (Courtesy of Teruo Harada, Fisheries Laboratory, Kinki University, Wakyamaken, Japan.)

FRY REARING

After grading, the larvae are stocked in floating nylon net cages, ranging from 2 to 50 m² in area and 1 to 3 m deep (Fig. 1). The use of cages, both for fry rearing and for growing for market, has been the key to the success of yellowtail farming and is now being tested or adopted in a number of other types of fish culture. Floating cages combine the advantages of small and large enclosures. When fish are crowded into small spaces, they burn fewer calories, consequently food conversion is more efficient. And of course the smaller the enclosure, the quicker and easier the harvest. However, in conventional pond culture the lower size limit of a pond to be stocked with a given number of fish is determined by chemical factors, principally the oxygen capacity of the water. Assuming any significant amount of circulation, this does not hold true for floating cages. While fish in cages are physically restrained in a very small area, the amount of clean, oxygenated water available is theoretically unlimited.

Yellowtail cages, the location of which is strictly regulated by the prefectural governments, are set out in parallel rows. They should be constantly tended, thus they are usually provided with platforms and a shelter for workers. One man can look after 10 to 15 of the larger cages.

As soon as the fry are ready to begin feeding, they are started on minced fish and/or shrimp. Shrimp and white-fleshed fish, such as sand eels and horse mackerel, which are low in oil and fat, are preferred to pink-fleshed, oily fish but if, as sometimes happens, such fish are in short supply, anchovies may be substituted. This is a dangerous practice, because yellowtail are not able to digest the unsaturated fatty acids in oily fish. Up to 100% mortality has resulted when fry were experimentally fed exclusively on anchovies. It is thought that some sort of dietary supplement could eliminate this danger, but until such a substance is developed, yellowtail culturists are advised to feed anchovies and the like as seldom and as little as possible. Some culturists stretch the supply of food fish by supplementing the diet of fry with zooplankton attracted into the net cages by lights suspended over them.

The supply of food for yellowtail fry could be stabilized if there were a suitable artificial food, but none is commercially available. Good results have been experimentally achieved at the Hiroshima Prefectural Fisheries Experiment Station with freeze-dried shrimp. Small, commercially valueless shrimp are dried, ground, sieved to remove bits of shell, mixed in a 1:1 ratio with water to produce a paste, and spread on a glass rod or plate for presentation to the fish.

A problem sometimes encountered in 7- to 10-day old fry is the development of calcium deposits in the form of urinary stones. There is no known cure for this condition, which may be symptomatic of nutritional deficiencies, and it may cause 80 to 90% mortality.

Fry sold for stocking in growing cages vary from 5 to 10 cm long and 8 to 50 g in weight, but those near the upper end of this range are preferred. Growth to this size usually requires 4 to 6 weeks in cages. Fry specialists are usually located no more than 100 km from their customers, and transfer of stock could probably be accomplished by truck, but live boats are presently used. An indication of the expansion of the yellowtail culture industry is the fact that in 1955 200,000 fry were sold and stocked, while in 1960 20 million were supplied to growers.

GROWING FOR MARKET

Growing for market is done in cages, 35 to 100 m^2 in area, 3 to 6 m deep, made of nylon or metal, and stocked with 40 to 100 yellowtail/m^2 (see Fig. 1). In choosing a location for cages, the following factors must be taken into consideration:

1. Circulation. There must be sufficient exchange of water through the cages to wash away feces and uneaten food and ensure that dissolved oxygen concentrations are at least 3 ppm.

2. Temperature. Yellowtail cease to grow below 15°C and die at 9°C. Culture is usually carried out at 18 to 29°C, and optimum temperatures are considered to be in the upper half of this range.

3. Salinity. Although yellowtail are pelagic fish, they do well at relatively low salinities. For best growth, cages should not be placed where the salinity drops below 16‰.

4. Pollution. When large-scale yellowtail farming was started, care was taken to select culture sites remote from sources of domestic, industrial, and agricultural pollution. It is clear, however, that pollution is going to be more of a problem in the future as Japan continues to grow and industrialize. Growers of piscivorous fish such as yellowtail have further cause for concern since it has been demonstrated that remoteness from agricultural lands does not completely protect their stock from contamination with pesticide residues via the food chain.

5. Protection from wind and waves. Cage culture sites should be located in sheltered waters.

6. Accessibility. It must be economically feasible to transport fry and food to the farm and ship marketable fish from it.

Feeding is as crucial in growing yellowtail for market as it is in fry culture, and the same general rules apply with regard to dry versus oily fish (see above). An indication of the importance of feeding in cage culture of yellowtail is the fact that 49.4% of the budget of the average farmer goes for food.

Artificial diets for the final phase of yellowtail culture are by no means perfected, but they are in more general use than is the case for fry culture. The best developed so far consists of at least 70% of fish meal, made from the same kinds of fish used to feed fry, with 5 to 10% of gluten as a binder. Other ingredients include a vitamin mix similar to that fed to chickens and catfish in the United States (see p. 170); minerals, especially iron and cobalt to prevent anemia; and the enzyme protease. This diet is prepared for feeding by mixing with water in a 1:1 ratio. The growth obtained on this mixture is not as good as that obtained by feeding fresh fish, but if the two foods are alternated, growth is better than if fresh fish alone is fed. Large-scale adoption of the artificial diet is presently retarded by its high cost, which will drop if yellowtail culture continues to expand. Another problem is that yellowtail do not respond well to food in pellet form. Research is now going on with the goal of finding a more suitable shape for food particles.

Considerable effort has been expended in determining how much and how often to feed. Japanese researchers, taking into account that growth, survival, and conversion are not all maximized at the same level of

PLATE 1.   Feeding small yellowtail in floating cage culture, Japan. (Courtesy Ziad Shehadeh, Oceanic Institute, Hawaii.)

feeding, have concluded that the optimal feeding regime for yellowtail consists of two feedings daily, at about 0100 and 1400 hours, together amounting to 10% of the body weight of fish fed.

Yellowtail in net cages grow remarkably rapidly. By December, when they are generally harvested, the 8- to 50-g fry which were stocked in June are 30 to 50 cm long and weigh 1.0 to 1.5 kg or, occasionally, more. A few fish may be left in the cages for several more months, perhaps until the start of the next farming season, at which time they are marketed as 40- to 60-cm fish weighing 2 to 3 kg. This is done only in the southern-most farms, where the water temperature is suitable for wintering, and accounts for only about 5% of the total production of cultured yellowtail.

PROBLEMS

It has been found that if the organisms which inevitably attach themselves to the walls of net cages are not controlled, growth is retarded. In addition, such organisms often harbor pathogens. Therefore the nets are periodically replaced—every ten days for nylon nets, or every three months for metal ones.

PLATE 2. One-kilogram yellowtail after harvest from floating cage.

Despite such precautions, diseases and parasites do occasionally occur. The commonest parasite is the flatworm *Bendenia seriolae,* which attaches to the skin. Infestation with *Bendenia* is prevented by dipping the stock in freshwater every 10 to 14 days. Since this worm matures very rapidly, to miss or postpone one of the treatments may allow it to get a foothold. If flatworms are found, they may be eliminated by exposing the fish to freshwater at 26°C for 3 min, or for 5 min at 16°C.

Axine heterocerca, a trematode sometimes found in the gills, which may cause a fatal anemia, may be cured with various drugs or, more simply, by placing the fish in strong salt water (100‰ salinity) for 2 to 4 min.

A greater danger is the bacterial disease vibriosis, which is similar to cholera. No cure for vibriosis has been perfected yet, though experimental results obtained by oral administration of sulfa drugs or antibiotics look promising. The best prevention is the usual—keep the fish well fed and in good condition.

PRODUCTION AND YIELD

Total production of cultured yellowtail has increased from 300 metric tons in 1958 to more than 30,000 metric tons in 1968. If only the area actually occupied by net cages is considered as being under culture, the per hectare yields achieved at yellowtail farms are astronomical. Estimates as high as 280 metric tons/ha have been made for farms which use a new rotary cropping system similar to that applied in culture of milkfish (*Chanos chanos*) in Taiwan (see Chapter 17). Even if the amount of unproductive water in any culture area is taken into consideration, the per hectare food yields of yellowtail farms must be among the highest ever obtained through aquaculture.

MARKETING

Yellowtail farmers, mostly family groups, find a ready market for their product, a staple food in Japan, and are able to realize middle-class incomes on the basis of less than one hectare of water. A competitive factor in favor of yellowtail culturists, as opposed to fishermen, is that, contrary to the usual situation with cultured fish, their product is judged to taste better than wild fish.

Fishery-caught yellowtail are marketed in two separate size categories. The larger fish, called "buri," have no counterpart in cultured yellowtail. If only the small "hamachi" are considered, cultured yellowtail constituted 36% of the total sold at the Osaka fish market (a principal outlet) in 1965. If problems of feeding and fry supply are solved, this percentage will probably increase.

SIGNIFICANCE OF YELLOWTAIL CULTURE

Several species of *Seriola* are widely distributed in the warm waters of the Atlantic and Pacific, and it seems inevitable that eventually fish culturists in other nations will follow the Japanese lead and institute culture of *S. quinqueradiata* or its congeners. But the significance of the Japanese achievement with yellowtail goes far beyond the future of *Seriola* spp. By successfully farming *S. quinqueradiata*, Japanese fish culturists have demonstrated two important points:

1. Marine fish can be intensively cultured. (Truly intensive culture would, of course, include propagation in captivity; this has not yet been achieved with yellowtail, but other marine fish are being bred in captivity in Japan, and yellowtail almost certainly will be eventually.) In Japan, practical culturists, spurred by this success, are increasingly interested in other species of marine fish. While yellowtail remain the principal marine fish crop, other species already account for nearly 25% of

the total production of cultured marine fish. Outside Japan, experimental culture of marine fishes is increasingly taking place.

2. At least some species of pelagic schooling fishes can be cultured in close confinement. Already, it is being advocated that other pelagic species, even such large predators as tuna, be cultured. (See Chapter 30 for a survey of ideas in this area.)

Thus, the development in the 1960s of large-scale culture of yellowtail in Japan may be seen as having profound implications in terms of increasing the world food supply. If, however, the more pessimistic ecologists' prophecies with regard to the fate of the oceans, and particularly piscivorous fishes, come to pass, then Japanese yellowtail culture will be seen in retrospect as nothing more than a temporary, isolated phenomenon.

REFERENCES

Interviews and Personal Communication

Fujiya, M. Japanese Fisheries Agency.

30

Culture of Pelagic Fishes Other Than Yellowtail

INTRODUCTION AND RATIONALE

Fish culture probably began with freshwater pond fish. When fish habitually or necessarily stay within a restricted area, it is reasonable to engage in feeding or other management activities with some expectation of realizing a return on the investment of time, labor, and capital. Culture of fish of streams, estuaries, and the inshore oceans requires somewhat more elaborate precautions to ensure that the stock stays around for the harvest but presents no insurmountable problems, and such fish also have a long history of culture. But management, let alone culture, of truly pelagic fishes is quite another thing. Confinement of such a fast-moving, wide-ranging animal as a tuna or a mackerel would seem to be unthinkable, and the efficacy of casting one's bread upon the open oceans in the hope that it will come back scaled is questionable indeed. Even if pelagic fishes could be "herded," as some biologists have suggested, the three-dimensional nature of the oceanic environment would permit but little control of food supply, breeding, predation, and so on. The maximum intensity of such management would be much less than that achieved with such relatively unrestricted beasts as, say, range cattle.

In the 1960s, however, Japanese fish culturists exploded all the preconceived notions about culture of pelagic fishes. When cage culture of the large, fast-swimming, piscivorous yellowtail (*Seriola quinqueradiata*), which had been occurring on a small scale since 1928, assumed major commercial importance, it was time for reevaluation of the potential role of pelagic fishes in aquaculture. (See Chapter 29 for details on culture of yellowtail.)

RESEARCH IN CALIFORNIA

The most extensive work to date on the special problems of culture of pelagic fishes has been carried out at the U.S. National Marine Fisheries Service Laboratory at La Jolla, California, where more than 20 species of marine fish have been hatched and reared to the adult form in captivity. Many of the species cultured experimentally are of little commercial value but, as most pelagic fishes have similar life histories and ecological requirements, much has been learned that will be of value in future attempts to culture pelagic fishes.

OBTAINING AND HATCHING EGGS

Most pelagic fishes and some other saltwater fishes lay pelagic eggs. Hatching such eggs is not particularly difficult, the crucial factor being the specific gravity of the water, which must be greater than that of the eggs, so that they will float during development. Once a suitable water supply is secured, it is merely a matter of regulating environmental parameters to suit the species being reared. The requirements of many species were worked out in the late nineteenth and early twentieth centuries by biologists of the U.S. Fish Commission.

Obtaining eggs is somewhat more problematical, and with most species biologists are still dependent on wild brood stock. However, it seems only a matter of time until suitable techniques of hormone-induced spawning are worked out for the commercially important marine fishes as they have been for so many freshwater species.

MAINTAINING EARLY LARVAE

More difficult problems are encountered in attempting to culture the larvae of pelagic fishes. At hatching, such larvae are in a prolarval or embryonic condition lacking pigmented eyes and have an incomplete digestive system. A second period of development is undergone during which egg yolk is gradually assimilated. Prolarvae are "passively active" while drifting in the sea. They do not have sustained powers of directed swimming but have the capability of sensing the approach of planktonic predators and avoiding the near surface water during periods of strong sunlight.

FEEDING OLDER LARVAE AND FRY

As egg yolk is used up in growth and metabolic processes, the retina of the eye becomes pigmented, a functional mouth and digestive system develop, and the prolarva approaches the true larval stage at which it is capable of directed swimming and capturing food organisms.

The onset of the true larval phase seems to occur at one of two distinct stages of egg yolk depletion, depending on the species. In one group of pelagic larvae, the feeding instinct manifests itself prior to complete utilization of egg yolk. These larvae have a built-in aid to survival in that an energy deficit, caused by lack of food, can be reduced by drawing on the remaining yolk reserve.

A second group of pelagic larvae do not begin feeding until all egg yolk has been absorbed. These larvae, to survive, must find food within a short time because the only energy source available for swimming and physiological processes comes from absorption of body tissues. A second noteworthy feature, which might be characteristic of this group of larvae, is that in all of the species reared to date, onset of feeding occurred in fish at a very small size. The food organisms eaten are also small. For example, the European plaice (*Pleuronectes platessa*), which begins feeding at approximately 7 mm in length, while some yolk remains in reserve, is able to consume food organisms as large as 0.3 mm in diameter. The Pacific sardine (*Sardinops caerula*), however, which does not begin feeding until all yolk is absorbed at a length of 5.5 mm, is able to consume organisms not much larger than 0.08 mm in diameter. Many of the species in the group which start feeding late, such as Pacific mackerel (*Pneumatophorus diego*), Pacific barracuda (*Sphyraena argentea*), Northern anchovy (*Engraulis mordax*), and jack mackerel (*Trachurus symmetricus*), range in size from 3 to 4 mm at onset of feeding and feed on organisms smaller than 0.05 mm in diameter. Also, being small larvae, their initial swimming endurance and range are limited.

Rearing these most difficult of pelagic larvae necessitated developing a technique for supplying very small food organisms, at the precise time of complete yolk absorption, and in sufficient quantities to allow larvae of low mobility to find food in all parts of the aquarium. The magnitude of the task can be assessed when it is learned that sardine larvae were observed to search only about 1 cm³ of water per hour at onset of feeding but required a minimum of four food organisms per hour to replace energy lost in swimming and body functions. The 2300-liter aquaria, found most suitable for rearing this species, contain approximately 1,800,000 cm³ of water. The staggering number of food organisms which must be supplied if high survival of larvae during the first 12 hours of active feeding is to be obtained is thus apparent.

A second major problem occurs in supplying food organisms in a series of increasing sizes. As the larvae grow, the optimum size of food organisms consumed increases, since larvae tend to seek out the largest organisms they can swallow.

The technical problems of supplying large quantities of food organisms in varying size ranges have been met, in part, by collecting them at night. A 1000-W underwater lamp connected to a submersible pump is suspended several feet below the surface of the sea. The strong light attracts copepods from a wide distance and concentrates them near the pump where they are sucked up with the water and transported to the surface. Plankton-enriched water then passes through a series of filters which further concentrates food organisms and the highly enriched filtrate is piped to a 900-liter storage tank. This process collects organisms with a cross-sectional diameter of 0.028 mm and larger. Prior to being fed to fish larvae, concentrated plankton is graded by filters to remove organisms larger than 0.10 mm. The portion containing large copepods, crab larvae, chaetognaths, and so on, may be fed to advanced fish fry and juveniles.

After larvae have grown large enough to consume zooplankters larger than 0.25 mm, supplemental feeding of brine shrimp (*Artemia*) eases the task of collecting an ever increasing supply of wild food organisms. *Artemia* eggs are hatched in the laboratory in a continuous series and provide a convenient but incomplete source of food for fish larvae.

GROWTH AND SUITABILITY OF VARIOUS SPECIES

Rate of growth in fish is governed by inherent individual and species gene potentials, which are, in turn, to a large degree moderated by water temperature. Quantity of food present in the environment plays a subordinate role as far as rate of larval growth is concerned; food must be present in sufficient amounts to offset the constant drain on body energy. Marine fish larvae do not seem to have much, if any, capacity for storing an energy reserve in the form of fat. Larvae held for a short time under suboptimal conditions rapidly develop lethal physiological disturbances. During the critical first days of active life, survival is a matter of balance between energy expended and energy obtained from food. The slight margin of energy obtained in excess of physiological needs is reflected in growth.

A slight increase in larval size is correlated to an increase in size of food particles eaten and in an ability to range further per minute of swimming in search of food. In short, efficiency of living increases with size. The mode this efficiency takes is, however, different in various types of fish larvae. For example, clupeoid species as Pacific sardine, northern anchovy, and a number of species of flatfish (Pleuronectidae and Bothidae) begin active life as long slender larvae. Other types of larva, such as Pacific and jack mackerel, Pacific barracuda, and tunalike fish, begin active life in a similar form, but subsequent ontogeny of the two body types and food capacities are very dissimilar.

Soon after feeding begins (about 3.5 mm), the Pacific mackerel has developed a large head and capacious mouth. The sardine larva at this time is about 5.0 to 5.5 mm long and has a small head and mouth. The relative capabilities of the two types of oral apparatus may be seen in the progressive increases in food consumed by the Pacific mackerel-type larva as compared to the sardine-type.

At a length of 4.5 mm, the Pacific mackerel might contain about 0.004 cm³ of food. Assuming that the volume of the Pacific mackerel increases roughly with the cube of its length, then doubling its length would result in an eightfold increase of body volume; the volume of food, however, increases about 200 times. The sardine, while increasing in length from 4.0 to 7.0 mm, theoretically increases in bulk by a factor of about 5.4 times. Its average food volume increases during this growth only by a factor of 1.25 times. The sardine-type larva requires approximately 30 days from onset of feeding to undergo metamorphosis to the adult body form, whereas the mackerel-type larva metamorphoses to the adult form in approximately 7 days after onset of feeding.

POSSIBLE APPLICATION TO COMMERCIAL CULTURE

The factors influencing growth, as outlined, point to the potential of culturing species having a mackerel-type development and growth. All of the pelagic, commercially valuable piscivorous species reared so far exhibit this type of development.

On the other hand, clupeoid fishes such as sardines and herring have the advantage of being able to digest carbohydrates, which more or less piscivorous pelagic fishes apparently cannot do. Thus advanced fry and adults can be fed on present commercial fish foods. This seeming advantage will probably be of less importance as new feeds, tailored to the needs of pelagic fishes, are developed to supplant the presently used "trout pellets."

The fastest growth rates among pelagic fry have been recorded for such typically solitary piscivores as the Pacific barracuda, which has been grown from 4.5 to 35 mm in about 8 days after commencing piscivorous feeding—a doubling of length every 24 hours. Solitary fishes are ill suited for intensive culture, however, as, aside from their space requirements, they will accept only live fish as food. Schooling piscivores such as mackerel, on the other hand, may be easily stimulated to a "feeding frenzy," at which time they will take ground anchovy, squid, or other prepared foods.

Of the species studied at La Jolla, the best results have been obtained with the Pacific mackerel. The chief problem encountered has been heavy cannibalism, commencing at about 9 days after hatching, when the fry are about 10 mm long. Up to 50% population reduction has resulted, but it is believed that much of this could be averted in a more spacious environment.

Pacific mackerel fry, kept in a 2300-liter tank and grown on a diet of large copepods, adult brine shrimp, larvae of other fish, and each other, attained lengths of 25 mm in 16 days after hatching. At this stage they were transferred to a pool 4 m × 3 m × 1 m deep, where they were reared to the age of 3 months before various problems of a nonbiological nature caused a 75% mortality. A second, nearly total, mortality 3 months later terminated the experiment. The 6-month-old mackerel averaged 0.11 kg in weight and 20 cm in fork length. Based on the 341 mackerel that survived to the age of 6 months, the prorated yield of this experiment is tremendous—31,130 kg/ha.

Schooling fishes, in addition to being easiest to feed, are best suited to intensive culture since they are preadapted to crowding. The nervous temperament of many of these fishes may be less of a problem than might be anticipated, as at least some species, when reared from early life under artificial conditions, do not exhibit the "wild" or excitable behavior of their naturally grown counterparts but behave more like trout cultured in a trout pond. A case in point is the successful rearing to the juvenile stage (80 mm) of the California flying fish (*Cypselurus californicus*). As far as can be learned, wild flying fish are extremely delicate and excitable and do not live long at any stage when placed in confinement. When cultured from the egg, however, a remarkable behavioral adjustment to life in a small area is observed.

THE JAPANESE TUNA CULTURE PROGRAM

Apart from yellowtail culture in Japan, growing of commercially important pelagic fishes to marketable size is largely in the realm of speculation. A notable exception is the tuna culture program of the Japanese Fisheries Agency. In addition to experimental breeding and hatching of bluefin tuna (*Thunnus thynnus*), this agency has inaugurated a tuna rearing experiment at Shizuoka Prefectural Fisheries Experimental Station. The immediate objective is to rear 0.3- to 0.5-kg 1-year-old tuna, captured in set nets, to marketable size (2 to 8 kg) 2-year-olds. The ultimate goal is to rear tuna to maturity and achieve reproduction under controlled conditions. The initial experiments will involve 200 tuna, divided between two octagonal floating pens, 5 m on a side and 9 m deep. It is not known whether adequate food can be provided or whether the fish will survive winter water temperatures, which sometimes drop to 12°C.

If tuna culture in floating cages proves unfeasible, the suggestion of M. Inoue of the Fishery Research Laboratory, at Tokai, that tuna be reared in atolls and lagoons in the tropical Pacific may yet be heeded.

RESEARCH OR SPECULATION WITH OTHER SPECIES

The only other pelagic fish which is remotely near the threshold of practical culture is the bluefish (*Pomatomus saltatrix*), which is being experimentally grown in tanks at the University of Rhode Island.

Artificial fertilization, hatching, and limited larval rearing of a number of pelagic fishes, including the Atka mackerel (*Pleurogrammus azonus*) and the Pacific saury (*Cololaibis saira*), has been achieved in Japan, but no attempts have been made to apply the results to practical culture.

Efforts have been made to rear artificially propagated yellowfin tuna (*Thunnus albacares*), but larvae have survived no longer than 20 days.

As mentioned, the suitability for culture of clupeoid fishes is reduced by the nature of their larval development. A number of species, particularly the oceanic herring (*Clupea harengus*), are commercially valuable, however, and some effort has been made to culture them. Eggs of the Baltic herring (*C. harengus membras*) have been artificially fertilized in the Soviet Union as a means of transporting this delicate fish from its native Baltic Sea to the Aral Sea for the purpose of introduction. So far as is known, more intensive culture is not contemplated.

About 1960, in Scotland, it was demonstrated that sperm of the oceanic herring could be preserved for long periods, so that hybridization of the many subspecies and races of this widely distributed fish would be possible. The problems involved in rearing the larvae have also been studied, and a few individuals have been reared past the point of metamorphosis to the adult form. This work has apparently not been followed by attempts at large-scale culture.

PROSPECTUS

It has been suggested that some of the coral atoll lagoons in the tropical Pacific might be converted into fish farms, an idea which at first appears attractive. However, schemes thus far proposed have involved organic enrichment of the lagoons. Recent experience in Hawaii has shown that sewage may be lethal to the tiny coral animals which construct and maintain the Pacific atolls. In some cases mortality has been attributed directly to the toxic effect of the sewage; in others excessive algal growth resulting from organic enrichment has been blamed. Addition of fertilizers undoubtedly would have a similar effect. Even the enrichment resulting from artificial feeding and high-density stocking of fish might eventually destroy the coral. Thus atoll farming, advocated as a means of increasing the world protein supply, might in fact only succeed in reducing the protein available to the peoples of the Pacific islands, while simultaneously destroying biological communities of great scientific and esthetic interest. Certainly no major efforts at atoll farming should be made unless preceded by careful long-term pilot studies.

With the problems of larval culture and controlled reproduction well on the way to solution, practical culture of pelagic fishes other than yellowtail seems more and more likely to become a reality. For the present, the Japanese method of cage growing seems most likely to succeed in enabling culturists to supply some of these fishes more cheaply and reliably than fishermen, who must hunt, corral, and capture them.

REFERENCES

ROEDEL, P. M. 1953. Common ocean fishes off the California coast. California Department of Fish and Game, Fisheries Bulletin 91.

ROLLEFSEN, G. 1938–39. Artificial rearing of fry of seawater fish. Rapp. et Proc. Verb. Cons. Internat. Explor. Mer. 109:133–134.

YUSA, T. 1960. Embryonic development of the Saury Cololaibis saira (Brevoort). Bulletin of the Tohoku Regional Fisheries Research Laboratory 7.

YUSA, T. 1967. Embryonic development and larvae of the Atka Mackerel Pleurogrammus azonus Jordan et Metz. Sci. Rep. Tohoku U., 4th Series, Biology Vol. XXXIII, 3–4.

Interviews and Personal Communication

SCHUMANN, G. San Diego Gas and Electric Co., San Diego, California.

31

Culture of Marine Flatfishes

The flatfishes (order Pleuronectiformes) include a number of valuable food fishes, marketed as plaice, sole, flounder, halibut, and turbot. As a group, they are among the most commercially important fish of North Temperate waters. Since they inhabit chiefly shallow waters, are generally hardy, and, unlike most of the marine commercial fishes, are somewhat sedentary, they were among the first marine fishes to be experimentally cultured.

A number of important flatfish fishing grounds showed evidence of slight environmental deterioration and/or overfishing as early as the last half of the nineteenth century. In an effort to maintain these fisheries, several species of flatfish were artificially propagated and larvae stocked along both shores of the Atlantic during the 1890s and the early part

PLATE 1.   Internal view of Port Erin flatfish hatchery showing racked incubation tanks. (Courtesy C. E. Nash.)

of this century. Although plaice (*Pleuronectes platessa*) have been thus cultured and stocked in Norway for 60 years, the efficacy of the practice is questionable, and it was long ago discontinued in most countries. Almost all of the recent progress in flatfish culture has been made in the United Kingdom, under the aegis of the White Fish Authority.

Research on more or less intensive culture of flatfishes in the United Kingdom began during World War II. No serious problems were encountered in most phases of culture, but, before 1962, no more than 10% of larvae in hatcheries could be reared to metamorphosis. In 1962, flatfish culture was revolutionized by James Shelbourne, then at the Fisheries Laboratory, Lowestoft, England. Shelbourne discovered that the main cause of heavy mortality of larvae was a form of bacteria. By adding a mixture of penicillin and streptomycin to the hatching tanks, he was able to control these bacteria and rear 60 to 80% of the larvae to metamorphosis. As cultured stocks become generations removed from wild fish, this antibiotic treatment may lead to lowered disease resistance, but there is no sign of such a phenomenon yet.

Since Shelbourne's breakthrough, the center of flatfish culture has shifted from Lowestoft to Port Erin, on the Isle of Man. A pilot hatchery built there by the British White Fish Authority was so successful that it functioned as a supply station for the other field stations which have been established (Plate 1). At present, flatfish culture in Great Britain has been restricted by limited financial support for development of hatcheries.

HATCHERY CULTURE IN THE UNITED KINGDOM

SPECIES USED

The species of flatfish chosen by the White Fish Authority for their initial culture experiments was the plaice, but sole (*Solea solea*) were subsequently found to be hardier and better suited to growing at high temperatures, as well as bringing twice the price, and both species are now cultured. Other flatfish, including turbot (*Scophthalmus maximus*), and lemon sole (*Microstomus kitt*) have been bred in captivity, using the techniques to be described here for plaice and sole, but to date suitable foods have not been found for the fry of these species. Both of the principal cultured species are cold water, benthic carnivores which require an oceanic salinity of about 35‰.

SPAWNING

Brood stock at Port Erin are kept in outdoor ponds and, for the spawning season (February to May), transferred to indoor tanks, 4.2 m square and 1.2 m deep, where spawning occurs naturally from February to May without any manipulation of the animals. The floating eggs are scooped off the surface and hatched separately. Spawning tanks, like all tanks at the hatchery, are supplied with recirculating, ultraviolet sterilized sea water, which has passed through plastic pipes only. A shallow layer of sand may be kept on the bottom of each tank, since captive flatfish are much less uneasy if they can partially bury themselves.

HATCHING AND FRY REARING

The hatching and fry rearing troughs, some of which have individual temperature control units, are 4.9m × 1.2 m × 1.2 m deep. Each tank is stocked with 30 to 40,000 eggs. Hatching requires 3 weeks at 6°C, and metamorphosis to the benthic form occurs 6 to 7 weeks later. Handling of all life stages is minimized by keeping the fish in the hatching tanks throughout culture. When it is necessary to transfer larvae or fry it may be done by means of a vertical overflow system, or by simply partitioning off the tank.

The newly hatched larvae are planktonic (otherwise there would be no need for the hatching tanks to be so deep) and commence to feed a few days after hatching. Provision of proper food at this time is very critical. The only known suitable food for young plaice and sole larvae is brine shrimp (*Artemia*), and even brine shrimp are not accepted by some other flatfish. For reasons which are not understood, but may have to do with pesticide residues, the source of the brine shrimp eggs is also critical. In 1966, when the White Fish Authority switched suppliers and purchased Utah eggs instead of the California eggs they had been using, only 3000 plaice and 5000 sole out of an expected crop of 100,000 of each species survived.

At first each young flatfish requires about 10 newly hatched brine shrimp a day, but larger larvae may take 200 daily. To cope with the almost continuous feeding required, the Port Erin hatchery is equipped with highly sophisticated automated brine shrimp incubator-feeder units.

Although brine shrimp are absolutely essential for very young larvae, British biologists attempt to wean plaice and sole larvae from brine shrimp as soon as possible, and certainly before metamorphosis. A good intermediate food has been found in the form of an oligochaete worm which inhabits the North Sea tidal zone, but automated techniques for its continuous culture have yet to be developed, and natural stocks are limited. After metamorphosis, the fry are fed on chopped mussels or fish until they are 3 cm long, at which time they are shipped to one of the White Fish Authority's other field stations for stocking in growing enclosures.

It has been found that shipping 3- to 4-day-old eggs is much more feasible than shipping young flatfish. Thousands of eggs can be shipped in a 3-liter flask, whereas at best hundreds of fry can be accommodated in a similar amount of space. It is expected that, once hatchery operations at other stations become as efficient as those at Port Erin, shipping of eggs and fry can be eliminated.

OBJECTIVES OF THE BRITISH WHITE FISH AUTHORITY IN CULTURING FLATFISH

The original objectives of the White Fish Authority in culturing flatfishes were threefold:

1. To investigate the feasibility of augmenting fisheries by stocking flatfish fry.

2. To investigate the use of enclosed natural areas for growing flatfish, with fertilization of the water and/or supplementary feeding.

3. To experiment with intensive culture in heated water.

STOCKING TO AUGMENT FISHERIES

Experience had already shown that stocking flatfish larvae accomplished nothing other than to provide a free meal for various predators. However, at the time when flatfish stocking was common practice, it was not possible to rear the larvae through metamorphosis, thus stocking of fry had never been tested. Accordingly, considerable numbers of fry were stocked in the Irish Sea. They fared no better than the previously stocked larvae. It was noticed that the behavior of hatchery-reared fry was altered, that they did not "know" how to cover themselves with sand, as wild fish do, to elude predators. Whether or not this behavior was a crucial factor in their high mortality, the stocking program was abandoned.

Actual stocking and yield data on the Irish Sea experiments are scarce, but the results of computer calculations made by Saul Saila of the University of Rhode Island's Aquatic Sciences Information Retrieval Center are of interest. Saila estimated the numbers of small winter flounders (*Pseudopleuronectes americanus*) which would have to be stocked to affect the fishery for that species in Rhode Island. Given the average annual fishery yield of 900,000 kg, it would require 22,500 kg of juvenile flounders, or several billion individuals, to increase the fishery yield by 10%. The additional fish would not come close to offsetting the expenses of hatchery culture.

CULTURE OF MARKETABLE SIZE FISH

The results of attempts to culture marketable size plaice and sole have been more encouraging. The first attempts at growing hatchery-reared plaice fry to marketable size were based on work done during World War II by Fabius Gross of Edinburgh University. Gross found that, by fertilizing an enclosed sea loch with superphosphate and sodium nitrate, it was possible to grow flatfish two to four times as rapidly as in untreated waters, but his work was restricted by the scarcity of young fish for stocking.

With hatchery techniques perfected, fry supplies ceased to be a problem, and in 1965 the White Fish Authority established another field station at Ardtoe, Scotland, where a 2-ha loch was dammed and stocked with 200,000 young plaice (Plate 2). In lieu of fertilization, the fish were fed with chopped mussels and trash fish. Problems were numerous the

PLATE 2.   General view of a Scottish loch, part of which is enclosed for flatfish culture.
(Courtesy C. E. Nash.)

first year: heavy rainfall and surface runoff diluted the sea water (plaice
are particularly sensitive to low salinities); rotting vegetation caused
oxygen depletion; and predators and competitors ran rampant. As one
worker put it, "We had found an excellent way of growing the shore
crab."

Nevertheless, the culturists at Ardtoe persisted, and in 1968 the first
crop of marketable fish was produced. It was decided that the original
enclosure was too large, and experiments are now carried out in smaller
enclosures, both within the original 2-ha loch and in the sea. Net cages
are also being used. Control of predation and salinity are still of concern,
but the principal stumbling block to commercial feasibility at this time
is conversion of food. It is estimated that it takes 5 kg of food to produce
1 kg of plaice or sole (wet weight). While high from an ecological viewpoint, such a conversion rate does not favor economical mass rearing.

CULTURE IN HEATED WATER

More intensive culture of plaice and sole, in artificially heated water,
began in 1966 at a nuclear power plant at Hunterston, Scotland (Plate
3). Sea water at Hunterston ranges in temperature from 9.5 to 16.4°C

PLATE 3. Experimental ponds for rearing plaice and sole in warm water discharge from Hunterston nuclear power plant, Scotland. (Courtesy C. E. Nash.)

in the summer and gets as cold as 3°C during the winter. At such low temperatures, growth of most fish, including plaice and sole, all but ceases. The power plant effluent temperature ranges from 10 to 23°C and, by mixing it with sea water, temperatures can be maintained such that plaice and sole will grow throughout the year.

The original installation at Hunterston consisted of no more than four cement tanks, each 14.4 m × 7.2 m × 1.2 m deep. In 1968, hatchery facilities were added, and plaice and sole are now spawned, hatched, and reared in heated water. The results have been remarkable. Growth in the early stages was accelerated and larvae metamorphosed within 3 weeks of hatching rather than taking 6 to 7 weeks as is required at the Port Erin hatchery. Growth of juveniles and adults was equally impressive (Plates 4 and 5). Sole stocked as 3.5-cm fish at 325 to 900/m² reached an average length of 15.3 cm and a maximum of 23.3 cm in 11 months. Similar growth was achieved by plaice. Minimum marketable length for both species is about 20 cm, which size they attain in 3 to 4 years in nature. The experiments at Hunterston have shown that it is possible to produce cultured fish of this size in 2 years or less.

The studies thus far carried out on culture of flatfishes in thermal

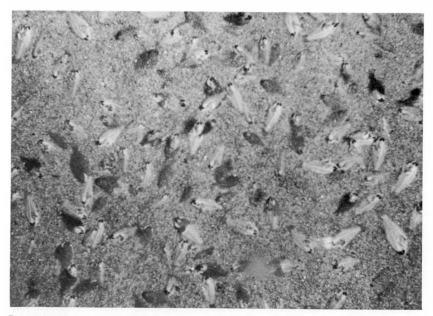

PLATE 4. Ten-week-old plaice (*Pleuronectes platessa*) which have completed meta-morphosis at Hunterston station. Note variety of pigmentation cover. (Courtesy C. E. Nash.)

effluent have by no means provided the last word on the subject. The maximum temperatures tolerated by plaice and sole are about 20 and 23°C, respectively; however, the optimum temperatures for culture of these fish are not known. By careful manipulation of the thermal regime, growth may eventually be further accelerated or thermal requirements may be altered by selective breeding and/or hybridization.

There was some concern about possible effects on flatfishes of the low concentrations of chlorine routinely introduced into power plant effluents to prevent fouling of pipes by sessile marine organisms. No ill effects have been observed, however, and, although the chemistry of chlorine in such situations is poorly understood, there are no indications that its use would in any way endanger human consumers of cultured flatfish. It may even have a positive effect by suppressing the growth of disease-causing bacteria. To avert another potential health hazard, fish reared in nuclear plant effluent should constantly be monitored for radioactivity.

CULTURE OF FLATFISH OUTSIDE THE UNITED KINGDOM

The achievements of the White Fish Authority have drawn the attention of fish culturists outside the United Kingdom, but so far no one has been

PLATE 5. Dover sole reach marketable size at 20 months of rearing in power station warm water discharge. (Courtesy C. E. Nash.)

moved to emulate the British effort. As mentioned, Norwegian culturists continue to breed plaice as well as the river flounder (*Pleuronectes flesus*), but they are still concentrating on improvement of hatchery techniques. In Japan, a number of commercially valuable flatfishes, including the mud dab (*Limanda yokohamae*), the starry flounder (*Platichthys stellatus*), *Lepidopsetta mochigarei*, *Limanda schrenki*, and *Xystreurys grigorjewi*, have been artificially spawned and hatched, but there are no indications that this achievement is to be applied to commercial culture.

Perhaps the next country to take up intensive culture of flatfish will be Denmark, That country's Limfjord, although it has no indigenous plaice, is ideally suited for their growth. Yearling and two-year-old plaice, harvested from the North Sea, have been stocked in the Limfjord at least since the 1920s, and perhaps as far back as 1910, and now support a sizable fishery. Since plaice for some reason do not spawn in the Limfjord, stocking is of necessity an annual affair. At the time the stocking program began, the North Sea was abundantly supplied, perhaps even overpopulated, with plaice. This is no longer the case, thus it may become necessary for Denmark to construct her own plaice hatcheries.

PROSPECTUS

Further developments are to be expected in the United Kingdom. It has already been mentioned that much remains to be learned about the thermal requirements of flatfish. The same applies to stocking densities and feeding.

It is apparent that plaice and sole are quite tolerant of crowding. Not only has good growth been achieved at initial densities of up to 900 fry/m² at the Hunterston station, but these fish have remained completely free of disease. It is likely that even better growth can be achieved at lower densities. This is probably not inconsistent with economic considerations if Shelbourne's estimate is correct. He calculates that if each flatfish were allotted 1 ft² (0.09 m²) of bottom, an amount equal to the entire British fishery catch could be produced annually in ponds covering 326 ha.

Important as thermal regimes and stocking densities are, it is probably feeding which is most crucial to the success of flatfish culture. Development of relatively inexpensive, high-protein, prepared foods in pellet form has been part of the key to success in culture of other carnivorous fishes and should be pursued in the case of flatfishes.

Other possibilities for improvement of flatfish culture include introduction of new species, hybridization, selective breeding, polyculture, and new developments in marketing and engineering. The last two will of necessity involve cooperation with personnel other than biologists and professional fish culturists. The success of the White Fish Authority so far already constitutes an outstanding example of profitable collaboration between biologists and engineers.

With respect to marketing, the situation with flatfishes in the United Kingdom is somewhat different from that in some other fish culture industries, where the culturist is in direct competition with commercial fishermen. The mandate of the White Fish Authority is to aid in maintaining a steady supply of flatfish for the British consumer; they seek to supplement the fishery, not compete with it.

Nevertheless, in the course of experimental culture of plaice, one development has occurred which could eventually provide culturists with an economic advantage over fishermen. Many cultured plaice show pigment abnormalities in the form of white patches. The cause of the discoloration is not known, but it is thought that handling of young fry results in the loss of patches of skin and that thereafter the fish cannot properly lay down melanin in the damaged areas. Such fry are of course at a disadvantage in a natural environment due to their high visibility but are otherwise not discernibly handicapped. If this condition could be deliberately induced in such a manner that the entire body would be white, it would be an advantage to processors, in that the fish would not need to be skinned. This would not only eliminate a step in processing but would bring about a considerable reduction in weight loss during processing.

Hybridization, improvement of strains by selective breeding, and introduction of new species to culture might all serve any of several purposes, such as increasing temperature tolerance, accelerating growth, or improving food conversion. As mentioned, intensive culture of flatfishes other than plaice and sole has been held back by the inability of culturists to feed the larvae. Selection of plaice and sole is just beginning, as is hybridization. Plaice × lemon sole crosses have been made in the United Kingdom, and in Norway the hybrid ♀ *Pleuronectes platessa* × ♂ *Pleuronectes flesus* has been produced, but in neither case have the characteristics of the hybrids been studied.

Some White Fish Authority biologists foresee a marine analog of Chinese pond polyculture (see Chapter 2) in power plant effluent, with flatfishes as the principal crop. Not only are there ecological niches to be filled with edible fishes, but addition of some kinds of fish might result in increased yields of flatfish. Of particular value would be scavengers to assist with the considerable problems of hygiene brought on by the use of such foods as chopped mussels, and algae feeders to take advantage of the luxuriant and potentially dangerous algal growths which often appear in heated sea water.

The flatfishes are chiefly inhabitants of North Temperate waters. For this reason, and because of their carnivorous feeding habits, commercial culture, if it ever becomes a reality, will probably be confined to the economically more fortunate countries. At present, White Fish Authority officials indicate that they do not know enough to predict the economic feasibility of a commercial flatfish farming adventure, but prospects look good. Nevertheless, fish culturists and government agencies outside the United Kingdom seem content to let the White Fish Authority more thoroughly explore the possibilities before committing their resources.

REFERENCES

LOFTAS, T. 1967. Fish farming for the future. Ill. London News (July 22, 1967):18–19.

NASH, C. E. 1969. Thermal aquaculture. Sea Frontiers 15:(5):168–276.

SHELBOURNE, J. E. 1964. Artificial propagation of marine fish. *In* J. T. Russell (ed.), Advances in Marine Biology, Vol. II. Academic Press, New York.

White Fish Authority. 1967. Research and Development Progress Report, 1967. Fish Resources.

Interviews and Personal Communication

BERTELSEN, E. Danish Ministry of Fisheries.

NASH, C. E. The Oceanic Institute, Waimanalo, Hawaii.

RICHARDSON, I. D. British White Fish Authority.

SAILA, S. University of Rhode Island (from a letter to Dr. S. Holt, Chief, Fisheries Biology Branch, FAO, Rome, Italy, January 17, 1964).

32

Shrimp Culture

Decapod crustaceans of the suborder Natantia (shrimps or prawns) are found in fresh and salt waters virtually all over the world and many of the larger species are highly valued as human food. In most countries, shrimp culture is nonexistent or at best in the experimental stages, but some species have been cultivated in southeast Asia for five centuries or more. The methods in use are crude, often consisting of no more than trapping and confinement of young shrimp in brackish water ponds for several months before harvesting. In most southeast Asian countries production of shrimp has long been incidental to the culture of brackish

587

water fish, but monoculture of shrimp, with concurrent technological advances, is emerging in a number of countries.

Truly intensive shrimp culture, with breeding in captivity, has yet to play a significant role in southeast Asia but has been part of the picture in Japan since 1934, when Motosaku Fujinaga achieved the first success in spawning and partial rearing of the kuruma shrimp (*Penaeus japonicus*). Fujinaga strove to perfect his techniques until 1959 when, with the financial assistance of several fishery companies, he was able to set up a pilot hatchery and farm. By 1967, some 20 operators were using his techniques to produce 4000 tons of shrimp annually from 8500 ha of water.

There is presently no successful commercial shrimp culture in the Western hemisphere, but scientists and entrepreneurs in the United States and several Latin American countries have noted the similarity between American commercial shrimps and the kuruma shrimp and have attempted to apply Japanese culture methods, to date with only partial success. Attempts have also been made in the United States to apply southeast Asian methods, but would-be shrimp culturists have been held back not only by biological and technological problems but by legal difficulties with regard to the ownership and use of estuarine waters.

Shrimp of various species are fished virtually from pole to pole, but commercial shrimp culture is, with the notable exception of Japan, confined to the tropical and subtropical regions. In recent years, however, experimental shrimp culture has been initiated in temperate zone countries. Nor is freshwater exempt. Experiments in Malaysia and elsewhere have demonstrated the potential for culture of freshwater prawns.

SHRIMP CULTURE IN SOUTHEAST ASIA

Despite the long history of shrimp culture in southeast Asia, little is known about the biology of the species cultured there. Up to thirteen species of Penaeid shrimps and seven species of Caridean shrimps, all edible, may be found in ponds used in culture, but none of the Carideans and only seven of the Penaeids (the banana prawn, *Penaeus merguiensis*; the Indian prawn, *Penaeus indicus*; the sugpo prawn or giant tiger prawn, *Penaeus monodon*; the green tiger prawn, *Penaeus semisulcatus*; the yellow prawn, *Metapenaeus brevicornis*; *Metapenaeus ensis* and *Metapenaeus burkenroadi*) are eaten. In India *Penaeus carinatus*, *Metapenaeus dobsoni*, and the Palaemonid species *Leander styliferus* may be added to the list of cultured shrimps. All the cultured species share the habit of entering shallow coastal waters when very young. There they settle to

PLATE 1.   Shrimp (prawns) cultivated in Taiwan. Species illustrated from left to right: *Metapeneus monceros;* two *Penaeus teraoi; P. japonicus; P. monodon;* four *P. semisulcatus.* (Courtesy I. Chiu Liao, Tungkang Marine Laboratory, Taiwan.)

the bottom, where they grow rapidly for several weeks then return to the sea. It is a matter for debate whether postlarval shrimp actively seek out brackish water or if only a small fraction of the total population find their way to the estuaries by chance. Whichever is the case, several species occur in estuaries in sufficient abundance to support a fairly significant shrimp culture industry. Selection for species occurs only at the harvest or marketing stages, except in the Philippines where the sugpo prawn is cultured exclusively.

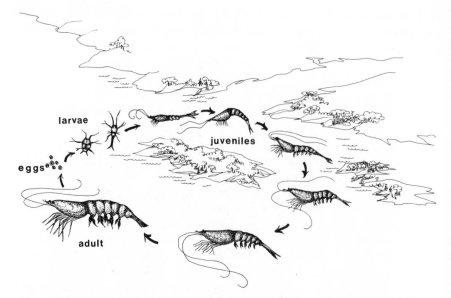

FIG. 1.   Life cycle of Penaeid shrimps. (After U.S. Bureau of Commercial Fisheries, Circular 312, Gulf of Mexico Shrimp Atlas.)

MALAYSIA AND SINGAPORE

Unlike most parts of southeast Asia, the Malay Peninsula has no tradition of brackish water pond culture. However, there is a large demand for shrimp by the predominantly Chinese population of Singapore. There is no local shrimp fishery, so shrimp had to be imported until the initiation in 1937 of shrimp culture in modified mangrove swamps in rural Singapore. Singapore thus became the first country to support shrimp culture exclusive of fish. The demand for shrimp remains high today, but shrimp culture is on the decline due to the rapid usurpation of coastal lands for industrial purposes. Consequently, shrimp culture has spread into neighboring areas of Malaysia.

Before attempting to modify a mangrove swamp for shrimp culture, its suitability for that purpose must be evaluated. The first step in site evaluation is sampling by cast net to determine the relative abundance of various shrimps. (The species list on p. 588 is in order of preference.) If large numbers of Caridean shrimps are captured, the site is likely to be rejected. There are seven additional factors important in site selection.

1. Tidal range and swamp elevation. Tidal inflow and outflow should ideally be such that the pond can be harvested every day, since premium prices are paid for very fresh shrimp. If the swamp is too high or high tides too low, no sea water will enter on neap tides, even at high tide, thus reducing the number of fishing days. In practice it is not possible to harvest any pond every day; culturists aim at 20 to 22 harvest days per month. Very low swamps and high tides provide more potential harvest days, but expenditures for construction and maintenance of dikes and sluice gates may be prohibitive under these circumstances.

2. Depth. It should be possible to maintain about 0.7 m of water in the pond at all times.

3. Location relative to the coast. If the pond is too near the beach, the dike may be damaged by wave action. Usually a fringe of at least 15 m of mangroves is left undisturbed between the shore and the pond for protection. On the other hand, if the pond is too far from the coast it may be less accessible to shrimp and there may be a danger of flooding with rainwater during the monsoon season.

4. Soil quality. The bottom should be mostly hard clay, perhaps mixed with sand and organic detritus. Soft, porous bottom materials are to be avoided, since the swamp bottom is the source of construction material for the dike. There should be no more than 50 cm of silt on the bottom since, for unknown reasons, heavy silting reduces shrimp catches. Large amounts of organic matter are also to be avoided, since their decay may produce anaerobic conditions leading to mass mortalities of shrimp.

5. Surface area. The smallest ponds require at least two workers daily for maintenance and operation. For this reason ponds of less than 12 ha are not economically feasible.

6. Salinity. Salinity of sea water entering a shrimp pond should be at least 18‰ at all times; 24 to 30‰ is considered optimal.

7. Drainage basin. Heavy freshwater run-off can reduce salinity below the preferred level for shrimp. For this reason swamps which receive large freshwater streams are not favored. If a swamp is otherwise suitable, however, it may be feasible to divert such streams.

Once a suitable swamp is located it is converted into a pond by surrounding it with an earthen dike, or bund. The bund must be high enough to prevent the highest tide from flowing over the top, strong enough to withstand water pressure differentials, and wide enough for operating personnel to walk on top. Construction is usually by manual labor. Given the soft soils of mangrove swamps, this is more efficient than dredges, tractors, bulldozers, and so on. Indeed, where machinery has been used in bund construction it has been necessary to make the bund wider and higher merely to support the machinery.

Clay for bund construction is taken from the swamp bottom in slabs about 30 cm × 15 cm × 10 cm thick. These slabs are laid side by side around the perimeter of the pond and each layer is sun baked before the next is added. The quality of workmanship at this stage will largely determine the amount of maintenance necessary in the future. Even a well-constructed bund will sink about 30 cm/year, but poorly constructed bunds may sink as much as 90 cm.

Clay slabs are excavated so as to create a pattern of channels radiating from the sluice gates to the farthest corners of the pond. Experience has shown that this practice increases yields. It may be that the channels serve to distribute young shrimp evenly so that they are less vulnerable to predators and make more efficient use of the food supply. Not only the channels but the entire bottom is cleared of mangrove stumps and leveled, since stumps and other irregularities provide hiding places for predators. The bottom should slope toward the sluice gates to permit total drainage if desired.

Sluice gates must be large and numerous enough to allow tides to enter and leave the pond rapidly, thus avoiding extreme pressures on either side of the bund. Sluice gates should be located so that the incoming tide will not generate eddies and whirlpools, which may trap the incoming shrimp. The gates themselves should be built with a concrete foundation and sides, and the bund adjacent to the gates should be reinforced by wooden pilings. A simple wooden windlass is used to operate the gates.

The channel connecting the pond to the sea should be straight or nearly so and kept clear of plants. It is believed that plants vibrating in the current frighten young shrimp.

Owing to the rapid turnover of water in such ponds, fertilization is not practiced. A few experimental attempts did not appear to produce any significant differences in yields. It is customary, however, to drain the pond completely at least once a year and let it sun bake for 3 to 4 days before refilling. This is believed to increase the rate of production of plankton and other microorganisms.

Next to normal sinking and erosion, the biggest problem in pond maintenance is the mud lobster (*Thalassina anomala*), which burrows in bunds, weakening them and causing leaks. Mud lobsters may be killed by placing a piece of calcium carbide about 25 mm in diameter into the funnel-shaped burrow. A small amount of quicklime poured into the funnel has the same effect.

Daily stocking of shrimp ponds is usually uncontrolled. When the incoming tide in front of the sluice gates reaches a height of 60 cm above the pond level, the gate is simply opened and shrimp and various other organisms are allowed to swim in. A 60-cm head is necessary to ensure that no shrimp swim out of the pond. At dead high tide a 0.13-cm mesh wire screen is placed across the gate to retain the young shrimp. Although there are undoubtedly seasonal spawning peaks for each of the cultured species, there is some reproduction of all species throughout the year, so that this method is practicable year-round.

This sort of stocking permits entry of numerous predators which, if not controlled, will substantially reduce the yield of shrimps. The most serious predators are fish, which may be effectively controlled by poisoning with teaseed cake every 4 months.

Ponds so stocked always contain shrimp at all stages of growth. Large shrimp, up to 4 cm in carapace length, are harvested at dusk or night on ebb tides as they seek to return to the sea. Daytime harvesting is not practiced since few of the shrimp are swimming about during the day. Shrimp are captured in a conical net about 9 m long with mesh decreasing from 2 cm at the mouth to 1 cm at the cod end. The mouth is constructed so as to fit in place of the sluice gate. The net is affixed directly behind the gate just before the tide ebbs. When the tide begins to ebb the gate is opened and shrimp attempting to leave the pond are captured. Damage to the net by the rushing current is averted by supporting it in a troughlike wooden raceway on the outer side of the sluice gate. When the water level in the pond is reduced to about 60 cm the sluice gate is closed and the net hauled up.

Reliable production figures are not available for this form of shrimp culture primarily because culturists are reluctant to reveal their incomes. Yields of 300 to 800 kg/ha are reported, but it is likely that well-managed operations yield up to 1200 kg/ha of edible Penaeid shrimps annually.

INDIA

Methods similar to that just described are used in India, but the usual sites for culture are 1- to 10-ha rice fields, which are stocked with shrimp only during the 6 months when rice is not grown.

Production varies greatly; figures from 300 to 1600 kg/ha for the 6-month growing season are cited in the literature.

INDONESIA

Even less intensive methods are used in Indonesia, where Indian prawns, green tiger prawns, yellow prawns, and *Metapenaeus ensis* find their way into tambaks used in brackish water fish culture (see Chapter 17) and are harvested at the rate of 25 to 400 kg/(ha)(year). Various tiny shrimps, notably Mysids, which are not used elsewhere, are also harvested to make trassi, a flavoring product used in Indonesian cookery.

THE PHILIPPINES

The traditional method of shrimp culture is also used in the Philippines, but Philippine shrimp farmers are increasingly adopting more sophisticated methods of growing the largest of the Penaeid shrimps, the sugpo prawn, alone or in combination with milkfish (*Chanos chanos*). Monoculture of the sugpo prawn of course depends on the availability of pure stocks of sugpo fry (postlarval stage). Eight- to fifteen-millimeter sugpo fry may be captured from rivers by dip-netting from a boat, or by wading fishermen, but more often they are taken by use of lures, called "bonbon." These lures consist of bundles of twigs or grass tied at intervals of 1 to 2 m to a line strung between two poles set in a river near the bank. Sugpo fry entering the river attach themselves to these bundles and are captured by being shaken off into a dip net. Both methods are practiced day and night on all tides and in all weather. At the peak of fry migration it is possible to collect 1000 fry in an hour.

These collection methods result in the capture of a number of species of shrimp, but the sugpo are easily distinguished by the band of dark brown pigment running through their otherwise transparent bodies. After sorting, they are kept in lots of 100 to 250 in earthenware pots provided with banana leaves for attachment. Later they are placed in plastic bags for sale and delivery to shrimp culturists. Sugpo fry collection is a fairly profitable business, but as in any business where the supply of the commodity cannot be controlled by the operator, prices vary widely with time and place.

As in Malaysia and Singapore, pond site selection is crucial. Philippine culturists concern themselves chiefly with two factors, soil type and elevation. Sandy clay is the best type of soil, because it lends itself to dike construction, produces good growth of algal food for the young shrimp, and facilitates the burrowing habits of the sugpo prawn. Clay loam is an acceptable second choice.

Ordinary tides should fill the pond to a depth of at least 0.9 m. Ideally, it should be possible to exchange half the water in the pond as often as is necessary to maintain a temperature of approximately 25°C and a salinity of 20 to 25‰. Owing to the tidal range in the Philippines, which averages only 0.3 m during neap tides, this is not always possible. Fortunately, the sugpo prawn is capable of withstanding temperatures as

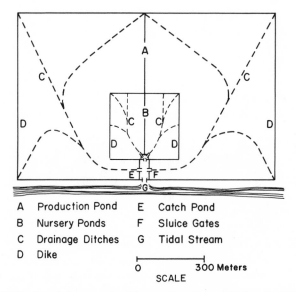

A   Production Pond   E   Catch Pond
B   Nursery Ponds   F   Sluice Gates
C   Drainage Ditches   G   Tidal Stream
D   Dike

├──────────────┤
0   300 Meters
        SCALE

FIG. 2.   Philippine shrimp pond. (After Delmendo and Rabanal, 1956.)

high as 30°C and salinities as low as 10‰ or as high as 35‰ for 1 or 2 days.

Pond construction in advanced Philippine shrimp culture is more elaborate than in the Malayan method. Figure 2 illustrates a rather elaborate but highly effective design for a 10-ha pond. Much larger ponds may be used, in which case they are usually subdivided by dikes into smaller units. Many ponds currently in use are much less sophisticated in design than the one illustrated. Dikes and sluice gates are constructed as in Malaysia and Singapore. The outer sluice gate opens into a catch pond, which is somewhat deeper than the rest of the pond. The entire pond bottom should slope toward the catch pond and sluice gates. A nylon mesh screen behind the sluice gates filters out predators and other undesired organisms. The drainage ditches in the bottom serve chiefly to facilitate harvest. Vegetation and large obstructions are removed from the bottom, but small branches and twigs are provided as places of attachment for the young shrimp.

After cleaning and leveling the bottom, the pond is dried for 2 to 6 weeks, depending on weather conditions. During this time agricultural lime should be applied at 400 kg/ha or more to absorb excess carbon dioxide and supply the calcium required by the shrimp during their moulting periods. Then the pond is filled with tidal water to a depth of 3 to 10 cm to induce the growth of the microbenthos complex known as

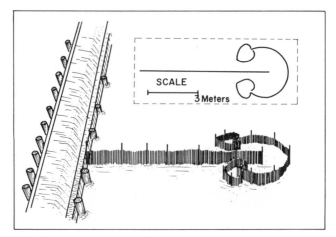

FIG. 3.   Philippine bamboo shrimp trap. (After Delmendo and Rabanal, 1956.)

lab-lab, which serves as food for the young shrimp. (For a more detailed discussion of lab-lab see Chapter 17.) After about a month the pond will be ready for stocking.

Sugpo fry are stocked at 300,000 to 500,000/ha, in the nursery section if the pond is provided with one. Newly stocked shrimp do not disperse rapidly, so stocking is best done by scattering them around the pond to lessen the danger of mass predation or severe competition for food. When sugpo are raised together with milkfish it is best not to stock until after the milkfish fry are large enough to be immune to predation by the sugpo. This sort of polyculture is becoming less popular since, although total yield in kilograms per hectare is higher than that for monoculture of shrimp, the dollar yield currently is lower.

Growth of floating, green filamentous algae, which might serve as food, is discouraged in nursery ponds since very young shrimp may become entangled in it. Lab-lab is the primary food but rice bran or dead fish and other animals ground into meal may be given as supplementary food. Such feeds are piled in the corners of the pond and serve as a source of fertilizer as well as food. Another supplementary food used with good results is small fish, boiled to soften the flesh and placed along the dike in late afternoon and early evening when the sugpo start feeding.

After 3 to 8 weeks in the nursery the shrimp are stocked in production ponds at 10,000 to 12,500/ha. In ponds set up as in Fig. 2 this may be accomplished by opening the nursery dike and driving them out. If separate nursery ponds are maintained, the shrimp are caught in bamboo traps (Fig. 3). After trapping there will still be numerous shrimp left in the nursery. Some of these may be captured by removing the branches and twigs to which they attach. The rest must be picked out of the mud after draining the pond.

Production ponds are operated similarly to nursery ponds except that, in addition to lab-lab, growth of filamentous green algae is encouraged. Sugpo prawns are reared for 4 months to 1 year. Throughout the nursing and rearing periods a continuous exchange of filtered tidal water is maintained in the ponds.

Average size at harvest, which is the chief determinant of the price the shrimp will bring, may be increased by lengthening the rearing period (Table 1). Supplemental feeding enhances not only growth but survival. Average survival from stocking to harvest for unfed shrimp is only 20%, but as many as 60% may survive when supplemental feeding is employed.

Harvest may be carried out by draining the pond and concentrating the shrimp in the catch basin or, if there is no catch basin, by attaching bag nets such as used in Malaysia and Singapore to the sluice gates. A more frequently used practice is to place bamboo traps similar to those used in nursery ponds (Fig. 3) in the pond. Lights are often used to attract shrimp into the traps at night. Whatever method of harvest is used,

TABLE 1.   AVERAGE RATE OF GROWTH OF CULTURED SUGPO PRAWNS (*Penaeus monodon*) IN THE PHILIPPINES

DURATION OF CULTURE	TOTAL LENGTH (MM)	BODY DEPTH (MM)	WEIGHT (G)
Fry	15.3	1.6	0.025
1 week	21.5	2.5	0.06
2 weeks	28.2	3.6	0.08
3 weeks	38.8	4.5	0.92
4 weeks	45.3	5.7	0.78
5 weeks	57.1	7.8	1.63
6 weeks	60.3	9.7	3.39
7 weeks	69.5	10.9	4.36
2 months	79.0	9.8	4.34
3 months	94.7	11.1	6.88
4 months	120.0	15.3	14.5
6 months	141.9	18.3	22.3
7 months	152.6	16.4	25.1
9 months	178.0	27.8	57.3
10 months	211.6	30.2	62.8
11 months	223.0	32.0	70.7
1 year	229.8	32.0	95.1

the burrowing habits of the sugpo prawn occasion some loss, which may be reduced by draining the pond and picking them out of the mud.

It is customary to refer to Philippine shrimp culture as more "advanced" than that practiced in other southeast Asian countries, but production per hectare lags behind the rest of the region; annual yields are from 250 to 900 kg/ha. It may be that monoculture of the sugpo prawn leaves open ecological niches which would normally be occupied by other species of shrimp. Philippine shrimp culture is also severely limited by the undependable supply of sugpo fry and the inefficiency of harvesting. Research on the natural behavior of the sugpo prawn might result in increased yields by providing the key to both artificial spawning and effective harvesting. Even the most dramatic improvement in breeding and harvesting techniques may not be adequate to counteract the damage done in recent years by industrial pollution of Philippine coastal waters.

TAIWAN

Methods of shrimp culture like those practiced in the Philippines are supplanting traditional methods in Taiwan as well. Unlike Philippine culturists, most Taiwanese shrimp producers continue to grow shrimp and milkfish together and seldom practice supplemental feeding. Nevertheless, the average annual yield of shrimp culture in Taiwan is 750 to 1500 kg/ha. This may have something to do with the ecological diversity of Taiwanese shrimp ponds, since Taiwanese culturists do not select a single species of shrimp but culture a combination of species, which may include green tiger prawns, kuruma shrimp, sugpo prawns, *Metapenaeus ensis, Penaeus carinatus,* and *Penaeus teraoi.* In 1968, fisheries officers at the Tainan Fisheries Experimental Station successfully spawned and reared several of these species in captivity, using the methods developed for kuruma shrimp in Japan (discussed later). Thus Taiwan may soon become the second country to support truly intensive shrimp culture.

Successful introduction of Japanese techniques to the southeast Asian countries would certainly increase production, but the rate of biological and technological advance is not the only factor that will influence the uncertain future of shrimp culture in the region. Shrimp is presently a luxury food, and it is unlikely that shrimp culture will become so efficient that the price will drop to a level competitive with that of brackish water fish. Thus, if increased human population or agricultural failure create a need for increased production of inexpensive protein foods in southeast Asia, the present trend toward shrimp culture and away from brackish water fish culture in such countries as the Philippines may be reversed.

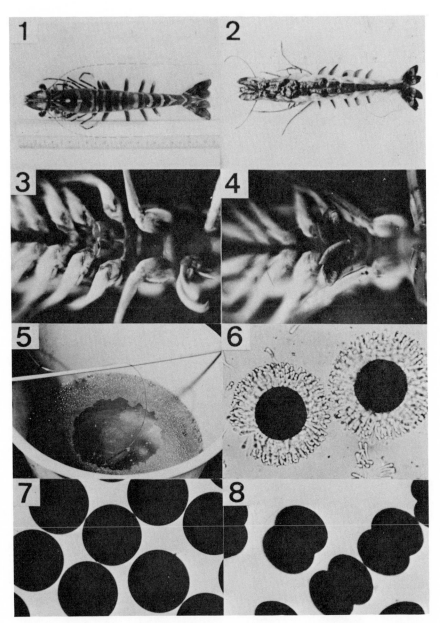

PLATE 2. Fig. 1. Spawner (*P. semisulcatus,* B.W. 90.1 g); Fig. 2. Mature ovary (O.W. to B.W. about 11–15%); Fig. 3. Before copulation, ventral view of thelycum (*P. japonicus*); Fig. 4. After copulation, ventral view of thelycum with stoppers (*P. japonicus*); Fig. 5. Bubbles on the water surface just after spawning; Fig. 6. Eggs just after spawning. (*P. semisulcatus*); Fig. 7. Eggs before absorption of water (*P. monodon*); Fig. 8. First cleavage (*P. monodon*).

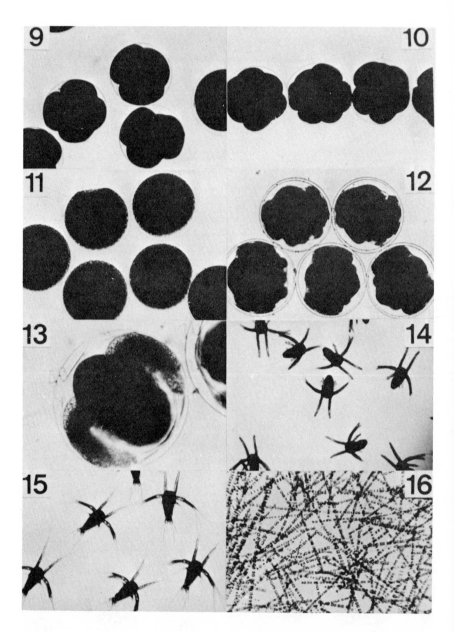

PLATE 2. Fig. 9. Second cleavage (*P. monodon*); Fig. 10. Fourth cleavage (*P. mono-don*); Fig. 11. Two hours after spawning (*P. monodon*); Fig. 12. Ten hours after spawning (*P. japonicus*); Fig. 13. Nauplius under the egg membrane immediately before hatching, 12 hours after spawning; Fig. 14. First Nauplius (*P. semisulcatus*); Fig. 15. Fifth Nauplius (*P. monodon*); Fig. 16. Cultured *Skeletonema costatum*.

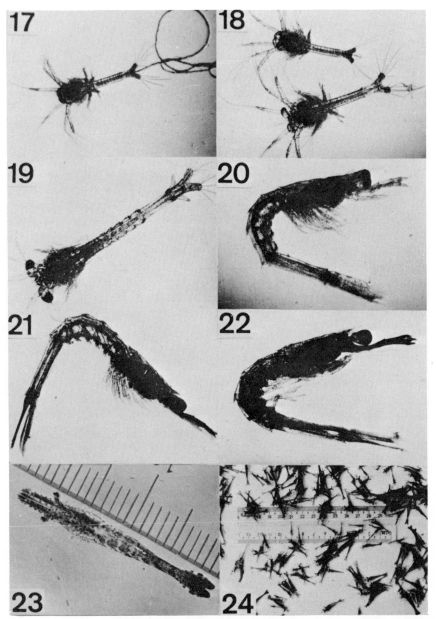

PLATE 2. Fig. 17. First Zoea (*P. teraoi*); Fig. 18. Comparison between first (above) and second (below) zoea (*P. teraoi*), Fig. 19. Third zoea (*P. teraoi*); Fig. 20. First mysis (*P. teraoi*); Fig. 21. Second mysis (*P. teraoi*); Fig. 22. Third mysis (*P. teraoi*); Fig. 23. Postlarva, 29 days after hatching (*P. monodon*); Fig. 24. Young prawns (*P. semisulcatus*). (Courtesy Dr. I Chiu Liao, Tungkang Marine Laboratory, Taiwan.)

INTENSIVE SHRIMP CULTURE IN JAPAN

The model for improvement of shrimp culture in southeast Asia and throughout the world is the Japanese method of farming the kuruma shrimp. Whereas in traditional methods of culture shrimp are allowed to pass through all the larval stages before human intervention occurs, in modern Japanese culture all life stages, from egg to adult, are passed in captivity. This not only provides a more dependable supply of shrimp but allows the culturist to grow his stock for a longer time, thus producing larger shrimp which can compete in the market with those taken by trawlers. At present, the kuruma shrimp is the only species so intensively cultured, although in recent years Japanese scientists have been able to adapt Fujinaga's methods to the green tiger prawn and *Metapenaeus ensis*. Yields attained by Japanese shrimp growers are far in excess of those achieved elsewhere, varying from 2000 to 6000 kg/ha, depending chiefly on whether running water can be supplied.

In nature, kuruma shrimp, like most Penaeid shrimps, mate and spawn at sea. Copulation between males and females normally takes place following each moult of the female. The sperm are encapsulated in spermatophores and inserted by the male into a special seminal receptacle of the female. This may occur at any time of year, the unused spermatophores being rejected with the shell at each moult. A supply of sperm is thereby available to the female when spawning occurs.

The spawning of the female lasts from mid-May to the end of September. During the spawning act, sperms deposited in the seminal receptacles are released and fertilization occurs as the eggs are discharged. The fertilized eggs are freely dispersed throughout the water and hatch to nauplius larvae 13 to 14 hours after spawning. The nauplii moult six times within the following 36 hours, passing into the protozoea stage with the sixth moult. The protozoeae, in turn, pass through three moults into the mysis stage during the following 5 days, and the mysis moult an additional three times during the next 5 days, then metamorphose into the first postlarval shrimp. At that stage, the animal ceases its planktonic existence and begins to crawl on the bottom, passing through an additional 20 to 22 moults, attaining adult characters and a body length of about 6 cm over a period of about 40 days.

The natural salinity at which larval development occurs ranges between 27 and 32‰. In culture this may exceed 35‰ without adverse effects. The larvae and postlarvae also withstand a wide temperature range of 15 to 33°C, though development is most rapid at 28 to 30°C.

Beginning with the first protozoea stage, the larvae begin to feed, taking unicellular algae, small crustaceans, and a variety of other planktonic microorganisms. The postlarvae begin immediately to feed upon small benthic organisms and plant and animal debris.

Some Japanese culturists engage in both spawning and rearing shrimp, but the majority of seedlings, as the juveniles are called, are produced by specialists. Some seedling producers grow their own brood stock, but most rely on fishermen to supply breeders. Only females are needed, due to their habit of storing sperm between moults. Spawners of 50 to 120 g are preferred to larger shrimp. Females may be easily distinguished from males by the dark grey ovaries visible between the carapace and the abdomen, but it is very difficult to determine the maturity of the ova.

Females suspected of being ready to spawn are placed, singly or in groups of two to five, in tanks 2 to 15 m² in surface area and 1 m deep, containing aged sea water, and left overnight. A mild water circulation is maintained to separate and aerate the eggs. If a female is ready to spawn, she will usually do so on the first night. The act of spawning is quite vigorous and may be detected by the presence of foam on the surface of the water. Sometimes the water takes on a pinkish hue from a jellylike material which coats the newly spawned ova. Kuruma shrimp produce an average of 300,000 eggs and sometimes as many as 1.2 million. The female is removed immediately after spawning to prevent her from eating the eggs.

Spawning can be carried out at temperatures of 22 to 33°C and salinities of 28 to 36‰, but 25 to 29°C and 32 to 35‰ are considered optimal. In nature, kuruma shrimp encounter a wide range of temperatures and salinities, but Japanese culturists prefer to keep these factors constant throughout the life cycle. This is considered particularly important during hatching, for which most culturists employ a temperature-controlled room. Hatching takes place in 13 to 14 hours at 25°C.

The newly hatched larvae may be transferred to larger tanks. The Shrimp Farming Co., Ltd., of Takamatsu uses 112- or 450-liter ceramic, tile-lined tanks stocked with 15,000 or 100,000 larvae, respectively. These tanks are maintained under greenhouse conditions at 26 to 30°C and vigorously aerated. In the early part of the season (April to June) supplemental heating by means of water-jackets in the tanks is required.

A newer and less expensive technique involves 10-m × 10-m × 2-m deep concrete tanks placed outdoors in full sunlight and used for both spawning and larvae rearing. Thirty to one hundred females per tank are stocked for spawning. Slow water circulation is maintained at all times, and the water supply is recycled after passing through 80- to 100-mesh/cm vinylon cloth filters.

Whatever type of tank is used for rearing the young shrimp, they

PLATE 3.   One of the larval rearing buildings at the Shrimp Farming Co., Takamatsu, Japan. Each 250-gal tank is capable of rearing 15,000 larvae. Fittings in foreground provide filtered sea water and air. Jackets at the base of each tank provide heating. Note greenhouse roof. (Photograph by J. H. Ryther.)

remain in it throughout the 10- to 12-day larval period and for the first 20 days of postlarval life. The season for rearing larvae and postlarvae lasts from early April to mid-September.

The nauplii do not feed, but thereafter it is essential that each life stage receive a special diet. The protozoeae are fed mostly on diatoms, occasionally supplemented by the flagellates *Isochrysis galbana* and *Monochrysis lutheri*. The latter are considered particularly valuable for very young fry. The original technique of diatom feeding, as developed by Fujinaga and still practiced by many culturists, involved pure cultures of *Skeletonema costatum* grown in separate tanks and pumped into the fry tanks. In 1964 Fujinaga developed a technique of culturing diatoms in the fry tank by merely adding 200 g of potassium nitrate and 20 g of potassium phosphate to each tank daily. This not only greatly reduces labor and expense but produces a mixture of species, which has been found to be preferable to a single species. In 1966 30 species of diatoms were identified from a single culture tank.

Diatoms are included in the diet of mysis larvae, but 90% of their food consists of newly hatched nauplii of brine shrimp (*Artemia*), cultured at a density of 10 g of eggs per liter. Five liters of such a culture per 15,000 larvae per day are fed. Clams and shrimp ground into 1-mm pieces may be used as a supplementary food, but brine shrimp nauplii have been found essential.

When kuruma shrimp reach the postlarval stage and assume a benthic life it is advisable to begin to circulate, as well as aerate, the water in the rearing tank. Brine shrimp may still be fed, but small pieces of annelids, nematodes, copepods, bivalves, and fish are the principal food. Cannibalism may begin to occur at this stage, so considerable amounts of food should be supplied, taking care not to overfeed and foul the water. The starting rate for postlarvae is 20 g of food/(day) (10,000 individuals). This is increased to 80 to 120 g/day over the 20-day growing period, during which the postlarvae undergo three or four moults. When they reach 15 to 20 mm in length and 10 mg in weight they are ready for stocking in production ponds or sale as seedlings. Total survival from egg to seedling stage averaged around 10% in the mid-1960s but rates as high as 50% have been reported in recent years.

Transportation of seedlings to production ponds is complicated by their cannibalistic tendencies. On short trips the seedlings are placed in 1-m × 0.7-m × 0.3-m deep boxes with fine mesh tops and bottoms. The boxes are placed in a tank with a water circulating system. At water temperatures above 25°C, losses to cannibalism of up to 30% are tolerated.

On longer trips (up to 10 hours) the seedlings are transported in trucks equipped with hermetically sealed refrigerated chambers to lower the metabolic rate of the shrimp and reduce cannibalism. The seedlings are placed in lots of 5000 to 10,000 in 20-liter polyethylene bags containing 6 to 8 liters of sea water and a supply of oxygen. The temperature inside the chamber is initially set at 1 to 15°C and gradually raised during transport to the temperature of the receiving water so as to prevent thermal shock. Thus 600,000 to 700,000 seedlings can be transported in a 2-ton truck at a cost of about 3 to 5% of the price of the seedlings.

There are two basic methods used in growing kuruma shrimp for market; pond culture and "river" culture. Up to three times as many kilograms of shrimp per area of water surface are obtained by the river method, but due to the difficulty of supplying large amounts of running water it is not suitable for large-scale operations. Attempts are being made to grow kuruma shrimp in net cages, a method which has to a large degree combined the virtues of running water culture and pond culture for many species of fish, but this is not being done on a commercial scale yet.

Kuruma shrimp production ponds vary greatly in size, from 0.01 to 10 ha or more. Small ponds, which are usually constructed of concrete, are more expensive to build but easier and more economical to maintain than the large earthen-banked ponds, which have a tendency to become choked with filamentous algae, which may suddenly die off and pollute the pond. Some shrimp-growing concerns use a combination of small and large ponds with a total area of perhaps 10 ha. Whatever the surface area, growing ponds are 1 to 2 m deep and have sand bottoms.

Interchange of sea water in pond culture is accomplished by means of one or two large (up to 1-m diameter) pipes which allow tidal filling and drainage. The pipes are fitted with fine mesh screens to keep the shrimp in and extraneous organisms out. About one-fourth of the volume of water in a pond is normally exchanged at each tidal cycle, but the precise amount is highly variable. If the amount of fresh tidal water entering the pond is not adequate to maintain a suitable concentration of dissolved oxygen, electric paddle wheels or some other means of aeration may be employed. If the tide level is such that interchange of water ceases altogether, pumping is recommended until natural circulation is restored. Pumps may also be used when some other factor, such as an overabundance of zooplankton, produces unusually low concentrations of dissolved oxygen.

The river-type culture facility consists of a series of oblong ponds, 0.5 to 0.7 m deep, supplied with water by an inlet pipe similar to that used in pond culture. Rapid exchange of water is achieved by pumping water out of the end pond rather than relying on tidal flow. Usually two pumps, with a total exchange capacity of 80,000 liters/hour, are used; both are operated at night when the shrimp are active, but only one is run during the day when they are buried in the sand, hence consuming little oxygen. The size of such culture systems is limited by the capabilities of available pumps. The Shrimp Farming Co., Ltd., operates a series of 28 ponds, each 100 m × 10 m (0.1 ha). Other operators use ponds as small as 0.02 ha.

The Shrimp Farming Co., Ltd., also grows shrimp in concrete or wooden tanks 50 m × 10 m × 1.3 m deep, with false bottoms made of fine screen. The false bottom is covered with about 2 cm of sand. Air is passed through a system of standpipes from beneath, forcing water up into the tank proper and creating a downward circulation through the sand, which acts as a filter. This sort of aeration and filtration greatly reduces the need for constant interchange of water, although some circulation is maintained.

The single most critical factor in stocking production ponds is the availability of dissolved oxygen; the minimum acceptable level is 3.5 ppm.

PLATE 4. Cement ponds 10 × 100 m through which sea water is pumped for culturing adult shrimp at the Shrimp Farming Co., Takamatsu, Japan. (Photograph by J. H. Ryther.)

Effective oxygen concentration can be increased by circulating the water more rapidly, thus more rapidly replacing the oxygen consumed by the shrimp. There seems to be no simple stocking formula based on dissolved oxygen concentration and rate of circulation, but perhaps an example will demonstrate the sort of relationship which obtains. If the water temperature is 28°C and the dissolved oxygen level is at 80% of saturation (6.3 ppm) and 10% of the volume of water in the pond can be replaced hourly, the number of seedlings stocked should be as many as can be expected to produce 0.11 kg/m² of shrimp in September, the last hot month, allowing for 30% mortality. If 33% of the water can be replaced per hour, the stocking density may be increased to whatever number will produce 0.50 kg/m² by September.

Even under optimum conditions, high population densities retard the growth of kuruma shrimp, so that culturists aiming for large shrimp may stock considerably fewer than the maximum permissible numbers. The decision to be made here is an economic one. Based on current prices, which will bring the culturist a better return, a large crop of small-to-medium shrimp or a smaller crop of jumbo shrimp? Another major factor in the growth rate of shrimp in Japan and other temperate zone countries is the date of stocking. Although kuruma shrimp can withstand temperatures as low as 4°C for short periods of time, they cease feeding below 10°C. Thus, if seedlings are stocked in the spring, they may be harvested after 6 months, but if stocking is carried out in the summer or fall, it may be necessary to hold the shrimp in ponds for as long as 9 months. This of course results in increased mortality.

One of the reasons shrimp and other crustaceans remain a luxury food is the poor efficiency of food conversion, necessitated largely by the great losses of energy involved in moulting. The kuruma shrimp culturist should expect to produce 1 kg of shrimp for every 10 to 15 kg of feed at the optimum temperature, 25°C. Table 2 is a feeding schedule which has been found adequate to prevent cannibalism and maximize growth of kuruma shrimp seedlings. Daytime feedings are discontinued after the

TABLE 2. FEEDING SCHEDULE FOR KURUMA SHRIMP (*Penaeus japonicus*) CULTURED IN JAPAN

WEIGHT OF INDIVIDUAL SHRIMP (G)	TYPE OF FOOD	AMOUNT FED (% OF TOTAL WEIGHT OF SHRIMP)	FREQUENCY OF FEEDING
0.1– 0.5 (first few days after stocking)	Ground clams and shrimp	200–300	2–3 times daily
0.1– 0.5	Ground clams and shrimp	50	2–3 times daily
0.5– 1.0	Small whole shrimp or ground clams	25	2–3 times daily
1 – 2	Small whole shrimp or ground clams	25	1 feeding daily, just before sunset
2 –10	Minced fish or, preferably, low-fat meat or ground clams	15	1 feeding daily, just before sunset
10 –20	Minced fish or, preferably, low-fat meat or ground clams	5	1 feeding daily, just before sunset

seedlings reach about 1 g in weight, by which time they are almost exclusively nocturnal.

Diseases and parasites present little or no problem in growing kuruma shrimp from seedlings to marketable size and cannibalism is not a problem once the shrimp are past the postlarva stage. Survival between stocking and harvest varies from 60 to 100%.

In Japan, as elsewhere, one of the principal problems faced by the shrimp culturist is the low efficiency of harvesting. Conventional methods of netting and trapping are only partially effective and inevitably necessitate draining the pond and picking up by hand as many of the buried shrimp as can be found. Better results have been obtained by use of a trawl with a high-pressure hose mounted in front to blast the shrimp from the sand, but the most effective piece of gear developed to date is an electric trawl. Some culturists simplify the harvesting problem by leaving the shrimp in ponds over the winter and harvesting a few at a time to take advantage of premium prices.

After harvest growing ponds are drained and dried for a few weeks and the algae are scraped off the concrete sides of small ponds. Sea water is introduced and allowed to sit in the ponds for a few days before restocking.

Most shrimp sold in Japan are destined for use in the famous Japanese dish tempura, for which only very fresh shrimp are satisfactory. Thus it is necessary to ship shrimp live. This is made possible by chilling them to 12°C. At this temperature they can endure long periods out of water, maintaining their greatly reduced metabolic rate solely on the basis of oxygen in the water trapped in their gill cavities. The chilled shrimp are packed in cedar sawdust, which is naturally repellant to insects. (Japanese cedar, *Cryptomeria japonica,* is not related to American cedars but is closely akin to the California redwood.) Due to the excellent insulating properties of sawdust, further refrigeration is not necessary for up to 2 days in the summer and 4 days in the winter. On arrival the shrimp may be revived by dropping them into warm water and sold live.

With the introduction in recent years of numerous technological improvements, Japanese shrimp culture has become so successful that cultured shrimp are being released in the Inland Sea to augment the shrimp fisheries. Local governmental units (prefectures) are responsible for stocking programs. Most of the shrimp for stocking are produced by federal Propagation Centers, but some prefectures maintain their own hatcheries.

Stocking is carried out in sandy-bottomed bays and inlets, geographically and oceanographically situated so that the shrimp are not likely to migrate out and are relatively free of predators. Most of these areas once supported large populations of kuruma shrimp which for one reason or another, usually overfishing, have been depleted. Final selection of sites to be stocked is determined by the prefectural governments in consultation with Fishermen's Cooperatives.

Postlarvae, such as might be sold for use in commercial shrimp culture, were originally used in this program, but it was found that stocking larger shrimp resulted in better survival. Thus propagation centers now rear postlarvae in running water tanks to the length of 10 to 15 mm before distributing them for stocking.

Stocked waters are opened for fishing when it is judged that the shrimp have reached harvestable size. There has been no biological proof of the success of the stocking program, but studies are being carried out by the Nansei Regional Fisheries Research Laboratory. Shrimp fishermen, however, report that in their opinion the stocking program has been effective in increasing local populations of kuruma shrimp.

EXPERIMENTAL SHRIMP CULTURE

UNITED STATES

The largest shrimp fishery in the world is that exploited by Cuban, Mexican, and American trawlers in the Gulf of Mexico. The brown shrimp (*Penaeus aztecus*), the white shrimp (*Penaeus setiferus*), the pink shrimp (*Penaeus duorarum*), and, to a lesser extent, the Caribbean brown shrimp (*Penaeus brasiliensis*) and seabob (*Xiphopenaeus kroyeri*) form the basis for that fishery as well as a similar one along the Atlantic coast of the United States as far north as the Carolina banks.

Though yields of these fisheries have remained high, they have been barely able to supply half the demand for shrimp in the United States. For this reason, and because the pollution and filling of tidal swamps and estuaries may eventually significantly reduce the available nursery areas, in recent years there has been considerable interest in shrimp culture in the United States.

American biologists have of course long been aware of Asian shrimp culture practices, but for some reason it was generally assumed that American shrimps, despite their close resemblance to Asian species of *Penaeus*, were not susceptible to culture. Recent years, however, have seen more or less successful experiments with both the Japanese and southeast Asian methods of shrimp culture, although there are as yet no commercially successful American shrimp farms.

It may be that there are estuaries in the United States which could be farmed using the Malayan method of stocking by natural recruitment, but thus far it has been found necessary to resort to artificial stocking to obtain satisfactory yields. The practicability of pond stocking is presently limited by the availability of postlarval and juvenile shrimp. Stock for culture may be obtained from trawlers, but shrimp taken in trawls suffer 85 to 100% mortality no matter how they are handled. Even if the healthiest appearing survivors are selected from trawl catches, a further mortality of 25% or so may be expected when they are stocked. The only capture device that does not produce high mortality is the cast net, but cast nets are practical only during the first few weeks of the year when shrimp are abundant in tidal streams.

The problem of recruitment ceases to be of importance if reproduction can be controlled. Attempts have been made to induce spawning of American Penaeids in ponds without success, and most authorities are in agreement that the future, if any, of American shrimp culture lies in adapting Japanese methods to American circumstances and species or, failing that, in introducing kuruma shrimp. One dissenter is John H. Knox, a business management consultant who has immersed himself in shrimp culture biology and technology and believes that, with suitable modification, the Malayan system of continual stocking and harvest would produce commercially feasible yields of brown shrimp, white shrimp, and possibly pink shrimp in the United States. Among the improved recruiting methods suggested by Knox are the construction of weirs to funnel postlarvae into a pond, driving shrimp by means of an electrical field, luring them by use of lights, and drawing them in with the type of bladeless impeller pump which is now routinely used in commercial handling of such fragile food items as eggs, tomatoes, and peaches. All of these methods have the common advantage of drawing in shrimp from a larger area than would be harvested by merely opening a sluice gate.

As in all forms of shrimp culture that depend on natural spawning and recruitment, the scheme envisioned by Knox depends largely on judicious site selection. The following factors appear to be crucial in growing American Penaeid shrimps:

1. Soil quality. The soil must be conducive to both the growth of shrimp and the construction of dikes. The latter factor of course implies little more than low porosity. Determining whether soil is chemically and physically compatible with shrimp is somewhat more complex. Complicating this decision are the preferences of individual shrimp species. Brown shrimp and white shrimp generally prefer softer, muddier sub-

strates than pink shrimp, which are commonly found on sandy bottoms. A more general requirement is the absence of metallic oxides, which are repellant, if not toxic, to shrimp.

2. Pollution. Domestic and industrial pollutants are of course detrimental to the welfare of shrimp, but a less obvious source of "pollution" is the mangrove tree. Although in Malaya shrimp ponds are built in mangrove swamps, the American species of mangrove at times impart to water a substance which is toxic to American shrimps.

3. Salinity. Postlarvae and juveniles of American Penaeid shrimps are tolerant of a wide range of salinities, but as they approach maturity it is important that excessive amounts of freshwater not enter the pond. The safe salinity range seems to be 22 to 37‰.

4. Depth. Since high production of benthic organisms is desirable, 0.5 to 0.7 m is considered to be optimal, although depths up to 1.3 m may be satisfactory if the water is unusually clear. In general, deeper holes are detrimental in that shrimp will congregate there rather than near the sluice gate when the pond is drained for harvest. One or two deep holes of known location may, however, be advantageous in the Carolinas, where sudden chilling may occur. Large-scale mortality of shrimp, particularly white shrimp, has been known to occur in shallow ponds when air temperatures drop to around 4 to 5°C.

5. Land configuration. This is primarily an economic consideration. Construction costs may be minimized by taking advantage of natural impoundments formed by tidal creeks, swamps, or lagoons which can be adapted for shrimp culture by merely placing a sluice gate across the mouth. Such sites, in addition to being rare, are often too deep or, if located on a creek, are subject to desalinization in the wake of heavy rainfall. In the majority of instances, a dike will have to be constructed. The prospective culturist should thus seek to reduce construction costs by using existing land features such as river banks and railway embankments as part of the dike system and by constructing dikes using soil excavated at or near the pond site. All ponds should be separated from the open sea by at least 15 m of marsh grass or other vegetation to break storm waves.

Further details of pond culture of American shrimps have not been worked out, but Knox felt confident enough of the commercial feasibility of the process to undertake surveys of some 5000 miles of coastline to locate suitable sites for a shrimp farm. Since the best type of marshland for this purpose is the least attractive for other forms of commercial development, he felt sure he could obtain the necessary property cheaply. Although he was successful in locating suitable sites, he found that con-

PLATE 5.   Feeding juvenile shrimp at rearing ponds warmed by power plant cooling
water, Turkey Point, Florida. (Courtesy Florida Power and Light Co.)

servation laws and restrictions on tidelands leasing made it virtually impossible to lease acreage suitable for shrimp culture in any of the eight
states where it would be feasible. These laws, plus an entrenched emo-
tional resistance on the part of the local citizenry to anything that would
restrict access by sportsmen to marshy estuaries, have until recently
thwarted all would-be shrimp farmers in the United States. This situation
has begun to change as state and local governments realize the economic
potential of estuarine aquaculture, but leases are still difficult to obtain.

The first tangible evidence of changing attitudes was the harvest, in
1970, of the first crop of shrimp (*Penaeus setiferus*) by Marifarms, Inc., a
1200-ha Japanese-American-owned farm near Panama City, Florida. The
1970 and 1971 harvests were experimental in nature, but in 1972 Marifarms moved into the commercial market. The 1000 ha of estuary
which comprise most of the farm are leased from the state of Florida
with the stipulation that the company release 20 million shrimp annually
into the natural environment. Marifarms has also moved to minimize
land use conflict by permitting permanent full public access to the area.

To most would-be American shrimp culturists, Knox's experiences
still seem to constitute another good argument for controlled spawning of
American Penaeids. White, brown, and pink shrimp and seabobs have all

PLATE 6. Dr. Durbin Tabb, University of Miami, holding 30-day postlarvae and 10-month adult cultivated shrimp. (*P. setiferus*)

been successfully spawned and reared in captivity, primarily at the United States Bureau of Commercial Fisheries Biological Laboratory, Galveston, Texas, and the University of Miami's Institute of Marine Science, using various adaptations of Fujinaga's original scheme. Figure 4 is a diagram of the type of setup in which shrimp are reared from egg to juvenile at Galveston.

None of the three species can yet be produced in commercial quantities,

TABLE 3. STOCKING RATES, ENVIRONMENTAL CONDITIONS, AND RESULTS IN EXPERIMENTAL CULTURE OF WHITE SHRIMP (*Penaeus setiferus*) IN THE UNITED STATES

LOCATION	NO. SHRIMP STOCKED PER HECTARE	WEIGHT OF SHRIMP STOCKED PER HECTARE (KG)	DATE OF STOCKING	DATE OF HARVEST	MORTAL-ITY (%)	WEIGHT OF SHRIMP HARVESTED PER HECTARE (KG)	PRODUCTION (KG/HA)	FOOD CONVER-SION	TEMPER-ATURE RANGE	SALINITY RANGE
Florida	118,000	341.6	July 15	About December 1	33	1,147.3	806.7	7.05:1	18–32°C	11.4–34.4‰
	180,000	214.0	July 15	About December 1	22	1,601.8	1,487.8	—	—	—
	60,905	1,059.0	November	May	20	1,266.0	207.0	—	—	—
Galveston, Texas	81,840	—	July 21	November 10	16	654.0	645.0	—	—	—

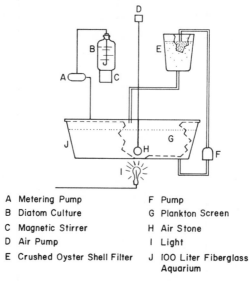

A	Metering Pump	F	Pump
B	Diatom Culture	G	Plankton Screen
C	Magnetic Stirrer	H	Air Stone
D	Air Pump	I	Light
E	Crushed Oyster Shell Filter	J	100 Liter Fiberglass Aquarium

FIG. 4. Setup in which shrimp are reared from egg to juvenile at National Marine Fisheries Service Laboratory, Galveston, Texas. (After Cook and Murphy, 1965.)

but the greatest degree of success has been achieved using the white shrimp. Thus far it has proven possible to rear up to 200 juvenile white shrimp per square meter in tanks 1.3 m deep, using sea water with salinities of 31.5 to 37.0‰.

White shrimp have also proven the most productive species for pond stocking. Table 3 lists the results of four experiments in which white shrimp were stocked in 0.05-ha ponds. In the Florida experiments, the shrimp were fed whatever amount of ground whole fish they would consume daily. At Galveston, fertilization, rather than direct feeding, was employed, 1 yard³ of chicken manure being deposited in a single location prior to stocking.

Brown shrimp have not been investigated as fully as white shrimp, but it appears that they do not grow quite as well in ponds. On the other hand, they are more hardy with respect to cold water. Pink shrimp, due to their preference for sandy bottoms, are less suitable for culture than white shrimp or brown shrimp in many areas.

It seems almost certain that American researchers will eventually devise techniques as biologically efficient as those currently used by Japanese shrimp farmers. Whether shrimp farming in the United States will ever become as economically rewarding as it is in Japan seems doubtful, although the interest shown by a number of large corporations argues in its favor. In addition to being confronted by high labor, site preparation, and equipment costs, American producers will be forced into more direct competition with shrimp fishermen than their Japanese counterparts. In Japan the demand for live shrimp for use in tempura can only be satisfied by cultured shrimp. Since they have a monopoly on this portion of the market, Japanese shrimp culturists are thus able to charge very high prices. In the United States there would be no such discrimination between cultured shrimp and the fishery product. The prospective American shrimp culturist must therefore ask himself how his anticipated production costs compare with the expense of outfitting, maintaining, and operating a trawler. It may be that intensive culture will never be able to compete with trawling in the United States and that the primary function of American shrimp culture will be to compensate for the destruction of estuarine nurseries by stocking shrimp for eventual harvest by fishermen.

LATIN AMERICA

Shrimp culture has yet to be accorded much consideration in Latin America, partly for economic reasons and partly because most Latin American shrimp fisheries have not been severely depleted. An exception occurs in the northern Gulf of California. Profligate use of Colorado River water for irrigation in California and Arizona has reduced that stream to a fraction of its natural volume by the time it enters Mexico and reaches the Gulf. The resultant drying up and salinization of tidal swamps and estuaries have virtually eliminated the important fishery for café brown shrimp (*Penaeus californiensis*), blue shrimp (*Penaeus stylirostris*), and white shrimp (*Penaeus vannamei*) (not to be confused with the white shrimp of the Atlantic) in that region. Mexican scientists are currently exploring the possibility of shrimp culture in ponds distant from the sea but connected with it by canals, as an alternative to this fishery.

Other experiments in culture of marine Penaeid shrimp in Latin America are those financed by U.S. capital in Mexico and Ecuador. In 1969 the first shrimp culture experiments were initiated in Honduras, using postlarvae supplied by the University of Miami.

AFRICA

Although significant shrimp fisheries are found in a number of African countries, commercial shrimp culture has not been attempted on that continent. Beginning in the mid-1960s pink shrimp were experimentally cultured in Nigeria, but the outbreak of civil war forced suspension of the FAO-sponsored project.

To date most practical and experimental shrimp culture has taken place in tropical and subtropical regions, where catastrophic die-offs of stock in shallow ponds during cold weather are less likely to occur. However, extensive shrimp fisheries occur in temperate and even subarctic regions and the highly successful kuruma shrimp industry of Japan demonstrates that shrimp can be successfully cultured well outside the tropics.

KOREA

In 1968, South Korea became the second temperate zone country where shrimp are grown commercially. The successful experimental culture there of *Penaeus orientalis* led to the establishment of about 50 ha of private shrimp ponds. Rapid expansion is foreseen.

EUROPE

Experimental culture of cold water species is currently under way in France, West Germany, and the United Kingdom. The French work is in the initial stages, kuruma shrimp having been imported from Japan in 1969.

West German biologists are further along, having succeeded in rearing the sand shrimp (*Crangon crangon*), which sustains an important fishery in the North Sea and the Baltic Sea, from the zoea stage to sexual maturity. At present no attempt is being made to apply this experimental success on a practical basis. This might prove to be quite difficult since the sand shrimp is much more pronouncedly metamorphic than the commonly cultured Penaeid shrimps. One might therefore expect that the environmental and nutritional requirements of the various life stages would be more diverse and difficult to satisfy than those of Penaeids.

Culture in the United Kingdom of the deepwater prawn (*Pandalus borealis*) and *Palaemon serratus* is also still in the laboratory stages but seems closer to commercial application. Substantial numbers have been reared through all stages and the optimum conditions of temperature, salinity, light, and population density determined. Researchers are presently seeking means to control diseases and a cheaper substitute for brine shrimp as an early food. A number of authorities, however, consider these species too slow-growing for practical culture.

FRESHWATER SHRIMP CULTURE

When shrimp are mentioned, one customarily thinks of a marine environment. But many of the largest and most desirable shrimps occur in freshwater. By far the best known of these is the giant freshwater prawn (*Macrobrachium rosenbergi*) of the Indo-Pacific region (Plate 7). Maximum length of *M. rosenbergi* is about 25 cm for males and 15 cm for females. Both sexes bring extremely high prices relative to other seafoods.

Attempts have been made, with some success, to induce *M. rosenbergi* to enter impoundments for growing to marketable size, but the rather low natural population densities characteristic of this species render commercial culture by the methods used in southeast Asia for marine shrimps less than practical. Accordingly, fishery workers in the Indo-Pacific countries have made efforts to spawn and rear *M. rosenbergi* in captivity. The first experimental success in this enterprise was achieved by an FAO-supported fisheries development project under the direction of S. W. Ling at the Fisheries Research Laboratory of the Fisheries Department of the Federation of Malaya at Glugor, Penang. Commercial culture there is not yet a reality but might well prove feasible with some improvement in the efficiency of the methods practiced at the Penang hatchery.

In nature, adult *M. rosenbergi* are found in virtually all types of fresh and brackish waters. Larval development, however, requires water of 8 to 22‰ salinity. Mating takes place a few hours after the female performs a premating moult in estuaries or in freshwater streams not far above tidewater. The male deposits sperm on the ventral thoracic legs of the female, but the sperm are not retained for long periods of time as in the marine Penaeids. Rather the entire complement of eggs is deposited and fertilized in brood chambers at the base of the thoracic legs within 6 to 20 hours after mating. Unmated females also deposit eggs after moulting, but these fall off after 2 or 3 days.

It is believed that, in nature, females may spawn 3 to 4 times a year, producing up to 120,000 eggs each time. The eggs remain attached to the female, who aerates them by beating her pleopods and carefully removes dead eggs and foreign matter, using the first pair of thoracic legs, up to the time of hatching. The process of incubation usually requires about 19 days at 26 to 28°C. From about the twelfth day the color of the eggs, originally bright orange, begins to fade to a pale grey. When this darkens to slate grey, hatching is imminent.

From the first minutes of life, *M. rosenbergi* larvae are active swimmers, but they are not initially strong enough to hold their own against a

Plate 7. Specimen of *Macrobrachium carcinus*. (Courtesy T. J. Costello, Tropical Atlantic Biological Laboratory, Miami, Fla.)

current. Thus river-hatched larvae are quickly swept downstream to a suitably saline environment. At first the larvae swim together in large groups, but after the tenth day of life they tend to separate. They feed primarily on zooplankton, but in the absence of an adequate supply of live animal food they will take minute bits of dead organic material or plants. Within 35 to 55 days the larvae have passed through 12 stages and metamorphose into juveniles.

Juveniles immediately adopt a benthic mode of life and commence to feed on benthic animals and organic detritus. Moulting occurs every 4 to 6 days. Juvenile *M. rosenbergi* are believed to crawl slowly upstream until, 2 to 3 months later, in the recognizable form of a young prawn, many of them have reached pure freshwater.

At this time young *M. rosenbergi* are 5 to 6 cm long (measured from eye socket to tip of telson) and weigh about 6 g each. The young prawns continue the migration begun as juveniles but swim and crawl at a much more rapid rate. Some may travel over 60 km upstream from the sea. From this time on, *M. rosenbergi* will eat almost any piece of living or dead organic matter of suitable size. When sufficiently hungry they may resort to cannibalism. Under favorable conditions, sexual maturity is attained in 9 months, after which time there may be a downstream migration.

Sometimes egg-bearing or "berried" female *M. rosenbergi* can be captured from the wild for use in culture. Three methods of capture are employed: trap, hook and line, and hand net.

Prawn traps are funnel-shaped structures of split bamboo or galvanized wire mesh. In use they are baited with pieces of fish, shrimp, or baked coconut and set overnight near the banks of rivers.

Hook and line fishing for prawns, using one or more barbless hooks baited with earthworms, small shrimp, or pieces of baked coconut, is practiced in both Malaysia and Thailand. A small sinker, just heavy enough to hold the baited hooks on the bottom, is attached about 25 cm above the uppermost hook. When a bite is detected a few seconds are allowed for the prawn to hook itself, then the line is drawn up steadily, without jerking. Berried female prawns caught on hook and line are kept for a day or so in submerged wire cages to ensure that they are not seriously injured.

Hand netting is done at night, when prawns move into shallow water to feed. A strong light temporarily stupefies the prawns and enables the fishermen to see the bright bluish reflection of their eyes. Specimens thus detected can be captured by careful netting from the rear.

Transportation of live prawns from the field to the culture site presents few problems. As long as they are kept moist, prawns will survive for several hours when packed in baskets between layers of soft plants. Longer journeys may require aerated tanks or plastic bags filled half-and-half with water and oxygen. If bags are used, the sharp tip of the rostrum must be cut off to prevent its puncturing the bags.

If berried females are not available in sufficient quantities, young prawns may easily be reared to maturity in captivity. Mature females are usually kept in groups in aerated 100- to 200-liter tanks. Such groups must be constantly watched for moulting. Newly moulted females must be screened off from the rest, else they will be attacked while the shell is soft.

Males may fight at any time, so they are usually kept individually in 60-liter tanks. When a female has moulted, 2 to 3 hours are allowed for the shell to harden, then she is placed in the tank of a male. Mating will occur within a few hours, followed by egg laying within 24 hours. If sufficient freshly moulted females are available, group spawning may be practiced. For this purpose 2 to 4 males are placed together with 8 to 20 females in a tank 2 to 3 m long, 1 to 1½ m wide, and 40 cm deep and mating and egg laying proceed as described. As in all stages of *M. rosenbergi* culture, vigorous aeration is essential. Both sexes are kept in freshwater through spawning.

Berried females, whether collected from the wild or mated in captivity, are transferred to individual 50- to 200-liter tanks. When the eggs begin to turn grey, sea water is added daily in small amounts so that by hatching time the salinity is 8 to 15‰. This is not only better for the larvae than freshwater, it seems to produce better rates of hatching.

An average female produces about 50,000 larvae, which are almost immediately transferred to a rearing tank 2 to 3 m long and 50 to 70 cm wide containing 16 to 20 cm of water. The bottom of the rearing tank slopes slightly to a collecting pit at one end. Larvae are captured for transfer by shading all but one corner of the hatching tank. The larvae are attracted to the lighted corner, where many of them may be captured by dipping with a cup. The remainder are siphoned into the rearing tank. They are quite hardy with respect to water quality, but every effort is made to maximize growth and survival by providing near-optimal conditions: 12 to 14‰ salinity, a temperature of 26 to 28°C, and a pH of 7.0 to 8.0. Sudden changes in these environmental parameters are scrupulously avoided. If the salinity of the water in the hatching tank is less than 12‰, the rearing tank is initially filled to less than its capacity with water of the same salinity as that in the hatching tank. Gradual upward adjustment of salinity is made by slowly siphoning a predetermined amount of sea water into the tank. An aerator and stirrer are used to keep the dissolved oxygen concentration near saturation.

The basic larva food used at Penang is brine shrimp nauplii, which are hatched in the rearing tank, using eggs imported from the United States. Brine shrimp eggs are placed within a floating ring at one end of the tank, which is shaded. The newly hatched nauplii are attracted to the lighted end of the tank and eaten. For the first few days of rearing the daily complement of brine shrimp eggs is about ¾ teaspoonsful per batch of 50,000 larvae. This is gradually increased to 3 teaspoonsful per day when the larvae are 30 days old, then held at this level until they metamorphose to juveniles. Some brine shrimp should be present in the rearing tank at all times to prevent cannibalism.

Since brine shrimp eggs are quite expensive, supplementary feeds are used as much as possible. A suitable natural food is fresh ripe fish eggs, thoroughly washed to eliminate ovarian tissue and other foreign matter. In 1968 workers at the Songkhla Marine Fisheries Station in Thailand succeeded in rearing *M. rosenbergi* larvae using mullet (*Mugil* spp.) eggs as the primary food with brine shrimp nauplii as an occasional supplement.

Prepared foods are also used, although brine shrimp nauplii appear to be essential during the first day of life. The current favorite at Penang is a vitamin-fortified mixture of fish flesh and egg custard steamed together, drained and passed through a screen to provide particles of an appropriate size (Table 4). Fish flesh or egg custard may also be fed

TABLE 4. PARTICLE SIZES OF AN ARTIFICIAL FEED FED TO LARVAE OF THE GIANT FRESHWATER PRAWN (*Macrobrachium rosenbergi*) IN EXPERIMENTS AT THE FISHERIES RESEARCH STATION, GLUGOR, PENANG, MALAYSIA

AGE OF LARVAE (DAYS)	FOOD PARTICLE SIZE (MM)
2– 4	0.4
5–10	0.5
11–20	0.9
20+	1.4

separately. Powdered dried chicken blood is successfully used in Thailand. *M. rosenbergi* larvae will also eat phytoplankton, and the Hawaii Division of Fish and Game has found it convenient to grow larvae in "green water" for the first 12 days.

Prepared foods are provided at the rate of approximately 30% of the total body weight of larvae per day. For convenience in feeding, most of the rearing tank may be shaded at feeding time to concentrate the larvae in one place. Feed is gently spread on the surface of the water with a medicine dropper. This should be done slowly enough that each larva has a chance to feed. If most of the larvae appear to be carrying food particles, they are being correctly fed.

Use of prepared feeds requires that great attention be paid to cleanliness. Uneaten food particles and fecal matter are siphoned off twice daily. Every 10 days or so a partial change of water is advisable. This is done by shading part of the tank and siphoning from that portion.

One of the disastrous consequences of overfeeding or lack of cleanliness may be an incurable fungus infection which manifests itself as small white patches on the tails and at the bases of the appendages of the larvae. Infected larvae should be removed and destroyed. If a large percentage of any batch of larvae are infected, it is best to sacrifice them all.

Larval *M. rosenbergi* are also subject to a serious protozoal infection which, if caught in the early stages, may be treated with 0.2 ppm of malachite green for $\frac{1}{2}$ hour daily, or for 6 hours with a single dose of 0.4 ppm copper sulfate.

Prawn larvae are sensitive to nicotine, so tobacco smoking should be prohibited in the vicinity of larvae rearing tanks.

Yet another source of mortality among late larvae is jumping out of the water. Stranding may be averted by shading the sides of the tank to keep the larvae in the center.

When the larvae are ready to metamorphose, small branches, stones, shells, and so on, are placed on the bottom of the rearing tank to provide shelter for the freshly moulted juveniles, thus deterring cannibalism. When 90% of the larvae have metamorphosed, the remaining slow growers are transferred to another tank for further rearing.

The metamorphosed juveniles must next be acclimated to freshwater. This is accomplished within 3 to 8 hours by gradual replacement of the brackish water in the larvae rearing tank.

Juveniles may be reared in the same tanks used for the larvae, but it is better to provide more spacious quarters in the form of larger concrete tanks or earthen ponds with cemented brick walls. These facilities vary in surface area from 5 to 50 m^2 and in depth from 15 cm to 1 m and are abundantly supplied with shelters for the young prawns. In addition to the usual aeration, it is desirable to maintain a slight flow of water through the tank or pond. Stocking rates for juveniles vary from 2 to 10/m^2.

The juvenile prawns are very catholic in their food habits, but growth is maximized and cannibalism minimized by feeding fresh animal material as often as is economically feasible. Fresh fish, mollusks, and earthworms cut into pieces according to the size of the juveniles are the principal food, although whole live aquatic worms and chironomid larvae are used when available. Supplementary foods include dried animal material softened in freshwater for $\frac{1}{2}$ hour before use and pieces of grains, peas, beans, and soft aquatic plants. Feeding is carried out four times daily; three daylight feedings and one at night. To avert cannibalism the amounts fed must be in excess of what the shrimp can consume. Insofar as possible all uneaten food is siphoned out after each feeding.

Juvenile *M. rosenbergi* are subject to the same diseases as larvae. As a preventive measure juvenile rearing ponds are completely dried, drained, and disinfected before and after each use. Disease treatment is the same as described for larvae.

With good management juveniles should grow to 2- to 3-cm prawns within about 30 days, with a survival rate of about 50%. When they reach about 4 cm in length the young prawns are suitable for stocking in production ponds. Almost any sort of pond over 200 m² in surface area and 50 cm in depth, with a water temperature of 22 to 32°C, can be used for growing *M. rosenbergi,* but larger ponds, 1000 m² or more in area and 1 to 1½ m deep, are more economical.

Ponds for prawn culture are prepared in the same general manner as fish ponds in southeast Asia: predators are eradicated, inlets and outlets screened, and aquatic plants removed. It is desirable to have a gentle flow of water through the pond at least a few hours each day.

Prawns may be stocked alone or in combination with fish. One finds in the literature references to growing *M. rosenbergi* in combination with "carp." It should be emphasized that this refers to various of the Chinese carps, not to the common carp (*Cyprinus carpio*), which competes for food with prawns and brings a far lower price. Fishes successfully used in culture with *M. rosenbergi* include such herbivores and/or plankton feeders as the big head (*Aristichthys nobilis*), grass carp (*Ctenopharyngodon idellus*), silver carp (*Hypophthalmichthys molitrix*), catla (*Catla catla*), rohu (*Labeo rohita*), milkfish (*Chanos chanos*), *Osteochilus hasselti, Barbus gonionotus,* grey mullet (*Mugil cephalus*), kissing gourami (*Helostoma temmincki*), sepat siam (*Trichogaster pectoralis*), and three-spot gourami (*Trichogaster trichopterus*). Suitable stocking rates depend not only on the numbers and kinds of fish stocked, but on the quality of soil and water. Table 5 is a general guide to stocking practices.

Natural production within the pond supplies most of the food for prawns at this stage. Productivity is enhanced by monthly application of 200 kg of cow dung and 10 kg of lime per hectare. Supplementary feeding is also practiced, using 75% animal material, including small pieces of fish, mollusks, earthworms, offal, live insects, and silkworm pupae, and 25% plant material, such as various grains and rotten fruit. Five percent of the total body weight of prawns is fed daily, half in the morning and half in the afternoon. Waste may be prevented by placing food in trays along the side of the pond.

Both food and shelter for the prawns may be provided by growing small patches of *Ipomoea* in the pond. The area covered by *Ipomoea* should not, however, exceed 10% of the area of the pond. Further shelter for moulting prawns may be supplied by placing branches on the bottom of the pond.

The most serious management problem encountered in this phase of

TABLE 5.   STOCKING RATES FOR GIANT FRESHWATER PRAWNS (*Macrobrachium rosenbergi*) CULTURED ALONE AND WITH FISH IN SOUTHEAST ASIA

POND CONDITIONS	STOCKING RATE (PRAWNS/HA)	STOCKING RATE OF FISH
	PRAWNS CULTURED ALONE	
Rich	15,000	
Medium	10,000	
Poor	6,000	
	PRAWNS CULTURED WITH FISH	
Rich	6,000	Full
	12,000	Half
Medium	4,000	Full
	8,000	Half
Poor	2,000	Full
	4,000	Half

prawn culture is oxygen depletion, to which *M. rosenbergi* is more sensitive than most fishes. When prawns migrate toward the edges of the pond and appear sluggish in their movements it is time to apply remedial measures to increase the supply of oxygen.

Under favorable conditions prawns in ponds should reach marketable size (15 cm and 100 g) in 5 months, at which time they may be harvested by draining the pond or seining.

Less intensive culture of *M. rosenbergi* in paddy fields is also practiced, but the future of this form of prawn farming is threatened by the increased use of insecticides on rice. The bunds of paddy fields used for this purpose should be slightly raised so as to retain at least 12 cm of water throughout the 4-month growing period of the rice. Inlets and outlets should be equipped with screens to prevent the escape of prawns and entry of predators. These screens should extend about 0.3 m above the water surface or prawns will climb over them and escape. One or two small sump pits 1 m × 2 m × 50 cm deep should be constructed near the outlet to trap the prawns when the field is drained.

Paddy fields should be stocked only after the rice seedlings are fairly well rooted. Due to the shorter growing season, older prawns 2 to 3 cm longer than those used in pond stocking are preferred. These are stocked at about 1 prawn/15 m² under average conditions. Supplementary feeding is not practiced.

Since culture of *M. rosenbergi* is still largely in the experimental stages, there are few data on the yields which may be achieved. However, a commercial, albeit pilot, operation on Oahu, Hawaii, under the supervision of the State's Department of Fish and Game can produce 3,000 kg/ha of giant prawns. Larval survival rates in a highly automated hatchery are satisfactory even though they could still be improved. At moderate flow-rates of water through the adult rearing ponds, limits on stocking are mostly conditioned by dangers of disease. But even lacking experimentation with a more vigorous flow, T. Fujimura, who has pioneered this success, is confident that 4,000 kg/ha are attainable with continuous stocking and harvesting schedules.

Large freshwater shrimps similar to *M. rosenbergi* are found in practically all tropical and subtropical regions, and *Macrobrachium rude* is coincidentally cultured along with *M. rosenbergi* in India, but the only other freshwater species thus far studied to any extent are *Macrobrachium malcolmsoni* and *Macrobrachium carcinus*. In Pakistan, the Directorate of Fisheries has succeeded in experimentally breeding and rearing *M. malcolmsoni* in a freshwater pond (size not given). Stocking with about 15,000 young yielded 560 kg of adult shrimp. Breeding was accomplished by confining pairs in very fine-meshed cages, 68.5 cm $\times$ 35.5 cm $\times$ 18 cm deep at the side of the pond and feeding with rice bran each morning.

Macrobrachium carcinus has been experimentally reared through all larval stages at McGill University's field station in Barbados, but mortality has been too high for effective culture. As far as is known, neither spawning nor hatching of eggs in captivity has been accomplished.

It is reported that in Peru larvae of the freshwater shrimp *Macrobrachium caementarius* are harvested from the lower end of rivers, raised in tanks for a few months, and used to stock both private and public waters.

PROSPECTUS

Further growth of the shrimp culture industry will result from expansion into new parts of the world, improvements and innovations in culture techniques, and introduction to culture of new species. Most of the countries which presently support commercial or experimental shrimp culture have been mentioned here. New additions to the list may be expected regularly, as virtually every country with a seacoast has some potential for shrimp culture.

NEW TECHNIQUES

A number of new and/or experimental techniques have also been mentioned here. At present the chief stumbling block in shrimp culture is poor survival of the larvae. The key to better survival is apparently a better diet. Japanese culturists have achieved the greatest success in supplying the nutritional needs of young shrimp, but their reliance on imported brine shrimp is costly and, if imitated by many other countries, might result in the depletion of brine shrimp stocks in the United States. Mention has already been made of the successful substitution in Thailand of fish eggs for brine shrimp in the diet of *Macrobrachium rosenbergi*. It has been suggested that amphipods or other locally available small crustaceans might be cultured as food for young shrimp in countries where brine shrimp are not available.

Improvement might also be effected in the feeding procedures for older shrimp. It would be particularly advantageous to culturists if a prepared food comparable to the pelleted feeds used in culture of carp, trout, and other fishes could be developed for shrimp. A suitable food pellet would have to incorporate a superior binding agent, since shrimp, rather than ingesting pellets whole as do most fish, hold them and pick them apart. In a British study *Palaemon serratus* were fed with pellets containing one of eight different protein foods. In each case the prepared feed produced much poorer growth than fresh food containing the same protein source. When, instead of a protein source, powdered polyethylene was added to the basic formula as a "nutritionally inert" filler, growth was better than that achieved using five of the other eight experimental feeds. Experimental work on pelletized diets for shrimp was also carried out at Florida State University, where a food conversion ratio of 3:1 in pink shrimp under experimental conditions was reported.

Other promising experimental techniques include growing shrimp in thermal effluent and monosex culture. The former method is being studied chiefly in the United States where numerous large power plants release large quantities of heated water used for cooling purposes. Conclusive results are not yet available, but indications are that growth and survival of Penaeid shrimps may be enhanced in "thermally enriched" environments.

Monosex culture has not yet been attempted, but its feasibility is suggested by the fact that female Penaeid shrimps of most, if not all species are larger than males of the same age; female sugpo prawns cultured in the Philippines average two to three times as large as the males at harvest. Research in West Germany has disclosed that female sand shrimp convert food more efficiently than males. On the other hand, male *Macrobrachium rosenbergi* grow much larger than females. It seems likely that similar sexual dimorphisms exist in other groups of shrimp as well.

NEW SPECIES FOR CULTURE

Nearly all attempts at shrimp culture to date have involved Penaeid shrimps or members of the palaemonid genus *Macrobrachium*. Only a minority of these species have been cultured commercially and few of the remainder have been adequately studied, so additional members of these groups may be expected to enter the picture. Of particular interest among the Penaeids are those exceptional species which are highly euryhaline. Notable among them are the Australian greentail prawn (*Metapenaeus benettae*) and *Metapenaeus ensis*, both of which spawn naturally in estuaries, a trait that could greatly simplify estuarine culture. Postlarval greentail prawns apparently prefer water of less than 20‰ salinity. Australian biologists are currently experimenting with the culture of these and/or similar species, locally called "greasy backs."

Edible non-Penaeid shrimps which possess the ability to reproduce in brackish water include *Caridina gracilirostris*, most species of *Macrobrachium*, *Palaemon*, and *Leander*, and some species of *Palaemonetes*. *Macrobrachium lanchesteri* of Malaysia will even reproduce in stagnant freshwater. In addition to possessing the considerable advantage of being very hardy with respect to high temperatures, low dissolved oxygen concentrations, and waters with low mineral content, *M. lanchesteri* is believed to be herbivorous.

In concentrating on members of the genera *Penaeus*, *Metapenaeus*, and *Macrobrachium*, shrimp culturists have overlooked species with less complex life cycles. Such species, if they could be cultured, would eliminate many of the difficulties now encountered in rearing young shrimp. These species may be roughly divided into four categories:

1. Species with no free-swimming larvae. In *Sclerocrangon boreas* of the North Atlantic and Pacific and *Sclerocrangon ferox* of the arctic Atlantic the young remain attached to the female's pleopods and do not feed up to the zoea stage. These two benthic species reach lengths of over 12 cm and appear potentially suitable for culture. Certain other members of their genus, however, produce pelagic larvae.

There are species of shrimp which bypass the larval stages entirely and emerge from the egg as juveniles, but most of them are too small for human consumption. An exception is *Bythocaris leucopia*, which exceeds 9 cm in length. *B. leucopia* is a deep water species, however, and thus might not be amenable to culture. A better possibility might be the Japanese *Pandalopsis coccinata*. The 15-mm shrimp which emerges from the egg of this species is classified as a zoea, but is benthic and probably has feeding habits similar to the adult.

2. Species with larvae which do not feed. There are species of shrimp which have no functional mouth parts as zoea and subsist on yolk until they become juveniles. Most of these are deep water pelagic species and would probably be difficult to maintain in captivity. A possible exception is the benthic *Glyphocrangon spinicauda*.

3. Species with large larvae. Much of the difficulty in rearing shrimp larvae stems from their small size. Larger larvae could be fed such readily available animals as copepods or amphipods or might even take prepared foods. Large larvae, however, mean large eggs, hence these species produce less eggs than the commonly cultured shrimps. Extreme cases such as *Richardina spinicincta,* which may produce as few as six eggs, are obviously unsuitable for culture, but species which produce several hundred eggs might prove satisfactory if extremely high survival of larvae were achieved.

4. Species with few larval stages. *Pandalus kessleri,* which is fished, although not cultured, in northern Japan, hatches as a very advanced 8-mm zoea and moults to become a postlarva within a few days. *P. kessleri* has the disadvantage of being very stenohaline, thus unsuitable for tidewater culture. The same advantages and disadvantages apply to *Pandalus platyceros,* which is currently being cultured experimentally in the United Kingdom. Other species that appear to have similar rapid development include *Argis lar* and perhaps other members of that genus; the holarctic *Lebbeus polaris,* and *Lebbeus groenlandicus,* found from northeast Asia to Greenland.

The species just mentioned represent only a tiny fraction of the known species of Natantia. Many of the rest are obviously unsuited for culture by virtue of size, habits, or habitat. Nevertheless, there are undoubtedly others of potential value to the culturist. It should be noted that almost all of the species discussed are boreal or tropical in distribution. It is thus particularly likely that additional cultivable species will be discovered in the South Temperate and Antarctic regions.

Whatever species or techniques are used in shrimp culture, it is unlikely that dramatic improvements will drive the price of shrimp down. The generally carnivorous habits of shrimp, along with the great amounts of energy lost in moulting, virtually guarantee that food conversion will be inefficient. Thus as long as there are ample markets for luxury foods the profit potential of shrimp culture will remain high. But if worldwide or, in some instances, local food shortages develop, shrimp culture may be overshadowed by biologically and economically more efficient forms of brackish water aquaculture.

REFERENCES

ANON. 1970. First cultured shrimp harvested at Florida farm. Am. Fish Farmer 2(1):7.

ANON. 1972. Shrimp harvest at Marifarms. Am. Fish Farmer 3(2):5–7.

BORJA, P. C., and S. B. RASALAN. 1957. A review of the culture of Sugpo, *Penaeus monodon* Fabricius, in the Philippines. FAO Fisheries Reports 57(2). FRm/R57.2 (Tri.), R/1, pp. 111–124.

COOK, H. L., and M. A. MURPHY. 1965. Rearing Penaeid shrimp from eggs to post-larvae. Proceedings of the Southeastern Association of Game and Fish Commissioners, 19th Annual Conference, pp. 283–288.

COOK, H. L., and M. A. MURPHY. 1969. The culture of larval Penaeid shrimp. Trans. Am. Fish. Soc. 98(4):751–754.

DALL, W. 1958. Observations on the biology of the greentail prawn *Metapenaeus mastersi* (Haswell) (Crustacea:Decapoda:Penaeidae). Aust. J. Mar. Freshwater Res. 9(1): 111–134.

DELMENDO, M. N., and H. R. RABANAL. 1956. Cultivation of "Sugpo" (Jumbo Tiger Shrimp) *Penaeus monodon* Fabricius, in the Philippines. Indo-Pacific Fisheries Council, Proceedings of the 6th Session, Sections 2 and 3, pp. 424–431.

FORSTER, J. R. M., and T. W. BEARD. 1969. Some experiments on the preparation of a compounded diet for the prawn *Palaemon serratus* Pennant. I.C.E.S., C.M. 1969/E: 6, Fisheries Improvement Committee.

FUJIMURA, T. 1966. Notes on the development of a practical mass culturing technique of the giant prawn *Macrobrachium rosenbergi*. Work Paper of the Indo-Pacific Fisheries Council. IPFC/C66/WP (47).

FUJINAGA, M. 1967. Kuruma shrimp *(Penaeus japonicus)* cultivation in Japan. FAO World Scientific Conference on the Biology and Culture of Shrimps and Prawns, Mexico City, 6/12–24/67. FR:BCSP/67/E/44.

HUDINAGA, M., and J. KITTAKA. 1967. The large scale production of the young kuruma prawn *Penaeus japonicus* Bate. Inf. Bull. Plank. Jap.

HUGUENIN, J. E. 1968. Commercial shrimp culture: an overview. Unpublished manuscript. 25 pp.

IDYLL, C. P. 1965. Shrimp nursery-science explores new ways to farm the sea. Nat. Geog. 127(5).

KNOX, J. H. 1966. A business man/investor's evaluation of the scientific problems facing shrimp farming. *In* E. A. Joyce, Jr., and B. Eldred, The Florida shrimping industry. State of Florida, Board of Conservation, Division of Inland Fisheries, Education Series, No. 15.

LING, S. W. 1961. Notes on the life and habits of the adults and larval stages of *Macrobrachium rosenbergi*. Proceedings of the 9th Indo-Pacific Fisheries Council (II).

LING, S. W. 1966. Feeds and feeding of warm-water fishes in ponds in Asia and the Far East. FAO World Symposium on Warm Water Pond Fish Culture. FR: III-VIII/R-2.

LING, S. W. 1967a. Methods of rearing and culturing *Macrobrachium rosenbergii* (de Man). FAO World Conference on the Biology and Culture of Shrimps and Prawns. Mexico City, 6/24/67. FR: BCSP/67/E/31.

LING, S. W. 1967b. The general biology and development of *Macrobrachium rosenbergi* (De Man). FAO World Conference on the Biology and Culture of Shrimps and Prawns. Mexico City, 6/24/67. FR:BCSP/67/E/30.

LUNZ, G. R. 1957. Pond cultivation of shrimp in South Carolina. Proceedings of the Gulf and Caribbean Fisheries Institute, 10th Annual Session, Nov. 1957, pp. 44–48.

MEIXNER, R. 1966. Eine Methode zur Aufzucht von *Crangon crangon* (L.) (Crust. Decap. Natantia). Arch. Fischereiwiss. **XVII**(1):1–4.

PANIKKAR, N. K. 1957. Osmotic behavior of shrimps and prawns in relation to their biology and culture. FAO Fisheries Reports 57(2). FRm/R57.2 (Tri.), R/1, pp. 527–538.

PANIKKAR, N. K., and R. G. AIYAR. 1939. Observations on breeding in brackish water animals of Madras. Proc. Indian Acad. Sci. (B) 9(6):343–364.

REEVE, H. R. 1969. The laboratory culture of the prawn *Palaemon serratus*. Fish. Invest. London Ser. II 26(1).

THAM, A. K. 1957. Prawn culture in Singapore. FAO Fisheries Reports 57(2). FRm/R57.2 (Tri.), R/1, pp. 85–94.

WHEELER, R. S. 1968. Cultivation of shrimp in artificial ponds. *In* Report of the Bureau of Commercial Fisheries Biological Laboratory, Galveston, Texas, Fiscal Year 1967. U.S. Bureau of Commercial Fisheries Circular 295, pp. 8–9.

WILLIAMSON, D. I. 1957. The type of development of prawns as a factor determining suitability for farming. FAO Fisheries Reports 57(2). FRm/R57.2 (Tri.), R/1, pp. 77–84.

Interviews and Personal Communication

BORGA, P. C. Chief, Minor Sea Resources Section, Marine Fisheries Biology Division, Philippine Fisheries Commission.

ESCRITOR, G. L. Shellfish and Crustacean Section, Estuarine Fisheries Division, Philippine Fisheries Commission.

FORSTER, J. R. M. Shellfish Culture Unit, Fisheries Experiment Station, Conway, England.

FUJIMURA, T. Hawaii Division of Fish and Game, Honolulu, Hawaii.

FUJIYA, M. Japanese Fisheries Agency.

HASEGAWA, I. Director, Inland Sea Fish Farming Association, Yashima Center, Yashima, Japan.

LING, S. W. Fisheries Biologist, FAO, Bangkok, Thailand.

MERICAN, A. B. O. Fisheries Biologist, Fisheries Research Station, Federation of Malaya, Glugor, Penang, Malaysia.

PROVENZANO, A. J., JR. Institute of Marine Sciences, University of Miami, Florida.

RABANAL, H. R. FAO, Rome, Italy.

SUGIURA, I. General Manager, The Shrimp Farming Co., Ltd., Takamatsu, Japan.

33

Lobster Culture

One of the most highly prized of all seafoods is lobster, and efforts have been made to culture this delicacy at least since the 1860s. From the biologist's or culturist's point of view "lobster" may indicate either of two very different types of animal. Seafood gourmets find considerable difference between the two as well, but many consumers are less discriminating. The animals most of us visualize when lobster is mentioned are the American lobster (*Homarus americanus*) and its European counterpart, *Homarus vulgaris*, both of which are restricted to the North Atlantic. However, much of the lobster tail of commerce (including that

633

marketed as rock lobster, Australian lobster, or South African lobster) comes from one or another of the spiny lobsters, which are found virtually throughout the oceans. There are many species of spiny lobsters belonging to several families, but all may be distinguished from *Homarus* by the lack of the formidable claws or chelipeds. A more important difference to the culturist is the much more complex larval development of spiny lobsters, which has kept spiny lobster culture lagging behind that of *Homarus,* although neither animal has yet been reared commercially.

CULTURE OF HOMARUS SPP.

EARLY STOCKING PROGRAMS

Early attempts at culturing *Homarus* concentrated on hatching larvae for stocking in the sea. In 1885 the U.S. Fisheries Commission began releasing newly hatched lobster larvae along the northeastern coast of that country and by 1907 numerous federal or state-supported hatcheries were established in five states. Canada and Newfoundland soon followed suit but found that hatcheries were neither biologically nor economically justifiable and discontinued stocking by 1917, although Canada still maintains a small experimental program. The early boom in the United States fell off for the same reasons, but the practice of lobster culture never completely died out and the state of Massachusetts still carries out a stocking program. A different sort of stocking program is contemplated in British Columbia and Oregon, where some biologists hope to establish naturally reproducing populations of *Homarus americanus.*

Though artificial propagation of *Homarus vulgaris* was achieved 20 years before the first success with *Homarus americanus,* lobster hatcheries were not established in Europe before 1921. Since then, experimental culture and stocking have been undertaken in Norway, Sweden, Denmark, Germany, France, the Netherlands, and the United Kingdom, with results no better than those achieved on the other side of the Atlantic. Large-scale experiments are presently being conducted only in Norway and the United Kingdom. At one time British researchers were also experimenting with the culture of another homarid species, the "Dublin prawn" (*Nephrops*), but this work has apparently been discontinued.

Commercial rearing of edible size lobsters has been the subject of much speculation and occasional experimentation but has never been achieved in Europe or North America.

PLATE 1.   American lobsters in mating position photographed at Woods Hole Oceanographic Institution. (Courtesy J. Atema.)

NATURAL HISTORY

As far as is known, the natural history of the two species of *Homarus* is virtually identical. Mating, which occurs in the summer months, follows the pattern for decapod crustaceans in that it is dependent on the moulting of the female. According to some authorities copulation can occur up to 12 days after the female moults, but chances of success are much higher within 48 hours. The success of mating depends partly on the relative size of the male and female; if there is great disparity in size, it proceeds only with difficulty or not at all.

Courtship is apparently initiated by a pheromone secreted by the freshly moulted female. When a male is drawn to within a meter or so he responds by advancing on the very tips of his walking legs, with continuous rapid movement of the maxillipeds and side to side movement of the antennae. Upon making contact, the two lobsters stroke each other with their antennae for up to 30 min (Plate 1). At the conclusion of this courtship, the male mounts the female, gently rolls her over, and they copulate in a head-to-head position. Males appear limited to mating with one or two females in a period of a few days. Females may on occasion mate with more than one male.

Sperm is stored in the female's seminal receptacle for 9 to 13 months, at which time the 5000 to 125,000 eggs are extruded, fertilized, and ce-

PLATE 2.   Female American lobster (*Homarus americanus*) bearing fully developed eggs. (Courtesy Biological Laboratory, National Marine Fisheries Service, Boothbay Harbor, Maine.)

mented to the nonplumose hairs of the swimmerets. The eggs are carried by the female for 10 to 12 months before hatching. Thus the total time between mating and hatching may be as much as 2 years, though cul-turists have succeeded in reducing this period, by means of temperature control, to as little as 11 months. During the period of incubation the eggs change color from nearly black to green to brown. Hatching has been recorded at temperatures as low as 9.4°C but usually takes place in spring or early summer at 15 to 20°C.

The newly hatched mysis larvae, which make their first moult almost

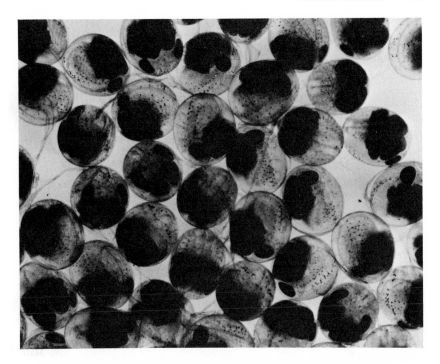

PLATE 3. Lobster (*Homarus americanus*) eggs fully developed and ready to hatch. (Courtesy Biological Laboratory, National Marine Fisheries Service, Boothbay Harbor, Maine.)

immediately, assume a planktonic existence near the surface. After a few more moults they acquire well-developed claws and other external appearances of tiny lobsters (Plate 4). At this time they are capable of swimming yet may still be found among the plankton, but soon they take up a benthic, nocturnal existence. Development to this stage takes 9 to 33 days, depending on temperature, but survival from hatching to the benthic form (fourth stage) is probably less than 0.1%.

Postlarval lobsters ordinarily undergo four more moults in the first season of life, from the date of hatching to the end of that calendar year, for a total of 10 moults, counting larval moults, in that season. Thereafter moulting becomes less frequent, until by the attainment of sexual maturity at about 6 years moulting occurs no more than once annually. Sexual maturity usually coincides with the attainment of commercial size (about 20 cm in length and 0.5 kg in weight), though the precise rate of growth is dependent on temperature. Data on lobster growth rates are scarce since all external indications of the individual identity of a

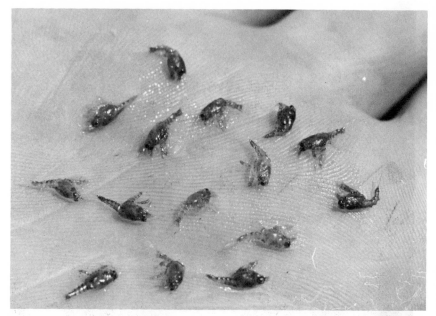

PLATE 4. Third-stage lobster larvae (*Homarus americanus*) eight days old. (Courtesy Biological Laboratory, National Marine Fisheries Service, Boothbay Harbor, Maine.)

lobster are lost with the shedding of the exoskeleton. Captive lobsters kept individually have been observed to grow 9.5 to 17.4% in length and up to 50% in weight at a moult. There appear to be no significant differences in rate of growth due to sex.

Lobsters in Massachusetts have been observed to moult as early as April 25 and as late as December 29 at temperatures as low as 3.3°C, but most moulting occurs from May to October at temperatures of 15 to 20°C. There are two seasonal moulting peaks, one in early summer when the water temperature first reaches 15 to 20°C and one in early fall. Fall moulting is probably a response to increase in volume of body tissue due to heavy feeding during the summer.

Homarus americanus has a life span of 50 to 100 years and attains weights as great as 19 kg.

HATCHING AND REARING THE YOUNG IN CAPTIVITY

Early lobster hatcheries relied on eggs taken from "berried" females and hatched separately with the aid of a device to agitate the water. It is currently believed that the female lobster is a more effective incubator than any man-made device. Either way, hatching lobster larvae has always been easier than rearing them and, although first-stage larvae were stocked along the Atlantic coast of the United States and Canada from 1891 to 1917, it has been the opinion of most lobster culturists that for a stocking program to be effective the young lobsters must be capable of assuming a benthic life upon release. Thus rearing of larvae has long been a preoccupation of lobster culturists.

The largest lobster hatching and rearing operation today is that carried on by the state of Massachusetts on the island of Martha's Vineyard, under the direction of John T. Hughes. Hughes and his associates have succeeded in mating lobsters in captivity, but the chief source of eggs for hatching continues to be the lobster fishery. Spring-caught females bearing brown eggs ("Brown eggers") are selected, since their eggs will hatch in a few months, whereas "green eggers" would have to be held over winter. The prospective mothers are placed in running water (salinity 30 to 31‰) in hatching tanks 274 cm × 91.5 cm × 30 cm deep, divided into compartments for individually holding up to 70 lobsters. (Berried *Homarus vulgaris* reportedly cannot be kept together because they will claw the eggs off each other, but this is not so with *Homarus americanus*.) They are fed on shellfish viscera and/or fish. Hatching usually commences in mid-May, when the water temperature reaches 15°C, and becomes intensive during a 2-week period in June and July.

The newly hatched larvae are swept away by the current and collected in a wire screen box then transferred to circular fiberglass tanks with a diameter of 400 mm and concave bottoms. Experienced workers estimate the number of larvae by eye and place about 3000 in each tank. Water is circulated through the rearing tanks to keep the larvae drifting in midwater. If they were to accumulate on the bottom, cannibalism would severely decimate their numbers. Water enters the tank from the bottom through an inverted, perforated plastic cup, which breaks up the flow and keeps water circulating evenly in all parts of the tank. The outlet is an overflow pipe equipped with a similar device.

The first fry feed used at Martha's Vineyard, and still used in Norwegian hatcheries, was finely ground beef liver, but ground clams and frozen adult brine shrimp (*Artemia*) have been found to produce much better survival. Crab viscera are another acceptable food. Ground clams or brine shrimp are mixed with sea water in a proportion determined by the demands of the larvae, and one tablespoon of the mixture fed every three hours throughout the larvae's stay in the hatchery. Heavy feeding helps avert cannibalism, but overfeeding may clog the inlet and outlet pipes, thus reducing circulation and increasing cannibalism as well as raising the threat of pollution.

Using these methods an average of 22% survival to the third moult has been attained, with a record season's survival of 42.6%. Next to cannibalism the major cause of mortality is a gas disease caused by supersaturation of the water with nitrogen, a condition brought on by air leaks in the circulating system. Lobster larvae are sensitive to certain metal ions, but this has become less of a problem at the Martha's Vineyard hatchery since lead piping was replaced by plastic.

In Norway, lobster fry in hatcheries are attacked by a parasitic suctorian (*Euphelota* sp.). Heavy infestations cause mass mortality by offsetting the larvae's buoyancy and forcing them to the bottom. Chemical treatment has proved futile, but *Euphelota* can apparently be excluded from hatcheries by filtering sea water through sand before it is used.

CURRENT STOCKING PROGRAMS

When the lobster fry reach the fourth or fifth stage (between third and fifth moults) they are planted at selected points along the Massachusetts coast. The assumption is that they soon settle to the bottom where natural mortality is much less than that for planktonic larvae. The repeated presence of fourth-stage larvae in plankton catches, however, indicates that they do not all settle immediately, or perhaps none of them do. The logical way to eliminate this cause of concern is to grow all of the larvae to fifth stage.

A more gratuitous assumption is that stocking lobster larvae contributes substantially to the fishery. There is circumstantial evidence suggesting that the first large-scale propagation program, put into operation by the state of Rhode Island in 1901, improved the fishery off that state. In 1904 the Rhode Island catch had declined to 169,747 kg or 20 kg/pot. In 1919, after 17 years of stocking, the catch was up to 736,422 kg or 35 kg/pot. These data were used as a rationale for further efforts on both sides of the Atlantic, but there is no evidence that any of the subsequent stocking programs paid dividends. Rhode Island eventually gave up on lobster culture as being "uneconomical."

The precise fate of lobster larvae released into the sea cannot be determined for the same reason that growth rates of wild lobsters cannot be studied, the impossibility of tagging (a satisfactory internal tag has reportedly been developed in Maine, but details are not yet available), but statistical inferences can be made as to the practical value of lobster culture as currently carried out. In 1950 Clyde C. Taylor of the U.S. Fish and Wildlife Service made the first critical evaluation of a hatchery program. He demonstrated that if 10% (a rather optimistic figure) of the lobster larvae reared and stocked in Maine in 1943 had survived to commercial size and were captured in 1949, they would have constituted only

0.015% of the fishery for that year. There is no indication that the situation is or ever was substantially different anywhere else.

RECENT APPLIED RESEARCH

Biologists at the Martha's Vineyard hatchery have succeeded in selectively breeding various odd-colored lobsters, which could serve as natural tags for growth studies and assessment of stocking programs. Although odd colors have persisted through ten moults in the laboratory, it is known that food affects the color of lobsters and it is conceivable that odd-colored lobsters released in the sea might change diets and therefore color. According to Hughes, however, certain genetically determined colors are permanent and so distinctive as to rule out confusion with any environmentally induced colors.

The rate of development of both eggs and larvae can be accelerated by raising the water temperature. Eggs held at or above 20°C will hatch about 3 months sooner than eggs started in sea water at normal winter temperatures, with no apparent harm to the resulting larvae. Similarly, the rate of moulting of larval lobsters may be increased by a factor of 2 to 5 by heating the water. The upper limit of temperature for first stage larvae appears to be 27°C, but third- and fourth-stage larvae can endure constant temperatures as high as 31°C.

A factor that might affect the feasibility of rearing larvae to fifth stage is population density. Density has been shown to affect survival to fourth stage, but little is known as to what constitutes the optimum density at any stage.

Adult lobsters in hatcheries are usually kept in very shallow water, for reasons of convenience. There has never been any problem in getting lobsters to mate under these conditions, but until recently females would jettison their eggs before fertilization. This problem was alleviated at Martha's Vineyard by the simple expedient of raising the water depth from 15 to 45 cm. No one knows just why this works, but it has enabled Massachusetts biologists to contain the entire life cycle of *Homarus americanus* within the laboratory and opened up the possibility of selective breeding. Among the goals of selection, in addition to the "natural tags" described, are higher growth rate, larger claws, more meat in relation to body size, and resistance to disease.

POSSIBILITY OF COMMERCIAL CULTURE

As intriguing as the possibilities of lobster breeding appear, the economic prospects for self-contained lobster culture in the near future do not look bright, at least not in the United States or the United Kingdom. It appears that at present commercial lobster culturists would be more likely to succeed by growing juvenile lobsters to marketable size. The principal source of stock for such an operation would likely be the lobster fishery. In some places there would need to be legislative changes to allow fishermen to retain small lobsters for this purpose. Before such legislation could be rationally considered it would be necessary to weigh the economic and gustatory benefits of lobster culture against the possible effects of reduction of wild breeding stocks.

If any method of growing marketable size lobsters in captivity achieves economic success, the state of Massachusetts can take much of the credit for laying the groundwork. Biologists at the Martha's Vineyard hatchery have succeeded in raising lobsters up to the age of 10 years and as large as 154 mm in carapace length. Fifth-stage lobsters to be retained for growth to maturity are stocked in individual 15 cm × 15 cm compartments in a trough containing 15 cm of water. After about 2 years it becomes necessary to provide larger compartments, but individuals must always be isolated or given abundant hiding places to prevent cannibalism on freshly moulted lobsters. The only alternative to isolation is to stock lobsters at very low densities. All of these methods present obvious drawbacks to commercial culture.

Captive lobsters are ordinarily fed fresh fish or shellfish. If either food is not fresh, it will be rejected. Each lobster at Martha's Vineyard is fed as much as it will consume daily, except during the winter when they are fed but once a week. It is estimated that it takes about 8 kg of food to produce 1 kg of lobster.

Lobster culture could be made more efficient if a pelleted feed, such as those used in culture of catfish and trout, were available. In addition to sinking, a satisfactory lobster pellet would have to be of a texture to accommodate the lobster's feeding habits without dissolving or crumbling. Unlike fin fish, which engulf pellets whole, lobsters grasp and chew their food. By using the seaweed gel carrageenin, pellets with the consistency of used chewing gum can be manufactured. Such pellets, incorporating commercial trout chow, dog food, or cat food as foods, are now being tested at Martha's Vineyard. These pellets are stable in water and accepted by lobsters, but whether they will produce growth comparable to that achieved with fresh foods is as yet unknown.

In nature it takes 5 to 7 years from hatching to produce a marketable size lobster. During much of this time little or no growth is going on, due to water temperatures below 10°C. By maintaining year-round high temperatures it is possible to greatly accelerate growth rates. Lobsters kept at a constant 15.6°C moult twice as often, hence grow twice as fast as those exposed to the full range of natural temperatures in the North

Atlantic. Further, lobsters will adapt to temperatures as high as 30°C—16 to 21° above normal summer water temperatures in their natural habitat. This knowledge is the basis for current work on *Homarus* culture on the Pacific coast of the United States, where they are not native. Lobsters are being experimentally grown at San Diego, California, in heated discharge from a San Diego Gas and Electric Co. power plant.

Growth rate is probably capable of further improvement, independent of thermal manipulation; great individual differences in growth rates have been observed in stock of both sexes kept under identical conditions, suggesting that selective breeding for rapid growth would be practicable.

By using heated water and other intensive care techniques biologists at Martha's Vineyard have in at least one instance produced a marketable size lobster in a little over 2 years after hatching, and marketable 3-year-olds are a commonplace.

It would also be propitious for the future of lobster culture if the public could be educated to accept smaller lobsters or lobster meat processed so that the original size of the lobster could not be determined.

Hughes believes that commercial lobster culture, using heated water, is now feasible. Perhaps even more optimistic are some members of the staff of a West German marine biological station on the island of Helgoland, in the North Sea. They have launched an ambitious project to investigate the possibility of culturing *H. vulgaris* from larvae to commercial size in cages on the sea floor, under virtually natural conditions. The optimism of Hughes and the German researchers notwithstanding, we feel that for the present any sort of commercial culture of either species of *Homarus* seems at best an intriguing gamble.

CULTURE OF PALINURID LOBSTERS

NATURAL HISTORY

There are 7 genera and about 30 species of spiny lobsters (superfamily Scyllaridea), of which at least 15, all members of the family Palinuridae, are commercially important (Table 1). So far as is known, all the commercially important species have similar but very complex life histories.

Mating occurs in inshore waters at various times of year, depending on species and age of females. Old females mate earlier than young and some of them are thought to mate twice a year. In mating, males extrude sperm in the form of a viscous fluid which attaches to the underside of the female. There the surface rapidly hardens to form a sperm sac. Some

TABLE 1.   THE COMMERCIALLY IMPORTANT SPECIES OF SPINY LOBSTERS AND THEIR RANGES

SPECIES	RANGE
Jasus lalandii	Australia, Juan Fernandez Islands (off coast of Chile), New Zealand, South Africa, Tasmania, Tristan da Cunha
Jasus verreauxi	Southern Australia, New Zealand, Tasmania
Palinurus elephas	Great Britain to the Mediterranean
Panulirus argus	Florida, the Caribbean area, the Atlantic coast of South America, Bermuda
Panulirus gracilis	Southern Mexico to Ecuador
Panulirus inflatus	Gulf of California to Southern Mexico
Panulirus interruptus	Southern California, Lower California (Mexico)
Panulirus japonicus	Japan
Panulirus laevicauda	Bermuda to northeastern South America
Panulirus longipes	Western Australia, Ryukyu Islands, Taiwan, Philippines, New Caledonia
Panulirus marginatus	Hawaiian Islands
Panulirus ornatus	Indian Ocean, Ryukyu Islands, Taiwan, Philippines
Panulirus penicillatus	Indo-Pacific, Korea, Lower California (Mexico), Galapagos, Costa Rica
Panulirus regius	Western Mediterranean, West Africa
Panulirus versicolor	Indo-Pacific

time later, the eggs are extruded and pass over the sperm sac. At that time the female breaks the sperm sac by scratching it with her legs, thus fertilizing the eggs. They are then passed on to the swimmerets where they are attached, to remain until hatching. The number of eggs produced by a female varies with size and species. Estimates as low as 50,000 and as high as 4,000,000 have been recorded. The time between mating and fertilization is probably dependent on water temperature. In general it appears to be less than that for *Homarus*.

Certainly hatching occurs much sooner than in *Homarus;* only 3 weeks are required in some species. During this time the eggs change color from bright red-orange to a dark brown which fades until they are almost colorless just before hatching.

If, from the culturist's point of view, the reproductive process of spiny lobsters is more efficient than that of *Homarus,* the spiny lobsters more than make up for it with protracted and complex larval development.

FIG. 1. Young spiny lobster (Phyllosoma). (After Dees, 1963.)

The newly hatched larva is a flat, leaflike, and unusually delicate animal 2 to 3 mm long, known as a phyllosoma (Fig. 1).

Phyllosomas are planktonic and float horizontally with the legs extended. At least some species are negatively phototropic and make fairly extensive diurnal vertical migrations. The larvae remain planktonic while passing through a large number of moults, most of which do not result in metamorphosis, although progressive changes in form may be detected. Usually 3 to 6 months and 6 or more metamorphoses are required to reach the puerulus stage, but in one laboratory experiment on *Panulirus japonicus*, 16 moults were passed through in 178 days without metamorphosis. It is possible that the number of larval moults is not fixed for a species and that frequency of metamorphosis is partially dependent on nutrition or some such factor.

The puerulus is superficially similar to the adult but is transparent, lacks lime in the skeleton, and may still be planktonic. Finally, the puerulus moults to become a juvenile about 2.1 cm long and settles to the bottom. After two or three more moults it acquires the reddish-brown color of adult spiny lobsters. Survival to this point is extremely low since predation is extensive at all stages.

Juvenile and adult spiny lobsters are also subject to predation but avoid it to some extent by hiding under rocks and other cover during the day and foraging at night. Muddy bottoms and strong currents are avoided. In nature adult spiny lobsters consume a wide variety of foods, including fish, worms, mollusks, and smaller crustaceans. Cannibalism may occur if foods containing calcium carbonate are not available. Spiny lobsters may act as scavengers but exhibit a marked preference for fresh food.

Frequency of moulting, and therefore rate of growth, varies with food supply, water temperature, and sex. Mature females moult but once or twice a year, before mating and sometimes after hatching or otherwise disposing of their eggs. Males grow larger than females and may moult at any time of year, thus are thought to moult more often. Growth between moults is about 5 to 10% of body length for *Panulirus argus*. Maximum size and age varies with species. *Panulirus interruptus* is said to reach a weight of over 13 kg, but such specimens are rare, and 3 kg would be considered large. In California this species is believed to take at least 7 to 9 years to reach the legal minimum size of 10½ in. (27 cm), although *P. argus* may attain this length in 3 years.

ATTEMPTS AT CULTURE

Obvious difficulties notwithstanding, many attempts have been made to rear spiny lobsters from the egg in captivity. Perhaps the first efforts were those of the California Department of Fish and Game, starting in 1911, with *Panulirus interruptus*. Sporadic efforts with that species have continued, but little progress has been made, though Margaret Knight at Scripps Institute of Oceanography in La Jolla, California, has been able to hold larvae in finger bowls for up to 4 months. Soviet scientists also report some success in holding and rearing *Palinurus elephas* in aquaria and nursery ponds.

Other members of the family Palinuridae which have been hatched and held in the laboratory for 5 to 10 moults include *Panulirus japonicus* in Japan, *Panulirus argus* in Florida, and *Jasus lalandii* in South Africa. Mention should be made of the confusion between the terms "stage" and "moult." There are descriptions in the literature of rearing *Panulirus japonicus* and other species as far as the "tenth stage," but when compared to natural larvae of *Panulirus interruptus,* for which the larval stages have been thoroughly described, the supposed tenth-stage larvae appear to be third or fourth stage. This confusion is occasioned by the lack of any direct correspondence between developmental stage and number of moults. Under unfavorable conditions of temperature or food supply, spiny lobster larvae may moult without appreciable growth or morphological change. The term "stage" should be used only to refer to morphologically distinct forms.

Other Palinurid species which have been hatched in captivity but not reared beyond the very early stages include *Panulirus inflatus* in California, *Panulirus longipes* in Japan, and *Panulirus polyphagus* in Malaysia. The necessity of unpolluted water free from silt or small bits of detritus on which the larvae may entangle their long legs and the need for nearly constant temperature would certainly constitute serious restraints to commercial culture, but the main problem at present is feeding. Early larvae of most species accept brine shrimp nauplii quite readily, but attempts to feed older larvae on adult brine shrimp, sea urchin eggs, and larval gobiid fishes have met with slight success. Complicating the situation is the fact that almost nothing is known of the natural food of spiny lobster larvae.

CULTURE OF SCYLLARID LOBSTERS

All of the commercially important spiny lobsters belong to the family Palinuridae. The related Scyllaridae are scarcely exploited. However, it should be pointed out that those species whose habits are most conducive to fishery exploitation are not necessarily those best suited for culture. Some Scyllarid lobsters possess the advantage for culture of having shorter and less complex larval development. In general they also appear to be hardier animals.

The Scyllarids *Ibacus ciliatus, Ibacus novemdentatus, Parribacus antarcticus,* and *Scyllarus bicuspidatus* have been hatched in Japan but have not been reared for more than a few days. Similar lack of success has been experienced in India with *Scyllarus sordidus.* However, Phillip B. Robertson of the University of Miami's Institute of Marine and Atmospheric Sciences has become the first to rear any spiny lobster to the juvenile stage. He reared the sand lobster (*Scyllarus americanus*) from the egg through 6 to 8 phyllosoma stages to metamorphosis in 32 to 40 days at 25°C and salinities of 23.2 to 38.6‰. Most important from the culturist's point of view, Robertson was able to achieve this on an exclusive diet of brine shrimp nauplii. Robertson's success notwithstanding, it will almost certainly be a long time before larval culture of the sand lobster or any other scyllarid lobster can be recommended for other than experimental purposes.

REARING OF SPINY LOBSTER PUERULI OR JUVENILES TO MARKETABLE SIZE

Clearly, if there is to be commercial culture of Palinurid or Scyllarid lobsters in the near future it will involve capturing pueruli or juveniles and growing them in confinement. This possibility has been considered in Australia where young *Panulirus longipes* can be captured in abundance on the west coast. To retain these young for culture would presently be illegal, but if it could be shown that culture was economically feasible and would not deplete natural stocks, the legal restraints might be removed. Most undersize *P. longipes* taken off Australia are 2 to 3 years old and would need to be held for only 1 to 2 years to reach marketable size. Experiments have shown that the food conversion ratio of such animals fed on fish or abalone is about 6:1, which is quite efficient for a crustacean. However, taking into consideration the costs of food, capture of stock, pond construction, maintenance, and so forth, it seems unlikely that such culture would be profitable in Australia at this time.

The type of culture suggested for Australia has reportedly been attempted with *Panulirus japonicus* by Japanese researchers. Young *P. japonicus* caught in traps have been held in ponds or shallow bays and fed trash fish. The results indicate that it may be feasible to produce salable lobsters at a profit.

There may be other countries where the practice of capturing and growing young spiny lobsters will prove profitable. The feasibility of such culture would be enhanced if growth rates could be increased as has been done with *Homarus americanus*. This might be done by improving the quality of food fed to spiny lobsters. Most Palinurids in captivity are fed fish, probably for reasons of availability, but crustaceans have been shown to produce better results. A better possibility for increasing growth rates is the use of heated water. Steven A. Serfling of San Diego State College in California has been able to increase the growth rate of young *Panulirus interruptus* in the laboratory by as much as 260% over average natural rates by maintaining them at 28°C. It is not known whether this would be feasible within the framework of commercial culture.

PROSPECTUS

Improvements will surely be made in culture techniques for Homarid, Palinurid, and Scyllarid lobsters but, in the near future at least, the vast majority of the lobsters consumed by man will be fishery products.

REFERENCES

ALLEN, B. M. 1916. Notes on the spiny lobster (*Panulirus interruptus*) of the California coast. U. Calif. Pub. Zool. 16(12):139–152.

CARLSON, F. T. 1954. The American lobster fishery and possible applications of artificial propagation. Yale Conserv. Stud. 3:3–7.

CHITTLEBOROUGH, R. G. 1968. Can we culture marine crayfish? Fish. Indust. News Service, W. Australia 1(3):16–17.

DEES, L. T. 1963. Spiny lobsters. U.S. Bureau of Commercial Fisheries, Fishery Leaflet 523. 7 pp.

DOTSU, Y., K. SENO, and S. INOUE. 1966. Rearing experiments on early phyllosomas of *Ibacus ciliatus* (von Siebold) and *I. novemdentatus* Gibbes (Crustacea: Reptantia). (In Japanese, with English abstract, tables, legends.) Bull. Fac. Fish. Nagasaki U. 21:181–194.

HUGHES, J. T. 1968. Biologists breed lobsters selectively. Comm. Fish. Rev. 30(10):20.

HUGHES, J. T., and G. C. MATTHIESSEN. 1967. Observations on the biology of the American lobster (*Homarus americanus*). Massachusetts Division of Marine Fisheries, Technical series No. 2.

INOUE, M. 1965. On the relation of amount of food taken to the density and size of food and water temperature in rearing the phyllosoma of the Japanese spiny lobster, *Panulirus japonicus* (V. Siebold). (In Japanese, with English abstract, tables, legends.) Bull. Jap. Soc. Sci. Fish. 31(11):902–906.

JOHNSON, M. W. 1956. The larval development of the California spiny lobster *Panulirus interruptus* (Randall) with notes on *P. gracilis* Streets. Proc. Calif. Acad. Sci., 4th Ser. 29(1):19–22.

JOHNSON, M. W., and M. KNIGHT. 1966. The phyllosoma larvae of the spiny lobster *Panulirus inflatus* (Bouvier). Crustaceana, 10(1):31–47.

KENSLER, C. B. 1970. The potential of lobster culture. Am. Fish Farmer 1(11):8–12, 27.

MOQUIN-TANDON, O., and J. L. SOBEIRAN. 1865. Establissements de Pisciculture de Concarneau et de Port-de-Bauc. Bull. Soc. Imp. Zool. Acclim. 2nd Ser., no. 2, 533–545.

ONG, K. S. 1967. A preliminary study of the early larval development of the spiny lobster *Panulirus polyphagus* (Herbst). Malaysian Agric. J. 46(2):183–190.

PRASAD, R. R., and P. R. S. TAMPI. 1960. On the newly hatched phyllosoma of *Scyllarus sordidus* (Stimpson). J. Mar. Biol. Ass. India 2(2):241–249.

ROBERTSON, P. B. 1968. The complete larval development of the sand lobster, *Scyllarus americanus* (Smith) (Decapoda, Scyllaridae) in the laboratory, with notes on larvae from the plankton. Bull. Mar. Sci. 18(1–4):294–342.

SAISHO, T. 1962. Notes on the early development of a scyllarid lobster, *Parribacus antarcticus* (Lund). Mem. Fac. Fish. Kagoshima U. 11(2):174–178.

SAISHO, T., and K. NAKAHARA. 1960. On the early development of phyllosomas of *Ibacus ciliatus* (von Siebold) and *Panulirus longipes* (A. Milne Edwards). Mem. Fac. Fish. Kagoshima U. 9:84–90.

SHEARD, K. 1949. The marine crayfishes (spiny lobsters) of western Australia, with particular reference to the fishery on the western Australian crayfish (*Panulirus longipes*). Council on Scientific and Industrial Research, Bull. 247. Division of Fisheries Rep. 18, pp. 1–45.

SMITH, F. G. W. 1958. The spiny lobster industry of Florida. Florida State Board of Conservation Education Series, No. 11. 36 pp.

TAYLOR, C. C. 1950. A review of lobster rearing in Maine. Department of Sea and Shore Fisheries, Research Bulletin 5.

TEMPLEMAN, W. Further contribution to the mating of the American lobster. J. Biol. Board Can. 2(2):223–226.

THOMAS, H. J. 1964. Artificial hatching and rearing of lobsters—a review. Scottish Fish. Bull. 21(6–9).

THOMAS, H. J. 1969. Lobster rearing. Scottish Fish. Bull 31:11–14.

WILDER, D. C. 1948. The growth of lobsters at Grand Manan, New Brunswick. Fisheries Research Board of Canada, Progress Report on Atlantic Stations, 41, pp. 14–15.

Interviews and Personal Communication

ATEMA, J. Woods Hole Oceanographic Institution, Woods Hole, Massachusetts.

HUGHES, J. T. Director, State Lobster Hatchery and Research Station, Vineyard Haven, Massachusetts.

KENSLER, C. B. U.N. Fisheries Research and Development Project, Mexico (1) D.F., Mexico.

MITCHELL, J. R. Woods Hole Oceanographic Institution, Woods Hole, Massachusetts.

SERFLING, S. A. Sea Grant Spiny Lobster Program, San Diego State College, San Diego, California.

34

Culture of Freshwater Crayfish

Among the best known and most highly esteemed crustaceans are the lobsters (*Homarus* spp.) (see Chapter 33), but they are rivaled as a delicacy by their smaller freshwater counterparts, the crayfishes or crawfishes (family Astacidae). (Not to be confused with the marine spiny lobsters, family Palinuridae, sometimes marketed as crayfish or crawfish.) True crayfish comprise more than 300 species and are found on all the continents except Africa. Although crayfish are esteemed as a gourmet food in several European countries and are the primary source of protein for certain tribes in New Guinea, they are generally underutilized by man.

Crayfish have attained importance as a commercial food product in parts of Europe and the United States. The most enthusiastic consumers

PLATE 1.   A bucket of Louisiana red crayfish, 5 to 6/kg. (Courtesy B. Glenn Ham, Photograph by Ray Utt, Cities Service Oil Co.)

of crayfish are the French, and crayfish farms have been in operation in France since 1880. But the most important crayfish producing area is Louisiana, the only American state with a history of French culture. In a good year more than 800,000 kg of "wild" crayfish may be caught and marketed in Louisiana; during bad years the yield may be less than half that figure. Fishery production is supplemented annually by 1.2 million kg produced on the 6000 to 7000 ha of crayfish farms in the state. Details are not available on techniques of crayfish culture in Europe, hence this report concentrates on practices in Louisiana.

NATURAL HISTORY OF THE PRINCIPAL CULTURED SPECIES

There are twenty-nine species of crayfish known to inhabit the waters of Louisiana, but only two are cultured. On most farms the dominant

PLATE 2.   A crayfish "boil" in the making. (Courtesy B. Glenn Ham, photographed by Ray Utt, Cities Service Oil Co.)

PLATE 3.   Crayfish burrow. (Courtesy James W. Avault.)

species is the red crayfish (*Procambarus clarkii*), but a few areas produce chiefly white crayfish (*Procambarus blandingi*). The natural history of the two species is similar.

Mating occurs in open water in the late spring, when the water level in the Louisiana swamps is high. At that time the male crayfish deposits sperm in an external receptacle on the female.

Shortly after the peak of the breeding season, female crayfish come out on shore and dig burrows near the water's edge. Since they are exposed to terrestrial predators at this time, areas well protected by emergent plants are preferred. Burrows are essentially vertical and usually 0.7 to 1.0 m deep, unless the water table is unusually high, in which case burrows half that depth may be found. By the end of July, all adult females have constructed burrows, which are then usually occupied by a single male, along with the female. Each burrow is capped with a plug of dirt. Young crayfish and unpaired males also seek shelter in the mud or in naturally occurring holes during the summer, but they do not burrow. Some additional mating may occur in the burrows, but the crayfish are thought to be essentially inactive until September, when egg laying occurs.

Eggs and sperm are simultaneously released by the female and the fertilized eggs attached to the underside of her tail until hatching, which occurs in 14 to 21 days with red crayfish and 17 to 29 days with white crayfish. The adults die soon after hatching. An average hatch is about 400 red crayfish (maximum 700) and somewhat less for white crayfish. Fertilization and egg laying often occur in the burrow, but growth and survival of the 25-mm, free-swimming young is greatly enhanced if open water is available. In years when the water is low in the fall, hatching may occur in the burrow, in which case the young suffer from overcrowding and lack of food. Or the female may crawl off overland in search of water and die of dehydration or be captured by a predator.

Young crayfish usually seek shelter in dense plant growth near shore, then move out to deeper water as they mature. Both red and white crayfish are annual animals, and maturity may be reached in less than 6 months after hatching.

Crayfish of all ages are omnivorous but prefer animal matter. Nevertheless, most crayfish, both in nature and in culture, must subsist on a predominantly vegetable diet.

POND CULTURE IN LOUISIANA

Crayfish culture is practiced in two types of water in Louisiana: rice fields and artificial impoundments. The latter, though its large-scale use dates back little more than 20 years, now accounts for more than 80% of the area devoted to crayfish farming.

SITE SELECTION AND CONSTRUCTION

As in most other forms of aquaculture, site selection is crucial in pond culture of crayfish. Due to the shallowness of crayfish ponds, it is essential that the land chosen be relatively flat. When the highest point in the pond bottom is covered by 0.3 m of water, depths of over 1 m should not occur in more than 25% of the remaining area. It is not, however, of any particular value if the pond bottom is level. In fact, small elevations in the bottom have the favorable effect of increasing the area available for burrowing of early breeders.

Densely wooded areas are not favored, since dense shrubbery and trees along pond banks hinder harvesting, shade out desirable aquatic plants, add leaves and debris which decay and reduce oxygen levels, and may keep temperatures in the shallows below optimum. It was in such environments, however, that pond culture of crayfish had its start in Louisiana and 2400 ha of densely wooded swampland are still in use in an area just west of the Atchafalaya Basin, where conventional pond construction would be extremely difficult. These leveed swamplands are also heavily used by waterfowl, which are extensively hunted in Louisiana.

Soil quality surely has some effect on crayfish production but, in Louisiana at least, the primary consideration is that the soil hold water.

Water supply is of course important, but crayfish are very tolerant of naturally occurring physical and chemical conditions. Aspects of water quality which are of particular importance to the crayfish farmer are temperature and hardness. Optimum temperatures for red crayfish are thought to be 21 to 29°C, but growth is not drastically reduced until the temperature falls below 13°C. Above 32°C, red crayfish burrow into the mud and become inactive. White crayfish do better at slightly lower temperatures and begin to burrow at temperatures as low as 27°C. Growth is the primary reason for the culturist to be concerned with water temperature, since neither species is likely to suffer significant thermally induced mortality at temperatures common in Louisiana. For the same reason, dissolved oxygen concentration, though unlikely to be a cause of mortality, should be kept as high as possible.

Cultured crayfish have been observed to do well at pH levels as low as 5.8 and as high as 8.2. Crayfish in acid waters tend to have thinner shells. Soft water also results in thin, soft shells as well as poor growth and survival. It appears that the water in crayfish ponds should have a total hardness of at least 50 ppm, and that up to 200 ppm is desirable.

In recent years interest in crayfish farming in the coastal swamps of southern Louisiana has increased, and some attention has been paid to the effect of salinity on crayfish. Preliminary experiments indicate that crayfish will reproduce and grow fairly well at salinities of 6 to 10‰.

Tolerant as crayfish are to natural conditions, they are extremely sensitive to synthetic chemicals and other substances which may find their way into agricultural water supplies. Such commonly used compounds as Pyrethrum, creosote, orthodichlorobenzene, sodium cyanide, turpentine, orthocresole, cresylic acid, pine oil, nicotine, carbon bisulfide, phenothiazine, calcium cyanamid, and chlorinated hydrocarbon pesticides have all been shown to be toxic to crayfish. The last-named group presents particular problems because of the possible danger to human health from dosages sublethal for crayfish and because the crayfish farmer may find it difficult to prevent contamination of his water supply by pesticides used on nearby agricultural land. It is recommended that crayfish culturists located in the vicinity of intensive terrestrial farming operations alert neighboring farmers and crop-dusting pilots of the location of their crayfish ponds and attempt to secure their cooperation in averting contamination. When spraying is going on nearby, pumping of water from streams and ditches into crayfish ponds should be suspended.

The danger of contamination by pesticides and other pollutants is greatly reduced if the water source is a deep well, but many farmers, for reasons of necessity or economy, use surface water. Whichever type of water is used, it is best if each unit of the farm is provided with its own pump. Units need not be discrete but may be portions of larger ponds separated by inside levees. The smaller the units, the greater the ease of water quality control and general management. On the other hand, the more small units are built, the greater the construction cost per hectare of water surface. A large farm might consist of one or two 100- to 300-ha ponds broken up into 10 or so units. One-family farms as small as 8 ha exist but do not provide a full family income.

Detailed advice on levee construction may be obtained from the Soil Conservation Service, county agricultural agents, or commercial contractors. In general, levees should be high enough to keep out flood waters and wide enough to permit access to vehicles.

Water circulation has been found to be a deciding factor in the success or failure of many crayfish farms. To facilitate circulation it is necessary to provide not only an adequate inflow of water but good drainage as well. Drain pipes, which must be screened with 12-mm or smaller wire mesh, should be located as far away from inlet pipes as possible to permit thorough mixing of water. Drains should be large enough to permit complete draining in 30 days.

STOCKING

Although some ponds contain natural populations of crayfish, stocking is nevertheless necessary the first year, after which the population should be self-sustaining. The usual procedure is to stock ponds with adults in May to July, at which time crayfish are "tough" and prices low. The pond should be flooded for at least 2 weeks before stocking and arrangements made to purchase freshly caught crayfish. Stock held in captivity for even so short a period as overnight should not be accepted. Large specimens, about 30 to 45/kg, are preferable. Suggested stocking rates for red crayfish vary according to the amount of cover in the pond (Table 1).

At least 80% of most lots of crayfish purchased for stocking in most parts of Louisiana will be red crayfish. If, as occasionally happens, a stock of predominantly white crayfish is obtained, the figures in Table 1 should be increased by 25%.

Crayfish of either species should be placed in the pond as soon as possible and kept cool and damp until that time. Predation on newly stocked crayfish will be reduced if they are released in densely vegetated areas or, if the pond is sparsely vegetated, in deep water far from shore.

TABLE 1.   SUGGESTED STOCKING RATES FOR CRAYFISH IN DIFFERENT TYPES OF
PONDS IN LOUISIANA

WILD CRAYFISH PRESENT	TYPE OF POND	STOCKING RATE (KG/HA)
Yes	—	20–25
No	Densely vegetated	40–45
No	Wooded	45–60
No	Open, sparse cover	45–60
No	Open, very little cover	60–100

J. G. Broom at Auburn University obtained experimental evidence
that stocking young red crayfish in March at 42,500 to 85,000/ha would
produce better yields than the conventional stocking system, but as far
as is known this method has not been put into practice, perhaps due to
the difficulty of capturing young crayfish for stocking. Broom suggests
the following stocking formula:

$$\frac{\text{desired yield (kg/ha)}}{\text{weight at harvest}} \times \frac{100}{\text{expected survival}}$$

$$\times \text{ number of ha to be stocked} = \text{number of crayfish to stock}$$

MANIPULATION OF THE WATER LEVEL

The most important management techniques in crayfish culture involve
manipulation of water level and quality. Each year ponds are drained
in late June or early July, when the females have started burrowing. Slow
draining is preferable, for fast draining will strand some crayfish which
are not ready to burrow and expose them to predators. Similarly, slow
draining allows young crayfish to seek hiding places for the summer.

Ponds are reflooded in September to ensure that the newly hatched
young will have ample water. If growth rates are normal, they should be
ready for harvest starting in late November or early December. The culturist should keep an eye on rainfall, which is the principal controlling
factor on wild crayfish harvests, and time reflooding so that he can begin
harvesting before the wild crop comes in. Early crayfish bring the best
prices and may spell the difference between economic success and failure.

Ordinarily, harvesting lasts until the pond is drained the following
summer. Between reflooding and the onset of the harvest season in ponds
having heavy cover, water levels 50 cm or so lower than those recommended above can be maintained to furnish more shallow water for
young crayfish.

During the harvesting season the water level should be kept fairly stable. This requires frequent pumping to replace water lost by evaporation. Pumping must also be resorted to to replace deoxygenated water. Any abrupt reduction in harvest is usually a sign of deoxygenation or some such condition and calls for partial replacement of water.

FOOD SUPPLY

Any animal food that cultured crayfish get is the result of natural production and their own foraging, but successful culturists encourage suitable food plants and discourage large, tough plants which interfere with harvesting, shade out smaller, more desirable plants, and are inedible by crayfish. Crayfish from ponds that do not contain adequate amounts of edible plants often have brown or black "fat" or livers, and tails which are not filled out, and are considered to be of inferior quality to well nourished crayfish with full tails and yellow livers.

Plants used as food and/or cover in crayfish ponds must be capable of survival both while the pond is flooded and when it is dry. The most desired food and cover plant in crayfish ponds is alligator grass (*Alternanthera phylloxeroides*). It can be seeded into a pond by raking it out of ditches and scattering it in the pond during drawdown in June and July. Alligator grass may, on occasion, grow too thick and hinder harvesting. If harvesting is done by boat, boat trails may be disked or raked on the bottom during the summer. Although cattle should generally be kept out of crayfish culture areas, grazing them on the dry pond bottom may help retard alligator grass.

Water primrose (*Jussiaea* spp.) is now considered as good as or better than alligator grass and may be planted together with it. Water primrose has the advantages of not growing as thick as alligator grass and being more tolerant of cold weather. A number of other plants which occur naturally in crayfish ponds, including pondweeds (*Potamogeton* spp.), *Elodea,* and duckweed (*Lemna*), are fair food and/or cover plants and should not ordinarily be discouraged. It is certain that fertilization of dry ponds would benefit all of these and other plants, but techniques to encourage desired species have yet to be worked out.

Crayfish will also eat almost any soft terrestrial plant. Some farmers have planted sorghum or millet as feed for crayfish and claim increased yields, but there is as yet no experimental evidence for this conclusion.

It is likely that supplemental feeding with animal matter would increase production, but it is doubtful whether it would be economically feasible. Experiments with feeding young red crayfish at Auburn University showed that growth was better on a diet of *Elodea* and ground

PLATE 4.   Hauling crayfish traps at the Cities Service Oil Co. crayfish farm near Fairbanks, La. Note vegetation, water primrose, which is an important food of the crayfish. (Courtesy B. Glenn Ham, photograph by John I. Fogelman.)

Tilapia than on either component alone. Research is currently being carried out at Louisiana State University with the aim of developing an economical, high-protein artificial feed for crayfish.

PROBLEMS

The most common pest plants in crayfish ponds are cattails (*Typha* spp.). They can usually be controlled only if caught early, at which time individual plants can be pulled out manually. If dense growths occur, disking and raking the pond immediately on drying may be helpful. Otherwise, it is necessary to call on state or federal agencies for assistance. Outside help is also usually necessary to control water hyacinth *(Eichornia crassipes)* which, unchecked, may completely cover the surface of a pond. A number of emergent plants other than cattails are usually present but seldom reach problematical densities in ponds which are drained and dried annually.

Filamentous green algae, which occur principally during the winter, also may hinder harvesting operations. Further, they are prone to die off suddenly in warm weather, creating a pollution problem. Partial removal of floating algal mats may be effected by vigorous circulation of the water.

Man is by no means the only animal which is fond of crayfish, and predation by fish, birds, raccoons, bullfrogs, snakes, turtles, salamanders, large water beetles, and others, is a problem confronting every crayfish culturist. The best protection against most of these predators is abundant cover in the pond, but fish, the most serious predators, may require additional measures. Annual draining and drying of course substantially reduce fish populations, but some of the worst predators, notably green sunfish (*Lepomis cyanellus*), bowfin (*Amia calva*), and bullheads (*Ictalurus* spp.), may survive in potholes until reflooding time. Rotenone at 2 to 3 ppm will destroy such fish. Great care should be exercised in its use, however, and a biologist consulted if possible, since 5 ppm of rotenone will kill some young crayfish.

It is standard practice to screen pond outlets to prevent fish from entering, but opinions differ as to the feasibility of screening inflow pipes. Many successful operators reason that most large fish are killed by the pump impellers and do not screen inflow pipes. In any event, the large 0.3- to 0.7-m pumps employed on most ponds move great volumes of water and make screening difficult.

It is recommended that, on new ponds supplied by surface waters, a temporary screen be installed on the inflow pipe and checked frequently during the first few days. If considerable numbers of large predatory fish are captured, a permanent screen should be employed. A suitable design consists of a cylinder of 25-cm weldwire about 1 m in diameter and 4 m long. Such a screen will allow young of predatory fish to enter the pond, but finer screens clog up very rapidly and require constant cleaning.

HARVESTING

One of the most difficult aspects of pond culture of crayfish is harvesting. The most frequently used harvesting device is an 0.8 to 1.0 m long funnel-shaped trap made of 21-cm mesh chicken wire and so constructed that it folds flat for hauling. Shorter traps are fully as effective for capturing crayfish, but dissolved oxygen concentrations may become very low in crowded, submerged traps, causing mortality. Long traps, however, may be propped up at an angle so that a small portion is above the surface of the water, and trapped crayfish can obtain air.

Traps are commonly baited with fish heads or chunks of gizzard shad (*Dorosoma cepedianum*), which is considered more effective than any other fish. Fresh fish may be scarce during the winter, and soybean cake,

PLATE 5.   Crayfish trap held by B. Glenn Ham, crayfish culturist. (Courtesy B. Glenn Ham, photograph by Ray Utt, Cities Service Oil Co.)

perforated cans of dog food, or almost any high-protein substance may be substituted. About 25 traps/ha are set out.

Most of the harvesting is done by professional fishermen hired by the culturist. This compounds the culturist's harvesting problems, since he is seldom able to maintain a crew of fishermen throughout a harvest season. But to realize maximum yields and fulfill contracts to buyers it is essential that harvesting be intensive throughout the season. The problem is particularly acute during cold weather, when fishermen may be reluctant to run their traps frequently, and during the height of the wild crayfish season, when they may be able to make more money harvesting wild stocks on their own. The demand for labor is particularly great at the start of the season, in late November and early December, when many culturists prefer to harvest as heavily as possible to beat the peak of the wild crop, to utilize adults which have spawned and will die shortly and reduce the danger of crowding of the young crayfish.

Some farmers circumvent harvesting problems by opening their ponds

TABLE 2. SCHEDULE OF PROCEDURES FOR POND CULTURE OF CRAYFISH IN
LOUISIANA

ACTIVITY	DATES
Construction of new ponds and flooding	By May 15
Stocking	June 1–30
Draining	June 15–July 15 (begin 2 weeks after stocking)
Fish control	July 15–August 1
Repairs, improvement, planting, and weed control	July 15–August 15 (as soon as possible after most of the pond is dry)
Reflooding	By September 15
Harvest	November 25–June 30 (begin 1 month later for new ponds)

to the public, either for a set fee or by charging substantially less than the retail price for the crayfish which are caught.

A summary and schedule of procedures for pond culture of crayfish in Louisiana is given in Table 2.

RICE FIELD CULTURE IN LOUISIANA

Crayfish have long been harvested as an incidental crop from rice fields in Louisiana, but it is only in the last 15 years that intensive cultivation has become common. Today, nearly 1000 ha of rice fields are operated in 2-year rotation as crayfish ponds and pastures. The approximate schedule for this sort of rotation is given in Table 3.

There are usually enough crayfish present in a Louisiana rice field to serve as broodstock. If not, adults may be stocked at 6 to 12 kg/ha during May. Some crayfish leave the drained fields in July and August, but most of them burrow into the moist soil. Rice stubble resprouts about a week after harvest, at which time the fields are reflooded to a depth of 15 to 45 cm. Crayfish in rice fields feed mainly on rice stubble and various aquatic plants which find their way into the fields. Rice-crayfish farmers should not use rice seed treated with aldrin, which may be toxic to crayfish.

Harvesting crayfish from rice fields, although tedious, is easier and cheaper than in ponds. Due to the more regular bottom, nets may be employed. The usual type, called a drop net or umbrella net, consists

TABLE 3.   SCHEDULE OF PROCEDURES FOR RICE FIELD CULTURE OF CRAYFISH IN LOUISIANA

ACTIVITY	DATES
Plowing	March 1–April 30
Replowing and planting rice	April 15–May 15
Flooding and stocking, if necessary	May 1–May 31
Draining	July 15–August 15
Harvesting rice, followed by reflooding	August 1–September 1
Harvesting crayfish	December 1–June 15
Drain and use as pasture	June 15–March 1

of a rectangular piece of netting, each corner of which is connected by a wire or rod to a ring located over the center of the net. A pole or other lifting device is inserted in this ring and the net laid flat on the bottom. Crayfish are baited into it with beef pancreas or other offal and captured by lifting the pole. Ordinary farm laborers rather than professional fishermen may be hired to do this chore.

Growth to the minimum marketable size of 10 to 15 g is rapid in either type of culture, requiring little more than 6 months. Larger 40- to 45-g crayfish, which bring the best prices, are harvested principally in the early part of the season and may be 8 to 14 months old. Well-managed ponds and rice fields commonly produce 400 to 700 kg/ha of crayfish, and yields of better than 1100 kg/ha have been achieved.

MARKETING IN LOUISIANA

Marketing is by no means organized. Crayfish may be sold live, boiled, or as peeled tails to wholesale dealers, restaurants, or individuals. Production of peeled tails may be increased if a mechanical crayfish peeler, currently being developed at Louisiana State University, is perfected.

DISEASES AND PARASITES

Disease is not a major problem to crayfish culturists in Louisiana. A bacterial rot causing decay of the rostrum was observed in red crayfish in Alabama but was readily cured by treatment with potassium permanganate at 3 ppm. An extremely heavy infestation of both red and white crayfish with larvae of the water boatman *Ramphocorixa acuminata* occurred at the U.S. Fish Farming Experimental Station, Stuttgart, Arkansas, but this insect has never presented serious problems in commercial culture.

CULTURE OUTSIDE OF LOUISIANA

Some gourmets find three large crayfish native to the Pacific Northwest of the United States superior to the Louisiana varieties. *Pacifastacus klamathensis, Pacifastacus leniusculus,* and *Pacifastacus trowbridgii,* sometimes marketed as "short lobster" (a term also used to refer to undersized lobsters taken illegally) are principally fishery products, but, in Whatcom County, Washington, they are beginning to be cultured. Most existing crayfish farms in Washington utilize ponds originally constructed for other purposes, such as irrigation, but ponds built specifically for crayfish are likely to become more prevalent.

Crayfish of several species are also raised for use as bait by sport fishermen in various parts of the United States. In the Finger Lakes region of New York, *Orconectes immunis* has been successfully raised in ponds with various species of bait fish. With supplementary feeding on soybean meal, fish meal, cracked corn, potatoes, and cut hay, 40 to 350 kg/ha were produced and higher yields are believed possible. Yields of up to 1000 kg/ha of *Orconectes rusticus* have been reported from fish ponds in Ohio.

Methods of raising crayfish for bait are by no means standardized. The essential difference from culture for human consumption is that young, rather than adults, are harvested. One successful grower in Missouri uses 12- × 6-m spawning ponds, 1 m deep. When spawning and hatching are complete, the adults are removed using a heavily weighted minnow seine. Some of the larger ones (20 to 28 cm long) are marketed as food, but this is of secondary importance. The young are fed on cornmeal mush supplemented with cut fish and commercial fish food pellets and harvested in small lots on demand.

As mentioned earlier, there is a considerable demand for crayfish in a number of European countries, and crayfish culture in Europe might have developed to a greater importance were it not for a recurring "plague" specific to *Astacus astacus,* the principal European commercial species. Crayfish plague is caused by the fungus *Aphomyces astaci* and is presently considered incurable. Nor does there appear to be any tendency on the part of *Astacus astacus* to develop a natural immunity. The history of the crayfish fishery in Finland illustrates the severity of the crayfish plague. In 1900 that country exported 15.5 million crayfish, but in the 1960s only 0.3 to 0.4 million were exported annually. The reduction is attributed solely to the plague. The plague-resistant American crayfish *Pacifastacus leniusculus* has already been introduced to Finland and is being artificially propagated there so that more extensive introductions may be made.

Recently, Soviet biologists developed a hatchery technique which is claimed to have saved *A. astacus* from extinction in Lithuania. It is not known if the crayfish plague had anything to do with the impending doom of the Lithuanian crayfish. In 1969, hatcheries in east Lithuania were expected to release about 500,000 young crayfish into lakes and rivers. The hatchery technique involves stripping the eggs from the females; up to 90 young have been produced from one female.

POSSIBILITIES FOR EXPANSION

There are a number of edible European crayfish other than *A. astacus,* and it is surprising that their culture has not been attempted. In fact, the nearly worldwide neglect of crayfish by aquaculturists is amazing. Opportunities for expansion of the crayfish industry would appear to be particularly plentiful in the United States. In addition to the species already discussed, a small commercial fishery for *Orconectes virilis* exists in Wisconsin, but otherwise crayfish are virtually neglected as a food resource in North America. A number of the species cultured as fish bait are large enough to be edible and surely it would be a relatively simple matter to adapt existing culture techniques for production of large, edible specimens.

Where expansion of crayfish culture outside of Europe has been attempted, it has usually involved introduction of red crayfish from Louisiana. It is likely, however, that in most countries there are indigenous species suitable for culture, and certainly these should be tried first. The giant crayfish (*Astacopsis gouldi*) of Tasmania, which reportedly reaches lengths of over 40 cm and weights of 3 to 4 kg is particularly intriguing in this regard. In Japan, where red crayfish were introduced in 1918 as food for bullfrogs, they have become a serious agricultural pest. Populations of up to 1200 kg/ha exist in some rice fields in western Honshu, where they nibble off rice shoots and weaken dikes by their burrowing activities. A similar situation exists in taro (*Colocasia esculenta*) fields in Hawaii. It is surprising that the Japanese, normally so quick to utilize aquatic and marine resources, have for more than 50 years failed to adopt rice-crayfish culture methods similar to those used in Louisiana or otherwise turn the presence of the red crayfish to their advantage. Perhaps the progress of crayfish culture has been impeded by rather unfortunate publicity implying that fortunes can be made by simply stocking any rice field or leveed tract of swampland with crayfish and harvesting it a year later. Nevertheless, there is money to be made, with hard work, from crayfish farming, and it would seem that crayfish are overdue to assume an important role in aquaculture. At present, they seem best suited for marketing as a luxury food, but if more intensive culture methods, particularly more efficient harvesting techniques, are developed, the tribesmen of New Guinea may not be alone in regarding crayfish as an important nutritional resource.

REFERENCES

ANON. 1970. Crawfish: A Louisiana aquaculture crop. Am. Fish Farmer 1(9):12–15.

AVAULT, J. W., JR., L. DE LA BRETONNE, JR., and E. J. JASPERS. 1970. Culture of the crawfish, Louisiana's crustacean king. Am. Fish Farmer 1(10):8–14, 27.

BROOM, J. G. 1961. Production of the Louisiana red crayfish *Procambarus clarkii* (Girard) in ponds. M.S. Thesis, Auburn University, Auburn, Alabama. 16 pp.

FIELDING, J. R. 1966. New systems and new fishes for culture in the United States. FAO World Symposium on Warm Water Pond Fish Culture. FR: VIII/R-2.

FORNEY, J. L. 1954. Raising bait fish and crayfish in New York ponds. Cornell University Extension Bulletin 986, pp. 23–27.

HAM, B. G. 1971. Crawfish culture techniques. Am. Fish Farmer 2(5):5–6, 21, 24.

LACAZE, C. 1970. Crawfish farming. Louisiana Wild Life and Fisheries Commission, Fisheries Bulletin No. 7. 27 pp.

PENN, G. H. JR. 1954. Introduction of American crawfishes into foreign lands. Ecology 35(2):296.

PHILIPS, R. H. 1970. Crayfish or crawdad, it's all one—small, mean and loved by Swedes. Nat. Fisherman (Nov. 1970): 16-A.

Interviews and Personal Communication

HEADLEY, A. A. Crayfish grower, Foster, Missouri.

35

Culture of Crabs (Brachyura)

Nearly all of the larger species of crabs (Brachyura) are edible, but the majority of the crabs of commerce are members of three families: Portunidae (swimming crabs), Xanthidae (mud crabs), and Cancridae (cancer crabs). A fourth family which should be investigated by culturists is the widely distributed but underexploited Potamonidae (freshwater crabs).

At least one of the swimming crabs, *Scylla serrata,* has long been an incidental product of brackish water pond culture in southeast Asia. In the Philippines and perhaps elsewhere young *S. serrata* are occasionally stocked in fish ponds but usually they enter of their own accord. They may even be discouraged as possible predators on small fish and shrimp and because they may burrow in the banks, but generally they are tolerated and harvested along with the fish and/or·shrimp crop. In this manner, with no management whatsoever, the production of *S. serrata* from brackish water ponds in Taiwan in 1966 was 168,102 kg—more than that of any other single species except the ubiquitous milkfish (*Chanos chanos*) and Java tilapia (*Tilapia mossambica*). Similarly, milkfish ponds in Java produce an average of 200 crabs/(ha)(year).

In recent years overfishing has threatened the fishery for *Scylla serrata* in some parts of southeast Asia and attention has been given to the possibility of its culture. The groundwork for culture has been laid by Ong Kah Sin of the Malaysian Department of Fisheries in his laboratory

668

studies of the life history of this species. The following outline may serve as a general description of the life history of portunid crabs.

Mating of *S. serrata,* and probably most portunids, occurs as early as the first year of life after the female undergoes the precopulatory moult typical of all members of the Crustacean subclass Malacostraca. The first such moult cannot be readily distinguished from ordinary moulting, nor is maturity attained after any particular number of moults following hatching.

Females preparing to moult exert a strong attraction on males, probably by means of a pheromone. A male approaching a female in premoulting condition climbs over her, clasps her with his chelipeds and the anterior pair of walking legs, and carries her around with him in what is termed the doubler formation. They may remain so paired for 3 to 4 days until the female moults. The male then turns the female over for copulation, which usually lasts 7 to 12 hours. Since the male is attracted before the precopulatory moult, mating usually proceeds immediately upon moulting, but successful mating can occur up to at least 3 days after the moult.

Sperm are retained by the female, and fertilization may not take place for many weeks or even months after mating. A single copulation may provide sperm for two or more spawnings before the female moults again. The fertilized eggs are attached to the female's pleopods, where they hatch within a few weeks. In captive spawners, as many as two-thirds of the eggs may fail to become attached.

As in all Brachyura, the larvae hatch as planktonic zoeae. A few eggs may hatch prematurely as "prezoeae," but these moult and become zoeae within $\frac{1}{2}$ hour. After passing through several zoea stages and a single megalops stage within about a month, the larvae metamorphose to benthic juvenile crabs.

Ong's laboratory spawn were fed newly hatched brine shrimp (*Artemia*) nauplii, but survival was poor. Ong suggested that brine shrimp nauplii were too large and too fast for *S. serrata* larvae. They might have fared better on phytoplankton; some portunids are known to feed mainly on phytoplankton as zoeae and switch to zooplankton in the megalops stage.

Ong's studies are already being put into application in Ceylon, where the Ceylon Fisheries Corporation reports "encouraging results" from the culture of young *S. serrata* fed on fish offal accumulated in the normal operation of the plant. The company is planning to intensify crab culture, which will presumably involve larval rearing and possibly breeding.

Perhaps the most famous of the portunid crabs is the blue crab (*Callinectes sapidus*) of the Atlantic coast of the United States. (It should be noted that there are at least four highly valued portunids commonly referred to as "blue crabs": *C. sapidus, Callinectes bellicosus* of the Gulf of California, and the Japanese *Portunus trituberculatus* and *Neptunus pelagicus*, both to be discussed later.) Like *Scylla serrata, Callinectes sapidus* commonly enters brackish water ponds, where it attains a good size. However, such ponds are seldom used for aquaculture in the United States, so the harvest of *C. sapidus* is almost exclusively a fishery operation, carried out chiefly in the Chesapeake Bay and Delaware Bay regions. Both of these estuarine fisheries are threatened by pollution, and as far back as the 1940s the Virginia Fisheries Laboratory was engaged in experimental culture of blue crabs.

In 1969, W. T. Yang of the Institute of Marine and Atmospheric Sciences of the University of Miami was the first to rear *C. sapidus* from the egg to marketable size (9 to 12 cm carapace width). The process took only 3 to 4 months from hatching, suggesting considerable potential for commercial culture. Survival was very poor, however—only about 200 (0.25%) of some 800,000 second zoeae metamorphosed to the crab stage. Clearly practical culture of *C. sapidus* is at least several years away. Research at Duke University's Marine Laboratory in North Carolina has shown that temperature and salinity factors are critical in the larval development of *C. sapidus* and perhaps this would be the most fruitful area for applied research in culture of this species.

Japanese biologists at the Kanagawa Prefectural Fisheries Experimental Station have experienced somewhat better survival to 2-cm juveniles with *Portunus trituberculatus*. Rearing beyond this size is not usually attempted, since the principal purpose of culture is to provide young crabs for stocking in the hope of augmenting the fishery. The prospect of growing *P. trituberculatus* to marketable size looks tempting; juveniles kept on a diet of fresh fish and shellfish reach 15 cm long and 300 g in weight within 8 to 10 months of metamorphosis, or about a year after hatching. However, to achieve this rate of growth requires that each crab receive 10 to 15% of its body weight in food daily. Further, *P. trituberculatis* of all ages are cannibalistic to the extent of preferring their own species to all other foods. The expense of holding and feeding each crab individually to avert cannibalism virtually precludes commercial culture.

The present source of eggs for larval culture is berried females taken by fishermen. Such females are placed in individual cement tanks 5 m × 13.3 m × 1.7 m deep and allowed to incubate their own eggs. Though the natural environment of *P. trituberculatus* is estuarine, it thrives in a variety of salinities, so the water supply for these tanks, which are used for the entire culture process, is natural sea water with a salinity of 33 to 34‰. Starting at hatching the larvae are fed marine *Chlorella*. Five gallons of dense culture are added to each tank daily for a week. Then the larvae are switched to brine shrimp nauplii at about 30 nauplii/(crab larva)(day). Newly hatched nauplii are fed at first, but the size of the brine shrimp is increased as the larvae grow and pass into the megalops stage.

Metamorphosis to the crab stage usually occurs within 25 days of hatching. After metamorphosis the juveniles are maintained for another 20 days. Though they are heavily fed on fresh anchovies, cannibalism takes its toll and only about 20% of the juveniles survive this period. These are sold to Fishermen's Cooperatives for stocking in bays and inlets.

Since the program was initiated in 1967, 90,000 crabs have been stocked. It is estimated that at least 5% of these must eventually turn up in the commercial catch in order to repay the cost of culture. It is too early to attempt an evaluation of this particular experiment but, based on similar practices with other crustaceans, it seems unlikely that it will appreciably benefit the fishery.

About a decade ago, experimental culture at the Usa Marine Biology Station of Kyoto University of the other Japanese "blue crab," *Neptunus pelagicus,* was cause for optimistic predictions that "practical application to the culture of young crabs can be expected," but apparently nothing has come of it. As with *Portunus trituberculatus,* the chief problem in culturing *Neptunus pelagicus* is cannibalism; average survival from hatching to the first crab stage was only 10%. The situation was complicated by the fact that feeding efficiency of the larvae depended not only on the density of the food organisms (brine shrimp nauplii and the early stages of the oyster *Ostrea gigas* and the barnacle *Balanus amphitrite*) but apparently on the population density of the zoeae; the more dense the zoea population, the better the growth attained.

Whether or not practical culture of *Neptunus pelagicus* ever becomes a reality, the experiments at the Usa station uncovered two phenomena that may have general application to crab culture: The rate of development of all stages of *N. pelagicus,* like that of many Crustacea, was greatly increased by raising the temperature. At 24 to 25°C hatching took one-third as long and growth of larvae proceeded significantly faster than in the case of larvae reared at the "natural" temperature of 18 to 19°C.

A less predictable finding had to do with the strong positive phototropism of the larvae. Not only are they attracted to light, they can be induced to feed by lighting their tank. Though periodicity in feeding was never completely eliminated, larvae kept under 24-hour illumination fed more heavily and grew significantly faster than larvae on a normal light regime.

There has been considerably less work done on xanthids than on portunids, but the extensive experiments of J. D. Costlow et al. at Duke University's Marine Laboratory on several species of small xanthids of no commercial value suggest that thermal effects are equally important in growth of xanthid crabs. They also reached the rather surprising conclusion that for at least one species, the small mud crab (*Rithropanopeus harrisii*), growth and survival were much greater at high temperatures which fluctuated daily within a 5° range than at any constant temperature.

Costlow is presently continuing this type of work with a larger species, the stone crab (*Menippe mercenaria*), which supports an important fishery in Florida and Cuba. T. S. Cheung of the University of Miami was able to carry this crab from the egg to the adult stage. Stone crabs spawn up to ten times a year, thus hatching may occur at almost any time. After five zoeal stages and a single megalops stage the larvae metamorphose to juveniles. In the laboratory this entire process consumed only 9 to 14 days, depending on temperature. Various foods were tried, but the best survival (about 50% to the first crab stage) was achieved with brine shrimp nauplii.

After attainment of the crab stage growth became much slower, and the smallest commercial size (about 100 mm carapace width) was reached no sooner than midway in the third year for crabs fed on shrimp. Stone crabs fed on trout pellets grew slightly more slowly but may have been less expensive to raise.

A principal source of meat in stone crabs is the very large claws. Fishermen commonly break off one or both claws and release the crabs in the expectation that they will regenerate and provide another crop. The possible application of this technique to culture was tested. Small crabs regenerated claws slightly larger than the originals within 8 to 10 months and two moults. Commercial size crabs, however, failed to moult or regenerate in captivity. It may be that with improved culture methods this practice will prove economically feasible. Or there may be other species with large claws which are better suited to culture. Certainly these possibilities should be explored, for the advantages to the aquaculturist of a meat crop which need not be slaughtered are potentially great.

Culture of cancrid crabs has scarcely been explored, though the family includes a number of edible crabs, among them the dungeness crab (*Cancer productus*) of the Pacific coast of Canada, the United States, and Mexico, generally considered the finest of all crabs by gourmets. In recent years the dungeness crab fishery in San Francisco Bay and elsewhere has declined. Accordingly, laboratory rearing is being attempted and the feasibility of commercial culture investigated in a recently begun project at Oregon State University's Marine Science Center.

Given the large number of edible species and the short (for a crusta-cean) life cycles of most crabs, it seems highly likely that commercial culture of at least a few species will become a reality in the not too distant future.

REFERENCES

COSTLOW, J. D. 1967. The effect of salinity and temperature on survival and meta-morphosis of megalops of the blue crab Callinectes sapidus. Helgoländer wiss Meeresunters 15:84–97.

IDYLL, C. P. 1969. Status of commercial culture of crustaceans. Paper presented at Food and Drugs from the Sea Conference, Kingston, R.I. August 24–27, 1969.

ONG, K. S. 1964. The early developmental stages of Scylla serrata Forskal, reared in the laboratory. IPFC, 11th Session, Kuala Lumpur. C64/Tech. 37.

ONG, K. S. 1966. Observations on the post-larval life history of Scylla serrata Forskal, reared in the laboratory. Malaysian Agric. J. 45(4):429–443.

OSHIMA, S. 1938. Biological and fisheries research in Japanese blue crab (Portunis trituberculatus). Japanese Imperial Fisheries Research Station 9.

PAYEN, G., J. D. COSTLOW, and H. CHARAIAUX-COTTON. 1969. Endocrinologie des inver-tébrés-mise en évidence expérimentale de l'independence de la realisation du sexe chez le Crabe Rhithropanopeus harrisii (Gould) à l'égard du complexe neuro-sécréteur organe de Hanströmglande du sinus. Compt. Rendu Acad. Sci. Paris 269:1878–1881.

YATSUZUKA, K. 1960. Studies on the artificial rearing of the larval Brachyura, especially of the larval blue crab, Neptunus pelagicus. Rep. Usa Mar. Biol. Stat., Kochi 9(1).

Interviews and Personal Communication

COSTLOW, J. D. Director, Duke University Marine Laboratory, Beaufort, North Carolina.

HASEGAWA, I. Director, Inland Sea Fish Farming Association. Yashima Center, Yashima, Japan.

KOBAYASHI, M. Head, Propagation Division, Kanagawa Prefectural Fisheries Experimen-tal Station. Nagai, Japan.

36

Oyster Culture

Marine aquaculture may well have begun with oysters, which were cultivated in Europe during Roman times. The origins of oyster culture, as opposed to harvesting "wild" oysters, may be rooted in the human urge for convenience, but there are at least two other historic rationales for the practice:

1. Many areas where adult oysters grow very well are not suited to their reproduction. These areas may be brought into production of edible oysters by simply transplanting excess young oysters from grounds where reproduction occurs.

2. Though oysters may be prepared for the table in any number of ways, many gourmets prefer them raw, on the halfshell. Regularly shaped oysters, with shells free of fouling organisms, are considered more desirable for this purpose. Such oysters are rare in nature but can readily be produced by the oyster culturist.

In this century, the trend toward oyster culture has been accelerated as a reaction to man's catastrophic impact on estuarine ecology. In the Orient, human population pressure has stimulated the development of aquacultural schemes for many organisms, including oysters, which are regarded not as a gourmet dish but a staple seafood. This has led to the development of various techniques of off-bottom shellfish growing which have increased production manyfold over natural levels.

The ever-increasing food demands of the exploding human population might in themselves have resulted in some depletion of natural oyster stocks. But, particularly in the "advanced" nations, population growth has been accompanied by destruction of natural oyster grounds, both intentionally in the course of development of seashore property for other purposes and unintentionally through domestic and industrial pollution. Pollution constitutes a double threat to shellfish industries: even where the animals themselves are not harmed they may, due to their habit of filter feeding, concentrate pollutants in their flesh and become unfit for human consumption.

Pollution effects on shellfish fisheries and culture are most severe in the more developed countries. By 1964 nearly 400,000 ha of shellfish grounds, or about 12% of the total active producing area in the United States, were closed to harvesting for health reasons. In some states all waters are closed. The net effect of pollution, overharvest, and generally poor management has been to reduce oyster production on the Atlantic coast of the United States from more than 43 million kg in 1920 to less than half that amount in 1964. In New York state, a principal producer of oysters but also a highly populous and industrialized state, oyster production over a 50-year period declined by 99%.

In recent years pollution and other factors have threatened oyster production in many countries. The town of Malebon, once the chief oyster port of the Philippines, has been virtually eliminated from the industry by pollution. Industrial pollution, siltation, and overharvest threaten future oyster harvests in several other areas of the country. In Australia, where oyster culture was initiated in 1896 to compensate for overharvest of wild oysters, the future of the industry in the principal producing area, the George's River estuary, is threatened by domestic pollution.

Some efforts have been made to reduce pollution in oyster-producing areas, but the principal response to the problem has been to seek biological and technological refinement of culture systems, including the construction of hatcheries where all or part of the life cycle can be carried out under controlled conditions, using water of known purity. D. H. Wallace of the National Marine Fisheries Service has suggested that this is the only hope for the oyster industry in the United States.

NATURAL HISTORY OF OYSTERS

Before efforts to conserve or culture oysters or any organism can proceed, the natural history and ecology of that organism must be understood. The American oyster (*Crassostrea virginica*) has probably been studied more thoroughly than any other marine organism, and the reader is referred to P. S. Galtsoff's (1964) monograph for details on the American oyster in nature. Here we briefly describe only those aspects of the oyster's life that are of particular interest to culturists.

SPAWNING

Spawning commences whenever the water reaches a certain temperature (the precise temperature depends on the species; see Table 1) and lasts as long as the temperature remains above that minimum. In the tropics this may result in year-round spawning. In oysters of the genus *Crassostrea*, which includes most of the cultured species, the genital products of both sexes are released into the water and fertilization and hatching take place outside the shell. In oysters of the genus *Ostrea*, the male releases sperm, but the female retains the eggs in her pallial cavity until hatching. Individuals of some species are alternately male and female.

HATCHING AND SETTLING

Hatching occurs anywhere from a few hours to more than a week after fertilization. The larvae are free-swimming for a few days to a few weeks, after which they metamorphose, settle on some solid object, and assume a sessile life habit for the rest of their lives. As the newly settled oysters or "spat" develop, mortality from predation, disease, and the effects of crowding may be very high. Development to sexual maturity takes about a year, but up to 4 years are required for wild oysters to attain marketable size. Both larvae and sessile oysters feed by filtering plankton and suspended organic matter.

ENVIRONMENT

Environmental requirements and tolerances for the various species are imperfectly known, as is oyster taxonomy, but Table 1 serves as a general guide to the habits and habitats of the most important commercial species. A number of species in addition to those listed in Table 1 are locally important, notably the Olympia oyster (*Ostrea lurida*), which brings the highest prices of any oyster in the northwest United States, and *Crassostrea rivularis* of Japan, which, due to its extremely high tolerance for turbidity, is the only oyster that can be cultured over some bottoms.

BASIC TECHNIQUES OF OYSTER CULTURE

HISTORY

In nature, oysters may settle on any hard substrate. Areas where conditions are well suited to their growth soon become covered with many layers of oyster shells. The most primitive methods of oyster culture involve little more than scattering cleaned oyster shells, called "cultch," on the bottom in such areas just before setting and letting nature do the rest until harvest some years later. Seventeenth-century Japanese culturists

TABLE 1. THE PRINCIPAL CULTURED SPECIES OF OYSTERS, THEIR DISTRIBUTION AND CHARACTERISTICS

SPECIES	COUNTRIES WHERE CULTURED	SPAWNING SEASON	SPAWNING TEMPERATURE (°C)	INCUBATION PERIOD	DURATION OF LARVAL PERIOD AND TIME OF SETTING	DEPTH AND TIDAL ZONE INHABITED
Crassostrea angulata (Portuguese oyster)	Portugal, Spain, Atlantic coast of France; experimentally in Japan, Tunisia, and California	Summer	20 or more	—	15–20 days	Intertidal; in estuaries where current is strong
C. commercialis (Sydney rock oyster)	Australia, from southern Queensland to eastern Victoria and New Zealand	Summer and fall	Peak at 21–23	6 hours	14–21 days; February-April	From intertidal to 3 m below low tide
C. eradelie (slipper oyster)	Philippines	Spring and summer, peaks during rainy season (July-August)	30–33	—	7 days	Intertidal
C. gigas (Pacific oyster)	Japan, Korea, Taiwan, Pacific coast of United States and Canada; experimentally in Australia, France. Netherlands, Portugal, Thailand, and United Kingdom	Peaks in Japan: May-June in Inland Sea, August-September in North Japan	Begins at 19–20, peaks at 23–25	5–6 hours	10–14 days; peak in August	Intertidal

TABLE 1. (*continued*)

SPECIES	COUNTRIES WHERE CULTURED	SPAWNING SEASON	SPAWNING TEMPERATURE (°C)	INCUBATION PERIOD	DURATION OF LARVAL PERIOD AND TIME OF SETTING	DEPTH AND TIDAL ZONE INHABITED
C. rhizophorae (mangrove oyster)	Experimentally in Cuba and Venezuela	Continuous, peaks May-September in Venezuela	—	—	Continuous setting, peaks in July-August in Venezuela, February-April in Cuba	0.5–3.0 m (intertidal)
C. virginica (American oyster)	Atlantic and Gulf coasts of United States, maritime provinces of Canada; experimentally in Japan and California	Long Island Sound: mid-July–early October; Chesapeake Bay: mid-June–mid-October; South Carolina: May-October; Gulf of Mexico: April-November	Begins at 20	—	10–21 days	Intertidal to more than 30 m; spawning and spat settling most successful in estuaries
Ostrea edulis (flat oyster)	Atlantic coast of France, Spain, Netherlands, Great Britain, Japan, United States (Maine and Pacific coast)	June-September in Morbihan area of France	20 or more	8 days	12–14 days	Little or no intertidal exposure; in estuaries where current is weak

TABLE 1. (*continued*)

680

SPECIES	TEMPERATURE TOLERANCE (°C)	SALINITY TOLERANCE (‰)	SUBSTRATE	SIZE MARKETED AND GROWING TIME TO THAT SIZE
Crassostrea angulata (Portuguese oyster)	About 15–25	Optimum 20–30; fails to reproduce above 34	Various; quite tolerant of turbidity	65 g (including shell); 3 years in France
C. commercialis (Sydney rock oyster)	Varies widely	Varies widely	Hard bottom, usually in the shade	85–100 mm (for consumption on halfshell), 2½ years in North, 3½ years in the South; 50–65 mm long (for shucking), 2–3 years
C. eradelie (slipper oyster)	25–33, possibly wider	Wide, up to 45, spawns at 15	Usually over mud on plants; cultured over sand; very tolerant of silt	75 mm diameter, 6–9 months
C. gigas (Pacific oyster)	15–30, optimum for larval development 23–25; best spat sets at 25 or more	Optimum for larval development 23–28 (varies with temperature) best spat sets at 15–18	Any hard substrate	30–60 g (including shell), 6–12 months (Inland Sea), 18 months (North Japan) halfshell size, 2 years

TABLE 1. (*continued*)

SPECIES	TEMPERATURE TOLERANCE (°C)	SALINITY TOLERANCE (‰)	SUBSTRATE	SIZE MARKETED AND GROWING TIME TO THAT SIZE
C. rhizophorae (mangrove oyster)	18.4–34.0	22–40 (extremes of short duration); optimum 26–37	Associated with mangrove roots; very tolerant of turbidity	75–100 mm diameter, 5–6 months
C. virginica (American oyster)	Larvae develop well at 17.5–32.2	Wide, to at least 32; best larval development above 16.5, some survive at 7.5, maximum for larvae 22.5	Hard	75 mm diameter, 4–5 years, Northern Atlantic Coast; 2–3 years, mid- to South Atlantic Coast and Gulf of Mexico
Ostrea edulis (flat oyster)	Wide—at least 4–22; optimum 15–20; 100% mortality at 26 sensitive to variation	Usually found above 25	Hard, not tolerant of turbidity	65 g (including shell), 75 mm diameter, 4 years

PLATE 1. Dr. and Mrs. S. Y. Lin holding bamboo slivers used for oyster culture in estuaries of Taiwan. (Courtesy Dr. Ziad Shehadeh.)

refined this method by using rocks, branches, and other objects of a size such that they could easily be moved from place to place with oysters attached.

In 1673 Gorohachi Koroshiya, a Japanese clam culturist, made the serendipitous discovery that oyster spat would settle on upright bamboo stakes anchored in the sea bottom. Not only are oysters so attached protected from benthic predators but the resulting three-dimensional "bed" offers many times more surface area for attachment than a flat bottom. The next logical step was the development of the many types of suspended collectors, hung from floats, which are in use today.

SPAT COLLECTION

Whatever method of capture is used, the time of spat collection is crucial to the success of oyster culture. The best time of year for spat collection varies greatly with species (Table 1), locality, and annual fluctuations in temperature, salinity, tide, and so on. If collectors are set out too early, large numbers of barnacles and other undesirable organisms with habits similar to those of oysters may be captured. In such major oyster breeding areas as the Gulf of Morbihan in France, Long Island Sound in the United States, and Miyagi Prefecture in northeast Japan, government biologists assist oyster culturists in selecting the best dates for spat collection. By examining the contents of plankton tows, biologists may determine the presence and abundance of larval oysters of various ages and thus forecast peak setting periods a few days in advance. In other regions, culturists must rely on past experience and their own test collections.

Proper placement of collectors is fully as crucial as timing. In many parts of the world, only a small percentage of the oyster grounds are suitable for spat collection. Thus a few fortuitously located culturists may specialize in the production and sale to other culturists of young oysters. Or oyster growers may collect spat on grounds far removed from their growing areas.

Yet another important factor in spat collection is the type of collector used. Dozens of materials and designs have been tested, but no one type has proven universally superior. The best collector for one species in one area may be a failure with another species somewhere else. Therefore the various collectors are discussed here under the descriptions of individual culture methods.

Spat are usually left on the collectors at least until they develop into "seed" oysters, 10 to 20 mm in diameter. Seed oysters may be cultured in place, moved to areas best suited to growing for market, or sold to other culturists.

GROWING FOR MARKET

Procedures used in growing adult oysters for market vary with species and locality and according to the use for which the oysters are destined. Oysters to be eaten raw on the halfshell must be given special care to insure the production of uniformly shaped shells if they are to bring the best prices.

The same principle that favors setting of spat on three-dimensional hanging collectors favors off-bottom culture of oysters for market. Nevertheless, bottom culture, with its inefficient use of the water column and the attendant dangers of siltation and predation, is still practiced in some countries, notably the United States, Canada, and France. In the United States and Canada it persists due to legal obstacles to leasing portions of the sea bed, as is done in other countries. In North America, most coastal waters are held to be public domain, and to place extensive

PLATE 2. (A) American oysters spawning at the Milford, Connecticut, NMFS Shellfish Laboratory. (Courtesy NMFS Shellfish Laboratory, Milford, Conn., permission *Trans. Am. Fish. Soc.*) (B) Larval rearing tanks in an oyster hatchery at Carantec, France. (Photograph by J. H. Ryther.)

floating structures in them would interfere with recreation as well as other commercial use, whereas bottom culture does not raise such problems.

OYSTER CULTURE ON THE ATLANTIC COAST OF NORTH AMERICA

SPAT COLLECTION

Though market size oysters are rarely grown off bottom in North America, spat collection, which takes up considerably less space than growing for market, may be carried out using collectors suspended from rafts. This method is particularly applicable where oyster spawners exist in brackish ponds or lagoons, which may not be considered public waters. The most notable example of this practice is the Ocean Pond Corporation, located on Fisher's Island in the New York waters of Long Island Sound, a body of water where oyster spawning is undependable at best. Ocean Pond is a 9.3-ha body of brackish water connected to the sound by a narrow channel. Annually, just prior to the oyster spawning season, this channel is blocked to prevent escape of oyster larvae and to accelerate warming of the water. The abundance of larvae is determined by continuous monitoring and when adequate quantities are detected collectors composed of scallop (*Pecten*) shells strung on 2.4-m-long galvanized wires are suspended from rafts. When setting is completed the pond is opened to the sea, but the juvenile oysters are not moved until the following spring, when they are sold to oyster growers throughout Long Island and New England. As of 1969, Ocean Pond Corporation had experienced successful sets for seven consecutive years, sustaining an annual production of 25 to 50 million 9-month-old seed oysters.

The success achieved by Ocean Pond Corporation is by no means typical in the northeast United States and eastern Canada, and inadequate supplies of seed oysters remain a problem for the industry in that region. Attempts have been made to import seed from the southeastern states, where spat setting is more reliable, but southern strains of C. *virginica* do not do well in the cold waters of the northeast.

Biologists in Maine have had sporadic success in spat collection and efforts to take advantage of natural spatfalls in northern waters continue in several areas. The most successful collectors in most cases have been strings of scallop shells. Cement-coated egg cartons proved a satisfactory substitute for shells in Prince Edward Island. In a few localities, chicken wire bags containing shells have proven more efficient than strings, but

PLATE 3. (A) Three-week-old flat oysters (*Ostrea edulis*) cultivated in a recirculation system in Conway, Wales. (Courtesy P. R. Walne.) (B) Twenty-liter continuous algal cultures used for the production of food for adult, juvenile, and larval bivalves in the experimental hatchery at Conway, Wales. (Courtesy P. R. Walne.) (C) Newly set oyster spat on shell culch are held in plastic bags for initial growth in a commercial hatchery on Long Island, New York. (Courtesy NMFS Shellfish Lab., Milford, Conn., permission *Trans. Am. Fish. Soc.*)

the opposite is generally true. Various kinds of artificial collectors have also been tested but have generally proven inferior. However, 3-mm-thick sheets of such nontoxic plastics as polyethylene or polypropylene have shown promise. Collectors made of these plastics are much lighter and easier to handle than strings or bags of shells and are cheaper to construct than shell strings. In recent years culturists in California, Virginia, and Canada have had good results using plastic mesh collectors similar to the "Netron" collectors used by Japanese culturists (see below). These have a special advantage for bottom culture in that the seed oysters can easily be removed.

Since most oysters grown north of New York are destined for consumption on the halfshell, crowding occasioned by bunching on the collectors is to be avoided. For this reason some culturists prefer bags of shells to strings since if an excess of spat is collected, it is a simple matter to break shells or separate them one from another. Synthetic collectors may be manufactured with the same requirements in mind by simply alternating bands of smooth material, from which spat may easily be displaced, with bands of roughened material.

HATCHERIES

Discussion of spat collection is academic, however, if adequate numbers of larvae are not present in nature. Since shortages of natural spawn are the rule in the northeast, the emphasis among successful culturists is on hatchery production of seed oysters.

Hatchery culture of oysters had its genesis in 1879 when W. K. Brooks of the Johns Hopkins University demonstrated that American oyster eggs could be hatched in the laboratory. Success in rearing larvae to the setting stage was not achieved until 1920 when W. F. Wells of the New York Conservation Commission opened the door for practical intensive oyster culture. In the mid-1940s a modification of Wells' technique, for the first time incorporating artificially induced spawning, was developed by V. L. Loosanoff and H. C. Davis at the U.S. Bureau of Commercial Fisheries Biological Laboratory at Milford, Connecticut. One or both of these techniques, known respectively as the Wells-Glancy method and the Milford method, are used at the five commercial oyster hatcheries currently operating on Long Island Sound. Figure 1 illustrates the main features of both systems by means of a flow diagram.

One of the most successful hatcheries, located at the time at Oyster Bay, New York, produced 100 million seed oysters in 1966, using essentially the Wells-Glancy method. That hatchery has since been relocated and expanded and its methods somewhat modified, but the essential elements of the practice remain the same and are described here as carried out in 1968. Sexually mature adult oysters are selected from the growing beds (the selection based on size, shape, and growth rate) and are held in the hatchery at 10°C. They are then conditioned for spawning by slowly raising the temperature to 18°C or more and holding the oyster at that temperature for 2 to 4 weeks, the duration depending upon the time of year. Spawning is then induced by raising the temperature to 25°C while holding the oysters in glass trays. By using this technique it is possible not only to spawn oysters on any day of the year but, more important, to spawn individual oysters twice a year or perhaps more frequently.

The fertilized eggs are transferred to one of a battery of 50 120-gal conical rearing tanks. Every other day the water is drained through the bottom of the tank, the newly hatched larvae being caught in a fine mesh screen. After each change of water the tanks are scrubbed with Alconox

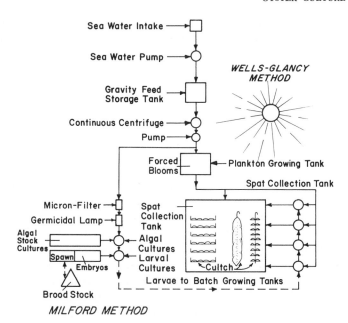

Fig. 1.   Wells-Glancy and Milford oyster hatchery methods. (After Gaucher, 1968.)

or Clorox to guard against disease organisms. The larvae are then graded by further screening, keeping only those individuals over 0.3 mm in diameter, or about 20% of the total, and discarding the remaining 80%. This process is believed to be very important in that it selects for a strain which retains the characteristic of rapid growth throughout life.

The selected larvae are then transferred to another tank containing water pumped into the hatchery from Oyster Bay, centrifuged through a battery of three Sharples centrifuges, and pumped into four 20,000-liter tanks in the upper floor of the building. A greenhouse-type roof admits sunlight and keeps the temperature above 30°C. The centrifugation of the water removes the animals and larger algal cells. If a sufficient number of small, unicellular algae suitable as food for the larvae are present and escape centrifugation, these are allowed to grow naturally for 24 hours; if such microorganisms are scarce, the tanks may be inoculated with 200 liters of algal culture and fertilized. The resulting culture of microorganisms, natural or inoculated, is then used to fill the larval rearing tanks.

After 10 to 15 days the larvae are ready to set, and when this is confirmed by microscopic examination they are transferred to one of six 3600-liter plastic settling tanks, each of which contains 10 bushels of specially selected, screened, and washed oyster shells, which may be treated with Clorox as an added precaution. It is most important to use clean shells in this operation or a bacterial bloom may develop in the tanks and kill the larvae or prevent setting. The clean shells are spread in the bottom of the tanks and the larvae introduced and left for 24 to 48 hours, during which time setting occurs.

Spat so collected are not as convenient to handle as spat set on hanging collectors. A proposed method of collecting spat in hatcheries involves releasing ready-to-set larvae into a long, narrow tank. After releasing the larvae, special collectors made of nontoxic plastic and almost the width of the tank would be passed through the entire length of the tank at a speed regulated by the rate of spatfall. With supervision by an experienced person, this method might prove highly efficient.

The shells with the spat attached are then transferred to half-bushel plastic mesh bags (Dupont Vexar) which are suspended from wooden beams into the nursing tanks. The 27,000-liter cement nursing tanks, each of which holds 200 bags, are located in a greenhouse where the temperature is kept at or near 30°C. These also are fed with water from one of the roof tanks in which an algal bloom is maintained, one-half of this tank each day being added to each holding tank. The high temperature and supplemental feeding results in rapid growth of the young seed oysters. The length of time they are kept in the holding tanks ranges from 4 days to a week or more, depending upon the availability of space and the timing of subsequent operations.

During this stage fouling by such animals as mussels, annelids, tunicates, gastropods, and hydroids may occur. At the Milford laboratory this was prevented by periodic treatment with Victoria Blue B, a biological stain. Clams, which customarily keep their shells closed during treatment, are unaffected by long-term exposure to low concentrations of this stain. Oysters, however, may open their shells and suffer mortality. Thus they are best treated by dipping them in a high concentration, say 200 ppm, for 15 min, which should kill most fouling organisms.

Eventually, the wooden beams with shell bags attached are transferred by chain-hoist and overhead rail outside to the dock where they are placed on floating rafts, each of which holds 200 bags. The seed oysters are suspended from the rafts into the bay water for 2 to 3 additional weeks until they reach fingernail size (1 to 2 cm), when they are ready for planting on the oyster beds. At full operation, 16 rafts or 1600 bushels (1 bushel = 35.2 liters) of seed-bearing shell are in use in the bay at one time.

Not all culturists introduce young oysters to bay water so soon after setting. Some growers are attempting to keep spat in large tanks for longer periods in order to enhance growth and reduce predation and fouling. Protection may also be afforded by placing the spat in natural or artificial ponds (it should be pointed out that artificial ponds may be relatively sterile, resulting in poor growth), intertidal waters, or the upper reaches of estuaries, where salinity is low, thus predation and competition are minimal. In such environments, mild sewage pollution may have the beneficial effect of producing food in the form of rich plankton blooms.

The entire operation from spawning to fingernail size seed requires 4 to 6 weeks. The hatchery can produce the small (3 to 4 mm) seed throughout the year, but the additional growth to 1 to 2 cm on the rafts cannot be done after the water temperature falls below about 10°C so the seed oysters are not planted on the beds in late fall or winter when they cannot grow and are at their most critical period for survival. Thus the hatchery operation has been restricted to late spring and summer.

As can be seen from Fig. 1, the Milford method is similar to the Wells-Glancy method in most respects but relies on pure algal cultures as a food source. (A recent modification, not yet perfected, utilizes freeze-dried algae.) Maintaining algal cultures is somewhat more expensive than centrifuging sea water but has the advantage of allowing the culturist to screen out such undesirable forms as *Chlorella*, which produces toxic metabolites, in high concentration fatal to oyster larvae, and to select for forms readily digested by the larvae. Among the latter are such chrysomonads as *Monochrysis lutheri* and *Isochrysis galbana*, which do not have thick cell walls. Another suitable species is *Scenedesmus obliquus* originally imported from Japan.

The advantages of pure algal cultures are not so clear-cut as they might seem, however, since we know little of the food habits of oysters in nature and even the best pure cultures are likely to be lacking important components of natural diets. We do know that no one species of algae in itself constitutes a satisfactory diet for oyster larvae.

Another point in favor of pure algal cultures as a food source is that the population density of larvae in culture may be maintained at a higher level. No matter what steps are taken to ensure an algal bloom, the amount of usable algae in a given volume of sea water is limited. Thus as the larvae grow they must be periodically thinned. If a pure algal culture is used as food, however, dense populations of larvae may be maintained simply by increasing the rate of feeding.

Some Wells-Glancy type hatcheries make occasional use of pure algal cultures when natural populations of algae are deficient and/or when light and temperature conditions are not adequate for the development of sufficiently dense cultures in greenhouse tanks. At one successful hatchery, algal culture is carried out in a constant temperature (about 20°C) room illuminated with fluorescent lamps. Sea water pumped into the room is filtered through wound-orlon filters and then passed through a bank of 4 2-ft ultraviolet lamps to kill any remaining microorganisms. The water is then enriched and used to fill the culture vessels, which consist of 11 227-liter and 8 136-liter translucent plastic drums. Presently in use as food organisms are the flagellates *Isochrysis galbana, Monochrysis lutheri,* and *Rhodomonas* sp.

As needed, these cultures are fed to the larvae to produce a total ration of 2000 cells/(larva)(day). The same concentration is used throughout the larval cycle but larger algae are used as the larvae grow larger.

The rationale for the construction of oyster hatcheries in North America is that in many years natural reproduction in certain areas does not provide enough larvae to sustain commercial culture, but development of hatchery techniques also opened the door for selective breeding of the American oyster (see p. 737).

The critical factor in determining the success or failure of an oyster hatchery is water quality. To insure good survival of larvae the water should have a salinity of 16.5‰ or greater, a pH of 6.75 to 8.75, and be relatively free of pollutants, particularly metallic salts, pesticides, and detergents. Areas where intensive, possibly toxic algae blooms occur frequently should be avoided. Silt may also have an adverse effect on oyster larvae, directly and by alteration of pH, but can be relatively easily removed by filtration. Even if the water quality is excellent, it may be necessary to treat incoming water with antibiotics, sulfa drugs, or ultraviolet radiation. To produce 5000 bushels of spat requires 32,000 to 40,000 liters of cleaned sea water daily.

A recent innovation which is finding increasing use by oyster culturists is the hatchery production of "cultchless" or "free" spat by inducing the oysters to set on sheets or screens of plastic or metal from which they may be removed, or on tiny particles of calcium carbonate. The free seed may then be grown in fine-mesh screen trays, in the hatchery or out-of-doors, until they are large enough to be planted on the bottom or, in some cases, until they reach marketable size. The advantages of this technique is that many more free seed oysters can be reared in the same space, they can be handled and transported far more easily and less expensively, and they grow into much more evenly and regularly shaped adult oysters than oysters which have set on oyster- or scallop-shell cultch.

GROWING FOR MARKET—BOTTOM CULTURE

Although hatchery production of seed oysters on Long Island Sound is highly sophisticated biologically and technologically, techniques used in growing seed oysters to marketable size in the United States barely qualify as culture. As with wild oysters, growth and survival of cultured oysters is largely determined by the quality of the beds, which should be located on a hard bottom in 1 to 12 m of water where tidal currents are strong. Once seed oysters are planted on the beds management is confined to predator and siltation control, and, when necessary, thinning and movement to new beds. One successful grower routinely thins oysters twice during the growing period. Seed oysters planted on the beds at about 200 bushels/ha (1 bushel = 35.2 liters) are redistributed over five times as much bottom at the end of the first year of growth, by which time they amount to 2000 bushels. At the end of the third year they have reached 4000 bushels and are distributed over twice as much bottom as at the end of the first year.

Even more primitive methods are used in maintaining public oyster grounds in the eastern United States. The various state governments initially place shells on the bottom in areas where reproduction occurs, then move them, with spat attached, to more saline waters for growing. Up to three subsequent moves may be undertaken, but little else is done by way of management except to regulate the season and method of harvest.

Accumulation of silt is particularly dangerous to young oysters, which may be smothered or prevented from feeding. Proper selection of beds plays a major role in discouraging siltation, but frequent transplantation and/or supplemental control measures are usually practiced by commercial culturists. The larger growers customarily clean oyster beds of silt using suction dredges or high-pressure water jets.

Predation is a problem wherever oysters are grown on the bottom, but it has been particularly severe in Long Island Sound, where the chief predators are oyster drills (*Urosalpinx cinerea* and *Eupleura caudata*) and the common starfish (*Asterias forbesi*). The upward trend in oyster production in that body of water since 1965, after 50 years of decline, is considered to be largely due to effective control of these animals. Long Island culturists, who formerly expected 10,000 bushels of seed to yield 10,000 bushels of oysters, now produce 100,000 to 150,000 bushels from the same amount of seed. Unfortunately, one of the basic techniques of predator control relies on chlorinated hydrocarbon biocides. These compounds, already beginning to be seen as ecological hazards on land, are even more frightening when applied to culture of aquatic filter feeders, which are extremely efficient in concentrating waterborne chemicals. But thus far, despite the history of DDT and other "safe" pesticides, the philosophy of the shellfish culturist seems to be "If it hasn't been *proven* dangerous to man, it's all right."

The principal biocide used in oyster culture in Long Island Sound is a potent mixture of chlorinated hydrocarbons marketed as "Polystream," which is illegal in most states. A week before stocking young oysters, drills (as well as starfish and other predators and competitors) are removed from the beds by suction dredging. Then Polystream is applied at 2000 kg/ha. Such treatment effectively curtails predator populations for 1 to 2 years and has been largely responsible for reducing annual losses to drills from 50% or more to less than 1%.

There are other, more laborious methods of effectively controlling oyster drills. One of the simplest involves use of an underwater plow, adapted from an agricultural disk plow. If the bottom is suitably soft, this device may be used to bury and kill drills, starfish, and perhaps limpets and mussels. In experiments conducted by the U.S. Bureau of Commercial Fisheries, drills buried under 6 cm of sand or mud suffered 92% mortality in 5 to 55 days. The higher the temperature, the more quickly they were killed.

Various chemicals less hazardous than the chlorinated hydrocarbons may be used as dips to kill drills and other predators and competitors, but their use entails the expense and labor of dredging up the oysters. Of these substances, rock salt is easily the cheapest as well as one of the most effective. By dipping the oysters in a saturated solution, then exposing them to the air for a while, most animals other than shellfish are killed, but oysters are unharmed. The time of immersion in the salt solution and exposure to air varies according to the type of organism to be controlled (Table 2).

The salt dip technique, with suitable modifications, is also likely to prove effective for controlling the predatory flatworm *Stylochus ellipticus* as well as many fouling organisms, including protozoa, hydroids, bryozoa, crustaceans, and algae. Small mussels are usually unaffected by salt treatment but may be controlled by similar treatment with a 0.5 to

TABLE 2.   IMMERSION TIME IN SATURATED NaCl SOLUTION REQUIRED TO KILL 5 TYPES OF OYSTER PREDATORS OR COMPETITORS

PREDATOR OR COMPETITOR	IMMERSION TIME (SATURATED NaCl) (MIN)	AIR DRYING TIME
Oyster drills	5	Several hours
Starfish	1	Several minutes
Boring sponges (*Cliona* spp.)	3	1 hour
Limpets (*Crepidula* spp.)	3–5	30 min–1 hour
Tunicates	1–3	1 hour

After Loosanoff (1958)

1.0% solution of copper sulfate. A quick dip, followed by air drying for 24 hours, is effective against not only mussels but a number of other undesirable organisms, notably drills. If the strength of the solution is increased to 2.0%, the drying time in air can be cut to a few hours.

Drills are among the deadliest enemies of untreated oysters, but starfish, which have been estimated to destroy more than 500,000 bushels of oysters annually in Long Island Sound alone, require more vigilance on the part of the culturist. Even when eliminated from an oyster bed, they may rapidly reinvade. To guard against starfish and other hazards, including siltation, scuba divers may be employed to continually monitor oyster beds. This precaution has played an important role in increasing the survival rate from seed to market size of one firm's oysters from 1 to 70%.

Constant treatment with suction dredges also plays a role in the fight against starfish. An older and more widely used device for the same purpose is the starfish mop, which consists of an iron beam 2 to 4 m long with a number of light chains attached. At the end of each chain is a bundle of rope yarn about 2 m long. As these mops are dragged slowly over the oyster beds the starfish become entangled in the yarn and are taken aboard and killed by immersion in boiling water.

Another popular control method involves quicklime which, when it comes in contact with a starfish, causes rapid disintegration and eventual death through its caustic action on the body membranes. Lime is usually applied at about 300 kg/ha, but up to 500 kg/ha may be used on beds located in deep water. To achieve complete coverage and to insure that the lime will adhere to the starfish very fine particles are used. This necessitates pumping it to the bottom so that slaking will not occur before the lime reaches the oyster bed. Though very effective, this practice is relatively expensive and may prove harmful to other commercially valuable animals such as lobsters.

Efforts have been made, without success, to institute biological control of starfish. The possibility of harvesting them for use in the manufacture of protein concentrates has also been explored. Starfish protein concentrate compares favorably with good quality fish meal, but the rate of natural production of starfish is not sufficient to support a profitable fishery. Until a cheap and efficient biological or mechanical method of starfish and drill control is developed, chemicals, however hazardous, are likely to continue to play a major role in bottom culture of oysters in North America.

The techniques used in oyster culture in the Gulf of Mexico and elsewhere along the Atlantic Coast are essentially the same as those used in Long Island Sound. From New Jersey south, however, oysters are customarily harvested after only 3 years growth. Formerly they were

PLATE 4. Starfish treated with quicklime for predator control disintegrate in about 24 hours. (Courtesy James White, Ed., *American Fish Farmer*.)

grown for the same length of time as northern oysters, but in the last 10 to 15 years a number of diseases have appeared which, until cures or resistant strains of oysters are found, necessitate early harvest to avert heavy losses.

In some places in the south, oysters have been successfully grown together with shrimp and various species of fish in fertilized brackish water ponds. It is not known whether the oysters derived any benefit from fertilization, but experiments in organic fertilization of ponds used primarily for oyster culture indicate that the American oyster derives little food from the type of algal bloom created by adding common fertilizers to sea water.

Crude as American oyster culture methods are, the yields achieved by

PLATE 5.   Oyster shells for culch on Long Island, New York. Unless employed for this or other purposes, disposal of shells becomes a problem. (Photograph by J. H. Ryther.)

culturists are many times higher than those recorded for public oyster grounds, which receive negligible management from government agencies. Prior to 1966, the better growers in Long Island Sound obtained yields of about 1000 kg/ha/year, whereas the average yield of public grounds was 10–100 kg/ha/year. Adoption since 1966 of improved methods of controlling siltation and predation has enabled Long Island Sound culturists to achieve a significant increase in production, so projected yields in that area now are of the order of 5000 kg/ha/year.

GROWING FOR MARKET—EXPERIMENTAL OFF-BOTTOM CULTURE

High production potential is the principal factor stimulating American oystermen to pursue the possibilities of off-bottom culture, despite the legal difficulties. The earliest efforts at off-bottom culture of American oysters were carried out in the middle and late 1930s at the Prince Edward Island Biological Station, Ellerslie, P.E.I., Canada. Seed oysters stored on the bottom over winter were carried through their first summer in floating trays then grown to market size on the bottom. The method was judged biologically effective but economically unfeasible. A similar technique, using strings of oysters suspended from rafts instead of trays, was tested by the U.S. Bureau of Commercial Fisheries in 1956 in Oyster Pond River, Chatham, Massachusetts. They succeeded in producing 70% marketable oysters in 2½ years as compared to the 4 to 5 years required by bottom culture. Further experiments by local shellfish officers in the

PLATE 6. Experimental oyster rafts at NMFS Shellfish Laboratory, Oxford, Maryland.
(Courtesy NMFS, Oxford, Md.)

same area had the extraordinary result of producing more uniformly
shaped oysters on suspended strings than were grown on the bottom.

The present center for experimental off-bottom oyster culture in North
America is the U.S. National Marine Fisheries Service at Oxford, Maryland. There various types of rafts, long lies, trays, and other suspension
methods are being tested in the hope of finding biologically, economically,
and legally feasible techniques for off-bottom culture of the American
oyster in North American waters.

OYSTER CULTURE IN JAPAN

ENVIRONMENTAL REQUIREMENTS

In contrast to American practices, perhaps the most sophisticated and
productive oyster culture in the world is that practiced in Japan. At
least eight species are involved, but by far the most important is the
Pacific oyster (*Crassostrea gigas*), which is also an important species for
culture in the United States, Canada, Korea, and Taiwan. Waters to be
used for culture of this species must meet the following requirements:

1. The farming area must be naturally sheltered against violent winds and waves or easily protected by inexpensive construction. (This is less important if long lines rather than rafts are used.)

2. Tides and/or currents must be sufficient to change the water completely and frequently.

3. Salinity and temperature must be suitable; optimal conditions are 23 to 28‰ and 15 to 30°C.

4. The water must contain adequate amounts of phytoplankton as feed for the oysters.

5. The culture area must be free of industrial and domestic pollutants.

GROWING FOR MARKET—RAFT CULTURE

Some Japanese culturists collect their own seed, but most purchase seed from seed oyster specialists located in the northern part of the country. Various types of bottom culture were formerly employed, but now all oysters grown in Japan are cultured off-bottom, usually suspended from rafts.

In raft culture, 1-month-old seed oysters, about 12 mm in diameter, attached to oyster or scallop shells strung 20 cm apart on No. 13 galvanized wires, with bamboo or plastic spacers, are suspended from the rafts for growth. The length of the strings of oysters, or "rens," depends on water depth in the culture area; in the Inland Sea, this varies from 10 to 15 m.

Rafts used in the Inland Sea are standard in size and construction, being made from 75- to 100-cm diameter bamboo (occasionally cedar) poles which are lashed together with wire in two layers at right angles to each other and with the poles 0.3 to 0.7 m apart. The standard raft is about 16 × 25 m in size and carries 500 to 600 wire rens.

The rafts are buoyed by hollow concrete drums, tarred wooden barrels, or specially constructed styrofoam cylinders. Styrofoam cylinders are replacing the other types of float and are now used in all newly constructed rafts. As the growth of oysters proceeds and the weight increases, additional floats are added as required. In some cases the styrofoam floats are encased in a large polyethylene bag to protect them from barnacles and other fouling organisms. The rafts float at the sea surface with no attempt made to keep them out of contact with the water. As a general observation, the Japanese rafts are relatively crude and do not compare with those used in Spain for mussel culture (see Chapter 39) with respect to size, durability, or care of construction.

Rafts are commonly laid out in lines 1.6 to 3.0 m apart, tied together with ropes, with two anchors at each end of the line. Ten or more rafts

PLATE 7.   Close-up of bamboo raft structure. (Photograph by J. H. Ryther.)

may be tied together in this manner. In other cases, the rafts are individually anchored fore-and-aft and placed randomly within a culture area
3 to 10 m apart. Rafts are left in the water permanently but may be
moved inshore during the spring and early summer for maintenance.
Average life of a raft is 5 years or more.

The disposition of rafts within a culture area is determined ecologically
with respect to currents, food, salinity, and so on, and legally by the appropriate prefectural government which allocates space for this purpose
to the local fishermen's cooperative association which, in turn, grants
the use of specific areas to the individual growers. The growers do not
pay for the privilege of using these grounds except indirectly through the
cost of membership in the cooperative association.

The foregoing description applies to the techniques used in the Inland
Sea. In the Sendai region of northeast Honshu, the scallop shell collectors
with the seed oysters attached are not strung on wires but are inserted between the strands of tarred, 13-mm diameter rice-fiber rope. This is not as
strong or durable as the wire ren, and consequently less oysters are grown
per ren. In addition, the rafts commonly used are smaller than those used
in the Inland Sea, measuring roughly 10 × 3 m.

PLATE 8. View of Kesennuma Bay, near Sendai, Japan. There are some 5000 rafts in this bay. (Photograph by J. H. Ryther.)

In the Inland Sea, the seed oysters are suspended from the rafts in July and August and by the end of December the individual oysters are 5 to 10 g wet weight shucked or 30 to 60 g including shells and are ready for market. Harvesting and marketing is begun in December and lasts throughout the spring. Thus the growing season of the oysters from spat collection to harvest is 6 to 8 months. Oysters harvested in the late spring are used primarily for canning because of the difficulty of shipping fresh oysters in hot weather. From 3 to 10% of the oysters are left out for a second year to produce a crop of exceptionally large (10 to 20 cm) specimens for the restricted halfshell market or other special purposes. Such oysters may be grown in special hanging cages (Fig. 2).

Formerly most Japanese oysters were grown for 2 years, but between 1945 and 1950 a 20 to 100% mortality of 2-year-old oysters occurred in western Japan. One-year-olds were not affected, and it became customary to harvest at that age. Today it is once again possible to culture 2-year-old oysters, but 70% of the total yield in Hiroshima Bay, the principal producing area in western Japan, is still composed of 1-year-olds. Better

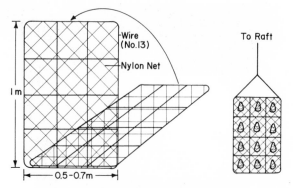

FIG. 2. Basket used in growing large single oysters in Japan. (After Fujiya.)

PLATE 9. A long line buoyed with tarred wooden barrels used for suspending oyster strings in exposed areas. (Photograph by J. H. Ryther.)

702

growth in culture of 2-year-old oysters is achieved if "hardened" seed (see pp. 711–713) are used.

In the oyster growing areas of northern Japan, the growth of oysters is slower in winter due to colder water temperatures. As a result, the oysters do not reach market size during their first year, as in the Inland Sea, but require 18 months from spat collection to harvest, which commences in November or December of their second year.

GROWING FOR MARKET—LONG-LINE CULTURE

The long-line method of oyster growing, first introduced to Japan about 1947, is becoming increasingly popular, particularly in the north. In addition to the lower initial expense and maintenance costs, long lines possess the advantage of withstanding winds, waves, and currents better than rafts. The long-line method has made it possible to grow oysters in exposed situations in the open sea where raft culture is not possible. The gradual increase in Japanese oyster production over the past decade (see Table 3) is due largely to the exploitation of such areas.

TABLE 3. PRODUCTION OF MARKETABLE OYSTERS IN JAPAN, 1957–1967

YEAR	OYSTER MEAT (METRIC TONS)
1957	19,000
1958	20,000
1959	24,000
1960	25,000
1961	22,000
1962	30,000
1963	36,000
1964	34,000
1965	35,000
1966	35,000
1967	45,000

Long lines are 45 to 75 m long and consist of a pair of 2.5- to 6-cm tarred rice ropes strung between two parallel pairs of metal, wooden, or styrofoam floats. Each end of the line, and sometimes the middle, is anchored. Floats are spaced 3 to 7.5 m apart, which distance is halved when the ropes are weighted down with oyster rens. Three rens, similar to those used in northern raft culture, are suspended from the ropes between each pair of floats. The length of the rens is usually 7.5 to 10 m, depend-

ing on the depth of the water. Rens should not be allowed to touch the bottom at any time.

PRODUCTION AND MARKETING

The production of both seed and market oysters in Japan varies from year to year by 5 to 10% but, on the whole, is rather constant for a given area. There has never been a failure in seed production nor an epidemic or mass mortality of oysters of all ages over a large area caused by disease, parasitism, predation, climatic extremes, storms, or pollution. All of these factors may reduce or eliminate production in restricted areas from time to time, but the industry as a whole is remarkably free from such risks.

Principal nuisances are fouling organisms, including mussels, barnacles, tunicates, algae, and sponges (listed in the order of their importance). Fouling of the northern Japanese oysters during their second winter is severe and may cause some reduction in the yield of oysters through crowding, smothering, or reducing the movement of water, and hence food, to the oyster. The weight of the fouling organisms also creates flotation problems in both raft and long-line techniques and necessitates

PLATE 10. (A) Oysters grown in raft culture are badly misshapen and heavily encrusted with fouling organisms. This is acceptable in Japan, where all oysters are shucked for market. (Photograph by J. H. Ryther.) (B) For the luxury halfshell trade oysters must be of even size and shape and free of fouling organisms, as in the case of these prime specimens of European flat oysters being sorted for market in Carantec, France. (Courtesy J. A. Madec.)

additional maintenance. However, because 90% or more of the oysters are shucked by the growers immediately following the harvest, the appearance of the oyster with respect to size, shape, and the presence of fouling organisms is not important. Most Japanese oysters would not, however, meet the standards of the halfshell market of Europe or the northeastern United States.

All of the shucked oysters which are marketed fresh or frozen for local consumption or export must be subjected to a purification process consisting of chlorination or ultraviolet radiation exposure of the shucked meat.

The Inland Sea of Japan is particularly well suited to oyster culture for a number of reasons. A heavy set may be consistently obtained in the estuarine waters of the Hiroshima region. Land drainage provides a moderate.to high level of organic production. Relatively warm water insures an early set and rapid growth throughout the year. Finally, the Inland Sea is extremely well sheltered and protected from the weather, permitting raft culture in almost any place including away from shore in relatively deep water (10 m or more) where the entire water column can be utilized. For these reasons, between 50 and 60% of the total oyster production of Japan (35,000 to 45,000 tons, shucked meat) comes from the Inland Sea.

In a typical raft culture operation in the Inland Sea, each ren produces an average of about 6 kg of oyster meat in the 6- to 8-month growing season, for a total of more than 4 metric tons of meat or 24 tons of whole oysters per raft of 600 rens. If it may be assumed that one-fourth of an area of intensive cultivation is covered by rafts, this is equivalent to five rafts producing 20 tons of meat or 140 tons of whole oysters, shells included, per hectare per year.

A medium-sized oyster business in the Inland Sea may consist of 100 to 150 rafts. Running this operation requires a staff of 10 to 12 regular employees and about 20 additional shuckers during the 5- to 6-month harvest season. The capital investment includes the rafts, two 6- to 9-m workboats, and a modest building which serves as a workshop, warehouse, and shucking house.

Many of the oystermen own no more than two or three rafts and the average is 10 to 15 rafts per owner, there being 8000 rafts operated by 700 oystermen in Hiroshima Prefecture.

Some culturists also maintain several rens near the workshop in shallow water unsuited for growth of oysters. These oysters are shucked on stormy days or whenever unforeseen demands for oysters arise.

In northern Japan, as mentioned earlier, the rafts are smaller, the rens are shorter, there are fewer oysters produced per meter of ren, and growth is much slower than in the Inland Sea. The yield of oysters per raft and per unit of effort and time is correspondingly less. In Kesennuma Bay, where there are 5000 rafts, the production per raft is only 0.15 metric tons for an 18-month season as compared to 4 tons per raft in 6 to 8 months in the Inland Sea.

The long lines are so variable with respect to length and depth (length of the rens) that no generalization can be made concerning their yield. One long line which was observed in Momono-Ura Bay on the Ojika

Peninsula (northwest of Sendai) was 60 m long, consisted of 11 tarred wooden floats and 300 7-m rens, and produced 1.2 tons of shucked oysters in the 18-month growing season. Roughly 44 long lines of that size could be located in 1 ha of water, representing a potential yield of 53 tons of oyster meat. Considerably larger yields could be expected from long lines operated further offshore, where rens may be as long as 10 to 15 m.

About 90% of the oysters produced in Japan are consumed locally. Less than 1% are sold in the shell; virtually all of the Japanese market being for fresh, shucked oysters which are rapidly shipped all over Japan and consumed raw or in a variety of cooked dishes. Roughly 10% of the oyster production, particularly those harvested late in the season, are canned or frozen for export, mainly to the United States.

As has been mentioned, the yield of oysters from any given area in Japan has remained almost constant for many years. Table 3 shows the annual oyster production for the whole of Japan for the past decade.

Production is currently levelling off due to the limited availability of new grounds for expansion of the industry. The market for oysters in Japan and elsewhere is expandable, but the prospects for a significant increase in production using present methods appear to be limited.

SPAT COLLECTION IN WESTERN JAPAN

As mentioned, most Japanese oyster growers do not collect their own seed oysters. Exceptions occur in areas where the salinity is 15 to 18‰. The Hiroshima region is particularly favorable in this respect since Hiroshima is a delta city receiving the effluents of seven rivers. The highly productive oyster industry of the Inland Sea is based largely on seed collected in this region, and the industry is centered around Hiroshima Prefecture. The breeding population consists of natural beds of oysters on the bottom of the estuaries and the bordering areas of the Inland Sea and is probably supplemented by the cultured oysters.

The collection of seed oysters in the Inland Sea is accomplished by suspending rens 1 to 2 m long bearing scallop shells from a bamboo framework driven into the bottom. The wires are hung across the framework by their middle, thereby forming a double collector string about 1 m or less in length. The wire is No. 16 galvanized made especially for oyster culture. About 40 scallop shells are pierced and strung on the wire about 2.5 cm apart with bamboo or plastic spacers between the shells. Although the plastic spacers cost three times as much as bamboo, they are becoming increasingly popular because they can be used repeatedly, whereas the bamboo must be replaced annually.

A new form of spat collector is being used on a large-scale experimental

PLATE 11. Seed oyster production near Hiroshima, Japan. Top: general view of area. Bottom: detail of above. Bottom right: close-up of young oysters on scallop shell several weeks after setting. (Courtesy *Science*, Vol. 161, September 13, 1968.)

basis, both by biologists at the prefectural laboratories and by some of the commercial growers. This consists of hexagonal plastic netting (Netron produced by Dupont) which has a rigid rubbery consistency and a mesh size of about 5 cm. This is cut into strips 15 cm wide and 10 m long which are suspended horizontally from the collecting racks. The set of oysters on this material appears to be satisfactory, but final evaluation of its effectiveness has not been made.

The Netron collecting strips cost about five times as much as standard shell collectors, which may make their use prohibitive to the smaller operators. However, they are extremely light and easy to handle in comparison with the shell rens and the same strip may be used both for collecting seed and subsequent growth of the oysters to market size, thereby eliminating the costly and time-consuming step of restringing the collector shells on new wire or rope rens. With the rising cost of labor in Japan, the latter consideration is of primary importance. Widespread adoption of this, or a similar labor-saving technique, is expected in the near future and should revolutionize the industry and its economics.

The shell collecting rens are suspended from the racks so that they will be 1 to 2 m above spring low-tide level and exposed for only a few days each month. The depth at which the collectors are placed is very important with respect to both getting the maximum set of oysters and avoiding a set of barnacles, the larvae of the two organisms occurring at slightly different depths. This practice varies somewhat from place to place depending upon the characteristics of the particular environment.

A good set is considered to consist of about 200 spat per shell (as many as 6000 is not uncommon) of which 50 to 60 survive to the seed oyster size of 1 to 1.5 cm. The oysters reach this size in about 1 month, following which they are removed from the collecting frames, unstrung from the wires, cleaned, and restrung on heavier (No. 13) wires with bamboo or plastic spacers 20 cm apart. The oysters are now ready to be suspended from the culture rafts for growth to market size.

THE SEED OYSTER INDUSTRY OF NORTHEAST JAPAN

The seed oysters used in other parts of Japan are produced by specialists, who also supply seed for export, the principal market being the United States which annually receives approximately 1 billion seed oysters or half the total Japanese production. It was until recently believed that the Pacific coast of North America was unsuited to reproduction of the Pacific oyster, but in the last few years substantial numbers of spat have been collected in the United States and Canada. Japanese seed oyster culturists, faced with the loss of half their market, have thus begun to promote *C. gigas* for culture in other countries, and experimental culture, using Japanese seed, is under way in Australia, France, the Netherlands, Portugal, Thailand, and the United Kingdom.

The Japanese seed oyster industry is concentrated in northeast Honshu, where the coastal waters range from brackish (<15‰) to oceanic in character (>30‰) depending upon rainfall, state of the tide, depth of water, and other factors. Because the coastal area is broken up by countless islands, embayments, and peninsulas, the tide, which has a 1- to 2-m average range, flows swiftly through the various cuts and channels and over the shallow, hard-bottom passages. It is in these numerous situations, and in the eddies formed by the tidal currents in particular, that oyster larvae are concentrated and the spat are collected.

In this relatively northern part of Japan the water temperature rises slowly, reaching the spawning temperature of 20°C only in late spring or early summer. Spat fall occurs from May through August with two or more peaks. The late season peak in August is normally favored by the growers, there being fewer fouling organisms in the water at that time and the survival of young oysters being higher than that of spat collected earlier in the summer. As elsewhere, the prefectural laboratory biologists forecast the setting and the fishermen put their collectors in the water at the prescribed time.

The seed collecting technique is essentially the same as that described for the Inland Sea operations but with some modifications. The collecting rens are similar to those used in Hiroshima Bay, except that no spacers are used between the shells. For Japanese use, scallop shells are employed. Because the seed oysters exported to the United States and Canada are used for bottom culture, scallop shell collectors are not used. The seed on the underside of a scallop shell lying on the bottom would suffocate or not have enough food for growth. Thus the seed grown for export are collected on oyster shells, the deep and irregular shape of which makes them suitable for bottom culture. A 1.8-m collecting ren consists of 70 to 80 oyster shells or 100 scallop shells.

The rens are suspended from two types of collecting frame. The more simple consists of crossed bamboo poles driven into the bottom at 1.5- to 3.0-m intervals with horizontal bamboo poles wired or tied into the crotch. These are usually placed at the edges of tidal currents. The more elaborate system consists of rectangular bamboo platforms 3 to 6 m long by about 1 m wide, nailed or wired to uprights driven into the bottom. The rens are doubled and dropped over the horizontal bamboo poles as closely together as they will fit. For spat collection they are placed just below the spring low tide level and are not exposed to the air at any time.

In a variation of this procedure, the collecting rens are suspended from

PLATE 12. The late Dr. T. Imai, pioneer in Japanese mollusk culture, holding string of oyster shells for spat collection. (Photograph by J. H. Ryther.)

long lines of the type used for growing market oysters. The same size and type of ren is used in this procedure, suspended the same distance from the surface as in the rack method.

One month or less after setting, when the seed oysters are 5 to 10 mm in diameter, they are moved to hardening racks. Those rens which were suspended from platforms are simply laid horizontally along the tops

PLATE 13.   Spat collectors (scallop shells) in a seed oyster farm in Miyagi Prefecture, Japan. (Photograph by J. H. Ryther.)

of the platforms. Those suspended from the X-type racks are moved to special hardening racks where they are suspended horizontally.

In the hardening process, the spat are so positioned that they are exposed to the air for 4 to 5 hours during each tidal cycle. The hardening process is indispensable for the export business, for which it was designed, since the seed oysters do not otherwise survive trans-Pacific shipment by sea.

Since 1966 it has been realized that hardened oysters suffer much lower mortality in the growth interval from spat to seed than those not so treated. During this critical period the greatest mortality occurs, normally about 80%. If, however, the young oysters are subjected to the hardening process, mortality at this stage is decreased to 20 to 60%. The reasons for this improved survival are not fully understood. Part of the explanation may lie simply in the increased attention given the young oysters at this important stage of their lives. In the transfer from the collecting to the hardening racks and the subsequent exposure to surface currents, waves, and so on, the oysters are thoroughly washed and cleaned of any silt or sediment, which is a major cause of mortality at that age and size. Perhaps the intertidal exposure itself has some unknown beneficial effect. Whatever the reason, the results are clear and the hardening process is now employed in virtually all of the seed production of Miyagi Prefecture, whether for export or domestic use. In some cases hardening is carried through two steps, moving the oysters to horizontal racks just below the low-tide level when they reach 5 mm in size and up into the intertidal range when they reach 10 mm.

Oysters moved from the collecting rens in September are left on the hardening racks until February, when they are 10 to 15 mm in size. They are then ready for export or for sale to domestic growers.

As mentioned, the greatest mortality in the postlarval life of the oyster occurs within the first few weeks from spat to seed oyster size. Normally, the spat fall is heavy enough that a considerable reduction in the set is not only expected but necessary, since there is not enough room on the collector shell for several hundreds or even thousands of larvae to grow to seed size. However, at times there are still excessive local mortalities in young oysters. A near total mortality occurred in 1962 in Matsuma Bay and reduced the seed production for the whole of Japan by more than 30% for that year.

The causes of such mortalities are usually not known. However, one of the recognized enemies of the young oyster is a flatworm which enters the shell in the larval form and destroys the oyster. This may cause mortalities as high as 50% in local populations, and there is as yet neither protection nor cure for the parasite.

The other major known risk to the industry is barnacles, which are considered a nuisance in culturing adult oysters but which may, in the seed business, completely dominate the set and physically almost eliminate the oysters. Again there is no cure for this danger other than to watch the plankton carefully and avoid putting collectors in the water when barnacle larvae are especially numerous, a situation which may unfortunately coincide with the peak setting time of oysters. The annual production of seed oysters for Japan over a 6-year period is given in Table 4.

Two million rens is roughly equivalent to 2 billion (2×10^9) seed oysters. The level of production is determined primarily by demand of both the foreign and domestic markets. However, there is a problem of space in any future expansion of the industry. This problem is becoming more severe as coastal waters receive more and more pollution, to which seed oysters are particularly sensitive. The seed industry was originally

TABLE 4.   PRODUCTION OF SEED OYSTERS IN JAPAN, 1960–1965

YEAR	SEED PRODUCTION (RENS)
1960	2,320,000
1961	2,238,000
1962	1,694,000
1963	2,857,000
1964	2,007,000
1965	1,972,000

centered in Matsuma Bay. A large fraction of the production still comes from that area, but this has been steadily decreasing within the present decade. The decline has been attributed to the heavy use of the area by the tourist industry and the resulting pollution of the inshore waters. The center of the industry recently has moved further up the coast to the less heavily populated Ojika Peninsula.

Seed oysters to be shipped overseas are packed under rattan mats in wooden cases. During transit they must be washed with sea water at least three times a day.

The process of oyster culture in Japan, from spat collection to harvest of 1- or 2-year-old oysters, is diagrammed in Fig. 3.

OYSTER CULTURE ON THE PACIFIC COAST OF NORTH AMERICA

GROWING FOR MARKET—BOTTOM CULTURE

Growers of C. gigas in the western United States and Canada generally rely on bottom culture for the same reasons as their counterparts on the Atlantic coast. The spat-bearing shells received from Japan are broken into small pieces and scattered on the beds. Growth to marketable size takes about three years, during which time they may be moved once or twice. Some culturists stock seed oysters in the intertidal zone for their first winter, then move them to just below low tide in the spring.

EXPERIMENTAL OFF-BOTTOM CULTURE

Experimental off-bottom culture of Pacific oysters, using rafts and trays, has been carried out at several places along the Pacific coast of North America, and the potential for such culture is considered to be great.

JUNE   ESTIMATION OF OPTIMAL TIME FOR SEED COLLECTION
JULY   PREPARATION OF SEED COLLECTOR AND SHELF
AUGUST   SUSPENDING SEED COLLECTION STRINGS

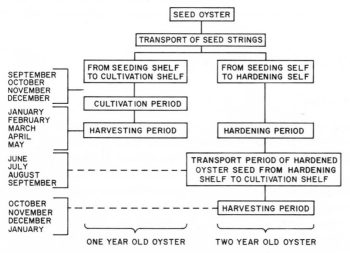

FIG. 3.   Schedule of procedures in oyster culture in Japan. (After Fujiya.)

As early as 1953, the Fisheries Research Board of Canada, in experiments conducted at Ladysmith, British Columbia, demonstrated that Pacific oysters suspended from rafts could be grown to market size in two years, and that oysters so produced were of higher quality than those grown for three years on the bottom.

SPAT COLLECTION

Spat collection is largely confined to Pendrell Sound, British Columbia, and Dabob Bay, Washington. At each of these locations about 200,000 2- to 5-m rens, each bearing 100 or more oyster shells, are annually suspended from racks or rafts. Commercially acceptable sets are obtained in most years.

HATCHERIES

Although hatchery techniques for oyster culture were first developed in the United States, their application on the American west coast is a recent phenomenon. Today, however, there are a number of hatcheries on the Pacific, including a commercial hatchery operated by Pacific Mariculture Co. at Pigeon Point, California, and a new experimental facility at

Oregon State University's Marine Science Center, Newport, Oregon. Pacific Mariculture, in addition to producing several subspecies of Pacific oyster, also supply seed of American, Olympia, flat, and Portuguese oysters, and were the first company to market "cultchless" seed oysters.

OYSTER CULTURE IN FRANCE

SPECIES USED

In France, oysters are grown on the bottom not because of social and legal sanctions but because bottom culture is the best way to produce the regularly shaped shells preferred by consumers of raw halfshell oysters. The gourmet's choice in halfshell oysters is the flat oyster (*Ostrea edulis*). Few waters are well suited to the growth of this species, thus most of the demand for halfshell oysters (which amounts to virtually the entire demand for oysters in France) is supplied by the hardier Portuguese oyster (*Crassostrea angulata*), which is less valued because of its incorrigibly irregular shell.

As the name implies, the Portuguese oyster is native to Portugal and Spain, but France, to which country it was accidentally introduced in 1868, is now the leading producer. Flat oysters are produced not only in France, but also in the Netherlands, Spain, and the United Kingdom, as well as on both coasts of the United States.

SPAT COLLECTION

In France, the two species are produced in different regions of the country. As is the case with a number of other oyster species, grounds suitable for spawning of flat oysters are limited. Virtually all of the seed production in Europe takes place in the Gulf of Morbihan and its tributaries on the southern coast of Brittany. Some growing of flat oysters for market is carried out in this region and along the southern Atlantic coast of France, but the main producing area is the north coast of Brittany. Both seed and market culture of the Portuguese oyster are centered on the southern Atlantic coast. Apparently it does not thrive in the cold, highly saline waters of northern France, but, as a precaution to protect the flat oyster industry, the Portuguese oyster is prohibited by law from being introduced to the Gulf of Morbihan.

The principal sources of flat oyster seed are 200 ha of natural beds in the central channels of the Trinity, Auray, St. Philibert, and Carnac rivers. These beds are protected by law and the introduction of other species of shellfish prohibited.

Spat collection in the Gulf of Morbihan occurs from June to September at temperatures of 20°C or more. Biologists employed by the Field Station of the Institut Scientifique et Technique des Pêches Maritimes at Carnac (on the Gulf of Morbihan) carefully monitor the plankton during the spring and summer and provide the growers with a forecast of the peak or peaks of setting. Equally important, they advise of the presence of large numbers of the larvae of barnacles or other fouling organisms. At the most propitious time, the growers put their collectors in the water, along the two banks of the four tributary rivers of the Gulf of Morbihan. Collectors are so placed that the young oysters are exposed for a few hours only during low tides.

The spat collectors consist of semicylindrical ceramic tiles about 0.3 m in length and 10.0 to 12.5 cm in diameter (originally roof tiles, now made specifically for the purpose). These are stacked in pairs, each pair perpendicular to those above and below. A stack consists of 5 or 6 pairs of tiles measuring about 1 m high and weighing 12 to 16 kg. The tiles in each stack are wired together so that they may be handled as a unit. Each tile is coated with a thin layer of lime which can easily be scraped off together with the attached oysters. The stacks of tiles are placed on wooden platforms in the river raising them 15 to 30 cm off the bottom.

Currently, experimentation is being carried out on a large scale with a new type of collector consisting of a rubberlike plastic mesh material of the same size and shape as the ceramic tile and coated with cement. These collectors are light and easy to handle and are flexible, which makes removal of the seed oysters easier. However, their lightness causes them to shift about in the rivers and results in some loss or damage. Further, they cost five times as much as tiles, can be used only once or twice compared to 10 years or more for tiles, and they do not catch as large a set of oysters. Thus it appears likely that the conventional ceramic tiles which have been in use for over a century will continue to be the popular form of spat collector for some time to come.

The collectors are left in position throughout the summer and fall, during which they may receive several successive spat falls. A successful set consists of 30 to 50 surviving seed oysters per tile, though 100 or more per tile is not uncommon. The seed oyster industry is normally a family operation involving three or four workers who put out 10,000 to 20,000 tiles, though some of the largest operators use as many as 100,000 tiles. All told, some 30 million collectors are set out each year, producing an average total of about 1 billion flat oyster seed. During the winter months,

PLATE 14. Intertidal parcs for culturing the European flat oyster in Brittany, France. (A) View of the parcs at low tide; (B) spreading young oysters. (C) Mature oysters are hand spread, turned, and cultivated. (Courtesy J. A. Madec.)

the tiles are brought in from the collecting grounds and the thumbnail sized seed oysters stripped off. Some of these are now transplanted to beds or "parcs" in southern Brittany, to be farmed by the same operators who were engaged in their production. The majority, however, are sold to growers on the north coast of Brittany.

There is little or no danger from storms to the oyster industry in Brittany since the bays and estuaries are well protected and the rivers carry little sediment. However, shifting sands caused mainly by tidal action may bury the spat or young seed oysters and must periodically be removed by hand labor when the beds are exposed. The most serious threat to the seed oyster industry is the occurrence of excessively low winter temperatures which may eliminate most of the set in some seasons, as in 1962–1963. Predation does not appear to be a serious problem on the collecting grounds.

Portuguese oysters are often collected on the same type of tile used for flat oysters, but these may be placed closer to shore, since the Portuguese oyster can withstand longer exposure to sun and the temperature extremes of the air. Increasingly, the use of tiles is being replaced by wire mesh bags of oyster shells, which are easy to obtain and less expensive. As with the tile collectors, the shell bags are placed in wooden racks or platforms 25 to 30 cm off the bottom. Commonly, the seed oysters are left in the collecting areas for 2 years instead of the 6 months for flat oysters.

GROWING FOR MARKET—BOTTOM CULTURE IN PARCS

The typical flat oyster grower in northern Brittany plants his oysters in specially prepared parcs in harbors and estuaries where they are exposed only during spring low tides. Because mortality is greatest during the first year, they are carefully tended during that period. During low-tide exposure, predators are removed, the young oysters are carefully moved about and spread evenly on the bottom, and excess sand is removed. Commonly the parcs, which vary from a few to several hectares in size, are protected from the incursion of sand and sediment by the construction of earthen and brush dykes 0.3 to 0.7 m high. The young oysters are left in the parcs for 1 year, or until they are 1½ years old. They are then picked up by hand, when the parcs are exposed, or by draggers, and moved to deepwater beds where they are always covered by 3 to 10 m of water. Here they are grown for 2 years until they reach nearly market size. They are then dredged up again and moved back to the shallow intertidal parcs where they are fattened and shaped for the halfshell market.

Deep water growing is a relatively new development in the French oyster industry, started some 20 years ago, and has involved the development of new techniques, such as the use of large draggers capable of dredging oysters from the deep water. Not only do the oysters grow more rapidly in deep water, but the practice has opened up new areas for oyster production, permitting some expansion of the industry. However, the additional year in the conventional shallow parcs is still necessary to smooth the rough edges of the shell and to "fatten" the meat. Thus the flat oyster in France is grown for an average of 4½ years and moved three times before it is ready for market.

Portuguese oysters are not given nearly such intensive care. Those collected on oyster shells are left on the shells until harvested for market, as is the common practice in American oyster culture. In some cases, the 2-year-old seed oysters are thinned and transplanted into new bags or wooden racks suspended off the bottom, since they grow faster and survive better when not in contact with the sediment. In other cases, the seed oysters are simply sown on the bottom. All of the culture of this species is done in intertidal waters within estuaries. Although they do not require the hard, sandy bottom needed by the flat oyster, those grown in soft sediments are periodically turned and brought to the surface by rakes to prevent their being suffocated by the mud and to ensure even growth of the shell.

FATTENING AND GREENING

An interesting variation on the conventional methods of French oyster culture is the "claire" method of fattening and greening of oysters as final preparation for market. This is done when the oysters have approached or actually reached full growth with respect to shell size. The term "fattening" is misleading, since the process involves primarily the deposition of glycogen. In the process, the oyster meat increases in size and weight, the color of the meat becomes creamy white and the flavor more sweet. The ecological conditions for this type of development are quite different from those favoring the most rapid growth of the oyster, and require calm, shallow, relatively warm, and rather brackish water rich in plankton or organic matter. Claires are small, shallow, artificial ponds 0.1 to 0.2 ha in size which are constructed on marshland adjacent to the sea. They are filled and drained during spring tides through a system of gates and channels connecting them to the sea. A single operation may include several hundred contiguous claires.

The claires are prepared in summer by draining and allowing them to sun bake for several weeks. Fertilizer is sometimes added at this stage.

PLATE 15.   Claires (small marsh ponds) for fattening and greening oysters in Brittanny. (Photograph by J. H. Ryther.)

They are then filled to a depth of about 25 cm and the water is exchanged twice a month during spring tides.

Oysters are stocked in the claires at a density of $4/m^2$ in late summer and fall, when they are $3\frac{1}{2}$ years old and average 40 g. In 6 months they roughly double in weight and represent the most desirable and highly priced oyster on the French market. In some of the claires, a special bluish-green diatom, *Navicula ostrearia,* develops naturally and imparts a green color to the oyster meat. This so-called greening apparently affects neither the taste nor nutritional value of the oyster but is nevertheless highly prized by the French gourmet and further increases the market value of the crop.

Although claire culture of the flat oyster is a declining practice, the fattening and greening of Portuguese oysters by this method is becoming increasingly popular as a means of producing a more attractive and valuable product which can compete on the market with the flat oyster. Hundreds of ha in the Marennes and Arcachon regions are now devoted to claire culture, in many cases utilizing ponds originally constructed for salt production. As in the case of flat oysters, the fattening and greening of Portuguese oysters requires a period of 4 to 6 months, during which the meat roughly doubles in size and weight. However, Portuguese oysters are stocked at 12/m².

PROBLEMS

Oysters of both species may be attacked on the growing beds by a number of predators, including fish, drills, crabs, and starfish. These are not generally considered to be a serious cause of mortality. Fouling organisms (barnacles, other mollusks, tunicates, algae) are considered mainly a nuisance, necessitating additional hand labor to prepare the oysters for market but not affecting their growth or survival. Excessive growth of algae on the oyster beds may, however, smother the oysters or create anoxic conditions leading to their mass mortality. In some cases snails (periwinkles) are introduced together with the seed oysters in late winter or early spring to control the growth of algae.

Flat oysters have not been subject to any serious disease or parasite problems since 1923, when they were nearly wiped out by an epizootic of unknown origin and nature.

According to figures obtained from two growers, the survival of flat oysters from seed to market size is as given in Table 5, starting with 1

TABLE 5.   SURVIVAL OF FLAT OYSTERS FROM SEED TO MARKETABLE SIZE IN FRANCE

YEAR OF GROWTH	AGE OF OYSTERS (YEARS)	SURVIVAL (%)	NO. OYSTERS
1	½–1½	50	5×10^8
2	1½–2½	80–90	4.5×10^8
3	2½–3½	80	3.6×10^8
4	3½–4½	75	2.7×10^8

billion 6-month-old seed. Thus the crop of 1 billion seed oysters could be expected to produce some 270 million market oysters. Survival after the fourth year declines sharply and the oysters at that age average 65 g (shell included), an ideal market size. With a few exceptions, to provide extra large oysters for the gourmet trade, the oysters are therefore harvested in the fall or winter of their fourth year.

The Portuguese oyster industry has been plagued in recent years by high mortalities, particularly in the Arcachon region. This has been variously attributed to pollution, disease, and predation but the actual cause is still not understood. For this and perhaps other reasons Portuguese oyster production, which has expanded rapidly since World War II, has now leveled off or actually begun to decline. To combat this trend, seed of the Pacific oyster were imported to southern France from Japan, beginning in 1967.

Both large- and small-scale oyster culture in France is largely a family business which has been carried on through many successive generations, using essentially the same methods, for over a century. Only within the past decade have new techniques and new approaches been cautiously introduced. These may, in the future, remove some of the risks and reduce operating costs, but they will probably not appreciably increase the production of oysters, already near its peak since most of the available and suitable bottom is already under intensive cultivation.

PRODUCTION AND PROSPECTUS

Flat oyster rearing parcs, which are leased from the government, vary in size from 2 to 20 ha. A total of 5200 ha are devoted to flat oyster growing as compared to 8000 ha for the Portuguese oyster. Average annual production of flat oysters is 9,000 metric tons or 1.7 metric tons/ha.* Portuguese oysters are produced at a rate of 7.5 metric tons/ha, for a total of 60,000 metric tons annually.*

There is little scope for increasing the productivity of the French oyster industry by any means other than switching to off-bottom culture and concentrating on shucked rather than halfshell oysters. The economics of the industry virtually guarantee that this will not happen in the near future. Improvements can be made in the production of seed oysters. Currently, the availability of seed is dependent on environmental conditions and in a bad year the price of seed may be ten times normal. Introduction of hatchery techniques would free the flat oyster grower from dependence on the seed oyster industry of the Morbihan region.

Experimental hatchery culture of the flat oyster was first carried out in the United Kingdom, where it was found that larvae cultured in water containing adsorbing agents, such as Fuller's Earth or magnesium trisilicate, or in artificial sea water containing no organic matter grew more rapidly than those reared in untreated sea water.

In 1968 the first oyster hatchery in France was constructed at Carantec, Brittany, by one of the larger commercial growers. The culture method used is essentially the Milford method. Though presently operated on an experimental basis, the hatchery is designed to supply all the flat oyster seed for the proprietor's growing operation.

---

* Total weight, including shells. Meat weight = approx. 1/6 total weight, but shucked oysters are not commonly marketed in France.

OYSTER CULTURE IN THE PHILIPPINES

HISTORY

In contrast to North American, European, and Japanese practices, oysters in southeast Asia are considered a staple seafood rather than a luxury. In the Philippines they bring a price no higher than that of some of the more common food fishes. Until about 1936, all oysters eaten in the Philippines were gathered from wild populations. Since World War II, industrial pollution and perhaps overfishing have greatly reduced wild oyster populations in the islands and oyster culture has been intensified.

Although Philippine oyster culture is essentially off-bottom culture, the techniques used are largely rather primitive. Recent slight increases in technological sophistication of culture methods may be attributed to biologists of the Philippine Fisheries Commission, who have established experimental growing areas in the midst of commercial culture operations to impress upon the conservative local oystermen the potential for improvement of their techniques. Progress, however, is slow and is retarded by poor financial support from the Philippine government.

Seed for oyster culture operations is obtained from wild stocks, which may contain any or all of several species, the taxonomy of which is not well understood, at least not by those involved in their exploitation. The most important species, in terms of abundance and size, is the slipper oyster (*Crassostrea eradelie*).

SPAT COLLECTION AND GROWING FOR MARKET

Since most Philippine estuaries have soft bottoms, setting usually occurs on fringing intertidal vegetation (predominantly mangroves). However, the slipper oyster will settle on hard bottoms, and it is the occasional areas having hard, sandy bottoms which are selected for oyster culture.

The simplest and commonest form of oyster culture in the Philippines consists of driving bamboo poles 5 to 10 cm in diameter into the bottom. The poles are thoroughly sun dried after cutting and the larger ones may be split. Originally, oyster shells were drilled and wired onto the poles as spat collectors. However, it was discovered that the bamboo itself served equally well as a collector, and the use of shells has been discontinued. The poles are set out in the intertidal flats in rows 0.3 to 0.7 m apart. There is no set pattern to the distribution of poles, which are simply arranged as closely together as possible, leaving only enough room between rows to permit access by small boat.

The new technique introduced by the Fisheries Commission involves the construction of a bamboo framework also driven into the bottom. Each of the so-called "plots" is 1 × 10 m in area and consists of three heavy bamboo horizontal stringers and 20 small-diameter bamboo crosspieces 0.3 to 0.7 m apart. The whole framework is wired to a series of uprights driven 1.0 to 1.3 m into the bottom. The "plot" is placed so that it is completely covered at each high tide.

From the bamboo crosspieces of each plot, 140 collector strings are suspended into the water. These consist of No. 12 or No. 14 galvanized wire of variable length (depending upon the depth of water or the height of the platform above bottom). Four to five oyster shells are strung on each wire 12.5 to 15.0 cm apart. Originally bamboo spacers were used between shells, but now a loop is made in the wire beneath each shell.

This technique has been adopted slowly by the commercial operators, since the plots are more difficult and costly to construct than the conventional poles. However, the yields are sufficiently greater than from pole culture that the method is finding increasing use in the Manila Bay region.

The oysters set on the bamboo poles or the shell collectors in early summer to midsummer. Obtaining of a successful set has never been a problem in Manila Bay, though the collection of too heavy a spat fall or of successive spat falls throughout the summer may cause overcrowding and require periodic thinning of the young oysters. Between 50 and 100 spat normally settle on each collector shell.

The oysters grow very rapidly in both pole and platform culture and reach a marketable size of about 75 mm in 6 to 9 months. Harvesting begins in January and lasts through the spring. Because the oysters mature sexually in their first year, care is taken to harvest them before they spawn in the late spring and summer of their second year. During the harvest, oysters less than 5 cm in length (those which set late in the year or whose growth was stunted by overcrowding) are removed from the collectors and broadcast on hard, sandy bottom areas for an additional 6 months of growth. Normally, about 20% of the oysters are left on the culture grounds to insure the presence of an adequate breeding population to sustain the industry.

In a variation of this technique, seed oysters are collected on bamboo poles and, at the end of the rainy season, moved into tidal estuaries and rivers for growth and fattening. Oysters transplanted into the rivers in January are harvested the following spring before the start of the next rainy season, since the oysters cannot withstand the freshwater dilution and the poles cannot be maintained in the swift current of the flooded rivers. Growth during the winter months in rivers is somewhat greater than that obtained in the bay, and the oysters are of much better quality.

PROBLEMS AND PROSPECTUS

There have been no documented cases of disease or serious parasitism in the Philippine oyster industry. There is some loss from excessive dilution of the bay waters during the rainy season, and some mortality from starfish, drills, and other predators occurs. The combined mortality from seed to market size oyster is estimated at about 20% in normal years. Blooms of toxic dinoflagellates which produce 50% or more mortality have occurred unpredictably and locally from time to time (about once every 10 years).

One of the major problems of oyster culture in the Philippines is the rapid silting of the bays and estuaries. This apparently has little or no direct effect upon the oysters, which can withstand remarkable amounts of suspended sediments, but is progressively filling in the available culture grounds within Manila Bay and elsewhere. It is estimated that the Manila Bay oyster industry will have to be abandoned for this reason within the next decade.

No attempt is made to monitor the culture area for domestic pollution. Despite the presence of a large U.S. naval base directly across Manila Bay from the oyster beds, no survey for coliform bacteria has been made in the last 15 to 20 years. However, there have been no known cases of illness which could be directly attributed to the eating of shellfish from Manila Bay or elsewhere in the Philippines.

In Manila Bay alone, 4000 ha of bottom are devoted to oyster culture but no figures are available concerning the production or value of cultured oysters anywhere in the Philippines. The average commercial operation in Manila Bay consists of one individual or family working 0.4 ha or less of bottom, perhaps with hired help at the busiest times of year. In most cases the operator is forced to build a small house on stilts over his grounds and sleep there to discourage poaching. The coastal towns retain the rights to oyster grounds adjacent to their shores and may rent parcels of bottom to oyster culturists, but more often the individual merely assumes squatter's rights.

No assessment has been made of the potential of oyster culture in the Philippines, but it is almost certain that production could be greatly increased by educating the oystermen to use more sophisticated methods and by expanding the industry into new waters. There are believed to be extensive undeveloped estuarine areas suitable for oyster culture, including some that do not have natural beds, but where breeding populations might be introduced. Philippine fishery biologists are not unaware of the possibilities for technological improvement and expansion but are handicapped by the lack of adequate personnel and funds.

OYSTER CULTURE IN AUSTRALIA

SPAT COLLECTION

Wooden collectors, rather than strung shells, are used in Australia for off-bottom culture of the Sydney rock oyster (*Crassostrea commercialis*). Collecting frames are constructed by nailing four 2-m long, 22- to 25-mm square hardwood sticks between two 1-m rails, so as to form a ladderlike structure. In some cases, the sticks are attached to the frame in pairs (Plate 16). The frames are tarred to protect the sticks from teredos and borers and to provide a clean, smooth surface for spatfall. About 50 of these tarred frames are stacked and securely wired together to form a "crate" about 3 m high. Crates are wired to very sturdy racks constructed by driving two parallel rows of posts into the bottom 1.0 to 1.3 m apart and nailing 50- × 25-cm hardwood rails along each row.

Although spat fall occurs in February to April, crates of sticks are usually laid out in December or January, so that any impurities in the

tar which might affect the spat are washed out. Sticks should not be laid out too early, however, or a set of barnacles or other undesirable animals may occur.

DEPOTING

The newly settled spat are left undisturbed until late winter or early spring, when they are transferred to the growing areas for a process known as depoting. Moving the sticks from the collecting areas before spring ensures against their receiving a second spatfall. During depoting, the frames of sticks are secured to racks in the same manner as they were in the collecting area. The placement of the frames in several layers serves to protect them from predation by fish.

PLATE 16. Rack culture of oysters (*Crassostrea commercialis*) in the George's River estuary near Sydney, Australia. (A) Tarred wooden racks (collecting frames) ready for placement; (B) view of culture site; (C) close-up of racks with juvenile oysters attached. (Courtesy Peter H. Wolf.) (D) Transfer of 2-year-old oysters for optional placement in trays where they are grown for an additional year. (E) Inspection of 3-year-old oysters in trays. (Courtesy Australian News and Information Bureau.)

731

GROWING FOR MARKET

"Nailing out" of sticks to grow oysters for market is carried out from June to August. Some culturists nail out sticks in early autumn, but this necessitates a wire fence to keep out predators. Nailing out involves breaking down the crates and laying each individual frame, oyster-bearing side up, at midtide across a rack identical to the ones used in spat collection and depoting. The frames are either nailed or wired to the rack. Wiring is more tedious but eliminates loss of oysters occasioned by jarring them off the sticks. Rows of racks are set 5 to 15 m apart, depending on the productivity of the growing area.

The nailed-out frames are left in the growing area for 1 to 2 years if the oysters are to be marketed shucked, or 1½ to 2½ years (depending chiefly on latitude) if they are to be eaten on the halfshell. Then the stick frames are removed and taken ashore and the oysters are detached by tapping them with a hard object. The harvest is cleaned of barnacles and graded, and oysters too small for marketing are placed in chickenwire trays for further growing. Some culturists routinely remove 2-year-old oysters from the sticks and keep them in trays for the final year of growing.

PROBLEMS

Spat collection presents no particular problem in Australia; there has never been a failure over a large area. However, disease, predation, and the rigors of the environment do pose threats to Australian oyster culture.

The only serious disease reported from Australia occurs in the southern estuaries during the winter. The diseased oysters show ulcerations attributed to an as yet unidentified microorganism. Low temperature, high salinity, overcrowding, and overexposure to the air are believed to be associated with high incidence of the disease. No treatment or cure has been developed.

Certain fishes, notably rays, may devour young oysters and in some estuaries wire fences are needed to protect even 1- to 2-year-old oysters. Oysters growing on or near the bottom of the rack may be infested by the mudworm (*Polydora ciliata*), which, even if not fatal, may lead to poor condition of infested oysters. While the Australian method of rack culture is similar to hanging culture in increasing yields by making full use of the water column, it does not protect the oysters from benthic predators which can climb. Thus during years of low rainfall when estuarine salinities are high, whelks and starfish may cause heavy losses.

The most serious dangers are environmental. All cultured oysters in Australia are grown intertidally and are thus potentially subject to killing winter frosts in the south and extreme summer heat in the north. Many oyster farmers have installed pumps to spray exposed oysters with sea water during these critical periods. In recent years an additional environmental threat has appeared in the form of domestic pollution.

PRODUCTION

Oyster culture in Australia is generally a family business; there are only a few major producers. There is no distinct seed industry and both large and small culturists generally collect their own spat. Oyster grounds are leased by the government to the culturist for 15 years with the right of renewal for another 15 years. As in most countries where oyster culture is practiced, suitable growing areas are more widespread than good spat-collecting sites, and some farmers maintain two widely separated leases, one for spat collection and one for growing for market. Salamander Bay, Port Stephens, New South Wales, is a particularly popular spat-collecting location.

The industry is centered in New South Wales, where 6600 ha are under lease. Total annual production of oyster meat for the state is about 1 million kg, evenly divided between shucked and halfshell oysters. This amounts to only about 150 kg/(ha)(year), but much of the leased area is not in active production and some is used exclusively for spat collection. A better idea of the potential of the techniques and the region may be obtained by examining the yields achieved by intensive cultivation. In the best and best-managed areas, annual yields of 2000 kg meat/ha are attained using the methods described; culture in trays has produced up to 5400 kg meat/ha. It is not clear whether production can be substantially increased. Virtually all available grounds are leased, but it would seem that at least some of the leased, inactive grounds could be brought into production. Another possibility is expansion into deeper water. This has been attempted by the simple expedient of using longer poles to support the racks, but this results in considerably increased cost of cultivation.

Recently, commercial culture of the Sydney rock oyster has begun in New Zealand, where it is regarded as a promising new industry. The methods used are the same as in Australia, but there is only a small area suitable for reproduction. Thus spat collection is carried out by the government, which sells sticks of caught spat to commercial growers.

EXPERIMENTAL OYSTER CULTURE IN TROPICAL COUNTRIES

Oyster culture in tropical and semitropical countries has not assumed much importance, except in the Philippines. It is likely that if oysters are to become an important crop in the tropics, methods will have to be found to culture native oysters rather than relying on introduction of exotic oysters as was done to initiate the important oyster industries of southern France and western North America. In one instance where introductions were attempted, flat oysters stocked in Bizerta Lake, Tunisia, experienced 100% mortality when the water temperature reached 26°C. Portuguese oysters were able to survive but could not reproduce in the highly saline water.

NIGERIA

Further south on the African continent experimental oyster culture in Nigeria is resuming after the civil war. Species under study are *Ostrea gasar* and *Ostrea tulipa*. Spat fall of one or the other species has been observed in 10 of the 12 months.

THE CARIBBEAN

Continuous spatfall is characteristic of *Crassostrea rhizophorae* of the Caribbean, at least in the southern portion of its range. Wooden collectors appear to be best for this species; investigators in Venezuela and Cuba tested several different types and found respectively that wooden planks painted with bitumen, and branches of the red mangrove (*Rhizophora mangle*), on which *C. rhizophorae* is most often found in nature, were most effective.

Biologists of the Marine Research Station of the LaSalle Foundation at Isla Margarita, Punta de Piedras, Venezuela, estimate that two harvests per year of *C. rhizophorae* are feasible there. Slightly earlier work by the Cuban National Institute of Fisheries indicated a potential production of 6,600 kg meat/(ha)(year) by hanging culture of this species.

Attempts in Cuba to produce a high-quality halfshell oyster by using French methods failed. In addition to problems created by extreme variations in salinity (22 to 40‰) and an abundance of fouling organisms, *C. rhizophorae* confined in shallow parcs produced such amounts of nitrogenous wastes as to severely pollute their environment.

PROSPECTUS OF OYSTER CULTURE

BOTTOM VERSUS OFF-BOTTOM CULTURE

As Table 6 indicates, hanging culture, either from rafts or long lines, is essential if the maximum potential of oysters as a source of food is to be

TABLE 6. AVERAGE YIELD OF DIFFERENT METHODS OF OYSTER CULTURE IN 5
COUNTRIES (MEAT WEIGHT, SHELLS EXCLUDED)

COUNTRY	SPECIES	GROWING METHOD	PRODUCTION (KG/HA /YEAR)
Australia	*Crassostrea commercialis*	Rack culture (best areas)	2,000
	Crassostrea commercialis	Tray culture (best areas)	5,400
Cuba	*Crassostrea rhizophorae*	Experimental raft culture	6,600
France	*Ostrea edulis*	Bottom culture in parcs	250
	Crassostrea angulata	Bottom culture in parcs	1,000
Japan	*Crassostrea gigas*	Long-line culture	26,000
Japan (north)	*Crassostrea gigas*	Raft culture	1,000
Japan (south)	*Crassostrea gigas*	Raft culture	20,000
United States (Atlantic coast)	*Crassostrea virginica*	Bottom culture, with intensive management	5,000[a]
	Crassostrea virginica	Public grounds— little or no management	10–100

[a] Projected yield using new methods.

realized. The infinitesimal yields from unmanaged public grounds in
the United States, where oysters once constituted the most commercially
valuable fishery, suggest that the contributions to fisheries of oysters in
highly developed countries are likely to be negligible; oysters must be
cultivated if they are to make a significant contribution to human
nourishment.

EXPANSION INTO NEW AREAS

Future growth of the oyster industry will involve both improvements in
efficiency of existing culture systems and expansion into new areas
of the world. Expansion will entail both development of culture methods

for new species, particularly in the tropics, and wider introduction in temperate waters of species for which culture methods have been worked out, notably the Pacific oyster and the flat oyster, which not only possess considerable advantages for culture but are backed by strong seed oyster industries in Japan and France, respectively.

Improvements may continue to be made in bottom culture for some time, and a few new areas may be brought into production using rather primitive methods. For example, in Florida, where American oysters, locally known as coon oysters, commonly form sizable reefs, both construction of artificial reefs made of oyster shells and cutting gaps in existing reefs which have grown above the water surface show promise of increasing oyster fishery yields. However, even in the United States and Canada, where bottom culture is traditional, most current research in oyster culture deals with off-bottom culture. With the exception of such specialized practices as the claire method of producing gourmet halfshell oysters in France, bottom culture appears to be on the way out.

MECHANIZATION OF PROCESSING

Improvements in processing as well as culture may contribute to the growth of the oyster industry. Notable in this respect is the development in the United States of a mechanical oyster shucker capable of processing 60 oysters/min as opposed to an average rate of 8/min by hand shucking.

GROWING IN HEATED WATER

Use of thermal effluent in oyster culture presents possibilities for both hatchery use and market production. Both possibilities are being tested at Northport, New York, on Long Island Sound, where a major grower has obtained use of a lagoon into which large volumes of heated water are discharged by a Long Island Lighting Company power plant. Since the American oyster feeds and grows only at temperatures above 10°C, the effect of the plant is to provide water suitable for growth on a year-round basis.

Experimental culture of young oysters in the lagoon produced more rapid growth than normal at all temperatures up to and including the midsummer maximum of 32°C, thus allaying any fears that the thermal effluent might be damaging to the oysters. Since the cooling water is taken from an area of high nutrient content, the high temperatures not only speed up the oysters' metabolism but provide an abundant food supply by inducing a luxuriant growth of microscopic algae. The shape and location of the lagoon are such that thermal effluent is distributed evenly, thus many rafts or trays, capable of supporting millions of seed oysters, may be set out. To further take advantage of the situation, a hatchery has been built beside the lagoon.

The size of the lagoon is probably prohibitive for large-scale production of marketable oysters but the company, looking ahead to the time when larger power plants will release greater volumes of thermal effluent in less restricted waters, is experimenting with growing oysters to market size in the lagoon. It is believed likely that growing time can be reduced by a year or more.

SELECTIVE BREEDING

Perhaps the most exciting recent development in oyster culture is the increased effort devoted to selective breeding, which is likely to result in oysters becoming the first truly domesticated aquatic invertebrates. This work is barely begun, but for the most part results are encouraging —early unsuccessful attempts to breed strains of *Crassostrea virginica* resistant to the fungus *Dermocystidium marinum* notwithstanding. Among the qualities sought are the ability to reproduce at low temperatures, disease resistance, more rapid growth, and better shape, texture, and flavor.

THE OYSTER RESEARCH INSTITUTE AT KESENNUMA, JAPAN

Selective breeding has only become possible since the perfection by Loosanoff and Davis of oyster hatchery techniques (see pp. 688–692). Another pioneer hatcheryman was Takeo Imai, former director of the Oyster Research Institute on Mohne Inlet at Kesennuma, Japan. In addition to native oysters, Imai experimented with the American oyster, the Olympia oyster, the Portuguese oyster, and the flat oyster. There are minor variations in culture of the different species, but essentially the same techniques are used for all. The implication for commercial oyster culture is that by using the standard methods developed by Imai almost unlimited numbers of juveniles of any species of oyster may be provided through hatchery rearing. (This statement applies to other types of shellfish as well. See Chapters 37, 38 and 40 for discussion of work at the Oyster Research Institute with scallops, abalone, and cockles.)

The facilities of the Institute consist of a small laboratory building, a workshop, and a field of 180 1000-liter tanks, which are located in Mohne Inlet. The tanks, which are the heart of the operation, consist of 0.1 mm-thick polyethylene bags 2 m × 1 m wide, about 75 cm deep and are set in a large floating platform approximately 30 m from shore. Seawater is pumped from the Inlet to the shore installation and passed through a deep sand filter. It is then run by hose out to the tank area and passed through a cartridge filter of diatomaceous earth at the hose nozzle before being used to fill the tanks.

The tanks are used for the spawning of adults and the rearing of larvae to the settling stage. Since the tanks are immersed in the sea, there is no temperature control and the organisms grow under the natural temperature conditions of the bay. Thus cultivation is a seasonal operation, with spawning and larval rearing confined to late spring, summer, and early fall. The water temperature of Kesennuma Bay ranges from 5 to 6°C in winter to 24 to 26°C in summer, the latter at the surface only. Much colder water is available in summer in the deeper waters of the bay, including the tank area itself which, though close to shore, is about 10 m deep. Thus a considerable range of temperature is available for different purposes in summer. Salinity of the bay water averages about 33‰. This may drop to half this value or less at the very surface during periods of heavy rainfall, but the deeper water maintains its high salinity, so the surface dilution does not present a problem. Water used for filling the tanks is taken from beneath the surface and is therefore relatively constant in temperature and salinity during the summer months.

Some degree of control in the spawning periods of the shellfish can be achieved by keeping the sexually mature individuals in small tanks with refrigerated water. This effectively postpones or prolongs the breeding period of those species which normally commence reproduction early in the season. Spawning in many of the species may be induced by temperature shock, that is, subjecting them to a sudden increase in temperatures of about 5°C for several hours. Frequently, spawning will proceed after the animals so treated are returned to their normal temperature. This treatment is very simply achieved by maintaining the animals to be conditioned in one of the tanks with slowly running sea water, from the supply described, which is first passed through a length of plastic garden hose coiled on top of the tank in an area exposed to the sun. Solar warming of the circulated sea water is adequate to provide the increased temperature for induced spawning.

The adults and larvae reared in the tanks obtain no food from the filtered sea water and must therefore be fed artificially. In the laboratory building there are two rooms devoted to the culture of unicellular algae in 100 or more approximately 14-liter pyrex carboys. The cultures are illuminated with banks of fluorescent lamps and are aerated. To facilitate their preparation, a large volume of water is enriched with culture medium and then passed slowly through a tube exposed to ultraviolet radiation and into the cleaned carboys. This does not ensure sterile conditions, but it is sufficient to kill the algae and animals in the filtered sea water and thus insure unialgal conditions. Organisms now in use include the flagellates *Isochrysis galbana* and *Monochrysis lutheri,* and the diatom *Chaetoceras calcitrans.* Stock cultures of other species are available and may be used from time to time. Usually, the larvae are fed the flagellates in the early stages and the diatoms or a mixture of the two in the later and larger stages. Feeding is calculated at a rate of 5000 to 15,000 algal cells per larva twice a day.

Support for the Institute originally came from local and federal fishery agencies and the fishery industry, but as operations become more efficient it is possible to finance more of its activities by sale of juvenile shellfish to commercial culturists. Eventually it is hoped that the Institute can become self-supporting.

The Institute is also doing some research on growing oysters, particularly flat oysters, to marketable size. By suspending flat oysters from rafts in 0.3-m × 1-m rectangular metal frames a marketable halfshell oyster can be produced in 2 years, or half the time required by commercial growers in France. At present, however, this method is prohibitively expensive for practical application in Japan.

It would be misleading to imply that the success achieved by the Oyster Research Institute in the rearing of young shellfish could quickly or easily be duplicated elsewhere, for its accomplishments reflect the unique knowledge, experience, and skill of Imai. Nevertheless, the techniques involved are not held in secrecy and there is no reason they cannot, with the help of adequately trained individuals, be adapted and applied elsewhere, as they must be if oyster culture is to progress rapidly.

INCREASING POLLUTION PROBLEMS

With continued improvement in hatchery techniques and off-bottom culture methods and the emergence of selective breeding, the short-range outlook for oyster culture is one of rapid growth. However, while fisheries and aquaculture in general are threatened by increasing pollution of aquatic environments, oyster culture and other shellfish industries are in a particularly precarious position. Not only are shellfish an estuarine crop, thus more often subject to great amounts of pollution than freshwater or pelagic animals, but as mentioned previously, they tend to concentrate pollutants. Thus levels of pollution far below those required to produce mortality or physiological damage may eliminate oyster culture by rendering its product unfit for human consumption. If pollution were to thus completely remove from production such major oyster growing areas as Long Island Sound, Manila Bay, or the Inland Sea of Japan, it would more than offset all the advances in oyster culture made since the Romans. Even thermal pollution, presently seen as a potential boon to oyster culture, might contribute to such a pollution disaster through synergistic effects with domestic and/or industrial pollutants.

REFERENCES

BRIENNE, H. 1967. L'huitre et l'ostréiculture en France. Tech. Sci. Munic. 62e Année (2):82–86.

BROOKS, W. K. 1880. The development of the American oyster. Stud. Biol. Lab. Johns Hopkins U. **IV**:1–104.

CALABRESE, A., and H. C. DAVIS. 1966. The pH tolerance of embryos and larvae of *Mercenaria mercenaria* and *Crassostrea virginica*. Biol. Bull. (Woods Hole) **131**(3): 427–436.

CLARK, E. 1966. The oysters of Locmariaquer. Vintage Books, Random House, New York. 203 pp.

Commonwealth Scientific and Industrial Research Organization, Division of Fisheries, Marine Biological Laboratory. 1945. The biology and cultivation of oysters in Australia. 47 pp.

DAVIS, H. C. 1953. On food and feeding of larvae of the American oyster *Crassostrea virginica*. Biol. Bull. **104**:334–350.

DAVIS, H. C. 1969. Shellfish hatcheries—present and future. Trans. Am. Fish. Soc. **98**(4): 743–750.

GALTSOFF, P. S. 1964. The american oyster. U.S. Fish and Wildlife Service, Fishery Bulletin Vol. 64.

GAUCHER, T. A. 1968. Potential for aquaculture. Electric Boat Div. Gen. Dynamics Rep. LL413-68-016. Groton, Conn. 22 pp.

HOUSER, L. S., and F. J. SILVA. 1966. National register of shellfish production areas. U.S. Department of Health, Education and Welfare, Public Health Service, Division of Environmental Engineering and Food Production, Shellfish Branch.

KESTEVEN, G. L. 1941. The biology and cultivation of oysters in Australia. CSIRO, Division of Fisheries. Report 5, pp. 1–32.

LOOSANOFF, V. L. 1958. Oysters: use of plastic for collecting oyster set. Comm. Fish. Rev. **20**(9):52–54.

LOOSANOFF, V. L. 1961. Biology and methods of controlling the starfish *Asterias forbesi* (Desor). U.S. Bureau of Commercial Fisheries. Fishery Leaflet 520.

LOOSANOFF, V. L., and H. C. DAVIS. 1952. Repeated semiannual spawning of northern oysters. Science **115**(2999):675–676.

LOOSANOFF, V. L., and H. C. DAVIS. 1963a. Rearing of bivalve mollusks. *In* Advances in marine biology, Vol. I. Academic Press, London.

LOOSANOFF, V. L., and H. C. DAVIS. 1963b. Shellfish hatcheries and their future. Comm. Fish. Rev. **25**(1):1–11.

LOOSANOFF, V. L., and C. A. NOMEJKE. 1958. Burial as a method for control of the common drill, *Urosalpinx cinerea*, of Long Island Sound. Proc. Nat. Shellfish Ass. **48.**

MACKENZIE, C. L. 1970. Oyster culture modernization in Long Island Sound. Am. Fish Farmer 1(6):7–10.

MEDCOF, J. C. 1961. Oyster farming in the Maritimes. Fish. Res. Board Can. **131:**1–158.

NIKOLIC, M., and S. A. MELENDEZ. 1968. El Ostion del Mangle. Instituto nacional de la pesca, Cuba, centro de Investigaciones Pesqueras, nota sobre investigaciones 7.

ROUGHLEY, T. C. 1922. Oyster culture on the Georges River, New South Wales. Technical Education Series, No. 25. Technological Museum, Sydney, Australia, pp. 1–69.

QUAYLE, D. B. 1956. The raft culture of the Pacific oyster in British Columbia. Prog. Rep. Pac. Coast Sta. Fish. Res. Board Can. **107:**7–10.

RYTHER, J. H. 1969. The potential of the estuary for shellfish production. Proc. Nat. Shellfish Ass. **59:**18–22.

SHAW, N. N. 1969. The past and present status of off-bottom oyster culture in North America. Trans. Am. Fish. Soc. 98(4):755–761.

Tohoku Regional Fishery Research Laboratory. 1965. Aquaculture in Tohoku regions of Japan. Part I: The production of shallow-water organisms from annual statistics. 85 pp.

WALLACE, D. H. 1966. Oysters in the estuarine environment. American Fishery Society Special Publication 3, pp. 68–73.

WALNE, P. R. 1964. The culture of marine bivalve larvae. *In* K. Wilbur, and Yonge, C. M., Physiology and Mollusca, Vol. I. Academic Press, New York, pp. 197–210.

WELLS, W. F. 1920. Artificial propagation of oysters. Trans. Am. Fish. Soc. **50:**301–306.

WINDHAM, D. 1967. Economic aspects of the oyster industry. Paper presented at the Oyster Culture Workshop, Sapelo Island, Georgia.

Interviews and Personal Communications

BORGA, P. C. Chief, Minor Sea Resources Section, Marine Fisheries Biology Division, Philippine Fisheries Commission.

ESCRITOR, G. L. Shellfish and Crustacean Section, Estuarine Fisheries Division, Philippine Fisheries Commission.

FRANCOIS, D. D. Director, Fisheries Branch, CSIRO, Sydney, Australia.

FUJIYA, M. Japanese Fishery Agency.

FURUKAWA, A. Director, Propagation Section, Tohoku Regional Fisheries Research Laboratory, Japan.

GOUZER, J. Oyster grower, Carnac, Morbihan, France.

HUMPHREY, G. F. Chief, Division of Fisheries and Oceanography, CSIRO, Sydney, Australia.

IMAI, T. Director, Oyster Research Institute, Kesennuma, Japan.

INGLE, J. B. Shellfish biologist, Biological Laboratory, U.S. Bureau of Commercial Fisheries, Oxford, Maryland.

KAN-NO, H. Oyster biologist, Tohoku Regional Fisheries Research Laboratory, Shiogama, Japan.

KOBAYASHI, M. Head, Propagation Division, Kanagawa Prefectural Fisheries Experimental Station, Nagai, Japan.

KOGANEZAWA, A. Director, Miyagi Prefectural Fisheries Laboratory, Ishinomaki, Japan.

MADEC, J. A. Oyster grower, Carantec, Finistere, France.

MARTEIL, L. Chief, Institut Scientific et Technique des Peches Maritimes, Laboratoire d'Auray, Auray-Morbihan, France.

MATTHIESSEN, G. C. Shellfish biologist, Wareham, Massachusetts.

RABANAL, H. R. FAO, Rome, Italy.

SAGARA, J. Shellfish biologist, Tokai Regional Fisheries Research Laboratory, Nagai, Japan.

SOUDAN, F. Institut Scientific et Technique des Pêches Maritimes, Paris, France.

TSUJI, R. President, Sanyo Fisheries Co., Ishiomaki, Japan.

VANDERBORGH, G., JR. Long Island Oyster Farms, Inc., Greenport, New York.

37

Culture of Clams and Cockles

Among the most widely distributed and utilized seafoods are clams (class Pelecypoda, order Eulamellibranchia). Their culture is second in antiquity only to that of oysters among aquatic invertebrates. Clam culture, however, has never become as widespread or highly developed as oyster culture, perhaps because some species of clams are so abundant and easy to harvest in nature.

743

CLAM CULTURE IN JAPAN

HISTORY

Clam culture originated hundreds of years ago in Japan. There are accounts in eighth-century Japanese literature of the transplantation of desirable clams to new areas. Although transplantation in itself does not constitute culture, it is frequently a prelude to culture. At exactly what time clams began to be cultivated in Japan is not known, but the practice was established before oyster culture began in 1673. Despite its even longer history, clam culture in Japan has not become nearly as sophisticated as oyster culture. Clam growers continue to use the ancient methods, for the very simple reason that they work.

EXPERIMENTAL METHODS

At least nine species of clam are cultured in Japan, including haigai (*Anadara granosa*), sarubo (*Anadara subcrenata*), tairagi (*Atrina japonica*), torigai (*Fulvia mutica*), hokkigai (*Mactra sachalinensis*), bakagai (*Mactra sulcataria*), hamaguri (*Meretrix lusorina*), agamaki (*Sinovacula constricta*), and asari (*Tapes japonica*). Takeo Imai at the Oyster Research Institute in Kesennuma experimented with the culture of a tenth species, the arc shell (*Anadara broughtoni*). He used the standard shellfish methods he developed for hatching and rearing the seed. Young arc shells were grown to marketable size in two years by placing them in a multichambered net bag suspended in the sea. This apparently is the first instance of off-bottom culture of a clam.

COMMERCIAL BOTTOM CULTURE

Commercial clam growers still depend on bottom culture. Techniques vary with the species raised, but we treat here only asari, the most commonly cultured clam.

Asari are dioecious and spawn twice a year, from April to June and again from October to December. From fertilization to development of the larval shell takes less than 24 hours, with the precise time depending on temperature. Growth of the larvae also depends on temperature and virtually ceases during fall and winter. Even during summer, the rate of growth is slow enough that not all larvae metamorphose before the fall spawning season begins. Thus there are some asari larvae in Japanese waters throughout the year.

For reasons not completely understood, in some areas which are quite suitable for growth of clams reproduction is poor. Japanese clam culturists take advantage of this fact by collecting seedlings from densely populated beds where extensive spawning occurs and transferring them to underpopulated beds where they grow much faster than they would on the original beds. On a good bed, asari planted at a length of 1.5 cm and a weight of 0.4 g may attain 4 cm or more and 13 g in 22 months.

Asari are quite temperature tolerant but grow best at 23 to 24°C. Eggs develop best at salinities of 17 to 30‰ but subsequently growth is favored by higher salinities; 24 to 32‰ is considered optimal while salinities below 10‰ are lethal.

A suitable bed for planting asari seedlings should contain 50 to 80% sand. Clams are cultured in very shallow water, but too lengthy exposure to air retards growth. As a rule of thumb, asari should not be planted more than 1.6 m above low tide or overexposure will result.

Beds are prepared for stocking by drawing a harrow over them with a tractor in order to loosen the bottom soil and eliminate weeds. The detrimental effects on tractors of salt water restrict Japanese clam culture to waters less than 25 cm deep at low tide. After harrowing, the bed is left undisturbed for a week before planting.

Seedlings may be planted in spring or fall, but spring planting is favored, as it results in immediate growth, thus shortening the time to harvest. The best times for planting are on slack tides, high or low, when the seedlings are less likely to be swept away by currents before they can settle. Seedlings are sown by hand, which requires a fair amount of care to ensure placement of a uniform number of clams per unit area. The density of stocking varies from 1 to 4 liters/m², depending on the size of the seedlings.

Among the problems besetting the Japanese clam culturist are wave damage, extreme temperatures, siltation, and predation. Ducks are the principal predators, but starfish, octopuses, drills, and crabs may also cause heavy losses.

Marketable size of about 10 g may be attained in as little as 1 year, and certainly within 2 years. Harvesting is carried on year-round, but the best harvests are achieved in fall and winter. Simple hand tools, such as hoes and shovels, are commonly used in harvesting. A more sophisticated device is a hand dredge consisting of a 60-cm-wide basket fitted with a series of long steel teeth and attached to a bamboo handle. The most efficient method of harvesting involves a 1-m-wide dredge attached by two ropes to a boat. The boat provides the power, but there is a long handle connected to the dredge, with which a fisherman in the bow can regulate the depth to which the dredge penetrates the bottom.

With such a dredge, two or three operators can collect about 85 liters of cultured asari daily.

No production figures are available for Japanese clam culture, but two of the Japanese species, *Anadara granosa* and *Sinovacula constricta,* are cultured in Taiwan, where they contributed 60,989 and 11,047 kg, respectively, in 1965. The principal clam for culture in Taiwan, however, is the blood clam (*Meretrix meretrix*), whose culture yielded 1,252,432 kg in 1965.

CULTURE OF "COCKLES" (ANADARA SPP.) IN SOUTHEAST ASIA

Anadara granosa is also cultured in China, the Philippines, Thailand, Borneo, and Malaysia. In Malaya *Anadara* spp. are known as cockles, a term which has no taxonomic significance. "Cockle" was originally applied to a single species of *Cardium* popular as food in the British Isles, but wherever English settlers have gone, they have taken the term and applied it to whatever local shellfish resembled their native cockle. *Anadara,* *Cardium,* and all the other creatures known as cockles belong to the order Eulamellibranchia, members of which are referred to in the aggregate as "clams" in this text.

The only southeast Asian country for which statistics on culture of *Anadara granosa* are available is Malaysia, so Malaysian practices are described here, despite the fact that they date back only to 1948, whereas elsewhere in southeast Asia *A. granosa* has been cultured for centuries.

HABITAT

In Malaysia, *A. granosa* is confined to the west coast of the Malay peninsula, where there are large estuarine mud flats, bordered landward by mangrove swamps. The salinity is highly variable in this environment since it is affected by land drainage during the wet season of the northeast monsoon. Normally near 30‰, it may fall to as low as 15‰ in restricted areas. However, unless there is continuous rainfall for several days, these low salinities seldom persist.

The cockle can withstand a wide range of salinities for short periods of time but does not thrive in areas which are brackish (i.e., < 18‰) consistently or for prolonged periods. Therefore its distribution is restricted from locations within estuaries or bordering mangrove swamps which receive large and continuous amounts of freshwater.

Of more importance to the distribution of the cockle is the nature of the substratum. The animal thrives and grows most rapidly in calm, undisturbed intertidal flats of soft mud containing up to 90% silt. This habitat preference is of particular significance for the future of cockle culture in southeast Asia, since few other shellfish can be grown in such an environment.

The peak of the spawning season of *A. granosa* occurs between June and October, with the exact timing depending on the weather. Spawning may occur at any time, however, as larvae are found in the plankton throughout the year.

The principle of the rather primitive techniques employed in culture of *A. granosa* in southeast Asia is the same one applied in Japanese clam culture. There are areas which, though well suited to growth of cockles, are little populated due to hydrographic conditions that prevent spat from settling. Similar, more favorably situated areas may accumulate spat at densities of 10,000/m² or more. Culturists thus collect seed from such areas and use them to bring otherwise barren areas into production.

CULTURE TECHNIQUES

Collection of the seed begins when they reach a size of 4 to 10 mm, about 4 months after they settle to the bottom. The equipment consists of a fine mesh wire scoop, a wooden trough or box, and sometimes a flat wooden board. With these implements the collectors set out for the beds in sampans with the ebbing tide and return 2 to 3 hours before high tide. On exposed tidal flats, the collectors propel themselves across the mud surface by kneeling on the wooden board with one knee and pushing it through the mud with the other foot. The scoop is swept several times through the surface layer of mud, rinsed in water, and the collected seed dumped into the receptacle. An experienced collector can gather as much as 121 liters of seed on a single tide, though the collection may consist of young clams and a variety of other small benthic organisms in addition to cockles. The collectors are generally not culturists themselves but sell their catch to the culturists. The price obtained depends on the number of extraneous animals; the purest catches of cockles are generally made early in the season.

Seed are sown from boats at high tide or half-tide. One person is required to propel the boat, while one or two others scatter seed with shovels, taking care to achieve an even distribution of 1000 to 2000 cockles/m². Sandy or weedy patches of bottom are avoided. Any other irregularities in distribution are compensated for at the next low tide. As the cockles grow, they may be thinned one or more times to achieve a final density of 300 to 600/m².

Growth is rapid and the crop may be harvested 6 to 12 months after stocking, at which time the cockles are 2.5 cm long and average 112/kg.

[The preceding is based on 1957 data. In 1958, to protect the fishery, the minimum legal marketable size was raised from 1 to 1¼ in. (2.54 to 3.18 cm), which may necessitate slightly longer growing periods.] Harvesting involves the same procedure used in seed collection, but a wider mesh scoop is employed. Using this technique, one man can harvest a ton of cockles in 5 to 6 hours.

There are no known cases of mortality of cockles caused by disease or parasitism. Predators, especially skates, may invade a newly planted bed and virtually wipe out the population, but this happens rarely. The boring mollusk, *Natica maculosa,* occurs in natural beds and has been transplanted with seed cockles to culture beds, but mortality from this cause is believed to be less than 1%. The gastropod *Thais carinifera* has on occasion caused mortalities.

PRODUCTION

In 1957 about 1400 ha of mud flats were leased by the Malaysian government to cockle growers. The culturists habitually stock many times the licensed area, thus estimates of the total area actually in use are virtually worthless, but it is believed that the area presently devoted to cockle culture is double that under cultivation in 1957. No more recent data are available. Total combined yield of *A. granosa* to Malaysian fishermen and aquaculturists in 1957 was 6,660,000 kg, compared to the fishery yield of 2,605,800 kg for 1938 (the last pre-World War II year for which data are available).

The potential of low-intensity culture of *A. granosa* in southeast Asia is quite good since the cockle can be raised cheaply with a minimum of labor (as few as three men can manage 10 ha of cockle beds, while twelve men can handle 100 to 120 ha), and occupies an environment which is little exploited by fishermen and aquaculturists. Cockle culture in Malaysia is particularly favored by the existence of a large market in Singapore, where 50% of the production is now shipped, either fresh or canned. Since *A. granosa* is not only cheap and tasty but richer in protein than most other shellfish, its culture should be encouraged as a means of partially supplying the increasing need for high-protein foods in southeast Asia.

CULTURE OF QUAHOGS (MERCENARIA MERCENARIA)

HATCHERIES IN THE UNITED STATES

The relative sophistication of clam culture in Asia and North America is exactly the opposite of the situation with respect to oyster culture.

Whereas North American oyster culture is quite primitive and Asian methods, at least in Japan, very advanced, the United States has taken the lead in clam culture. Not that anything approaching the efficiency of off-bottom culture of oysters has been achieved with clams, but American biologists have pioneered in the development of hatchery methods for clams, and fairly intensive clam culture is a possibility in the near future.

By far the most popular clam in the United States is the hard clam or quahog (*Mercenaria mercenaria*), which is especially important commercially in New England and Long Island. The possibility of hatching and farming quahogs was suggested as early as 1931, but early experimental shellfish culturists concentrated on oysters. Eventually, Victor L. Loosanoff and his associates at the U.S. Bureau of Commercial Fisheries Biological Laboratory at Milford, Connecticut, on Long Island Sound, perfected techniques for rearing quahogs under hatchery conditions.

The basic hatchery procedure used in quahog culture is the "Milford method," developed for culture of the American oyster (*Crassostrea virginica*) (see Chapter 36 for details). By using this method and controlling the water temperature, biologists were able to extend the spawning season of quahogs, which normally breed only during the summer. It is now possible to spawn these clams any day of the year.

Adaptation of a hatchery method to a new species of course entails altering the environmental conditions. Quahogs are more tolerant of varying temperatures than oysters but not as tolerant of low salinity. Although swimming larvae can withstand salinities as low as 15‰, at least 22‰ are required for successful hatching and development of larvae to the crucial straight-hinge stage (Fig. 1). The optimal salinity for development of quahog eggs from Long Island Sound is 27‰. Larvae are similarly plastic with respect to pH, but normal development depends on maintenance of a pH of 7.0 to 8.5.

Larvae develop normally at temperatures of 18 to 30°C, but the optimum temperature, and certainly the temperature that will produce the most rapid development, is usually considered to be near the upper limit of this range. At 30°C, setting may occur in 7 days after hatching, while at 24°C it may require 10 days to reach this stage. Above 30°C, survival drops off abruptly (Fig. 1).

Dissolved oxygen concentration does not seem to be a subject for great concern in quahog culture. However, the culturist should endeavor to provide well-oxygenated water, as experiments at the New Jersey Agricultural Experiment Station have shown that pumping and respiratory rates of quahogs, which were formerly thought to be regulated by feeding, are directly related to oxygen demand. The less oxygen is available, the

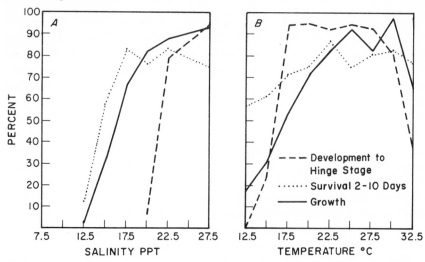

Fig. 1.   Growth of quahogs at different salinities and temperatures. (After Davis, 1969.)

harder the clam must pump, thus the less energy is available for growth.

Hatchery-reared quahogs are fed in essentially the same way as oysters reared in hatcheries which use the Milford method. The Japanese alga *Scenedesmus obliquus,* sometimes used by oyster culturists, has been particularly helpful to quahog hatcherymen. At Milford, quahog larvae fed on dried *S. obliquus* grew nearly as well as those given the best live food available.

To discuss "optimum" values for environmental parameters or "best" foods is of course to oversimplify, since all these factors interact. A particularly interesting interaction between feeding and temperature was noted for quahogs at Milford. At 10°C, quahog larvae will ingest algae, but digestion and assimilation do not occur. At 15°C, algae with thick cell walls were not digested, but those with thin cell walls were. At 25°C, larvae can digest most algae.

GROWING FOR MARKET IN THE UNITED STATES

Quahogs are grown for market using the same primitive methods that characterize American oyster culture. Seed clams, sometimes as small as 3 mm in length, but preferably at least 12.5 mm long so as to be less vulnerable to predators, are scattered on the bottom in shallow water and allowed to grow naturally. They may be thinned occasionally, and various techniques of predator control may be employed, but otherwise they are left undisturbed until they are dug up for market.

At present, the center of the clam industry in the United States is in the northeast, with New York as the leading producer. Growth of quahogs to marketable size in these northern waters requires 5 to 8 years, and there may be a southward shift in the industry to take advantage of the faster growing rates possible in southern waters. It is worth noting in this connection that the world's first commercial clam hatchery, slated to produce 300 million seed annually, was set up not in the northeast but in Virginia.

EXPERIMENTAL CULTURE

A development of potential importance for quahog culture in the southern United States was the hybridization in 1954 of *Mercenaria mercenaria* with the southern quahog (*Mercenaria campechiensis*) and successful rearing of the progeny. The southern quahog grows faster than *Mercenaria mercenaria,* and is better suited to culture in warm water, but does not keep nearly as well out of water, so that shipping is restricted. The hybrids appear to be intermediate with respect to keeping qualities.

Most experiments in clam farming in the southern United States have to date involved *M. mercenaria* seed obtained from Milford. Small Milford quahog seed, stocked in Alligator Harbor, Florida, in 1961 and 1962 reached marketable size (50 to 63.5 mm long) in 2 years. Seed stocked at lengths of 33 to 44 mm reached marketable size in 8 to 10 months.

These large seed were stocked in plots 9 to 76 m² in area. Previous work suggested that to successfully grow quahogs in Florida, the salinity must be at least 25‰ and the bottom firm. Practical considerations made it necessary to conduct the experiments in water not over 0.7 m deep at low tide, but it was found that if *M. mercenaria* seed were placed in water which was too shallow, high mortality at low tide might result from the very high summer temperatures.

Density of planting was found to be critical in achieving good growth. Seed stocked at 750/m² grew only 0.6 to 0.7 mm/month, but at concentrations of 100 to 500/m² growth was uniformly good, ranging from 1.4 to 1.7 mm/month.

Fencing was absolutely necessary to exclude predators from the experimental plots. Clams stocked on unfenced plots suffered 100% mortality, chiefly due to blue crabs (*Callinectes sapidus*), within a few months. Fences 2 m high, of 12.5-mm plastic coated wire or nylon netting sunk several centimeters into the substrate kept out most predators, although a few blue crabs did gain entrance. These were watched for and periodically removed with baited traps. Overall mortality in fenced plots ranged from 5 to 18%, with an average of 10%. Predation was apparently the sole cause of mortality, with blue crabs responsible for 90%. The only other important predator was the lightning whelk (*Busycon contrarium*).

Researchers at the Virginia Institute of Marine Science have successfully thwarted blue crabs and other predators in experimental quahog beds by simply spreading crushed oyster shell, crushed stone, or pea gravel over the surface of the beds.

It is encouraging to note that the researchers in Florida and Virginia did not dip into the Pandora's box of poisons used to kill predators on oyster beds in Long Island Sound. A rather disquieting bit of information apropos clams and pesticides is the fact that quahog larvae reared at Milford reacted positively to some pesticides and showed optimum growth at 5 ppm of lindane. It is to be hoped that clam growers will not conclude that what is good for the clam is necessarily good for the consumer, or for the future of clam culture.

The fences used to exclude predators retarded the circulation of water so that after several months a layer of soft sediment 12 to 25 mm deep was built up. This "mud," which was believed to be caused by the biological activities of the clams, did not appear to be detrimental in any way. On the other hand, siltation may cause up to 100% mortality of quahogs, depending on the amount of silt deposited.

A considerable obstacle to the commercial scale application of hatchery techniques for quahog culture is the expense involved in keeping seed in the hatchery until they reach optimum planting size. From the hatcheryman's point of view, it is desirable to sell 1- to 3-mm seed, whereas the grower would prefer 12.5-mm or larger quahogs. The use of dried algae as food may reduce the expense of feeding, but it seems impossible to do anything about the considerable space requirements of large seed.

Researchers at Florida State University have found it possible to grow *M. mercenaria* from less than 5 mm long to 10 to 15 mm at concentrations of up to $5,000/m^2$ in sand-filled, wire mesh-covered boxes placed in open water. It is doubtful if this method would be commercially feasible without modification.

Quahog seed produced at Milford have been shipped not only to growers in the Atlantic and Gulf Coast states of the United States but to Ireland, France, Japan, the Netherlands, and the United Kingdom as well. It is possible that in some of these countries culturists would do better to concentrate on developing hatchery methods for native species similar to those used on quahogs at Milford but, in Ireland at least, the quahog was imported for the specific purpose of stocking on mud bottoms where the cockle (*Cardium*), favorite shellfish of Irish consumers, will not grow.

Of perhaps greater significance is the culture of quahogs in Poole Harbor, Dorset, England, as part of an attempt to utilize human domestic and industrial wastes in a new ecosystem designed to produce food for human use and simultaneously reduce pollution. The potential pollutants involved are sewage, hot water, and flue gas, the latter two produced by the Poole Electric Generating Station. The flue gas, after scrubbing to remove traces of such impurities as oxides of sulfur, is used as a carbon dioxide supply for mass culture of the alga *Phaeodactylum tricornutum*. Sewage diluted with sea water serves as the nutrient source. The dense culture of algae thus produced is fed to larval quahogs, which are stocked in portions of Poole Harbor warmed by the thermal effluent from the power plant. The process, as compared to natural growth of quahogs stocked in nearby unheated waters, has been found to produce more rapid growth of all life stages as well as earlier and more prolific spawning.

CULTURE OF THE SOFT CLAM (MYA ARENARIA)

Many species of clams other than quahogs are eaten in the United States, but the only other species which is cultivated to any extent is the soft clam (*Mya arenaria*). Soft clams are of greatest importance in Maine, where they were nearly wiped out in the late 1940s, when a population explosion of the green crab (*Carcinides maenas*) occurred. The species has come back, however, and 36 seashore communities are presently cooperating with the Department of Sea and Shore Fisheries in enforcing a system of rotation of clam flats, with stocking policies determined by population distribution, survival, and growth rate. The clams harvested are mostly 3- and 4-year-olds and, in the better growing areas, produce annual yields on the order of 1500 kg of meat per hectare.

M. arenaria has been artificially propagated at Milford and has yielded to thermal manipulation of the reproductive cycle, but the practice has not become nearly as important as hatchery spawning of quahogs.

EXPERIMENTAL CLAM CULTURE IN FRANCE AND TUNISIA

Recently, French fishery authorities have become concerned over the greatly reduced stocks of clams in that country. Researchers at the Zoological Laboratory of the Faculty of Science at Brest have succeeded in rearing the native species *Venerupis aurea, Venerupis pullastra, Venerupis rhomboides, Venus fasciata, Venus striatula,* and *Venus verrucosa* to the benthic stage in 3 to 4 weeks in the laboratory, and, at least from a biological standpoint, prospects for future culture look bright.

A recent development in clam culture is the stocking and fattening of *Tapes decussatus* and *Venus verrucosa* in the brackish-water Bizerta Lake, Tunisia. The former species yielded 85 metric tons in 1967.

FRESHWATER CLAMS

Freshwater "mussels" (family Unionidae), despite their common name, are classified in the order Eulamellibranchia with clams and cockles. Most Unionidae are of little commercial importance, except as producers of pearls, but *Lampsilis claibornensis* of the Tennessee and upper Mississippi River valleys of the United States, was formerly sought by fishermen for its shell, used in the production of buttons and ornaments. The prevalence of plastic buttons and the siltation of large American rivers have drastically reduced the importance of the fishery for *L. claibornensis*, but the species is still sought as a source of food. Recently, *L. claibornensis* shells, cut or ground into small pellets, have been found to be ideal "seed" for culture of pearls, and their shells are now exported from the United States to Japan for that purpose.

Experiments conducted at the Auburn University Agricultural Experiment Station in Alabama by H. S. Swingle indicate that *L. claibornensis* has great potential value for culture in ponds, alone or with fish. An 0.8-ha pond was stocked with a few mussels, along with 3125 bluegill sunfish (*Lepomis macrochirus*)/ha, 1565 redear sunfish (*Lepomis microlophus*)/ha, and 312 largemouth bass (*Micropterus salmoides*)/ha. The mussels multiplied rapidly and were harvested annually, along with fish, over a 5-year period, yielding an average of 1010 kg/ha of unshelled mussels (318 kg/ha of meat). The sixth year, the pond was drained and 1271 kg/ha of mussels recovered (400 kg/ha of meat). The standing crop of fish was 464 kg/ha. In a control pond, stocked with the same numbers of fish but no mussels, the standing crop of fish was 317 kg/ha. It is believed that the difference in fish production was a function of the filtering action of the 49,440 mussels/ha which were found on the pond bottom. As the methods used in Swingle's study were far from intensive, all of these yields could undoubtedly be improved.

Larvae of *Lampsilis* spp. are parasitic on freshwater fishes, but infestations are seldom heavy and the results not usually serious. Thus it appears that *L. claibornensis* and its relatives could make a valuable contribution to aquaculture, particularly in ecologically integrated polyculture systems, where a host fish species might be stocked along with the main fish crop. In addition to the benefits already described, freshwater mussel exometabolites, commonly called "mussel mud," are a superior fertilizer for vegetables. The recycling capacity of a mussel-fish community is further enhanced by the possibility of periodically crushing excess small mussels and feeding them back to the fish.

PROSPECTUS

In attempting to formulate a prospectus for the culture of clams, one finds oneself dealing with two very different situations. In North America and Europe, clams are still a luxury food, whereas in most Asian countries they are presently or potentially one of the cheapest and most important sources of protein. The economic situation in North America and Europe, plus the fact that destruction of the estuarine environment is proceeding most rapidly on these continents, dictates that this is where research in intensive clam culture, especially hatchery methods, will be centered. Of particular value would be the continuation of the genetic work on *Mercenaria mercenaria* which was begun around 1960. It would also be advantageous if someone were to develop a method for growing uniform size clams, so that mechanical harvesting could be employed.

Some of the results obtained by American and European researchers may be of value to Asian culturists, but clam culture in Asia (outside of Japan, where the importance of shellfish hatcheries is likely to increase) will probably continue to be characterized by low-intensity methods for some time, particularly if the present high yields are sustained.

It should be borne in mind that there are hundreds of species of edible clams, including many which are unexploited by fishermen, let alone culturists. One or more of these species may be found to possess characteristics such that it will revolutionize the economics of clam culture. A number of presently uncultured species of clam, including the transverse arc clam (*Arca transversa*), Morton's cockle (*Laevicardium mortoni*), the small clam (*Pita morrhuana*), the razor clam (*Ensis directus*), and the surf clam (*Spisula solidissima*) have been artificially propagated experimentally at Milford, and this work should be extended.

At present, clam growers must compete, on the one hand with gatherers of "wild" clams, which are still abundant and easy to collect along many coasts, and on the other hand with such established growers of luxury foods as oyster culturists. Except for areas where there is a large established market for clams, for example, the market for quahogs in New England and New York or that for cockles in Singapore, the immediate future does not appear to hold the prospect of great expansion in the culture of clams.

REFERENCES

ANSELL, A. D., K. F. LANDER, J. COUGHLAN, and F. A. LOOSEMORE. 1964. Studies on the hard shell clam *Venus mercenaria* in British waters: growth and reproduction in natural and experimental colonies. J. Appl. Ecol. L:63–82.

BELDING, D. L. 1931. The quahaug fishery of Massachusetts. Commonwealth of Massachusetts, Boston. 41 pp.

CALABRESE, A., and H. C. DAVIS. 1966. The pH tolerance of embryos and larvae of *M. mercenaria* and *C. virginica*. Biol. Bull. (Woods Hole), 131(3):427–436.

CHANLEY, P. E. 1963. Inheritance of shell markings and growth in the hard clam *Venus mercenaria*. Proc. Nat. Shellfish Ass. 50:163–168.

DAVIS, H. C. 1969. Shellfish hatcheries—present and future. Trans. Am. Fish. Soc. 98(4):743–750.

LOOSANOFF, V. L., and H. C. DAVIS. 1963a. Rearing of bivalve mollusks. *In* Advances in marine biology, Vol. I. Academic Press, London.

LOOSANOFF, V. L., and H. C. DAVIS. 1963b. Shellfish hatcheries and their future. Comm. Fish. Rev. 25(1):1–11.

MENZEL, R. W. 1963. Seasonal growth of the northern quahog *Mercenaria mercenaria* and the southern quahog *M. campechiensis* in Alligator Harbor, Florida. Proc. Nat. Shellfish Ass. 52:37–46.

MENZEL, R. W., and H. W. SIMS. 1963. Experimental farming of hard clams, *M. mercenaria* in Florida. Proc. Nat. Shellfish Ass. 53.

PATHANSALI, D. 1961. Notes on the biology of the cockle, *Anadara granosa* L. Proc. Indo-Pac. Fish. Comm. 11(11):84–98.

PATHANSALI, D., and M. K. SOONG. 1958. Some aspects of cockle (*Anadara granosa* L.) culture in Malaya. Proc. Indo-Pac. Fish. Comm. 8(110):26–31.

SWINGLE, H. S. 1966. Biological means of increasing productivity in ponds. FAO World Symposium on Warm Water Pond Fish Culture. FR: V/R-1.

Interviews and Personal Communication

FUJIYA, M. Japanese Fishery Agency.

PATHANSALI, D. Director, Fisheries Research Station, Federation of Malaysia, Glugor, Penang, Malaysia.

38

Culture of Scallops

There is virtually no end to the list of edible invertebrates in the sea. Some, because of their scarcity or for other reasons, are rarely exploited commercially. Others, although abundant and commonly caught by fishermen, are not cultured on any appreciable scale, as is the case with scallops, one of the most popular sea foods. However, the situation may be changing.

It is reported that the deep sea scallop (*Patinopecten yessoensis*) is already being cultured commercially on a small scale in Japan, but details are not known. It is known that Takeo Imai of the Oyster Research Institute in Kesennuma succeeded in spawning and rearing this species, using the standard hatchery methods he developed for oysters (Plate 1).

The primary purpose of the Oyster Research Institute is to refine hatchery techniques, but experiments in growing shellfish to marketable size are also carried out there. *P. yessoensis* is grown by sandwiching seed scallops between layers of loose plant fiber in 0.3 m × 1 m rectangular metal frames, covered with a grid of nylon thread and suspended in the sea. When the animals are large enough, they are placed in similar frames without the fiber filling. Alternatively, scallops may be tied to hanging ropes by a piece of nylon thread strung through the "ear" of the shell. Both methods have produced marketable size scallops in 2 years, but the labor and materials required render them of doubtful commercial feasibility.

PLATE 1. Two-year-old Japanese scallops (*Patinopecten yessoensis*) grown in nylon-mesh sandwich type frame in suspended culture. (Photograph by J. H. Ryther.)

Much less success has been achieved with the European scallop, *Pecten maximus*. Researchers at the Zoological Laboratory of the Faculty of Science at Brest, France, have thus far not been able to rear this species past the veliger stage.

A decade before the establishment of the Oyster Research Institute, it was recognized that spawning of the deep sea scallop could be induced by raising the temperature. The same was later found to be true of the bay scallop (*Argopecten irradians*) of the Atlantic coast of North America. A. N. Sastry, in experiments started at Florida State University, Tallahassee, Florida, and continued at the Duke University Marine Laboratory, Beaufort, North Carolina, showed that food supply is also a controlling factor in gonadal maturation of bay scallops. By manipulation of these two factors, Sastry was able to spawn *A. irradians* out of season, even during months when they are never ripe in nature.

Sastry's work also involved successful rearing of the larvae to the sub-adult state (Plate 2). Although Sastry's methods have yet to be applied on a large scale, artificial propagation of scallops in the United States is on approximately the same level of development as was hatchery oyster

PLATE 2. Bay scallops spawned in the laboratory and grown in anchored floats. (Courtesy M. Castagna, Virginia Institute of Marine Science.)

culture in the mid-1950s, after the perfection of the Milford method for hatching and rearing the American oyster (*Crassostrea virginica*). It may be that commercial growing of hatchery-reared scallop seed would not be feasible, but the possibility should be explored.

REFERENCES

SASTRY, A. N. 1963. Reproduction of the bay scallop *Aequipecten irradians* Lamarck: influence of temperature on maturation and spawning. Biol. Bull. **125**(1):146–153.

SASTRY, A. N. 1965. The development and external morphology of pelagic larval and post-larval stages of the bay scallop *A. irradians concentricus* Say reared in the laboratory. Bull. Mar. Sci. **15**(2):417–435.

SASTRY, A. N. 1968. The relationships among food, temperature and gonad development of the bay scallop *A. irradians* Lamarck. Physiol. Zool. **41**(1):44–53.

YAMAMOTO, G. 1951. On acceleration of maturation and ovulation of the ovarian eggs in vitro in the scallop *Pecten yessoensis* Jay. Sci. Rep. Tohoku Imp. U. **19**:161–166.

39

Culture of Mussels

The widely distributed mussels (*Mytilus* spp.) are among the hardiest and most easily gathered seafood organisms, but their culture in most countries has begun only recently, if at all. Recent advances in technique, however, have established mussel culture as the most productive form of saltwater aquaculture, and its proliferation is virtually a certainty. At present, mussel culture, which is largely confined to Europe, is virtually unique among aquacultural practices in the Western world in producing a staple food rather than a luxury item. (So-called freshwater mussels, which have been experimentally cultured in the United States, are actually clams, and are discussed in Chapter 37.)

NATURAL HISTORY OF MYTILUS EDULIS

Almost all of the world's commercial production of cultured mussels consists of the common edible mussel (*Mytilus edulis*) and the Mediterranean *Mytilus galloprovincialis*, which may or may not be a separate species. *Mytilius edulis* reaches sexual maturity in its first year and spawns with rising temperature in the spring and summer. Eggs and sperm are shed separately and fertilization occurs in the open water. The free-swimming larvae remain planktonic for 10 to 15 days depending upon temperature, food supply, and the availability of settling sites. Young mussels are remarkably sensitive to light and require a relatively high light intensity in the region where they settle, a fact which may explain their natural occurrence in the intertidal zone. Although the peak of spawning occurs in midsummer, larvae may be found in the plankton from early spring to late fall and, in some parts of Europe, throughout the year.

Mussels normally occur in large banks or shoals. The reason for this highly gregarious behavior and discontinuous distributional pattern is not fully understood. It may be the fortuitous result of a hydrographic concentration of larvae at the time of their settlement, or it may be caused by a behavioral pattern of the mussels themselves, such as a mutual attraction through chemical mediation.

The mussels attach to the substratum by means of a byssus which they secrete. Their preference is to attach to fibrous or threadlike materials, but they commonly settle on or later move to other mussels and shellfish, rocks and stones, and algae. The mussel does not remain attached permanently by its byssus but may discard it and either crawl about or, it is said, secrete a bubble of gas and drift with the current or tide to find a more suitable niche where it may attach itself by a newly secreted byssus. This ability to move about is an important factor in mussel culture. Care must be taken in transplanting mussels to duplicate the conditions under which they were formerly living with respect to light intensity, temperature, salinity, and so on, or there is danger of their detachment and migration.

MUSSEL CULTURE IN FRANCE

HISTORY

The only country with a long history of mussel culture is France. The culture of mussels in France began with the shipwrecking of an enterprising Irish sailor named Patrick Walton at the point of Escale near the

PLATE 1. (C) Bouchet-type mussel culture in Arguenon Bay, near Mont St. Michelle, France. (A) Juvenile mussels wrapped on poles. (B) Mature mussels ready for harvest. (photos J. H. Ryther)

port of Esnodes in the fall of 1235. Faced with the problem of survival on this inhospitable shore, Walton first fashioned a wooden foot-boat (*acon*) with which he could negotiate the soft tidal flats. He then proceeded to design and build a large net (*alluret*) suspended on the flats by poles driven into the mud, with which he hoped to catch land and sea birds abundant in the area. Walton soon discovered that the supporting poles quickly became encrusted with mussels and that the shellfish on the poles grew much more rapidly than those living on the bottom. Further experimentation and exploitation of this discovery by Walton eventually led to the so-called *bouchet* system of mussel culture which is still in practice today with only minor modifications.

Although mussel culture has been practiced continuously since Walton's time along the southern Atlantic coast of France, the north coast of Brittany has become a chief producing area and is the basis for the description of culture techniques which follows.

TECHNIQUES

The environment along the northern coast of Brittany is characterized by a salinity of 32‰, dropping to about 29‰ in the surface water of estuaries during the winter rainy season. The annual temperature range is 5 to 20°C. The most striking feature of the North Brittany coast is the tide, which has a normal monthly range from 9 to over 15 m. This has a decided beneficial effect upon the mussels, whose growth appears to be directly proportional to the current strength. The extreme tidal range, advantageous to the mussel grower, is prohibitive for natural setting, since the larvae are carried out to sea before they reach setting size. Thus growers in Brittany collect seed from southern France, notably the La Rochelle region.

The collection technique, practiced mostly from May to July, but sometimes as late as September, involves suspending loosely woven ropes, 13 mm in diameter and 3 m long, in the intertidal regions near natural mussel beds. Within 2 weeks a set of seed mussels, each 5 to 10 mm long, will be found concentrated in the crevices between strands of the ropes.

The ropes are brought back to Brittany in May to August and spirally wrapped around 15- to 20-cm diameter oak poles 4 m long, which are driven for half their length into the sediment in the intertidal regions of harbors and river mouths. Poles may be wrapped from boats at high tide or by workers on foot and in oxcarts at low tide; in either case careful timing is necessary because of the tides.

Poles are normally arranged in rows perpendicular to the coastline. The precise pattern of placement of poles has been determined empirically for each culture site to achieve maximum growth and conform to a few legal restrictions (e.g., to not block navigational channels).

The bottom 25 to 50 cm of each pole are sheathed in smooth plastic to prevent crabs and other predators from climbing up. Sometimes a wooden framework is extended between poles and strung with twine to keep out predatory birds.

GROWTH AND PROBLEMS

The growth of mussels is rapid and they quickly cover the entire pole. Within a few months, the clusters extend several inches out from the poles, and the outer layer or layers must be broken off to ensure good growth of the inner core and to prevent their breaking off from their own weight and falling to the bottom. These removed mussels are placed in long, thin string netting and either wrapped in spiral patterns around new poles or slung between the poles. This process is repeated two to three times a year as growth proceeds.

PLATE 2. Experimental raising of mussels in plastic net bags in Norway. (A) A net bag filled with mussel spat. (B) Spat penetrating the meshes. (C) Mussels have now nearly all penetrated the mesh bag. (Credit Bjørn Bøhle, Fiskeridirektoratets, Bergen, Norway.)

In Norway, loss of excess mussels is prevented by placing spat in polypropylene net bags with expandable mesh. The young mussels penetrate the mesh and attach to the outside of the bag, where they grow more rapidly and evenly than mussels on conventional collectors (Plate 2).

Growth is seasonal (virtually none occurs during the winter) and averages 0.5 cm/month, so that 1 year from the time they are caught as seed, mussels average 5 cm in length and are ready for market. Some may be left on the poles for a second summer and fall of growth, but they are rarely left for a second winter. Ideally, cultured mussels should always be submerged, and those at the deep end of a field grow more rapidly than those inshore. Expansion into deeper water is restricted, however, by the greater hazard and expense of deep water culture. On the other hand, culture of mussels in the upper reaches of estuaries would expose them for too much of the tidal cycle.

Whereas mussels growing naturally on the bottom are subject to a variety of predators, off-bottom culture and the plastic sheaths just described are so effective in preventing predation that it is not considered a major problem. More serious problems in France are occasionally caused by the parasitic copepod *Mytilicola intestinalis*. More than 20 copepods per mussel may occasionally be found. Such a concentration is likely to cause death, while 10 to 20 severely retard growth. Since there is no known treatment, it is fortunate that such heavy infestations are rare. Biologists of the U.S. Bureau of Commercial Fisheries have recommended the use of Sevin and chlorinated benzenes, which are used to control various predatory crustaceans of the Pacific oyster (*Crassostrea gigas*) in Washington and Oregon, but the problem of concentration of these pesticides by such efficient filter feeders as *Mytilus edulis* would certainly offset the benefits of treatment. The only other common parasite of *Mytilus edulis* is a species of trematode which does not measurably affect production. Fouling organisms occur, as in all forms of shellfish culture, but it is rare that they seriously reduce production or necessitate great amounts of labor for their removal.

The hardiness of the mussels and the sheltered nature of the estuaries where they are cultured combine to minimize the danger of losses due to weather or climatic extremes. Mussels are at least as hardy with respect to pollution as to other environmental variables, but pollution does represent a problem to the culturist, since pollutants may be concentrated in the flesh of mussels and render them unfit for consumption. Government inspectors continually monitor French mussel grounds for pollution; areas found to be polluted are closed to mussel culture. This has become a rather serious constraint to the industry, terminating production in some areas and preventing expansion into new ones.

PRODUCTION AND ECONOMICS

As with the oyster industry, mussel grounds in France are owned by the government and leased annually for a renewable term of up to 25 years. Mussel farming is also similar to oyster culture in being largely a family business. The largest operator, located near Mont St. Michel, owns 480 rows of 170 poles each, for a total of 81,600 poles. An average farm consists of 10,000 to 25,000 poles.

No data are available as to the total area devoted to mussel culture in

France, but in 1965, 46,860 metric tons (shells included) were produced. More recent production figures are not available, but the industry appears to be holding its own. In the Mont St. Michel region, production from a single pole averages 10 kg/year (5 kg/year of meat) and the per hectare yield is about 4500 kg/year (2250 kg/year of meat).

At present the demand for mussels in France is great enough that almost all of the production is sold, in the shell, within the country. The remainder of the 80,000 metric tons consumed annually comes from other European countries, notably Spain, Belgium, Italy, and the Netherlands. This situation is likely to continue, since there is little opportunity for expansion of mussel culture in French waters. Adoption of the Spanish raft culture method (described below) might result in increased per hectare yields, but there are no suitable unexploited grounds to move into. In fact, the total area available for mussel culture in France is diminishing as a result of the increasing pollution of estuaries.

MUSSEL CULTURE IN SPAIN

HISTORY

Mussel culture in Spain is much less ancient than in France; its entire history spans little over a century, and it is only since World War II that it has become an important industry. The original center of Spanish mussel culture was on the Mediterranean coast near Barcelona, where the techniques used were the same as those practiced in France. The industry there was marginal, however, partly due to the fact that no natural setting of mussels occurs near Barcelona, and seed had to be obtained on the Atlantic coast, but primarily because growth is slow in the biologically poor waters of that region.

In 1946, the first attempt was made to grow mussels in the Galician Bays on the Atlantic coast of northwest Spain. The Galician bays are sunken valleys 7 to 11 km long, 2 to 5 km wide, and as much as 60 m deep—reminiscent of the Norwegian fjords. As the character of the environment is considerably different from the Mediterranean coast of Spain, new techniques were required for mussel culture. The solution was to borrow from the Japanese the raft method which has long been used for oyster culture with great success in the deep fjordlike embayments of Japan. Raft culture was an almost immediate success, so much so that Spain is now the world's leading producer of mussels.

ENVIRONMENTAL FACTORS

The temperature of the Galician bays ranges from 12 to 18°C at the surface and much lower in the deeper waters. The water is oceanic in character, salinity averaging 35‰ except at the very surface where heavy rainfall, particularly in winter, may result in a thin layer of nearly freshwater. Every 5 to 10 years this may be serious enough to produce some mortality of mussels within the upper 1 to 2 m, but it affects a minor fraction of the crop and does not constitute a serious threat. The freshwater flooding for some reason also does not appear to represent a danger to the natural, intertidal mussel beds upon which the culture industry is dependent for larvae and, to some extent, for seed. In fact, following periods of heavy rain, unusually heavy natural sets of mussels occur along the shore, possibly due to elimination of competitors and predators by dilution of the surface water.

The tidal range in the bays averages 4 m and is accompanied by very strong tidal currents. The bay waters are exceptionally rich in plankton and other suspended organic matter, partly because of turbulent mixing and the resulting nutrient enrichment of the water entering the bays from the sea, and partly through contributions from the land via rivers, runoff, and domestic pollution. The combination of high biological productivity, strong tidal currents, low and relatively constant temperature, and high salinity provides an ideal environment for the growth of mussels.

The bays average 30 m in depth and exceed 60 m in some places. For logistic reasons (anchoring of rafts, protection from storms, the maintenance of navigation channels) raft culture is largely confined to areas where the depth is about 10 m. The bays generally have narrow entrances and, on the whole, are very well protected from storms and strong winds (Plate 3).

TECHNIQUES

Mussels spawn in the Galician bays from early spring through late fall, but there are decided peaks in April and September. During the spring, seed is collected by lowering loosely woven and heavily tarred ropes, 12 to 25 mm in diameter, made of spart grass (a local product) or nylon, from the same rafts used for growing mussels for market. Every 30 to 45 cm, a wooden peg, about 12 mm square, is inserted between the strands of each rope to prevent the clumps of mussels from sliding down. The ropes are about 10 m long, so as to fall just short of the bottom. If they are allowed to touch the bottom, starfish take a heavy toll of the mussels.

For unknown reasons, mussel larvae do not set well on ropes in the fall, and seed must be collected from natural beds, which typically occur along the rocky shores of the numerous small islands located at the mouths of the bays. Seed mussels obtained in this manner are tied onto ropes in clumps, using a very fine, large mesh, rayon netting which disin-

PLATE 3.   Mussel culture rafts in one of the Galician Bays near Vigo, Spain.

tegrates within 24 hours of being placed in sea water. By that time, the mussels have secreted new byssuses and attached themselves to the rope. Transplantation, at this or any other time, is a critical process, due to the extreme sensitivity to light of mussels. If transplanted mussels receive too much light, they tend to migrate up or down the ropes seeking their optimum or accustomed conditions, and many fall off and are lost in the process. For this reason, transplantation is always carried out on cloudy days, and not usually during the summer.

The first mussel rafts used in Spain were old ships to which a wooden framework, bearing the ropes, was attached. Most rafts today are built on a platform of heavy wooden timbers, supported by four or more 3-m cubical floats made of wood wrapped in wire mesh and coated with concrete or, more recently, of molded fiberglass. Atop the platform is a crossed lattice of 50 × 100 mm or 50 × 150 mm sticks placed 0.3 to 0.7 m apart. The ropes are suspended from this lattice, which is supported

PLATE 4.   Close up of a large mussel raft of modern design which suspends 1,000 ropes.

so that none of the wood is exposed to sea water. An average raft of this type measures 20 × 20 m and will accommodate 500 ropes.

Within the past few years, some of the larger operators have constructed much more rugged rafts, about 700 m² in area, for use in deeper, more exposed waters (Plate 4). Such rafts, which can support 1000 ropes, are streamlined like a ship hull to withstand strong currents and may incorporate a shack to shelter workers.

In protected areas, the rafts are anchored only at one end by a single 20-ton concrete anchor, and are thus free to swing with the tide. This movement ensures a more even growth of the mussels suspended from all parts of the raft. In exposed areas, the rafts are anchored at each end.

GROWTH AND PROBLEMS

Growth of mussels in the Galician bays is minimal in summer and greatest during the winter. This seemingly paradoxical situation is probably due not to temperature effects but to the relative paucity of plankton in the stratified summer water. The result is that the spring set collected on the ropes and the fall set transplanted to the ropes are about the same size by the end of their first winter. As in France, growth is rapid and periodic thinning is necessary (Plate 5). This is done by stripping off the outer layer of mussels and binding them to new ropes in the manner described for transplantation of fall seed.

Normally mussels in Spain are left on the ropes through their second summer and harvested during their second fall, at which time they are 7.5 to 10.0 cm long. They are large enough for harvest by late spring but are in poor condition, being mostly spawned out, and do not ship well in the summer heat. Additional growth could be achieved by leaving

PLATE 5.   Mussels thinned from other ropes are bound to new ropes with thin rayon netting. Note wooden cross pieces to prevent mussels from sliding down ropes. (photos 3, 4, 5—J. H. Ryther)

them out for a second winter, but this is seldom done, since not only is fouling especially severe at that time, but many of the clumps of mussels are so large that they tend to break off of their own weight.

The risk of mortality of stock is remarkably low for an aquaculture operation. The design of the rafts and/or the sheltered nature of the culture sites minimizes losses due to weather and climatic extremes. Predation and disease are virtually nonexistent. *Mytilicola intestinalis,* which occasionally does severe damage to French mussel farmers' crops, occurs in Spain, but seldom are more than two of these parasites found on a mussel—a negligible problem. Fouling by algae, tunicates, barnacles, and so on, which may crowd out mussels on ropes, necessitate hand labor to prepare them for market, or reduce the circulation of water, is perhaps the most frequent problem but seldom exceeds nuisance proportions.

PRODUCTION AND ECONOMICS

As in France, mussel farming in Spain is principally a family business. Raft-culture areas are owned and controlled by the government and leased to the operators for a very nominal fee. The lease is for 10 years and is renewable but may be terminated by the government without notice. The operators therefore exist in a highly lucrative but singularly insecure atmosphere.

An average operation consists of two or three rafts and involves two or three people, but large operators may own up to 25 rafts. The small rafts previously described are capable of producing 60 metric tons of unshelled mussels/year. At an average of 10 rafts/ha, this amounts to a yield of 600,000 kg/ha of mussels, or 300,000 kg/ha of drained meat per year, in regions of intensive cultivation. The large rafts used in the outer bays can produce about double the amount of mussels grown on the small rafts. Since they are anchored at both ends and do not swing with the tide, they may be placed more densely than the small rafts. On the other hand, swinging with the tide has a beneficial effect on growth. The net result is that the per hectare yield of large, deep water rafts is about the same as that of small rafts.

About 94% of the 150,000 metric tons of mussels annually produced in Spain come from the five Galician bays. Half of these are canned and sold throughout Europe, 30% are consumed in Spain, and 20% are shipped alive to France. The high productivity of raft culture has caused rapid expansion in the past decade. In 1962, there were 1327 rafts in the Galician bays; in 1968 there were over 3000. The resultant increased production cut the price of mussels almost in half in those 6 years. In 1967, the government further decreased the margin of profit by requiring mussel culturists to subject all mussels to be sold alive (about half the crop) to chlorination for 24 hours. The installation or rental of the necessary equipment has added about 5% to the cost of production. Nevertheless, mussel farming in the Galician bays remains highly profitable.

The market for Spanish mussels in Spain and elsewhere in Europe is by no means saturated, but production appears to be levelling off due to the scarcity of room for expansion. Some expansion may be achieved, however, by installing more of the new large rafts in deep, exposed areas.

MUSSEL CULTURE IN THE PHILIPPINES

HISTORY AND RATIONALE

Though mussels are found virtually throughout the world, and are gathered as food over most of their range, commercial mussel culture is almost unknown outside of Europe. One might think that in southeast Asia, where a wider variety of seafoods are harvested than in Europe, mussels would be cultured for human consumption, but until the 1960s, when experimental culture began in Cambodia, the Philippines, Taiwan, and Thailand, this was not the case. The greatest success has been achieved in the Philippines, where the green mussel (*Mytilus smaragdinus*) has long been gathered for food but is not as popular as a number of other mollusks and has never been cultured.

Mussels are usually considered a fouling organism by Philippine oyster culturists, but now it appears that they have the potential to supplant oysters as the principal shellfish crop in the country. The impetus for mussel culture came about as a result of the rapid siltation in the past 15 to 20 years of the Manila Bay oyster grounds. Oystermen who attempted to collect spat in less silty offshore waters obtained poor sets of oysters but did get exceptionally heavy and almost pure sets of mussels. Upon being thus forcefully reminded of the potential of the area for mussel culture, biologists of the Philippine Fisheries Commission began in 1962 to experimentally grow green mussels at the Binakayan Demonstration Oyster Farm. The commission's intention was to demonstrate to the local oystermen the feasibility of mussel culture. At the same time they began to educate the public in ways of preparing mussels for the table. They have been sufficiently successful in both these tasks that a market for mussels has developed, and in one area near Bacoor, Cavite, a number of oyster culturists have switched to mussel farming.

NATURAL HISTORY OF MYTILUS SMARAGDINUS

The life history of the green mussel is similar to that of *Mytilus edulis* described earlier. In the Philippines, the green mussel spawns from early spring through late fall. There appears to be a peak of spawning and setting of seed mussels in April and May and again in September and October. Although it occurs within bays and estuaries in brackish water (15‰), it grows slowly in such environments. Optimal conditions, as far as is known, appear to be oceanic salinity (30‰), temperatures of 20 to 25°C, and water rich in plankton or organic matter but not in suspended sediments. However, these statements are merely inferences drawn from observations concerning the distribution and growth in culture of the mussel. Little or no basic biological research has been carried out on the species in the Philippines.

TECHNIQUES

Mussel culture in the Philippines so far differs from that in Europe in that the "seed" are young mussels which have already attached. These are removed from their natural substrate and reattached for culture. It has been found that if the byssus is cut with a sharp knife, it regenerates and the mussel is not injured, whereas pulling mussels off is usually fatal. Reattaching is accomplished by placing the young mussels in a 2-m × 1-m tray with a quantity of cultch in the form of oyster shells or bamboo stakes. The tray is suspended on poles so that it is about 0.3 m off the bottom and continually submerged.

Oyster shells with attached mussels are strung in groups of five on No. 10 wire. Strings of shells are about 1 m long, provided with loops as spacers between shells, and are suspended from bamboo platforms so positioned that the lowest shells are about 0.3 m off the bottom. About 1000 strings can be suspended from a 1-m × 10-m platform.

Bamboo stakes used as cultch consist of the whole tip of spiny bamboo (*Bambusa spinosa*). Such stakes are 5 cm in diameter at the base and 2 m long, but are placed in the tray so that only the upper half is exposed to the mussels. After attachment, the bare half of the stake is driven into the bay floor.

Whichever type of cultch is used, attachment of mussels occurs within a few hours, though some shells or stakes may take longer to fill than others. It is essential that green mussels be submerged throughout the growing period, so mussel culture is carried out farther offshore than oyster culture; usually where the water is 2 to 8 m deep at mean low tide.

POTENTIAL AND PROBLEMS

Though the platform method was the first attempted by the Fisheries Commission, the 60 or so commercial growers in the Bacoor area employ pole culture because the necessary materials can easily be procured in the vicinity. Culturists there now grow and market 3- to 8-cm long mussels within 4 to 10 months of collection. Growth to a larger size is conceivable, but 3- to 8-cm specimens are most acceptable to the Philippine consumer. Further, if the mussels are not harvested within their first year, clumps tend to fall off the poles of their own weight and perish in the mud.

The market for mussels in the Philippines is centered in Manila, but their popularity is spreading. About 8 ha of the potential culture area were in use in 1966 and produced about 2000 metric tons of unshelled mussels (1000 metric tons of meat), for a yield of 125,000 kg/ha. It is believed that hanging culture would be more productive, and annual yields well in excess of 250,000 kg/ha of meat are considered perfectly feasible.

The only serious problem thus far experienced by Philippine mussel farmers is caused by *Teredo* sp., a boring mollusk, which does not attack the mussels but may weaken bamboo platforms and cause their collapse.

A more serious problem may be encountered in the future if natural sets cannot be obtained in the culture area. Repeated removal by a growing industry of young mussels from natural beds could eventually result in depletion of breeding stocks. To ameliorate this situation, plantings of adult mussels are being made in Manila Bay and, to a limited extent, in other parts of the Philippines. It is hoped that, with plantings and the establishment of new demonstration centers, mussel culture can be spread about the country without damage to natural populations.

Outside of France, mussel culture is a fledgling industry but, as the success achieved in Spain demonstrates, potentially one of the most valuable forms of aquaculture. Not only are mussels widely distributed, hardy, prolific and easily adaptable to culture, but they may be among the most nutritious shellfish. The meat of Spanish mussels contains as much as 13% protein, 8% glycogen, and 2.4% fat. By contrast, clams and oysters normally contain 1 to 5% protein, 1 to 2% glycogen, and 0.1 to 0.6% fat.

PROSPECTUS

The major drawback of mussels as a culture organism is that since, unlike clams and oysters, they do not close their shell tightly when removed from the water, they die quickly and ship poorly. Since mussels do not command high enough prices to justify air shipment, this is particularly restrictive of the expansion of individual culture operations.

The next major advance in mussel culture may be the development of hatchery techniques, as has already been done for some other shellfish, notably oysters. *Mytilus edulis* was successfully spawned many times at the Milford, Connecticut, Biological Laboratory of the U.S. Bureau of Commercial Fisheries, but work on mussels was discontinued in favor of the more commercially valuable oysters and clams. The nerve locus causing the discharge of sex products in *M. edulis* is located in the adductor muscle, and stimulation of that muscle by any of various means was 90 to 100% effective in inducing spawning. The larvae were easily reared, even when fed only on such algae as *Chlorella,* which are completely unsuitable for oyster and clam larvae. Concurrent with any mussel hatchery program should be the investigation of the poorly understood life cycle of *Mytilus* spp.

It remains to be seen whether or not there will be sufficient economic or nutritional stimuli to provoke creation of large-scale hatchery programs for mussels, but, with or without hatcheries, mussel culture seems likely to be one of the fastest growing forms of aquaculture for some time.

REFERENCES

ANDREU, B. 1967. Biologia y parasitologia del mejillon gallege. Rev. Ciencias Madrid 30(2):107–118.

BØHLE, B. 1970. Forsøk med dyrking av blankjell (*Mytilus edulis* L.) ved overføring av yngel til nettingstrømper. (Experiments with cultivation of mussels by transplanting spat to net bags.) Fiskets Gang ar. 13–14, 2, April, 1970.

ESCRITOR, G. 1966. Tahong. Philippine Fisheries Yearbook, 1966, pp. 36–45.

FIELD, I. A. 1922. Biology and economic value of the sea mussel, *Mytilus edulis*. Bull. U.S. Bur. Fish. 38:127–259.

FRAGA, F. 1956. Variacion estacional de la composicion quimica del mejillon (*Mytilus edulis*). I. Invest. Pesquera 4:109–125.

LOOSANOFF, V. L., J. E. HANKS, and A. E. GANARES. 1957. Control of certain forms of zooplankton in mass algal cultures. Science 125:1092–1093.

LOOSANOFF, V. L., and H. C. DAVIS. 1963. Rearing of bivalve mollusks. *In* Advances in Marine Biology. Academic Press, London.

Interviews and Personal Communication

ANDREU, B. Director, Laboratorie de Investigaciones Pesqueras, Barcelona, Spain.

BORDE, J. Inspector, Institut Scientific et Technique des Pêches Maritimes, Laboratoire de St. Servan, St. Servan, France.

BORGA, P. C. Chief, Minor Sea Resources Section, Marine Fisheries Biology Division, Philippine Fisheries Commission.

ESCRITOR, G. L. Shellfish and Crustacean Section, Estuarine Fish. Division, Philippine Fisheries Commission.

FIGUEROS, A. Shellfish biologist, Laboratorie de Investigaciones Pesqueras, Vigo, Spain.

FRAGA, F. Chemist, Laboratorie de Investigaciones Pesqueras, Vigo, Spain.

LOSADA, L. Mussel grower, Villagarcia de Arosa, Spain.

MARTELL, L. Chief, Institut Scientific et Technique des Pêches Maritimes, Laboratoire d'Auray, Auray-Morbihan, France.

RABANAL, H. R. FAO, Rome, Italy.

40

Culture of Marine Gastropods, Especially Abalone

EXPERIMENTAL CULTURE OF ABALONE

When mention is made of edible gastropods, one usually thinks first of the land snails so esteemed by French gourmets, but many other marine gastropods are delicacies. In Florida and the Bahamas, various conchs are highly prized as food. Limpets and periwinkles, despite their small size, are eagerly collected by knowledgeable beachcombers. But the most valuable marine gastropods from an epicurean point of view are certainly the abalones.

There are about 100 species of abalone, distributed throughout the oceans, but the 10 or so large commercial species are for the most part confined to temperate waters. Important abalone fisheries occur only in

777

Japan, China, South Africa, New Zealand, southern Australia, Mexico, and the Pacific coast of the United States.

Abalones are slow-moving inhabitants of rocky intertidal areas and are easy prey for fishermen. Thus their numbers have been decimated in all the countries mentioned. It is somewhat surprising then, considering the scarcity of abalone and the high price they command, that their culture has been undertaken only in Japan and California.

JAPAN

In Japan, the effects of overfishing abalone were first felt in the heavily populated central and southern parts of the country. Attempts were made, with no success, to transplant seed abalone from Hokkaido and northern Honshu in the 1930s. After World War II, emphasis was shifted to hatchery propagation in the south, still with the intention of stocking to augment the fishery. Experimental work was initiated at the Tokai Regional Fisheries Research Laboratory at Nagai in 1959 and has since been taken up at several other laboratories. This account deals chiefly with work at the Tokai Laboratory and the nearby Kanagawa Prefectural Fisheries Experimental Station.

Research in all phases of culture continues to be a major activity at both of these stations, but research to date has been so successful that large numbers of young abalone are routinely grown to lengths of 1.5 to 2.0 cm and sold to fishermen's cooperative associations for stocking. Programs of hatchery rearing and stocking of invertebrates to support fisheries have usually been unsuccessful, but Japanese fishermen claim to recapture 10% of their seed 2 to 3 years after stocking, for a total annual yield of 100,000 hatchery-reared abalone, each 12 cm long or more. At this rate of return, the stocking program would be an economic success.

At least eight species of abalone are found along Japanese shores, but the most important commercially are *Haliotis discus, Haliotis diversicolor, Haliotis gigantea,* and *Haliotis sieboldi.* Of these four, the greatest success has been had in rearing *Haliotis discus,* a northern species.

H. discus normally spawns at temperatures of 15 to 20°C, during the period August to October. *H. gigantea* and *H. sieboldi* prefer the same temperature range but, being southern species, spawn later, from mid-October to December. *H. diversicolor* is believed to spawn in much warmer water during June to November in southern Japan.

In nature, spawning is preceded by a short migration into shallow water. Normally the male spawns first, and the presence of sperm stimulates the female to release her 10 million or so eggs. Fertilization occurs in the water, and the zygotes sink to the bottom, to hatch into trochophore larvae in about 13 hours at 16 to 17°C. The larvae enter the veliger stage 13 to 14 hours later then, after 6 to 11 days, sink to the bottom and metamorphose to juvenile abalone. As might be expected, development is faster in *H. diversicolor* than in the other three species. *H. diversicolor* may assume a benthic life within 43 to 46 hours of hatching. Larval abalone feed on planktonic diatoms and/or flagellates. Upon settling and developing a radula, the young switch to benthic diatoms. At a size somewhere between 40 and 100 mm, they begin to feed on macroscopic algae. From that time on, they are quite selective in feeding, which may be one reason why Japanese culturists have been rather hesitant to attempt to rear abalone to marketable size.

Hatchery stocks are comprised of adult abalone taken from nature. Abalone are captured by inserting a bar under the foot and prying them up. Considerable force often must be exerted, and in the process the foot may be injured. Bacterial infection resulting from such injury may be successfully treated with antibiotics and sulfa drugs.

Adult *H. discus* are maintained at low temperatures to inhibit spawning until it is desired. Both sexes will mature fairly rapidly if the water temperature is raised to 20°C. Maturity may easily be determined by examination of the gonads, which are green in mature females and milky white in males. The most active specimens are usually the best spawners. For purposes of breeding, males and females at a ratio of 1:4 are placed in 2 m × 2 m × 0.5 m deep outdoor concrete tanks and stimulated by thermal shock. Raising the temperature 3 to 7°C for 30 to 60 min, followed by a return to the original temperature (usually 20°C), is normally effective, but the procedure may need to be repeated. Care should be taken not to expose most species of abalone to temperatures above 28°C. If it is not convenient to heat the water in the spawning tank, a short exposure to air which is warmer than the water or an abrupt change in *p*H may be used to induce spawning.

The fertilized eggs are placed in inside hatching tanks, 2 m × 1.4 m × 1.4 m deep, at a density of about 100,000 eggs/tank. The larvae do not begin swimming immediately upon hatching but float at the surface for the first 4 to 5 hours. During this time they are collected and stocked in rearing tanks at 50 to 300/liter. The water in the rearing tanks is not changed during the free-swimming period. Swimming larvae at the Tokai and Kanagawa laboratories are not fed but nourish themselves on plankton occurring naturally in the sea water. At other laboratories, including the Oyster Research Institute at Kesennuma, swimming larvae are provided with specially cultured flagellates and diatoms.

When the young abalone are ready to settle and metamorphose, 50-cm square corrugated plastic sheets, which have previously been immersed

PLATE 1. One-year-old abalone, 3–5 cm in length cultivated at the Kanagawa Prefectural Fisheries Experimental Station, Japan. (Photograph by J. H. Ryther.)

in running sea water to provide a film of benthic diatoms, are placed vertically into the tanks. The abalone, which are about to commence feeding on benthic diatoms, settle on these collectors. After settling, the collectors, each bearing about 10,000 young abalone, are placed in lots of 1000 in 10 m × 10 m × 2 m deep outdoor tanks supplied with running sea water. Running water is not absolutely necessary but results in much better growth.

Diatoms remain the principal food until the abalone reach lengths of 2 to 3 mm (usually in about 50 days), after which time the soft brown alga *Undaria* should be fed every 2 to 3 days. *Undaria* is often scarce and finely chopped pieces of the green alga *Ulva* must be substituted. Satisfactory growth has also been obtained on an artificial diet consisting of a protein source in an insoluble calcium or sodium gel (Table 1).

The rate of growth is variable, but 1.5- to 2.0-cm seed, suitable for stocking, should be obtained in 4 to 8 months. Table 2 shows the growth of *H. discus,* fed on chopped *Ulva* and reared in running water, at the Kanagawa station.

TABLE 1. COMPOSITION OF AN ARTIFICIAL DIET FED TO EXPERIMENTALLY CULTURED ABALONE IN JAPAN

Dried *Undaria*	10%
Sodium Alginate	45
White fish meal	40
Starch	5
Vitamins	
(Crude protein	27.1%)

Survival from larvae to juveniles suitable for stocking is only about 1%. No recognized disease or parasite problems exist, and the major cause of mortality is believed to be lack of food, particularly in the period immediately after settling. Nevertheless, hatchery production of young abalone in Japan is increasing. For example, the Kanagawa station, which produced 120,000 young abalone in 1967, was expected to produce 1 million in 1970.

As far as is known, the only place in Japan where abalone are being experimentally grown to marketable size is the Oyster Research Institute. There young *H. discus* are reared to a length of about 5 mm using the methods described, then transferred to plastic wastebaskets covered with cloth netting to exclude predators, and suspended in the sea. They are fed chopped *Ulva* until they reach a length of 1 cm, at which time they are uncovered, released to the bottom and sides of the wastebaskets, and fed on fresh *Laminaria*. In this manner, *H. discus* can be grown to market size of 10 cm in 4 years. Similar techniques are being attempted with the red abalone (*Haliotis rufescens*), imported from California. It may be that the diet described could be substantially improved, as adult *H. discus* are reported to select *Undaria* from other seaweeds. *Melosira* and *Navicula*

TABLE 2. AGE AND GROWTH OF THE ABALONE *Haliotis discus,* EXPERIMENTALLY CULTURED AT THE KANAGAWA PREFECTURAL FISHERIES EXPERIMENTAL STATION, JAPAN

SHELL LENGTH (CM)	DAYS
0.42	21
0.73	40
0.88	50
1.25–1.40	100
2.30–2.50	130
3.00–5.00	1 year

PLATE 2. Plastic waste paper basket suspended in cloth bag for hanging culture of abalone, Oyster Research Institute, Kesennuma, Japan. (Photograph by J. H. Ryther.)

are said to be acceptable. Whatever is fed, the best time to feed adult abalone is 2 to 3 hours after sunset or before sunrise.

CALIFORNIA

More interest has been shown in the culture of marketable size abalone in California, for good reason. Overfishing of California abalone stocks started with the arrival of the first Chinese laborers following the 1849 gold strike. Abalone had long before become scarce in Chinese waters, and the immigrant Chinese not only took abalone for their own use, but dried great quantities for shipment back to China. New arrivals from the eastern United States took a liking to abalone as well, and California was eventually forced to severely restrict the harvest. Not only are there size limits and closed seasons, but it is illegal to dry or can abalone or ship it in any form. At present, large abalone are more abundant in California than anywhere else in the Western hemisphere, but the abalone industry is severely limited by these laws. The restrictions on processing and shipping abalone would, however, not apply to cultured abalone. Thus anyone who successfully grows large numbers of abalone to marketable size will be able to tap a new market.

The only large-scale effort to achieve these goals thus far has been going on for the past two years at Morro Bay, the principal abalone-producing area of California. California Marine Associates, the corporation which operates the Morro Bay hatchery, predicts little difficulty in producing an abundance of red abalone (*Haliotis rufescens*) seed for stocking and 2- to 3-year-olds for canning, but there is some doubt as to whether red abalone can profitably be reared to sizes suitable for "abalone steak." Even though culturists can market steaks from animals as small as 12.5 cm long (the minimum size which may be taken by fishermen is 7¾ in. or approximately 20.5 cm), it takes 4 to 5 years to grow red abalone to that size.

The most severe problem experienced in culturing red abalone is the difficulty of obtaining fertilized eggs. Males will respond to manipulation of the shell, temperature change, or almost any strong stimulus by shedding sperm, but females are reluctant to spawn. (One of the reasons California Marine Associates chose the red abalone as the species for culture may have been that females of other American abalones are even more difficult to spawn.) To date, the method of tank spawning described for Japanese abalone has been most effective. At Morro Bay, as in Japan, nutrition of the larvae is also a problem, and survival to 1 month of age and 5 mm in length of the best hatches is 10 to 20%.

Spawning and larval rearing takes place in indoor concrete tanks, and young abalone which have metamorphosed are kept inside for the first year of life, of until they are about 30 mm long, at which time they are transferred to outside reservoirs. All ages in culture are fed on kelp of a variety of species and supplied with sea water filtered through sand.

Stock on hand at Morro Bay presently consists of about 200 abalone from the first spawning in 1968, 10,000 1-year-olds, and 200,000 very young animals. The numbers of older abalone are expected to increase, since survival of the young has greatly improved since the first spawning. Further expansion will be facilitated by construction of an aerated growing tank with a capacity of over 4 million liters and a flow rate of about 4000 liters/minute. When construction is completed, the hatchery will accommodate several million abalone. Hopefully the economies of scale thus achieved, along with further improvement of techniques, will permit the commercial culture of red abalone to 12.5 to 15.0 cm.

CULTURE OF MUREX

The only other edible gastropod mentioned in the aquaculture literature is *Murex trunculus,* which is reportedly collected off the Tunisian coast and fattened in the brackish water Lake Bizerta. This seems strange, since Lake Bizerta is also used to fatten clams, mussels, and oysters. *Murex* spp. are among the most voracious predators on bivalve mollusks, and one would think it desirable to exclude them from bivalve culture areas. Nevertheless, 80 metric tons of *M. trunculus* were reportedly harvested from Lake Bizerta in 1967.

PROSPECTUS

There are undoubtedly a number of presently overlooked gastropods awaiting fishery exploitation and/or culture, but interest continues to center on abalone. Abalone has a number of advantages for commercial culture. Nearly everyone who tries abalone likes it and, since it has the reputation (well-deserved) of being a gourmet food, consumers expect to pay a high price. Abalone fanciers have remarked that the flavor of abalone steaks is actually improved by a few days in the refrigerator, which suggests that processing and shipping may be less problematical than they are with some seafoods. On the other hand, the slow growth rate is a definite drawback to intensive culture.

At present, the greatest obstacle to commercial culture, and one which might well be encountered in attempting to culture other gastropods, seems to be feeding of all life stages, but particularly juveniles. Would-be abalone farmers need to know what foods are favored in nature and, if gathering or growing these foods is not feasible, what artificial substitutes are satisfactory. Feeding methods must also be developed so that foods intended for captive abalone are accessible to them in adequate quantities. Whoever solves these problems may find himself engaged in a lucrative business.

REFERENCES

OBA, T. 1964. Studies on the propagation of an abalone *Haliotis diversicolor supertexta* Lischke. II. On the development. Bull. Jap. Soc. Sci. Fish. **30**(10):809–817.

Interviews and Personal Communication

FUJIYA, M. Japanese Fisheries Agency.

IMAI, T. Director, Oyster Research Institute, Kesennuma, Japan.

KOBAYASHI, M. Head, Propagation Division, Kanagawa Prefectural Fisheries Experimental Station, Nagai, Japan.

LEIGHTON, D. L. California Marine Associates, Cayucos, California.

SAGARA, J. Shellfish biologist, Tokai Regional Fisheries Research Lab, Nagai, Japan.

41

Culture of Squid

Squid and octopus, regarded with revulsion in some societies, are greatly appreciated as food animals in the Far East and the Mediterranean countries of Europe. Commercial suppliers of edible squid presently rely entirely on fisheries as a source, but *Sepioteuthis lessoniana* and the related but less commercially valuable cuttlefishes *Euprymna berryi, Sepia esculenta, Sepia subaculeata,* and *Sepiella maindroni* have been experimentally cultured, with considerable success, in Japan since 1960 or before.

Japanese biologists are reportedly able to propagate *Sepia subaculeata* and *Sepioteuthis lessoniana,* at least, in captivity, but have not produced a second generation in culture and find it more economic to use eggs or young squid collected from the sea as a source of stock. Eggs occur naturally in clusters but must be separated before being placed in the hatching tank, else very poor survival will result. The process of separation must be carried out very carefully and gently, as stimulation of the eggs may cause premature hatching.

At the Fukuoka Prefectural Fisheries Experimental Station in Japan, separated eggs are stocked at densities of up to $3300/m^2$ in a 195 cm $\times$ 115 cm $\times$ 60 cm deep tank supplied with sea water from a shower nozzle at 6 to 7 liters/min. Eggs in the tank are held in single layers in plastic mesh baskets.

It is best to keep squid eggs in darkness during incubation, to prevent

786

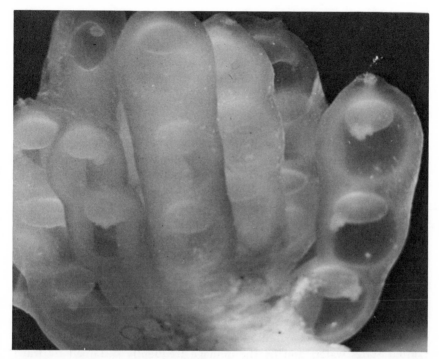

PLATE 1. A cluster of squid eggs (*Sepioteuthis sepioidea*). (Courtesy Rosenstiel School of Marine and Atmospheric Science, University of Miami.)

diatoms or green algae from growing on the surface and interfering with the eggs' development. If this precaution is observed, and if a good supply of fresh sea water is maintained, a hatching rate of about 95% may be achieved.

Young wild squid used for experimental culture are caught in the western part of the Inland Sea during July to September, at which time they weigh 20 to 80 g. Attempts have been made to rear small squid captured by trawlers, but the extremely high rate of injury and consequent mortality rendered this method impractical. Small set nets are now used, but even so, 20 to 30% of the catch must be discarded.

Larval and young squid are best maintained in glass tanks supplied with flowing sea water saturated with dissolved oxygen. Larvae commence to feed 16 to 48 hours after hatching, and for the next 40 days small live crustacea, particularly *Mysis*, are the best food. The initial stocking rate is about 1 squid/5 cm^2; after about 20 days this is reduced to 1 squid/10 to 15 cm^2. Under good conditions, the young will attain mantle lengths of 20 to 40 mm in 30 to 40 days, or somewhat longer in the case of *Euprymna berryi.* Japanese culturists emphasize the importance of proper feeding during this stage but supply few details, except that live food is essential to achieve the maximum survival rate of about 80%.

The 20- to 40-mm squid are stocked, at 30 to 50 g/m², in large tanks or ponds, and fed on shrimp or pieces of fish. Fish, apparently the better of the two foods, is converted at 2 to 4:1. The size of the food particles is determined by the average size of the squid. Ideally, each squid should receive a chunk of food weighing 8 to 10% of its own weight twice daily. Much remains to be learned about optimum stocking densities, feeding, and so on, but even now it is possible, starting with a 4-g animal, to produce marketable near-adult squid weighing 0.5 to 0.7 kg each within five months. Apparently, results such as just described cannot be obtained consistently, because, despite the fact that squid is the fifth most valuable fishery product in Japan, it has not been commercially cultured in that country or elsewhere.

In the United States, squid is usually consumed only by certain ethnic groups, but even so there is a sizable demand for the imported, canned product. An added demand is generated by the expanding field of neurological research. Neurobiologists use squid for many laboratory studies because of their giant nerve fibers—the largest found in any animal. To date supplies have been limited by the seasonal occurrence of squid in United States waters and by the damage frequently sustained by trawl-caught squid.

Recent work at the University of Miami's School of Marine and Atmospheric Sciences has made it possible to culture squid for research purposes, though rearing large numbers to edible size is not presently feasible. The cultured squid, which are hatched from eggs in the laboratory, have the added advantage of being adapted to aquarium life. Wild squid, placed in aquaria, frequently injure themselves by swimming into the tank walls.

The species of squid most successfully cultured at Miami is *Sepioteuthis sepioidea,* which has been reared from egg to maturity in less than 5 months. Hatching and subsequent culture are carried out in well-aerated, slightly agitated water of at least 30‰ salinity and 20 to 30°C in temperature. Once the eggs have hatched, the excitable nature of squid must be taken into account if maximum survival is to be realized. Among the precautions taken are the use of tanks made of opaque, textured materials; provision of natural-appearing substrates incorporating gravel, rocks, and artificial plants; and ultraviolet lighting. Even when these measures are taken, the tanks must be covered to prevent the squid from jumping out.

The most crucial factor in culture of *S. sepioidea* is feeding; the most frequent cause of death in young squid at Miami is starvation. While *S. sepioidea* will consume up to 180% of their body weight daily, they are very selective feeders. The best food for young has been found to be the mysid shrimp *Mysidium columbiae*. Older specimens are fed 3- to 7-cm fish, supplemented by penaeid shrimp. A daily ration of about 45% of the body weight of squid resulted in conversion ratios of 5 to 10:1. Maximum size attained in a 146-day growing period was 105 mm in length and 77 g in weight.

In addition to being considered a delicacy in countries where they are eaten, squid are an excellent source of protein. From a food processor's point of view, they are virtually a perfect animal, consisting of close to 100% edible matter. It is surprising, then, that commercial culture has not been at least attempted. Feed costs, the difficulty of rearing the larvae and possible shortages of eggs in nature would appear to be the principal restraints to commercial culture. The former, it would seem, could be substantially reduced by economies of scale in large operations, at least in Europe and Asia, if not in the United States.

The propagation and culture of octopus, while having been attempted in several laboratories and having met with some preliminary success, particularly in Japan, has not yet been sufficiently advanced to warrant treatment here.

REFERENCES

CHOE, S., and Y. OHSHIMA. 1963. Rearing of cuttlefishes and squids. Nature **4864**:307.

LaRoe, E. T. 1970. Squid raised to adult size at University of Miami. Am. Fish Farmer 1(7):16–17.

Interviews and Personal Communication

LaRoe, E. T. Institute of Marine and Atmospheric Sciences, University of Miami, Miami, Florida.

42

Culture of Seaweeds

Many of the numerous species of large marine algae, popularly known as seaweeds, are high in nutritional value, but, with a few minor exceptions, their use as human food is restricted to the Orient. They find wider use in animal feed, as fertilizer, as food additives, and in a number of industrial processes. Although seaweeds are harvested throughout the world, their culture is almost entirely confined to the Orient and reaches its peak of sophistication in Japan and China.

RED ALGAE

CULTURE OF *Porphyra* SPP. IN JAPAN

The most important algae commercially are members of the class Rhodophyceae, the red algae. Several red algae are eaten by man; by far the

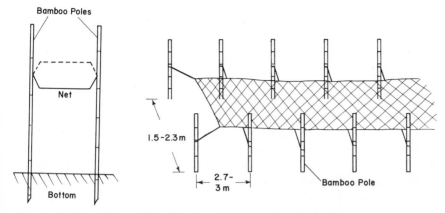

FIG. 1.   Rack used in growing *Porphyra* in Japan. (After Fujiya.)

most important of these are members of the genus *Porphyra*, known as "laver" in European and American countries and as "nori" in Japan, where it is the single most commercially valuable marine product.

Culture of nori reportedly began in the seventeenth century in Tokyo Bay. The original culture method, now obsolete and almost extinct in Japan, but still practiced in Korea, involves placing bundles of leafless branches of bamboo, oak, or other trees at or just above the mean water level in areas located well away from brackish water during September to October. Within 2 to 4 weeks the nonmotile monospores of *Porphyra* settle on the branches and develop into thalli (the portion of the plant which is eaten) about 1 cm in diameter.

The branches and their attached vegetation are then moved inshore, preferably to an area such as around the mouth of a river, where high concentrations of dissolved nutrients occur. There the thalli grow and are periodically harvested by plucking or cutting throughout the winter.

In Japan, man-made structures, usually nets, have supplanted branches as monospore collectors. The nets are made of palm fiber or synthetic twine 3 to 5 mm in diameter, with 15-cm openings, and suspended from poles driven into the bottom in such a manner that the flat surface of the net is parallel to the water surface (Fig. 1). Such nets are generally 1.2 m wide and vary from 18 to 45 m long. In place of nets, "hibi," made of parallel bamboo poles connected with rope at intervals of 10 to 15 cm, may be used, or nets may be suspended perpendicular to the water surface.

Whatever collecting device is used, catches of monospores are best when the water temperature is 22 to 23°C. After a storm is considered to be a particularly good time, as are the second to fourth days after the first or fifteenth of each lunar month.

Thalli can survive in temperatures of 3 to 20°C but the best growth occurs at 5 to 10°C. By April, when the water begins to warm, the thalli start to die off, and by June they have usually disappeared. In addition to temperature, a fungus disease, caused principally by *Pythicum* sp., may be lethal to *Porphyra* thalli, particularly at temperatures above 10°C. When prevalent, it may wipe out 50% or more of the nori crop. Fungus may be at least partially prevented by exposing the algae to air during part of the tidal cycle. On the other hand, too much exposure may reduce growth and toughen the thallus, making it unsuitable for food. The most satisfactory compromise is to expose the algae for approximately 4 hours/day. Most important is the occasional regulation of the level of the beds, since the best time and duration for emergence changes with the season and developmental stage of the alga. Also avoided is prolonged immersion at the very sea surface, since the surface layer in the estuarine areas may be almost pure freshwater, which is harmful if not lethal to the alga. Regulating the level of the bed is also important in controlling a green alga which sometimes fouls the nets but grows only in a narrow band of the water column.

Though the cultivation techniques just described have been practiced for several centuries, it was not until 1949 that nori growers or anyone else knew where *Porphyra* monospores come from, or what becomes of *Porphyra* during the summer. Then the English botanist K. M. Drew discovered the missing links in the life cycle of the British laver *Porphyra umbilacilis*. Not surprisingly, her description was subsequently found to apply to the Japanese species, including the five commonly cultivated species (*Porphyra angusta, P. kuniedai, P. pseudolinealis, P. tenera*, and *P. yezoensis*). As can be gathered from the preceding description of primitive methods of culture, active growth of the large, leafy *Porphyra* thallus occurs during the period from November to early April. Beginning in April the thalli become gradually smaller in size and eventually disappear by the end of July. In the meantime, beginning in late fall, some plants develop carpogonia, in the form of little-differentiated cells of the thallus, while others produce spermatia. Sexual fusion between the spermatium and the contents of the carpogonium then occurs, and the latter divides to form four carpospores. The liberated carpospores then drift and sink in the sea to germinate when they settle onto mollusk shells. The germinating spores now give rise to the so-called conchocelis phase, originally described as a separate shell-boring alga, *Conchocelis rosa*. The conchocelis continues to grow as a dark red incrustation on the shell through the spring and summer, releasing monospores in early autumn which, on settlement, grow into thalli. Some species of *Porphyra*, including all the cultivated species, carry out accessory reproduction by means of asexual spores released by young plants.

Knowledge of the complete life history of *Porphyra,* and particularly discovery of the conchocelis stage, has led to refinements of culture techniques which have revolutionized the nori industry. Today about 70% of the commercial production of nori in Japan is based on artificial production of monospores by various prefectural and municipal laboratories. Artificial seeding is carried out by placing oyster shells, either loose or strung on wires, in indoor concrete tanks in the early spring and adding chopped thalli. The carpogonia and spermatia thus liberated fuse to form carpospores which attach to the shells and burrow into the pearl layer, to become the conchocelis stage. *Porphyra tenera* is most commonly cultured in this manner, but other species or a mixture may be used.

The conchocelis plants are then cultured in the tanks until fall. The most important factor in growing the conchocelis stage is temperature, which should be kept above 25°C to prevent premature liberation of monospores. The maximum temperature permitted is 29°C. In many cases, both artificial heating and cooling of the culture tanks is required. Illumination is kept bright but not as strong as direct sunlight until the plants reach visible size: after that, it is reduced to about 500 lux. Salinity also varies according to the age of the plant; early and late in the culture period it is maintained at 30 to 31‰, but the rest of the time a salinity of 20‰ or slightly less is preferred to discourage disease.

The space requirements of conchocelis culture are modest. A tank 2.4 m × 1.8 m × 0.9 m deep will accommodate 250 strings of 10 oyster shells each, or enough to supply 125 of the small 18-m collecting nets.

Monospore collection may be carried out in the sea by placing conchocelis-bearing shells beneath the collecting nets. More often, however, the monospores are attached to the nets in the laboratory. In the fall, when the sea temperature is low enough to permit nori culture to begin, the water in the culture tanks is allowed to fall below 23°C and the monospores are liberated. Liberation can be accelerated by suddenly lowering the water temperature to 17 to 20°C. The yield of monospores may be augmented by reducing the daily photoperiod to 8 to 10 hours.

At the start of the nori culture season, growers bring their own nets to the laboratory and, for a small fee, are allowed to immerse them in the tanks for a few minutes to several hours, depending on the type of collecting apparatus used. The four types of collector employed are illustrated in Fig. 2; all have the purpose of keeping the nets and/or monospores in motion so that contact is maximized.

Development of artificial methods of monospore production has eliminated the uncertainty of obtaining a good set of spores in nature, which depends upon a variety of uncontrollable and largely unknown environmental factors. It also ensures a uniformly heavy set of spores on the nets.

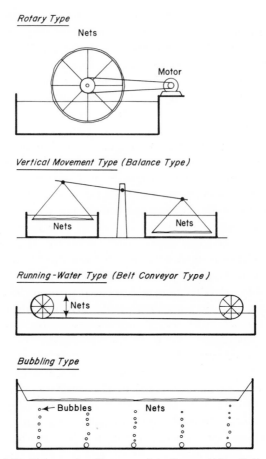

Rotary Type

Nets

Motor

Vertical Movement Type (Balance Type)

Nets   Nets

Running-Water Type (Belt Conveyor Type)

Nets

Bubbling Type

Bubbles   Nets

FIG. 2.   Types of *Porphyra* seed collector used in Japan. (After Fujiya.)

Finally, it serves one other very important function. For reasons un-known, the monospores are released in nature at a lower temperature than in the culture tanks. This natural spawning occurs in many of the southern areas of Japan, such as the Inland Sea, as late as mid-October, permitting the first harvest of nori in January. However, the tempera-ture of these same waters is usually low enough by early September to permit growth of the alga. Artificial spawning of the monospores can be achieved in early September or before, by manipulating the water temperature and the photoperiod, thus permitting the first harvest by early November and a considerably greater annual crop.

Still another method of artificial seeding takes advantage of the acces-

PLATE 1. *Porphyra* (Nori) culture in the Inland Sea of Japan. (Courtesy Japan Inland Sea Fisheries Laboratory, Hiroshima, Japan.)

sory reproduction of *Porphyra*. This is accomplished in the sea, by placing a second net over the main collecting net to catch the asexual spores given off by the young thalli.

Growing nori for market proceeds essentially as in older systems of culture (see above). Development of artificial seeding has, however, permitted one important advance in this phase of culture. Ordinarily, cultivated *Porphyra* become 10 to 15 cm long in one tidal phase (about 15 days), at which time they must be harvested, or the long leaves will be torn off by the waves. The plants on a single net may be thus harvested three or four times, after which the net should be replaced. Formerly, replacement was not always possible, particularly late in the growing season. Now, however, after spore collection, whole nets may be frozen at -20 to $-24°C$. Spores so frozen remain viable for several months.

After harvest, nori is washed in freshwater and cut into small pieces (10 to 15 mm square). About 4 kg of these pieces are placed in a 100-liter barrel of freshwater and stirred. The resulting suspension is spread within a wooden frame 17 to 21 cm on a side, resting on a mat made of fine stems of bamboo. The sheets of material thus formed, which weigh about 2.5 to 3.0 g each, are dried in the sun, scraped off the mat, and marketed as "hoshinori." Often the entire process of chopping, forming the sheets, and drying is done mechanically, either commercially or on a cost basis, by a prefectural or municipal organization.

One 18-m culture net produces 1000 to 3000 (average 2,500) sheets, or 35 to 105 kg of hoshinori annually. If it is assumed that one-fourth of an area of intensive cultivation is covered by nori net, this amounts to 100 nets/ha, producing a total of 750 kg during the 6- to 8-month growing season.

Since 1960, the annual production of nori for all of Japan has been relatively stable at about 120,000 metric tons (4×10^9 sheets)/year, with the exception of 1963 when unusually high temperatures and disease reduced the crop to about half that figure. Prior to 1960, the annual production was more irregular but averaged about half the value since that time, the difference being attributed to the introduction and immediate adoption of the artificial spore collection technique.

Nori brings a very high price, but the capital investment is small and, once the nets are mounted in position, the labor involved is relatively light. Clearly, the 70,000 or so nori growers find the industry extremely profitable.

How much nori contributes to the nutrition of the Japanese people is another matter. Hoshinori is usually eaten in small quantities or used as a condiment. It is estimated that the per capita consumption in Japan is only 50 g/year. It is, however, highly nutritious and digestible. Protein comprises 30 to 50% of the dry weight of nori, which makes it superior to rice and comparable to beef in that regard. It is also a good source of several vitamins (Table 1).

The supply of nori in Japan seldom equals the demand. Despite this fact and the relative ease with which *Porphyra* may be cultured, expansion of the industry does not seem likely. The principal reason is that most of the shallow areas suitable for nori growing are already under cultivation. And, unfortunately, as Japan grows and industrializes, many

TABLE 1.   NUTRIENT COMPOSITION OF NORI, A FOOD PREPARED FROM THE RED ALGA *Porphyra* IN JAPAN (CONTENTS PER 100 G OF ALGAL SHEETS)

WATER	PROTEIN	FAT	CARBO-HYDRATE	ASH
11.4 g	35.6 g	0.7 g	44.3 g	8.0 g
PROVITAMIN A	VITAMIN B_1	VITAMIN B_2	NIACIN	VITAMIN C
44,500 IU	0.25 mg	1.24 mg	10.0 mg	20 mg

of these areas are becoming severely polluted. Since it does not seem likely that per hectare yields can be appreciably increased, the prospectus is for a slow decline of the industry unless steps are taken to halt inshore pollution. A possible if less esthetic alternative would be completely artificial culture in tanks or ponds.

Outside of Japan, *Porphyra* is cultivated only in Korea, but the Chinese and Taiwanese people consume considerable amounts, and supplies are short throughout East Asia. Accordingly, Japanese interests are exploring the coasts of New Zealand and South America for stocks of *Porphyra,* while the Taiwanese are similarly engaged off California. If suitable areas are found, the first step in exploitation will probably be harvesting of wild stocks, but culture may eventually develop in these areas.

CULTURE OF OTHER RED ALGAE IN JAPAN

A number of other red algae are cultured in Japan, although crude methods, not approaching the sophistication of nori culture, are used. The most important edible alga so cultured is funori (*Gloiopeltis*), which is also used in the manufacture of glue. Production of funori has been increased by simply placing boulders or chunks of concrete in the sea to provide a substrate for settlement in otherwise suitable locations.

GREEN ALGAE

The only green algae commonly used as food by man are *Enteromorpha* and *Monostroma* (aonori). Both are sometimes encouraged to grow together with *Porphyra*. *Monostroma* commands the highest price of any seaweed in Japan. In addition to being grown as a byproduct of *Porphyra* culture, *Monostroma* is sometimes the object of monoculture. Details of the culture method are not available, but it is similar to *Porphyra* culture in that the crop is grown on nets suspended horizontally at intertidal levels near river mouths. Usually three crops are harvested from a net

TABLE 2. PRODUCTION OF THE GREEN ALGA *Monostroma* FROM ONE 36-M × 1.5-M NET IN DIFFERENT LENGTHS OF TIME

CROP	DAYS TO PRODUCE	WET WEIGHT PRODUCED (KG)
1	100	10
2	45	10
3	30	6

before it is replaced. Table 2 shows the production of *Monostroma* on one 36-m × 1.5-m net.

Before powdering, aonori is made into sheets like hoshinori. Between 1955 and 1960, an average of 254,701,000 sheets of aonori were produced annually in Japan, as compared to over 2 billion sheets of hoshinori.

BROWN ALGAE

Several brown algae (class Phaeophyceae), mostly the large species collectively known as kelp, are of value to man in the same ways as red algae, and a few of them are cultured. In Japan, the most important are wakame (*Undaria pinnatifida*) and konbu (*Laminaria* spp.). Of the two, wakame is by far the more intensively cultivated, having undergone a revolution in culture methods at about the same time as *Porphyra*.

CULTURE OF *Undaria*

Unlike *Porphyra, Undaria pinnatifida* is a cold water alga, occurring naturally in northern Japan. Until the late 1950s, methods of wakame culture were extremely primitive, consisting of no more than anchoring plastic floats at suitable depths for attachment of young plants. About 1956, however, a technique of artificial spore production was worked out at Onagawa, in the Sendai region of northern Honshu. The practice spread rapidly to all parts of Japan suitable for growing *Undaria,* and today the majority of wakame sold comes from cultured plants.

The life history of *U. pinnatifida,* as of other brown algae, is rather complex and involves an alternation of generations between sexual and asexual forms. The asexual form, or sporophyte, is the obvious macroscopic plant of which the fronds of the main thallus, or body, are used as food. The sporophyte phase grows during the winter months when the temperature is between 10 and 15°C.

At the base of the sporophyte is an area of specialized tissue known as the sporophyll. During the winter months, this area develops asexual zoospores which are released in the spring and early summer when the water temperature rises above 14°C. After a brief planktonic life, the zoospores settle and adhere to a solid surface (stones, shells, etc.), there germinating to produce the microscopic sexual plant, the gametophyte. Germination of the zoospore and growth of the gametophyte occurs between temperatures of 15 and 20°C.

Gametophytes develop male and female sexual organs during the summer months and in September the sperm are shed and fertilize the egg within the female sex organ or oogonium. The fertilized egg then develops into the young sporophyte at the same point of attachment as that of the gametophyte.

The optimal and extreme temperatures and salinities differ at these different stages of the life cycle, so it is difficult to generalize. For the whole life cycle a temperature range of 10 to 20°C is suitable, though both higher and lower temperatures may be tolerated at certain stages. Salinity tolerances of *Undaria* are rather narrow, neither gametophytes nor sporophytes developing below a salinity of 27‰ with best growth occurring between about 30 and 33‰. Thus the alga is restricted from estuaries and other brackish-water regions and typically occurs in open sea situations.

In culture, the mature sporophyte plants are brought into the laboratory and placed in concrete or plastic tanks. Into these tanks are placed rectangular plastic, wooden, or metal frames roughly 1 m × 0.3 m in size, around which is wrapped 90 cm of approximately 1.6-mm diameter braided cotton string. The tanks are kept in a cool, shaded area, but require no special attention.

The zoospores are released from the plants and attach to the string in late spring or early summer, the gametophytes and ultimately the young sporophytes developing on the string during the summer and early fall. This process may be carried out in the central to southern parts of Japan where the summer water temperature exceeds 25°C and precludes the natural distribution of the alga. All that is needed is a cool, shaded area to hold the gametophytes at or below 23°C over the warmest part of the summer.

In autumn, when the temperature falls below 20°C in the sea, the string-frames are taken from the tanks and the string containing the young algal plants is cut into short pieces and placed in the ocean. This is done in a variety of ways, either from bamboo rafts or from buoyed long lines. In either case, loops of tarred 6- to 12-mm rice rope are suspended from the bamboo rafts or long-line floats. To these, secondary weighted dropper lines are attached, and short, 5.0- to 7.5-cm pieces of the algae-bearing string are inserted and tied between the strands of the larger line at intervals of about 15 cm. The length of string used depends upon the density of the sporophytes, an attempt being made to include about 10 *Undaria* plants on each piece of string.

The alga grows rapidly under these conditions, young sporophytes put out in September being ready to harvest by midwinter. Harvesting is accomplished by either cutting pieces off the plants or by removing the whole plant, this depending upon the rate of growth, the total supply, and the market price.

PLATE 2. Wooden frame with strings for collection of *Undaria* zoospores. A good set may result in over 1200 *Undaria* plants from this one frame. (Photograph by J. H. Ryther.)

Culture of *Undaria* may be done at any depth down to about 6 m depending upon the clarity of the water. Since relatively high salinities and low temperatures are required, the most favorable locations are in fairly open areas. As in the case of oyster culture, the long lines are able to withstand rather rigorous conditions with respect to winds, waves, and currents.

There is a disease problem in the culture of *Undaria* which causes detachment of the young fronds and can result in a serious reduction of the crop. At present the origin and nature of the disease is not known. Fouling organisms, chiefly barnacles, setting on the algae cause some difficulties but do not result in any appreciable losses.

Artificial collection of *Undaria* spores is done both by private growers, for their own use, or as a service by municipal or prefectural laboratories.

A small and very simple operation belonging to one fisherman, which was observed at Nagai (south of Tokyo), consisted of no more than a 1000-1 canvas pool set in the woods and containing 20 plastic frames, each with 100 m of string. Assuming a good set of sporophytes (10 plants every 10 cm of string) as many as 250,000 *Undaria* plants could be produced from this one operation. At the other extreme, at the Kanagawa Prefectural Station, there are several large (about 1.8 m × 3.7 m × 0.9 m) concrete tanks devoted to *Undaria* culture. These handle a total of 500 frames or 50,000 m of collecting string. The sporophyte-bearing string is sold to culturists through the local cooperative association.

One bamboo raft, 36.6 m × 1.8 m in size with hanging ropes as described can produce on the average 1 ton wet weight or about 112.5 kg dry weight of *Undaria*. On the Ojika Peninsula there are 1333 such rafts in operation, which annually produce a total of 145 metric tons dry weight of the alga.

If it is assumed that 0.1 ha under intensive cultivation is covered by rafts, the resulting 6.8 rafts/ha produce 0.8 tons (dry weight) of *Undaria*.

Undaria production for all of Japan has been highly variable from year to year but there has been an upward trend from about 40,000 tons (wet weight) or less in the early 1950s to an average of about 60,000 tons since 1960. Most of the increase may be attributed to the introduction in 1960 of the technique of artificial spore collection.

Because *Undaria* grows in much the same type of environment as oysters, there is some conflict between the two industries. Particularly in the more northern parts of Japan, where oyster production is marginal and requires two years, there is some tendency for the fishermen to shift from oyster to *Undaria* culture at the expense of an expansion of the oyster industry.

Preparation of wakame for market is simple, involving no more than drying the fresh plant. The dried product is chopped into small pieces and consumed raw as a vegetable, salad green, or garnish.

CULTURE OF *Laminaria*

Methods of *Laminaria* cultivation in Japan are primitive, much like those used for red algae other than *Porphyra* (see above). In the past, a substantial amount of the *Laminaria* grown in Japan, as well as part of the Korean harvest, was shipped to China, where it is in great demand as a food and a source of iodine. Only a few of the many species of the order Laminariales are edible, and none of these is native to Chinese waters. In 1927, however, one of the edible species, *Laminaria japonica,* was found growing off the coast of Dairen in northern China. It is theorized that zoospores attached to lumber rafts and to the bottom of cargo boats used to tow them from northern Japan, and settled off Dairen.

For the following 20 years, *L. japonica* was harvested, but until 1943 cultivation was confined to the practice of throwing stones bearing spores or young sporophytes into the water. Attempts at more intensive culture, both on rafts and on the bottom, were made by the Japanese during the last three years of their occupation of China, but with little success. After the end of World War II, their efforts were continued and, following the Communist revolution, greatly accelerated. The first success in experimental culture was achieved in 1949. The results saw immediate practical application, with remarkable results. In 1959, total production of *L. japonica* was 387 times the peak production in pre-Communist years, and, although reliable data are lacking, production has apparently increased further in the past decade.

As is the case with any cultivated organism, a clear understanding of the life history of *L. japonica* was necessary before truly efficient culture methods could be developed. The gametophyte generation of this biennial species lasts only about two weeks, while the rest of the life cycle is spent as a large, edible, sporophyte, known as "haidai" in China.

Asexual reproduction involves the formation of sori on mature sporophytes. Reduction division occurs within the sori, resulting in the release of zoospores which swim about for about two hours before attaching and germinating into male and female gametophytes. Gametophytes eventually release sperm and eggs which combine to form sporophytes and complete the cycle.

The basic technique involved in artificial propagation of *L. japonica* consists of collecting zoospores from sporophytes in late autumn and lodging them on short "ladders" made of bamboo splints and hung from floating rafts. Young sporophytes appear on the ladders by January and are removed for use in one of the forms of culture to be described.

Research on the ecology of *Laminaria* has disclosed a considerable amount of valuable information about the reproductive cycle, particularly with respect to temperature and light. Under natural conditions, production of new sori and release of zoospores by existing sori are discontinued at temperatures above 20°C. The duration of motility of zoospores is also affected by temperature, ranging from a scant few minutes or less at 20°C to more than 48 hours at 5°C. The lowest temperatures are not necessarily optimal, however, since germination of spores and formation of gametophytes proceeds best at 10 to 20°C. The fruit of these findings has been the maturation of sporophytes, formation of sori, and collection of zoospores indoors at controlled temperatures, usually 10 to 15°C.

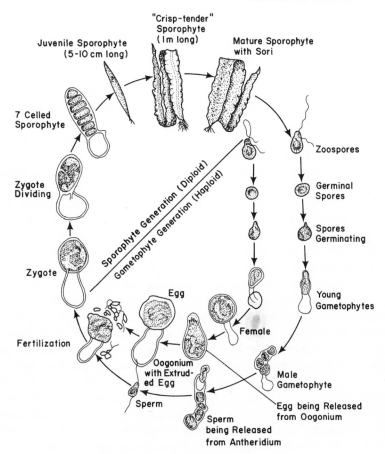

FIG. 3. Life history of *Laminaria japonica*. (After Cheng, 1969.)

Formation of sori has been found to occur only on sporophytes growing at depths of 0.66 m or less. However, this does not appear to be related to light conditions, for production of sori proceeds normally in the turbid waters of the Choushan Archipelago in Chekiang Province, where visibility is limited to about 10 cm. Attachment and germination of zoospores also appears to be independent of light; these processes have been shown to occur under conditions ranging from darkness to an illumination of 4000 m-candles. Spore collection is usually carried out in the shade, simply to avoid high temperatures in the surface waters.

An especially fast and efficient method of zoospore collection is known as the "dark dry" technique. Sori-bearing sporophytes are placed in open baskets, without water, and covered with cotton cloth or rush mats. The baskets are hung in a dark room for 3 to 6 hours at a constant temperature below 20°C, then placed in sea water. Absorption of water by the partially dehydrated sporangia ruptures their coverings, and the zoospores are discharged within a few minutes.

Sexual reproduction by the gametophyte generation is affected by light conditions. The intensity of light is not as important as adequate exposure to darkness. Release of eggs and sperm proceeds satisfactorily under conditions of periodic or continual darkness. Darkness is necessary for fertilization, since sperm released in the light may turn away from eggs or lose their motility.

Since *L. japonica* is a green plant, prolonged periods of total darkness would of course be lethal to sporophytes. Illumination has also been shown to be desirable for gametophytes; thus, in indoor propagation facilities, periods of illumination at 500 to 4000 m-candles should be alternated with periods of darkness.

Sporophyte growing practices, as well as artificial propagation, have been enhanced by studies of the ecology of *L. japonica*. Illumination and temperature are important, as might be expected, but the chief determining factor in the success or failure of *Laminaria* culture is the supply of nitrogen. Chinese algologists observed that, whereas *L. japonica* flourished naturally in the waters around Dairen, which are continuously enriched by sewage, it seldom reached marketable size in the waters of other parts of northern China, which generally do not contain more than 5 mg/m³ of nitrate nitrogen. It was determined that young sporophytes, 1 to 2 m in length, require 6 mg of nitrogen daily if their growth is not to be stunted. Accordingly, fertilization with such highly soluble and nitrogenous compounds as ammonium nitrate, ammonium sulfate, and sodium nitrate has become an integral part of current culture methods.

Studies of the light requirements of *L. japonica* sporophytes have disclosed that, while the best growth appears to occur 1.5 m or more below the surface, it is in fact to the culturist's advantage to keep his plants nearer the surface. *L. japonica* grown near the surface do not compare favorably with plants grown in deeper waters in terms of length and fresh weight but exceed them in dry weight by 10 to 20%. In quiet water at 10°C, young sporophytes less than 2 cm in length grow fastest at illuminations of 1000 to 2000 m-candles; larger plants require less illumination.

L. japonica is native to subarctic waters, and though it grows well in the relatively cold waters of northern China, high temperatures were long thought to be a severe limiting factor in the spread of its cultivation. Recently, however, it has been shown that the annual exposure to near-freezing temperatures experienced by *Laminaria* in Japan is not necessary for successful culture, and that temperatures as high as 21.5°C are suit-

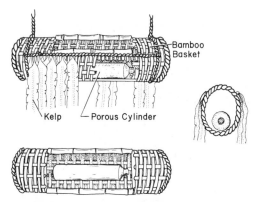

FIG. 4. Three views of bamboo basket with cylinder of fertilizer used in culture of kelp in China. (After Cheng, 1969.)

able. Again, dry weight is a better measure of the effectiveness of culture than length or fresh weight. In large plants, the latter two cease to increase above 13°C (plants less than 2.5 m long continue to increase in length at temperatures of up to 20°C), whereas dry weight continues to increase, reaching a peak growth rate somewhere between 13 and 20°C.

Discovery of the temperature requirements of *L. japonica,* along with the development by algologists at the Institute of Oceanology, Academia Sinica, of a new, more temperature-tolerant strain of the plant, called "Hai-Ching #1," have made it possible to extend *Laminaria* culture as far south as the subtropical waters of Kwangtung province.

As mentioned, the earliest attempts at culture of *Laminaria* sporophytes involved growing the plants on the sea floor. Bottom culture is still employed in China as well as in Japan and Korea, but most of the tremendous increase in the Chinese production of *L. japonica* has come as the result of various forms of raft culture. Rafts, which are generally anchored to stakes driven into the bottom in water about 10 m deep, are of three types:

1. The basket raft consists of five or more cylindrical bamboo baskets about 1 m long and 17 cm in diameter, tied together in a row. Each basket contains a porous, tightly stoppered, earthenware cylinder which holds 1.5 kg of fertilizer. An opening on the upper side of the basket facilitates removal of the fertilizer cylinder, which must be replenished monthly. A rope is tied along each side of the basket, and young sporophytes attached to it by inserting the basal end of the stipe between the strands of rope (Fig. 4). As the liquid fertilizer seeps through the cylinder, the nutrients are absorbed by the sporophytes.

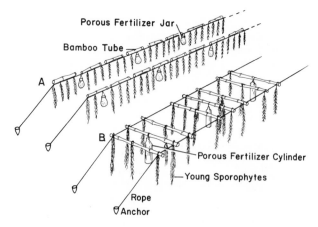

A Single-line Bamboo Rafts
B Double-line Bamboo Rafts

FIG. 5. Bamboo tube rafts used for kelp culture in China. (After Cheng, 1969.)

2. The single line tube raft is constructed by tying bamboo or rubber tubes end to end to form a single line perhaps 60 m long. About every 6 m along the raft, a 2.8-liter jar of porous earthenware containing fertilizer is suspended to a depth of 1.3 m. Young sporophytes attached to ropes, as described for the basket raft, are similarly attached at intervals of about 50 cm. The distance between plants on a rope varies between 1.7 and 3.0 cm, and the length of the ropes may be 1 to 3 m. Rafts are usually placed 3 to 4 m apart. The precise spacing of plants, ropes, and rafts depends on water transparency, temperature, currents, and custom. Normal densities of *L. japonica* sporophytes cultured on single line tube rafts are from 72,000 to 134,000/ha.

3. The double line tube raft is a ladderlike structure composed of 1-m-long bamboo tubes tied across two 12- to 15-m ropes at intervals of about 1 m. The tubes are arranged in groups of three, with a porous 2.8-liter earthenware cylinder of fertilizer suspended from the middle tube of each group. Ropes bearing young sporophytes are attached to the tubes in the same manner described for the other types of raft. Up to four double line tube rafts may be linked, 5 m apart, to form a large unit. Fig. 5 illustrates the two types of tube raft.

Productivity is generally reckoned in terms of the amount of algae produced per kilogram of fertilizer. Although the basket raft requires somewhat less labor to operate, and may produce higher quality *Laminaria*, it yields only about 1 kg of algae per kilogram of fertilizer, whereas the

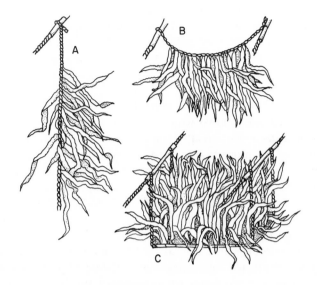

A   Perpendicular
B   Parallel
C   Upright

FIG. 6.   Arrangements of kelp suspended from culture rafts in China. (After Cheng, 1969.)

single line and double line tube rafts are capable of producing 3.75 kg and 3 kg of algae, respectively, per kilogram of fertilizer. Accordingly, the single line tube raft is most popular.

The conventional manner of attachment of ropes bearing sporophytes to tube rafts is perpendicular to the water surface, but parallel and upright arrangements have been found to be preferable. The upright arrangement, which involves fastening the ropes parallel to each other in a 3-m × 1-m bamboo bed suspended between two rafts, has the advantage of allowing the blades of *Laminaria* to point upwards, thus permitting greater access to sunlight (Fig. 6).

Some culturists give young sporophytes a head start by immersing them in a solution of fertilizer, usually ammonium nitrate. Absorption of nitrogen by plants so treated has been found to increase with concentrations up to 1 g/m^3 when the time of immersion does not exceed 6 hours. Stronger solutions may be used if the immersion time is reduced accordingly; 15 min is ample in a solution containing 400 g/m^3 of fertilizer. Immersion, accompanied by aeration and constant stirring, may be carried out weekly. The fertilizer solution may be reused several times.

Bottom culture of *L. japonica* persists in China largely because of the lower cost in terms of labor, material, and capital investment. Bottom culture is also customarily carried out in conjunction with raft culture, to take advantage of excess fertilizer. In addition, growth of *Laminaria* on the sea floor is encouraged because the plants provide shelter for various edible fishes and invertebrates, which in turn help fertilize the algae.

The classical method of bottom culture, used both in conjunction with raft culture and alone in areas which are too shallow at low tide to permit the use of floating rafts, is called "stone-casting." Originally, stone-casting consisted of no more than placing large stones, averaging 16 kg in weight, in the water to provide an attachment surface for naturally produced zoospores. Today, zoospores may be collected and allowed to attach to the stones before deposition. Stones are usually placed in rows 2 m apart, with each stone separated by 20 cm from the next one in its row, resulting in densities of 3800 to 4800 stones/ha.

Where the bottom is soft, round baskets made of bamboo or willow twigs may be substituted for stones. This method is particularly suitable for growing young sporophytes for subsequent use in raft culture. The baskets, generally 50 to 66 cm in diameter and 20 to 50 cm high, are deposited on the sea floor at densities determined by the transparency of the water and strength of the current, but averaging about 480/ha. Spores may be attached to baskets before they are placed in the sea, or sori-bearing sporophytes may be tied to them.

Laminaria grows poorly, if at all, on rocky beaches exposed to heavy tidal action. The Chinese have brought even such areas into production by construction of "tiered farms." These culture areas consist of pools, varying in depth from 40 to 150 cm, bounded by rock and cement dams of graduated height. Stones are placed in the pools for attachment of zoospores.

A cheap, simple method of culturing *L. japonica* would be to attach young sporophytes directly to ropes anchored on the sea floor. This technique is being tested in China but at last report was still in the experimental stages.

Although, as indicated, *L. japonica* is now being grown in southern China, the main centers of culture are still in the north, and periodic shipment of both young and mature sporophytes is necessary. Accordingly, techniques for long-distance transport have been developed. Success in shipment hinges on judicious lowering of the metabolic rates by controlling temperature and oxygen supply. This not only reduces consumption of stored energy but discourages harmful microorganisms which thrive on the metabolic wastes of *Laminaria*.

Both young and mature sporophytes are shipped in sea water in metal-lined boxes. The water temperature, which is checked every 6 hours, is maintained at 5°C by means of packing the boxes in bags of ice. The boxes are aerated once or twice a day and inspected to determine whether the buildup of metabolic wastes requires a change of water. The water must not be allowed to reach a state where it appears milky, or death of the algae will be imminent. *L. japonica* may be thus transported in complete darkness for periods of up to a week. The time in transit may be extended by occasional exposure to diffused light; direct sunlight is detrimental. Young sporophytes may receive immersion treatment in fertilizer solution upon arrival at their destination.

Sporophytes with sori must be treated more carefully in transit. Neither immersion in water nor dryness is likely to be lethal to such plants, but both conditions may cause premature shedding of zoospores. Sori-bearing sporophytes are shipped, without water, in the same type of boxes used for other sporophytes, but with the addition of a wooden rack inside each box. Enough space is allowed between tiers of the rack to ensure adequate oxygenation. An alternate method is to eliminate the rack and spread wet cotton on the bottom of the box and wet cheesecloth over the surface. In either case, shipments in transit are sprayed with small amounts of sea water once or twice a day. Care must be taken not to spray directly on the sporangia or the zoospores may be washed away. Temperature control is the same as in shipment of young sporophytes. Upon arrival, sori-bearing sporophytes are washed with sterilized sea water, any dead zoospores are removed, and the plants are placed in sea water at 10 to 15°C to await discharge and collection of zoospores.

With proper care, cultured *L. japonica* sporophytes will reach lengths of over 3 m in 4 to 5 months, at which time they are large enough for harvest. The maximum length attained in raft culture is over 6 m. Raft-cultured *Laminaria* are harvested from boats, but harvest of bottom-grown plants necessitates the services of trained divers.

The traditional harvest season is mid-June, just prior to the time when rising water temperatures cause the fresh weight of the algae to decline. Once again, however, dry weight is a better indicator for the culturist. It has been found that the dry weight of *L. japonica* continues to increase at water temperatures of 20.0 to 21.5°C—long after fresh weight has begun to decline. From this, Chinese algologists conclude that production of *Laminaria* could be increased by 11% by delaying the harvest until early July.

A harvest technique employed by some growers and recommended for all is periodic removal of the distal ends of the sporophytes before they slough off. In addition to salvaging what would otherwise be wasted

PLATE 3.   Harvest of full-grown *Laminaria* attached in parallel arrangement from a single-line raft, Liaoning Province, Northeast China. [Courtesy Tien-Hsi Cheng in *Economic Botany*, **23**(3), 1969.]

portions of the plants, this is believed to facilitate current flow and enhance photosynthesis. It has been shown experimentally that such partial harvesting does not lead to infection, and it is believed that if this technique were universally adopted, a net gain in production of 15% would result.

Laminaria is relished as a food not only in the coastal provinces but throughout China, particularly at times of year when green vegetables

TABLE 3.   NUTRIENT COMPOSITION PER 100 G OF DRIED *Laminaria japonica* (NOT TREATED IN BRINE)

Carotene	0.57 mg
Thiamine	0.09 mg
Riboflavin	0.36 mg
Niacin	1.60 mg
Protein	8.20 g
Fats	0.10 g
Carbohydrates	57.00 g
Coarse fibers (roughage)	9.80 g
Inorganic salts	12.90 g
Calcium	1.18 g
Iron	0.15 g
Phosphorus	0.22 g

are scarce. Unfortunately, it is customary to soak the algae in brine before drying it for sale, a practice which seriously lowers its nutritional value. Table 3 summarizes data published by the Department of Nutrition, Chinese Academy of Medical Sciences, on the nutrient composition per 100 g of unsoaked, dried *L. japonica*.

In addition to its food value, *L. japonica* is prized for medicinal and industrial uses. Among the commercially valuable constituents isolated from the alga are iodine, potassium chloride, algin, mannitol, laminarin, bromine, and chlorophyll.

A number of diseases, predators, and fouling organisms, including various rot diseases, bryozoans, ascidians, amphipods, and undesirable algae, are found on *L. japonica*. No effective means of control have been found for any of these.

Perhaps more serious than these biological problems are the economic problems faced by *Laminaria* culturists. On the one hand, culture of *L. japonica,* as an important source of food for the Chinese people and an economic asset to the country, represents one of the greatest achievements of the Communist regime. On the other hand, algal culture in such sterile waters as the Yellow Sea greatly increases the demand for chemical fertilizers, which are already in high demand and low supply for use on terrestrial crops. To what extent organic fertilizers could be substituted in either *Laminaria* culture or terrestrial agriculture is not known, but it appears that, in the latter case at least, some such substitutions would be feasible. Chinese authorities are currently considering another substitute source of fertilizer for *Laminaria*—large-scale introduction of nitrogen-fixing algae and bacteria into the Yellow Sea.

Further problems are incurred by the excessive amounts of manual labor required by present culture methods, which would be economically prohibitive in many countries today and may eventually prove so in China.

Economic problems aside, culture of *L. japonica* is currently a valuable industry in China. It is difficult to estimate the average yield or total production of *Laminaria,* since virtually all sources of information have been closed since mid-1966, and even before that, data were spotty and of questionable accuracy. Nevertheless, a few figures seem in order.

As early as 1962, average yields of 2 metric tons/ha were reported from Fukien province. Bottom culture in fertile areas of Shantung province reportedly yields 2.4 metric tons/ha. Yield data for raft culture are not available, but yields certainly average higher than for bottom culture.

Total production appears to have risen steadily since 1949. According to a prediction made in 1962 by Tseng Cheng-Kuei of the Institute of Oceanology, Academia Sinica, who has done as much as any one man to improve and expand *Laminaria* culture in China, the country should by now be producing 100,000 metric tons, dry weight, annually, as compared to the peak pre-Communist output of 62 metric tons. This is impressive until one considers that to meet the total demand for *Laminaria,* ten times that much is needed. Such a production total is, however, within the realm of possibility. China has an estimated 50,000 ha of coastal waters suitable for culture of *L. japonica.* As of 1960, only 3000 to 5500 ha were being utilized. A considerably higher proportion is presently cultivated, yet the area is almost certainly still far from totally exploited. If it were, and if an average yield of 2 metric tons/ha could be sustained, the required 1 million metric tons could be produced.

Given the economic problems cited, and the absence of recent data, it is impossible to predict the future of *Laminaria* culture in China, though the potential is clear. It is the opinion of knowledgeable algologists and aquaculturists that further applied research and a great deal more basic research on the biology of *L. japonica* are necessary if this potential is to be realized. The effect of recent political events in China on such activities is unknown, but may not be as severe as some observers have speculated. At the very least, barring catastrophe, culture of *L. japonica* will continue to be an important industry in China.

There appears to be small likelihood that the *Laminaria* culture methods developed in China will be adopted by her neighbors, with the possible exception of Korea. The labor-intensive techniques currently employed would likely be economically prohibitive in Japan or Taiwan, while the further southward spread of *L. japonica* would appear to be blocked by high temperatures. The latter obstacle might eventually be removed by further improvement of the temperature-tolerant strains already developed by Chinese researchers.

CULTURE OF *Caulerpa* AND OTHER SPECIES

The methods used in the highly successful culture of *Gracilaria cornopifolia* (used only in the production of agar, but not as human food) in Taiwan have been adapted to farming of *Caulerpa* sp. for the fresh vegetable market in the Philippines. *Caulerpa* is grown in brackish water ponds seeded with chopped fresh pieces of the mature plant. Grazing mollusks of the genus *Cerithium* must periodically be removed from such ponds. (*Caulerpa* has recently been found to be an excellent source of certain medical products, thus the emphasis in *Caulerpa* culture may shift away from food production.)

Because of the use as food additives of colloids derived from the tropical brown algae *Eucheuma* sp. and *Hypnea* sp., mention should be made here of the advanced experimental cultures for these species in the Philippines. The techniques are based on suspending thalli under protected and other optimal growing conditions.

PROSPECTUS

The only seaweed commonly eaten in the West is the red alga *Rhodymenia palmata*. "Dulse," as it is known, unlike most seaweeds, is completely digestible by man and is eaten in dried, raw form or used in cooking in the Maritime Provinces of Canada, Ireland, Scotland, and the Mediterranean countries. The feasibility of culture has not been explored.

The course taken by seaweed culture in the future is likely to depend on the degree of economic development of the country in question. In the poorer countries, labor-intensive methods may continue to be applied not only to algae currently cultured but to species new to culture. Increases in production may result from improved technology, intensification of culture, introduction of new species to culture, and increases in the total area devoted to culture. Labor-intensive cultivation of algae in Japan may continue for some time, but the immediate future of seaweed culture in the wealthier countries would seem to involve largely low-intensity methods.

Although cultivation of seaweeds for industrial purposes may spring up anywhere, by far the greater part of the demand for edible seaweeds will probably continue to be in Asia, at least in the near future. As mentioned, the red alga known as dulse is widely eaten in Canada and Europe, and certain large kelps are made into pickles in Alaska, but, in general, Westerners do not find seaweeds very palatable or digestible. It has been suggested that Oriental peoples have developed a special intestinal flora to enable them to digest seaweed foods. It is doubtful whether, under ordinary circumstances, most Westerners would take the trouble to cultivate such a flora. If, however, contamination of the environment, and particularly the oceans, with pesticides, PCB's, and so on, proceeds at the alarming rate predicted by some sources, man may find it increasingly advantageous to move lower on the food chain. If this happens, the importance of algae, including seaweeds, in Western diets may increase, and their culture may be undertaken.

REFERENCES

CHENG, T. H. 1969. Production of kelp—a major aspect of China's exploitation of the sea. Econ. Bot. 23(3):215–236.

DREW, K. M. 1956. Reproduction in the Bangiophycidae. Bot. Rev. 22:553–611.

KUROGI, M. 1963. Recent laver cultivation in Japan. Fish. News Inter. (July-September, 1963).

SAITO, Y. 1962. Fundamental studies on the propagation of Undaria pennatifida (Harv.). Sur. Contrib. Fish. Lab., Fac. Agric. Tokyo U. 3:81–101.

TAMIYA, H. 1959. Role of algae as food. Proceedings of the Symposium on Algology. New Delhi, India. December 1959.

Tohoku Regional Fisheries Research Laboratory 1965. Aquaculture in Tohoku regions of Japan. Part I. The production of shallow-water organisms from annual statistics. 85 pp.

Interviews and Personal Communication

ARASAKI, S. Department of Fisheries, Faculty of Agriculture, University of Tokyo, Japan.

DOTY, M. University of Hawaii, Honolulu, Hawaii.

FUJIYA, M. Japanese Fishery Agency.

SAITO, Y. Algologist, Naikai Regional Fisheries Research Laboratory, Hiroshima, Japan.

43

Culture of Edible Freshwater Plants

As mentioned in Chapter 42, man has scarcely begun to utilize the large marine algae or seaweeds, and their culture and harvest as food is almost entirely confined to the Orient. If man has neglected marine plants, he has virtually ignored freshwater plants. Freshwater algae are small, usually microscopic, and would seem to hold little interest as food for so large an animal as man. However, many fishes, including some of the largest species, derive much of their nourishment from phytoplankton, and it is at least theoretically possible that man could do the same. Thus in the 1950s, when it became evident that human population growth was outstripping the food supply, there was a flurry of interest in the culture of unicellular freshwater algae, particularly *Chlorella* spp. Research continues, but it is increasingly clear that this form of aquaculture is an extremely speculative venture. It is not beyond the realm of possibility that

815

it will someday be economically feasible to prepare a nutritious, palatable food from unicellular algae, but that day does not appear to be near. There are, however, a number of higher aquatic plants which can be harvested and prepared by conventional means for human consumption, though they are seldom eaten and even less frequently cultured.

WATER SPINACH (IPOMOEA)

In Thailand, Malaysia, and Singapore, water spinach (*Ipomoea reptans*) is sometimes grown in very shallow ponds or, in lesser concentrations, in deeper ponds where the primary crops are fishes and/or freshwater shrimps. *I. reptans* is occasionally harvested as a green vegetable, though it is more often fed to livestock. It has been reported to produce more than 100 kg/(ha)(day) of green matter, which sounds exciting until one considers that the average plant contains 92.5% water and only 2.1% crude protein and 2.9% carbohydrates. Quite logically, then, human consumption of water spinach is confined to the lowest economic classes, who can afford no other vegetable. Before completely dismissing *I. reptans* one should take time to calculate that, based on the data above, a 1-ha pond, in a tropical region with a year-round growing season, planted with nothing but this species, could annually produce 766.5 kg of protein and 1058.5 kg of carbohydrates, or considerably more usable food than many commercially profitable fish ponds. If nothing else, this should serve as an indication of the potential productivity of freshwater plants.

WATERCRESS

High water content vitiates the food value of more than a few other freshwater plants, but it is by no means a universal problem. An aquatic plant which is highly nutritious and particularly rich in vitamins and minerals is watercress (*Nasturtium officinale*). In North America, where it is considered a gourmet food and commands a high price, watercress is usually gathered from the wild, but in Europe and Hawaii it is commercially cultivated. A few individuals in the continental United States do culture it for personal use, and seeds are sometimes available from garden supply houses.

All that is needed to grow watercress is a year-round source of cool, running water. For health reasons the water should be free of domestic pollutants, since human pathogens may be transmitted by the plant.

Planting entails no more than soaking the seed overnight so it will sink, and scattering it in 2 to 15 cm of water with a moderate current.

The preference of watercress for very small streams has made large-scale harvest of the wild plant economically unfeasible, but large amounts might be cultured in a small area by means of digging a parallel grid of ditches. Watercress possesses the distinct advantage for culture of maintaining its growth throughout the year in all but the coldest of climates.

SAGITTARIA

Another edible aquatic plant found nearly everywhere in the North Temperate Zone is the arrowhead (*Sagittaria* spp.). The fleshy tubers of this emergent plant were once commonly eaten by American Indians during the winter but have fallen into disuse. Arrowhead is reportedly cultivated as a food plant in China, but details of this practice are not available.

Aquarium hobbyists culture *Sagittaria* spp. as submerged ornamental plants but are generally unaware of their edible qualities. Indeed, since submerged *Sagittaria* do not develop the characteristically shaped leaves which give the genus its common name, most aquarists would not recognize the wild plant. Aquarists and those few Americans who make a practice of gathering arrowhead tubers are in agreement that *Sagittaria* spp. respond positively to periodic thinning, and it seems that cultivation would be feasible.

The tubers are high in starch, though not in protein, and are used in many of the same ways as potatoes. The critical food needs in most parts of the world are for protein, but it is possible that in some areas cultured arrowhead tubers could become popular as a winter substitute for potatoes and the like.

WILD RICE

Rice might be considered an aquaculture crop, but since rice farming is a traditional form of agriculture, we shall leave its discussion mainly to writers on that subject. Rice does concern us here inasmuch as fish, shrimp, and crayfish are sometimes cultured in rice fields in Japan, southeast Asia, and the southern United States. This form of aquaculture has become less prevalent with the spread of insecticide use in rice farming, but perhaps the second thoughts about hard insecticides which are currently being entertained throughout the world will reverse this trend. Rice-fish and rice-invertebrate culture are discussed in Chapters 5, 8, and 34.

Mention should be made of "wild rice" (*Zizania aquatica*), which is more truly aquatic than true rice. As the name implies, wild rice is not generally cultured. Most of the commercial supply of this nutritious but expensive grain comes from the upper Great Lakes region of the United States and Canada, where certain Indian tribes have exclusive rights to harvest it. The distribution of wild rice is not limited to that area, however, and there is potential for its culture as a source of food and revenue. In recent years, a few small farmers and communities in northern California have planted it for private use, but little or none has reached the market.

CATTAILS

One of the most versatile of edible aquatic plants is the widely distributed cattail (*Typha latifolia* and *Typha angustifolia*). No less than six parts of this common and prolific plant can be used as vegetables or made into flour. As early as 1919 the possibility of cattail culture was discussed but nothing has come of it, and cattails remain virtually unused.

POSSIBILITIES FOR DEVELOPMENT

The preceding are but a few of the edible plants found in the world's freshwaters. If, in the future, some of these plants are to be farmed, development will have to proceed along lines quite different from those followed in fish culture. Although such practices as fish culture in rice fields will persist in some areas, it can be said in general that the presence of large rooted plants is undesirable in bodies of water used in fish culture. There are, however, in most parts of the world, considerable expanses of swampland and shallow ponds totally unsuited to most forms of fish culture. Man's present policy toward such areas is to "develop" them for industry, residential construction, or conventional agriculture. If, as seems likely, the pressing needs of the growing human population dictate that we alter remaining wetlands at all, surely it is esthetically more appealing and ecologically wiser to use them as wetlands than to "reclaim" them for uses which are not in accord with their nature. One possible use of wetlands is for the culture of aquatic plants. At present, there is virtually no interest in this sort of aquaculture, but research should be begun.

REFERENCES

CLAASEN, P. W. 1919. A possible new source of food supply. Sci. M. (August, 1919).

Ho, R. 1961. Mixed farming and multiple cropping in Malaya. Proceedings of the Symposium on Land Use and Mineral Deposits in Hong Kong, South China, and Southeast Asia. Hong Kong University Press.

LeMare, D. W. 1952. Pig-rearing, fish-farming and vegetable growing. Malayan Agric. J. 35:156–166.

Interviews and Personal Communication

ANDERSON, E. Department of Anthropology, University of California, Riverside.

Appendix

Pond Siting and Construction

Aquaculture occurs in a variety of enclosures and structures: floating cages and rafts, racks attached to pilings, raceways, troughs, artificial streams, recirculating systems, and aquaria and their uses are all described in this book. Nevertheless, the most common and versatile structure, particularly for culture of freshwater fishes, remains the oldest—the pond. Many special features have been incorporated in ponds designed for specific situations, so that the reader who has a particular animal or

821

form of aquaculture in mind should consult the appropriate chapter before engaging in pond construction. Nevertheless, there are certain principles of location and construction which obtain in all cases. Thus, although the following comments on aquaculture ponds are largely based on the experience of catfish farmers in the United States, they are applicable to pond culture throughout the world. Since ponds used for spawning and hatching of eggs are almost always highly specialized, we discuss mainly ponds used for maintaining and growing young and adult animals. However, many of the factors involved, for example, water quality, are of equal importance to the breeder as to the market grower or subsistence farmer.

SITE SELECTION

As in any form of agriculture, it is of crucial importance to choose a suitable site for pond culture. The main considerations in siting are topography, water supply, and soil quality.

The ideal location for a fish pond is on land that is flat or nearly so. In hilly areas, or wherever ponds are built on sloping ground, they must often be irregular in shape, which reduces the efficiency of harvesting operations. An irregular bottom contour is as undesirable as an irregular periphery, since it interferes with all operations which are dependent on draining, including harvesting.

In large river basins, flooding sometimes presents a serious danger to aquaculture operations. (There are of course low-intensity systems of aquaculture in which flooding is not only tolerated but desired. Here, however, we refer only to those types of culture in which the culturist selects the precise kinds and numbers of animals which inhabit his pond.) The best sites for ponds are in small, stable drainage systems where the possibility of flooding is remote. Often, however, considerations of climate, topography, availability of stock, or access to markets dictate that ponds be built in regions subject to flooding. In such cases, it should be possible to construct levees higher than the historic high water mark for the pond site. In areas where aquaculture is of major economic importance, government assistance may be sought in construction of levees or other measures to prevent or control floods.

Access to the pond at all times is of course essential. A pond that can be reached only by a strenuous journey or a pond that is inaccessible for part of the year is not likely to be a well-managed pond. The type of access needed depends largely on how the crop is to be disposed of. For a family or village fish pond used in subsistence aquaculture, such as those found in many parts of Africa, all that is necessary is that it be located conveniently to the family or village it is to serve. On the other hand, commercial aquaculture in, for example, the United States, necessitates access not only to the operators but to processors and/or markets. This usually means that the pond must be served by a good road, so that the time between pond and table can be minimized.

A negative aspect of easy access is that it may encourage poachers. In the United States, it is thus considered desirable that the topography permit ready detection of trespassers. In the Philippines, oyster culturists find it necessary around harvest time to construct temporary dwellings on their culture sites. In other parts of the world different precautions are practiced. The prospective pond culturist should be familiar with the likelihood of poaching in his region and plan accordingly.

A final crucial factor in pond location is drainage. Topography, aided by engineering, should permit the pond to be drained completely dry. This can of course be done by pumping, but this adds an expense which can be avoided where adequate natural drainage exists. In some parts of the world, legal problems with drainage may arise if ponds are not adjacent to a natural waterway and must be drained through other private or public lands, or if government agencies regulate or prohibit drainage into streams.

All ponds in an aquaculture complex should be constructed so that each may be drained individually, with no water passing from one pond to another. This prevents the spread of disease and the passage of wastes from pond to pond and also makes it possible to drain any pond for harvest or repairs without affecting other ponds.

WATER SUPPLY

QUANTITY

The amount of water required for freshwater pond culture is of course considerably less than that needed for, say, culture of trout in raceways. Nevertheless, a constant supply of high-quality water must be available throughout the year to replace losses due to evaporation, seepage, and drainage during management operations. The prospective pond builder should base his considerations on conditions prevailing during the hottest and/or driest part of the year, when evaporation is greatest and oxygen depletion in the pond most likely. A rule of thumb adopted by catfish farmers in the south central United States, where the climate is classified as "humid subtropical" and annual evaporation loss from ponds is approximately 1.0 to 1.5 m, is that a flow of 4500 to 5000 liters/min is adequate for a 16-ha operation. Culturists in other parts of the world must of course adjust for local differences in temperature and rainfall.

WELL WATER VERSUS SURFACE WATER

By far the best source of water for a pond is a well, although spring water or even surface runoff may be successfully used. Well water is preferred because it assures more dependable flow and is usually free of disease organisms, parasites, predators, trash fish, pesticides, silt, and other contaminants and pollutants. Since fry are more susceptible to most of these dangers than older fish, well water is particularly to be preferred in fry rearing ponds. Other than the expense of drilling a well, the only problem normally encountered in using well water is that it is often low in dissolved oxygen and high in carbon dioxide and nitrogen. This situation can readily be reversed by spraying it into the pond, splashing it in off a flat surface, or passing it over a series of baffles so that harmful gases may be exchanged for atmospheric oxygen. (Such precautions probably need not be taken where the only animals cultured are Anabantids, Clariid catfishes, or other creatures with accessory air-breathing organs, but even then high oxygen content might favor good growth.) A prospective pond culturist considering locating where aquaculture is not common, or where wells are not the usual source of water for ponds, might profit by drilling a test well before committing himself to any other source of water.

FILTERS TO EXCLUDE UNWANTED ANIMALS

Most surface water supplies contain fishes and/or invertebrates which may act as predators or competitors in culture ponds, so it is desirable, and usually necessary, to install a filtering device on the inlet pipe. The filter material should have at least seven meshes to the centimeter. One of the best materials for this purpose is Saran screen, Style MS-904, available from the National Filter Media Corp., New Haven, Connecticut. Saran screen is inexpensive, durable, and easily fabricated into various shapes. The simplest and most common Saran filter is the "sock" type (Fig. 1). The Saran sock, which may be attached to the inflow pipe by means of clamps, rubber bands, or a drawstring, should be at least 1.8 m long for a pipe with a diameter of 15 cm, and should extend well below the surface of the water to reduce stress on the screen.

Where water is discharged at high velocities, it may be necessary to construct a rigid-frame filter. As with the sock filter, stress may be re-

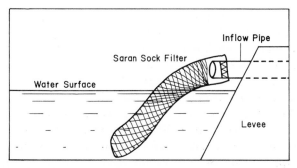

F ig. 1.   Saran sock to filter inflow to fish pond. (After Sills.)

duced by submerging the screen. One way of doing this is by constructing a floating box filter, as illustrated in Fig. 2. All filters must be cleaned periodically.

POLLUTANTS AND CONTAMINANTS

Pond water supplies may also contain organic and chemical pollutants and contaminants. The most common and easiest to cope with is silt. When the water supply is temporarily muddied, addition of 15 to 22 bales of hay per hectare or a 3:1 mixture of cottonseed meal and super-phosphate at 110 kg/ha will clear it. These organic materials should not be used in hot weather, when they may induce heavy algal blooms. A substitute treatment for such times is gypsum, applied at 200 to 900 kg/ha and repeated at 7- to 10-day intervals. These are intended only

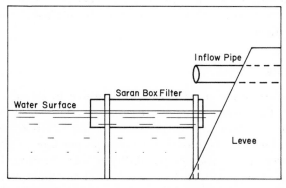

F ig. 2.   Floating saran box filter below fish pond inflow pipe. (After Sills.)

as emergency measures not as substitutes for a well-vegetated watershed, which will minimize soil erosion and siltation.

Iron is a common constituent of pond water supplies but causes problems only in small ponds. It may be precipitated out by pumping the water into a settling pond before it reaches the culture pond or by passing it over a bed of rocks or gravel. Some other pollutants and contaminants may be settled out, and aeration helps to remove all gaseous pollutants, but in general the culturist dependent on surface runoff has little control over water quality. Before essaying aquaculture with runoff as a water supply, one should ascertain whether the water supply is sufficiently free of contaminants for the animals he wishes to raise, whether it is safe from a human health standpoint, and what is the likelihood of future contamination.

CHEMICAL FACTORS

Chemical characteristics of water which may affect aquatic animals are innumerable. Among the most important and easily measured are dissolved oxygen content, pH, and akalinity. Optimum values for each of these parameters vary with the species of animal involved. The culturist should know the requirements of the animals he proposes to grow before stocking a pond and should continually monitor his ponds for the characteristics just mentioned. In brackish water ponds, the same applies to salinity. Sources and techniques of water supply for such ponds are very specialized and are discussed in the text, particularly in Chapters 16, 17, and 32. If pH is too low, it may be adjusted by adding 550 kg/ha or more of lime to the pond. The precise dosage must be determined empirically, so it is best to lime ponds a little at a time until the desired level is reached. There is no comparable practicable treatment to correct high pH; adding acidic substances is not effective. Alkalinity may be increased by treatment with lime or reduced by application of sulfate of ammonia fertilizer.

TEMPERATURE

The temperature of the pond must also be suited to its inhabitants, but in some cases the culturist may exert control over this aspect of the environment. Where two sources of water having different temperatures are available, they may be mixed in various proportions to provide intermediate temperatures. Another possibility is solar heating. Up to now, solar heating in aquaculture has consisted of no more than exposing water to sunlight in a broad, shallow pool before admitting it to culture ponds, a technique which is wasteful of both water (through evaporation) and space. At present, however, there are plans to incorporate solar heaters in aquaculture projects planned by the New Alchemy Institute in Massachusetts and New Mexico.

Although solar heating has been neglected, much attention has been focused of late on other sources of heated water in aquaculture, and culturists increasingly will be exploring the possibilities of thermal enrichment. It is true that growth of many aquatic organisms can be enhanced by increasing the temperature, but caution must be exercised. The characteristic curve of temperature versus growth ascends slowly and descends sharply; the optimum temperature for growth is often very close to the upper tolerance limit. Thus the culturist who intends to heat his water should either have extremely accurate control over temperature or allow a fair margin of error, taking natural summer temperatures into consideration.

SOIL QUALITY

FERTILITY

Two aspects of soil quality are important to pond culturists throughout the world; fertility and water retention. Fertility is not always a major consideration. In such systems as classical Chinese carp culture or brackish water culture of milkfish, where the fish derive all or most of their nourishment from naturally occurring organisms, it is of paramount importance and is usually augmented by addition of manure or chemical fertilizers. On the other hand, in such operations as catfish culture in the United States, or Japanese eel culture, in which growth of the stock is dependent on supplementary feeding, soil fertility is of little value, and excessive fertility may in fact be detrimental. Such forms of aquaculture are thus sometimes profitable in areas which are of poor quality for terrestrial agriculture.

WATER RETENTION

Water retention is of universal importance for obvious reasons. Leaky ponds may be repaired by treatment with various sealants but only at considerable expense. If possible, it is best to locate ponds over soils which are known to retain water well. Opinions differ as to what constitutes a good soil in this regard. Some authorities recommend soils with a high clay content; others cite a mixture of 70% sand and 25% clay, with enough silt to fill the spaces between the sand particles, as ideal. The discrepancy may be due to differences in the types of clay present or to the treatment the pond bottom and levees receive during construction (see below). In any event, before building a pond it is wise to seek professional advice. In the United States, this may be obtained free of charge from the Soil Conservation Service. If no such aid is available, the culturist may draw on the experience of other pond owners in his vicinity but should bear in mind that soil composition is highly variable, even within small areas.

Whatever type of soil underlies a pond, proper construction plays a large role in preventing leakage. Of particular importance is compaction of the soil in dams and levees, and preferably throughout the pond. A sheepsfoot roller is best for this job; otherwise the wheels of a heavy tractor are satisfactory. If the soil is judged to retain water well, it should be ripped and disked to a depth of 25 to 30 cm before compaction. Ponds built on lower quality soils may be improved by addition of 30 to 45 cm of good soil before compaction. For each soil there is a certain moisture content for optimum compaction. A good rule of thumb is that a soil compacts best when it is too wet to plow but not wet enough to discharge water during the process of compaction.

If compaction is not judged to be sufficient to eliminate leakage, one of a variety of sealants may be used. Perhaps the most common is bentonite clay, though its use is restricted by the huge amounts required; it is not expensive or in short supply, but if the pond is not located near a source of bentonite, shipping costs may be prohibitive. Bentonite has the property of absorbing large amounts of water and expanding to 8 to 20 times its original volume, thus plugging the pores in the soil. It is applied, preferably to the dry pond bottom, at 0.5 to 1.5 kg/m² (a laboratory analysis of the soil should be made to determine the precise rate of application), spread evenly, mixed with the soil, moistened, and compacted as described above.

Bentonite appears to be particularly effective on sandy or silty soils, but is not satisfactory in ponds which are subject to partial drying; under such circumstances it will crack. Cracking while the pond is being filled may be averted by applying a mulch of straw or hay to the pond bottom just before the final passes of the sheepsfoot roller (Plate 1).

Ponds constructed on fine-grained soils may also leak, in which case treatment with a chemical sealant may be indicated. Such sealants, which act by dispersing clay particles, are effective in soils composed of more than 50% particles finer than 0.74 mm in diameter and which contain less than 0.5% soluble salts (based on dry soil weight).

Although many soluble salts may function as chemical sealants, the

PLATE 1. This pond is being lined with clay and the sheepsfoot roller is being used to compact the clay. This process is used to prevent leakage through the bottom of the pond in porous soils or bedrock with fissures. (Photograph by D. C. Tindall, courtesy U.S. Department of Agriculture, Soil Conservation Service.)

most commonly used are sodium chloride (which does not increase water salinity when properly applied) and various sodium polyphosphates, notably tetrasodium pyrophosphate (TSPP) and sodium tripolyphosphate (STPP). Particle size of such sealants should be such that 95% will pass through a No. 30 sieve and be retained by a No. 100 sieve.

Application rates for chemical sealants are much lower than for bentonite; sodium chloride is usually applied at 0.04 to 0.17 kg/m^2 and polyphosphates at 0.01 to 0.02 kg/m^2. A laboratory analysis of the soil is essential to determine which sealant and rate of application will be most effective.

The sealant is first mixed with the soil and then compacted, as described, to form a blanket 15 to 30 cm deep, depending on water depth. For good compaction, the soil should be neither too moist nor too dry. The area of the pond near the normal water line should be further protected with 30 to 45 cm of gravel.

A newer, and increasingly popular method of pond sealing involves

PLATE 2. Installation of a polyethylene pond liner. Light areas are sand used to hold liner in place during application. Dozer building burm around pond with excavation from watercourse along edge. (Photograph by G. F. Clark, courtesy U.S. Department of Agriculture, Soil Conservation Service.)

lining the pond with a flexible membrane of polyethylene, vinyl, or butyl rubber (Plate 2). These materials are structurally weak and must be handled very carefully, but when properly installed they are easily kept intact and watertight. Minimum thicknesses for pond liners used over materials no coarser than sand are 2 mm for polyethylene and vinyl and 4 mm for butyl rubber. Where the bottom contains gravel, these thicknesses should be doubled.

Ponds to be lined should be allowed to dry until the surface is firm. Stony areas should be cushioned with a layer of fine material. Liners are carefully laid in strips with a 15-cm overlap for seaming with appropriate cement or tape. The lining is anchored by excavating a trench, 20 to 25 cm deep and 30 cm wide, completely around the pond above the water level, and burying the edges of the liner in the trench with compacted backfill. To protect against punctures and tears, all pond liners should be covered by at least 23 cm of soil no coarser than silty sand. The effects of this sort of sealant on pond fertility are not known.

PESTICIDE CONTAMINATION

In major agricultural areas, yet another aspect of soil quality must be taken into consideration—contamination by pesticides. This is particularly problematical where cotton is grown, for cotton farmers normally use massive doses of chlorinated hydrocarbon insecticides. There has been at least one documented case in Arkansas of a minnow farmer who attempted to convert cotton land to fish production and found concentrations of endrin and dieldrin in the soil so high that, when the land was flooded, the water became toxic to fish. Toxaphene had been used in the same locality, but it was not possible to detect it in the water, so it may be that toxaphene-treated areas are not objectionable for fish culture. While pesticide concentrations high enough to kill fish are still rare, these compounds may be concentrated, by direct uptake or via the food chain, to the point where eating fish grown where pesticides are used could endanger human health. Similar situations have already occurred in relation to certain marine fishery products. In assessing possible pesticide pollution, the farmer must consider not only the soil at the prospective pond sites but also the likelihood of contamination by air or water-borne pesticides intended for use on adjacent lands.

CONSTRUCTION

DUGOUT VERSUS LEVEE PONDS

There are two general types of pond which may be constructed, the dugout pond and the levee pond. The former type, as the name implies, is excavated, and the bottom lies below the surrounding ground surface. In levee ponds, the water is retained by a dam and levees or their natural topographical equivalents. Levee ponds are preferable for all forms of aquaculture, since, unlike dugout ponds, they may be drained completely dry without pumping. In Africa and Asia, ponds used for fish culture are sometimes created by damming natural watercourses (barrage ponds), but, even where surface runoff is the source of water, it is preferable to divert water into culture ponds, because all but the most stable natural streams are subject to occasional destructive flooding.

BOTTOM PROFILE

To facilitate drainage, the pond bottom should slope toward the outlet at about 0.2%. All obstructions should be removed, and irregularities smoothed out. If soil used in levee construction is taken from the pond site, the borrow ditches should be outside the pond. In general, the smoother the bottom, the better, but some Asian milkfish and shrimp ponds incorporate a series of ditches to facilitate drainage. If such ditches are not well constructed and maintained, however, they may in fact impede drainage and allow stock to avoid capture during harvest. Other specialized features of pond bottoms for certain types of aquaculture are described in the text.

An essential feature for economic harvesting is a harvest basin or catching pond, in which stock may be concentrated after the pond is drained. American channel catfish culturists prefer harvest basins which are 0.6 to 0.75 m deeper than the rest of the pond. Such basins usually account for at least 10% of the pond surface area, so that the animals are not overcrowded to the extent that they suffer mortality. The size of basin used in culture of other species should take into account the relative hardiness of the species in question. A cross section of an American catfish pond is included in Chapter 6 (Fig. 1). The shape of harvest basin is not standardized, but some culturists prefer circular basins where fish may easily be surrounded with a seine.

If a separate catching pond is substituted for the harvest basin, the drainage system must be designed to accommodate animals as well as water. Several such designs are described in the text. If properly located, one catching pond may serve two or more production ponds.

DRAINAGE

The outlet should be so designed that the pond, exclusive of the harvest basin, can be drained in 48 hours or less. Computing the right diameter of drainpipe for a given pond is a bit of an engineering problem; in the United States the Soil Conservation Service can help. One of the best drainpipe designs is an L-shaped structure often used in American catfish ponds. When the pond is full, the short leg of the L is perpendicular to the bottom with the end a foot or so above the water surface. In this position it acts as an overflow device in case of high water. Total or partial drainage is made possible by a swivel joint at the apex of the L. By rotating the short leg of the L until the end is at a given level, the surface of the pond can eventually be brought to that level. When both legs of the L are parallel to the pond bottom, the pond will drain completely. Often, especially during hot weather, it is desirable to discharge oxygen-deficient water from the bottom of a pond. The drain pipe just described may be adapted for this purpose by means of a double sleeve overflow (Fig. 3).

The end of any drain pipe must of course be screened to prevent fish from escaping. It may be helpful to gravel the pond bottom in the im-

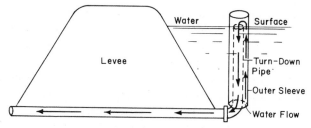

FIG. 3.   Double sleeve overflow on turn-down fish pond drain pipe.

mediate area of the drain pipe so as to prevent its being clogged with silt.

If there is any danger whatever of flooding, the pond must be equipped with a spillway, preferably a concrete structure. Spillways should be screened so that fish cannot escape over them. If the spillway opens near a natural stream, screens mut be provided to prevent the entry of wild fish during high water. Details of spillway structure should be determined in consultation with engineers or successful pond operators in the vicinity of a proposed new pond.

DAMS AND LEVEES

Dams and levees literally hold a fish farm together and are thus of paramount importance in construction. These structures should of course be made of nonporous materials and be of sufficient size to withstand the pressure of the water in the pond. Where mechanized equipment is to be used in pond management, levees are made considerably wider than is structurally necessary—at least 3 m on top—to permit access to vehicles. About 0.6 m of freeboard should be left between the water surface and the top of the levee. Recommended slopes are at least 4:1 on the inside and 3:1 outside. The bank should drop off sharply at the water surface to prevent the establishment of emergent plants, which permit siltation, provide shelter for unwanted small fish, and may interfere with management measures.

Levees, and in particular their bases, should be cleared of all roots, limbs, and other obstructions so that a water-tight bond between fill and base can be made. A further safeguard against leakage is the incorporation of a "core trench" in the center of the levee. Before actual levee construction begins, a trench about 1.2 m wide and deep enough to reach the first layer of impervious material is excavated around the periphery of the pond. The trench is then filled with impervious soil, thus providing an unbroken water-tight band from the top of the levee to far below the pond bottom.

Immediately after construction, levees should be planted to grass to impede erosion. In the United States and other affluent countries larger plants are discouraged, because they interfere with mechanized management operations and because they provide shelter for snakes and other semiaquatic predators, a factor of doubtful importance in most types of aquaculture. In southeast Asia, on the other hand, fruit and vegetable crops are often planted on levees. Tree-size plants serve the additional functions of shading out emergent plants and providing a windbreak to impede erosion of the bank and cooling of the water, to say nothing of their esthetic value. In general, it can be recommended that where the culturist is faced with severe economic competition, levee planting should be confined to grass, but otherwise crop plants are a desirable addition.

SIZE OF PONDS

Ponds of all sizes have been successfully used in aquaculture. Even after the choice of size has been narrowed down by consideration of biological, topographical, and economic factors, the culturist usually must weigh the relative advantages of small and large ponds.

Advantages of small ponds
1. Easier and quicker to harvest.
2. Can be drained and refilled more quickly.
3. Easier to treat disease and parasites.
4. If for any reason all or part of the stock in one pond is lost, it represents less of a financial loss.
5. Less subject to dam and levee erosion by wind.

Advantages of large ponds

1. Less construction cost per acre of water.
2. Take up less space per acre of water.
3. More subject to wind action, therefore less susceptible to oxygen deficiency.
4. More conducive to rotation with rice or terrestrial crops.

Pond depth depends partly on the type of animals to be cultured but primarily on the climate. Where freezing is a possibility, at least 1.8 to 3.0 m of water should be provided in part of the pond to eliminate the possibility of "winterkill." In warmer climates, or in ponds where stock is not overwintered, deep water is superfluous, for it is low in productivity and often virtually devoid of dissolved oxygen. In most forms of aquaculture, rooted aquatic plants are considered to be detrimental, so nowhere in the pond should there be water less than 0.75 m deep.

SHAPE OF PONDS

A final consideration in pond construction is shape. Many if not most aquaculture ponds are oblong and rectangular, but square ponds require less levee construction to achieve the same water surface area. If the oblong shape is used, large ponds should if possible be constructed with the long axis perpendicular to the prevailing winds, to reduce levee erosion. Conversely, small ponds should be constructed with the long axis parallel to prevailing summer winds to take advantage of wind aeration.

The shape, profile, and construction features of most aquaculture ponds being built today are extremely regular, for reasons of human efficiency. In regions of the world where augmentation of the human food supply is of great importance, the need for efficiency is obvious. In more fortunate parts of the world, economic competition dictates regularity as part and parcel of intensive monoculture. Unfortunately, this regularity all too often makes for esthetic monotony. It may be, however, that true polyculture, taking into account not only the pond ecosystem, but interactions between the aquatic and terrestrial communities, would benefit from irregularity. Ultimately we may find that the most productive system is one that is laid out irregularly, in conformity with the land it occupies. The possibility should be explored. For the present, the limitations of our understanding of ecosystems combine with economic factors to enforce the trend toward regularity.

REFERENCES

DILLON, O. W., JR. 1970. Pond construction, water quality and quantity. Paper presented to the California Catfish Conference, Sacramento, Calif. Jan. 20–21, 1970.

GRIZZELL, R. A., JR. 1967. Pond construction and economic considerations in catfish farming. Paper presented at 21st meeting, Southeastern Association of Game and Fish Commissioners, New Orleans, La., Sept. 25–27, 1967.

HADDEN, W. A. no date. Sealing ponds. Louisiana Cooperative Extension Service. 3 pp.

HADDEN, W. A. no date. Planning and constructing catfish ponds. Louisiana Cooperative Extension Service. 5 pp.

HUET, M. 1960. Traité de Pisciculture, 3rd Edition, Chide Wyngart, Brussels, 369 pp.

McLARNEY, W. O. 1970. Pesticides and aquaculture. Am. Fish Farmer 1(10):6–7, 22–23.

RENFRO, G., JR. 1969. Sealing leaking ponds and reservoirs. U.S. Soil Conservation Service, SCS-TP-150. 6 pp.

SILLS, J. B. no date. Saran screen fish barriers. U.S. Fish Farming Experimental Station, Stuttgart, Arkansas. 3 pp. (Mimeographed.)

Texas Agricultural Extension Service and Department of Wildlife Science, Texas A & M University. 1967. Proceedings of the Commercial Fish Farming Conference. College Station, Texas, Feb. 1–2, 1967.

Index of Names of Animals
and Plants

Index of Persons, Places
and Institutions

Subject Index